Foundations of Engineering Mechanics

V. A. Svetlitsky, Statistical Dynamics and Reliability Theory for Mechanical Structures

AF789409

Springer
Berlin
Heidelberg
New York
Barcelona
Hongkong
London
Mailand
Paris
Tokio

Engineering ONLINE LIBRARY
http://www.springer.de/engine-de/

V. A. Svetlitsky

Statistical Dynamics and Reliability Theory for Mechanical Structures

Translated by N. L. Reshetov

With 189 Figures

Springer

Series Editors:
Vladimir I. Babitsky
Loughborough University
Department of Mechanical Engineering
LE11 3TU Loughborough Leicestershire
United Kingdom

J. Wittenburg
Universität Karlsruhe (TH)
Institut für Techn. Mechanik
Kaiserstr. 12
76128 Karlsruhe
Germany

Author:
V. A. Svetlitsky
Bauman Moscow State Technical University
The Department of Applied Mechanics
2-nd Baumanskaya St. 5
107005 Moscow
Russia
e-mail: voronov@rk5.bmstu.ru

Translator:
Nikolay Reshetov
25-1-56, 13-th Parkovaya St.
105215 Moscow
Russia

ISBN 978-3-642-53657-1 ISBN 978-3-540-45826-5 (eBook)
DOI 10.1007/978-3-540-45826-5

Cataloging-in-Publication Data applied for

Bibliographic information published by Die Deutsche Bibliothek
Die Deutsche Bibliothek lists this publication in the Deutsche Nationalbibliografie; detailed bibliographic data is available in the Internet at <http://dnb.ddb.de>.

This work is subject to copyright. All rights are reserved, whether the whole or part of the material is concerned, specifically the rights of translation, reprinting, reuse of illustrations, recitation, broadcasting, reproduction on microfilm or in other ways, and storage in data banks. Duplication of this publication or parts thereof is permitted only under the provisions of the German Copyright Law of September 9, 1965, in its current version, and permission for use must always be obtained from Springer-Verlag. Violations are liable for prosecution act under German Copyright Law.

Springer-Verlag Berlin Heidelberg New York
a member of BertelsmannSpringer Science + Business Media GmbH

http://www.springer.de

© Springer-Verlag Berlin Heidelberg 2003
Softcover reprint of the hardcover 1st edition 2003

The use of general descriptive names, registered names, trademarks, etc. in this publication does not imply, even in the absence of a specific statement, that such names are exempt from the relevant protective laws and regulations and therefore free for general use.

Typesetting: Camera ready by author
Cover-Design: de'blik, Berlin
Printed on acid free paper 62/3020/kk - 5 4 3 2 1 0

Preface

The monograph text is based on lectures delivered by author during many years for students of Applied Iechanics Department of Bauman Ioscow State Technical University. The monograph includes also analitical results of scientific research obtained in collaboration with industry.

Progress in developing new equipment has called for a better understanding of the physical peculiarities pertaining to the action of designed structures in real conditions. This is necessary for increasing the accuracy of the analysis and making these structures more reliable.

It has been found that classical determined perturbations are not principal and that determinism-based methods of classical mechanics prove insufficient for understanding and explaining physical effects that arise at the operation of instruments located on moving objects, the vibration of rocket engines, the motion of a vehicle, and the action of wind and seismic loads. Therefore the necessity arose for devising a new physical model to analyze these dynamic processes and, in particular, for creating a new mathematical apparatus that would allow us to take into account non-deterministic external excitations. The theory of random processes that had been developed well enough as applied to problems of radio engineering and automatic control, where the effect produced by random excitations appeared to be commensurable with that of deterministic excitations and where the ignoring of the random excitations would bring about incorrect results, became such an apparatus. Therefore the theory of random processes began to be used for the solution of specific problems relevant to radio engineering, in particular, earlier than for the analysis of mechanical systems where the random excitations were frequently being ignored. In many applied problems such ignoring of random excitations, the more so really small ones, is quite allowable and their solution does not require the use of statistical mechanics. If, however, random excitations are comparable (in probability characteristics) with known forces and especially when only random excitations act on the system, the classical methods of analysis become unacceptable and the obtainment of numerical results demands the use probability-based methods. Therefore introducing statistical methods of analysis in practical design is a necessary condition of making reliable structures.

It is impossible to increase the reliability of new equipment designed for operation in extreme conditions and meeting the most stringent requirements to operation accuracy and reliability, and at the same time to achieve a reduction in the material input per unit of consumption, without the use of statistical mechanics.

The first four chapters are devoted to the fundamentals of the theory of probabilities and random processes. Consideration is given to random quantities and functions and their probability characteristics; probability density distribution functions; and mathematical expectations and variances. Different types of distribution laws encountered in practical problems are expounded. Non-stationary and stationary random processes that are of practical importance in analyzing the vibrations of mechanical systems are examined. Principal results of the special theory of stationary random functions and the implementation of spectral presentation of stationary random functions at steady-state vibration analysis are described, and the theory of Markovian processes is discussed.

Chapters 5 – 8 are concerned with the random vibrations of systems with a finite number of degrees of freedom and of systems with distributed parameters. The theory of random vibrations is presented similarly to the classical theory of vibrations that allows us to establish in the most obvious manner in what way these sections of mechanics (determined vibrations and random vibrations) are related and where they differ. The methods of analyzing random vibrations presented in the textbook make it possible to investigate dynamic processes that arise in mechanical systems (i.e. to determine the probabilistic characteristics of the generalized coordinates of a system and of their derivatives for a system with a finite number of degrees of freedom, and to obtain the probabilistic characteristics of the state of stress and strain of a system with distributed parameters).

The ninth chapter deals with the fundamentals of the theory of reliability and analyzes numerical methods of determining the probability of no-failure operation of mechanical systems under a single or small number of loading cycles and at a confined time of the process.

The use of methods of statistical mechanics in practical design becomes a possibility only when statistical information about random perturbations is available, but there are cases where we cannot obtain such information to the extent required for the design. In our presentation of the theoretical fundamentals of statistical mechanics we assume that the necessary information about random perturbation is known. In engineering practice, however, the situation may be different due to a very large volume of experimental investigations involved that sometimes turns out to be unfeasible because of technical difficulties or inadequate financing. The problem of obtaining the probability characteristics of perturbations is much more difficult than that of finding a subsequent solution to equations of the state of a system. Therefore this textbook contains a chapter expounding the theory and numerical methods of analyzing problems of the dynamics of mechanical systems when the available information about random perturbations is insufficient for a design that uses statistical mechanics.

Moscow, June 2002 *Valery Svetlitsky*

Contents

Introduction

When investigating dynamic processes that arise, for example, in mechanical systems, we often have to deal with the analysis of possible actions, the nature of which is not completely clear to us. These actions can be caused by both external uncontrollable (random) excitations and uncontrollable variations in the geometry and parameters of a system.

For instance, actions of this kind include the irregularities of a road or an airfield pavement that cause vibrations of objects moving on them (Fig. 0.1), the gasdynamic and technological misalignments of a jet engine thrust (Fig. 0.2), and the scatters of the parameters of amortization systems (Fig. 0.2). When these uncontrollable actions of external forces and scatters of the parameters of a structure do not produce any essential troubles in the behaviour of a system and can be neglected, we obtain a practically exact solution. This point of view is characteristic of classical mechanics and, in particular, of the classical theory of vibrations, where it is usual to assume that at the given initial conditions (known with certainty) and known forces for every instant of time there is a unique state of the system and their time sequence is developed in a uniquely determined "trajectory". It is common practice to call such one-to-one relationship between the state of the system and time as deterministic. A deterministic relationship between the state of a system and time completely excludes the presence of various uncontrollable actions in nature. The classical theory of vibrations is confined to deterministic processes, practically not focussing attention on the analysis of random processes.

Fig. 0.1.

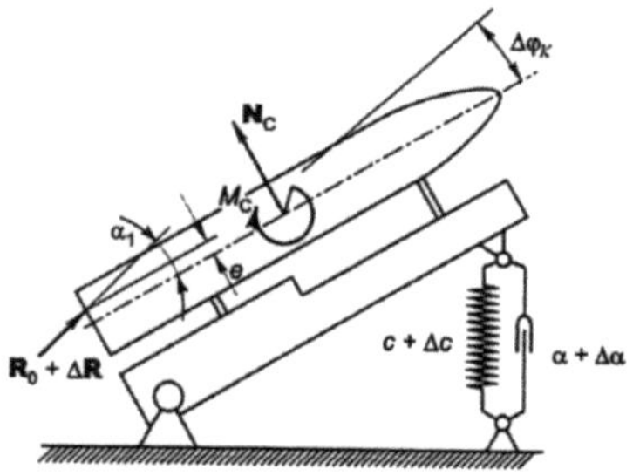

Fig. 0.2.

There was a period in the development of classical engineering when scientists and designers considered randomness an annoying obstacle, which, theoretically, could be avoided by a more careful conduct of the experiment. They also believed that the scatters of the initial data and forces were of no important significance and could be made arbitrary small using more precise instruments. The microworld was the only domain where they admitted the true role of randomness (works of Maxwell and Bolzmann on the theory of gases), whereas in the macrouniverse it was regarded as a consequence of our superficial knowledge of the laws of nature. Laplace gave an exact wording to this approach: "An intelligent being that at each given instant would know all driving forces of nature and have a complete picture of its state could, provided his mind were able to make a sufficient analysis of these data, express by one equation both the motion of the world's largest bodies and that of the smallest atoms. Nothing would remain unknown to this creature capable of taking a simultaneous view of both past and future" (Laplace. The Analytical Theory of Probabilities). Some difficulties arise, however, when one tries to put this statement in practice. They are connected with the obtainment of information about the position of the bodies of a system and their velocities at a given instant of time. It is practically impossible to find out these parameters. For example, the motion of the molecules of a gas can be described by differential equations, but to solve them it is necessary to have the initial conditions at a given instant of time, i.e. we must in a trice get information on the position of the molecules in space, which is practicable only during its transmission at an infinite velocity, and this cannot be achieved, because no signals can be transmitted at a velocity greater than that of light. Therefore the problem of obtaining information about the position of the molecules takes on fundamental importance. Another tacitly used assumption to the effect that absolutely exact measurements are theoretically possible (we can obtain absolutely exact values of the molecules' coordinates and of their first derivatives). This contradicts the Heisenberg uncertainty principle. Thus, it is in essence impossible to obtain accurate information on the initial state of the gas molecules, and hence we cannot predict the behaviour in time. A strictly one-to-one relationship between cause and effect requires continuous copying of the surrounding world, which absolutely excludes the occurrence of a new, qualitatively different phenomenon. If it were really so, the evolution of the world with its continuous change and qualitative variation in life forms would be impossible. All this suggests that randomness is not a consequence of our ignorance but an objective reality. The Gauss principle of least constraint to derive which we use the methods developed for the analysis of probability problems can be a good example proving the reality of randomness.

The Gauss principle of least constrain is formulated in the following way [12, 43]: the actual motion of a system under the action of forces and imposed boundary constraints without friction differs from motions being executed under the same initial conditions by the property that for the actual motion the measure of a deviation from a free motion, i.e. the constraint is a minimum. By the constraint at a given instant Gauss means a measure of deviation of a system moving under the action of external active forces (with due account of holonomic and nonholonomic constraints imposed on the system) from a free motion, which it would have since the considered instant under the action of the same external forces, if the constrains imposed on it were eliminated beginning from that instant. Let us first recall the mathematical formulation and the notation of this principle. Then we consider the positions of a material point mass m_i at the instants of time t and $t + \Delta t$ (Fig. 0.3). Now let us expand the function $\mathbf{r}_i\,(t + \Delta t)$ in a series and confine ourselves by the square-law part of expansion

$$\mathbf{r}_i\,(t + \Delta t) = \mathbf{r}_i\,(t) + \dot{\mathbf{r}}_i \Delta t + \frac{1}{2}\ddot{\mathbf{r}}_i \Delta t^2. \tag{0.1}$$

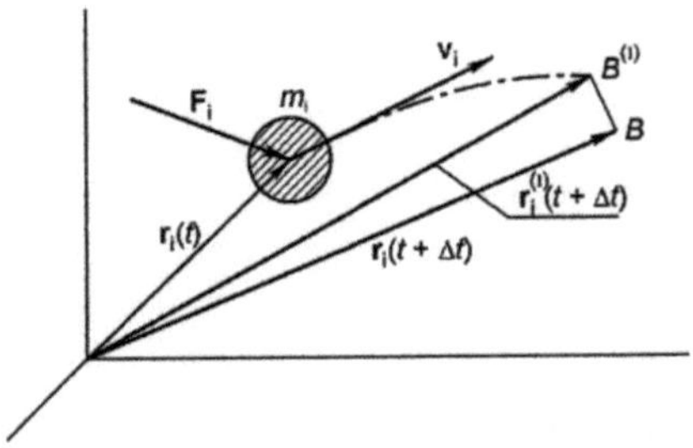

Fig. 0.3.

The vector $\mathbf{r}_i\,(t + \Delta t)$ characterizes the position of the mass m_i with due account of the constraints imposed on its motion. If the constraints were absent, the position of the point at the instant $t + \Delta t$ would be defined by a vector $\mathbf{r}_i^{(1)}\,(t + \Delta t)$, which can be presented within the third order infinitesimals as

$$\mathbf{r}_i^{(1)}\,(t + \Delta t) = \mathbf{r}_i^{(1)}\,(t) + \dot{\mathbf{r}}_i^{(1)} \Delta t + \tfrac{1}{2}\ddot{\mathbf{r}}_i^{(1)} \Delta t^2,$$

or (as $\mathbf{r}_i^{(1)}(t) = \mathbf{r}_i(t)$, $\dot{\mathbf{r}}_i^{(1)}(t) = \dot{\mathbf{r}}_i(t)$)

$$\mathbf{r}_i^{(1)}(t + \Delta t) = \mathbf{r}_i(t) + \dot{\mathbf{r}}_i \Delta t + \frac{1}{2}\ddot{\mathbf{r}}_i^{(1)} \Delta t^2. \tag{0.2}$$

In the absence of constraints the equation of motion of the point takes the form

$$m_i \ddot{\mathbf{r}}_i^{(1)} = \mathbf{F}_i, \tag{0.3}$$

therefore we have

$$\mathbf{r}_i^{(1)}(t + \Delta t) = \mathbf{r}_i(t) + \dot{\mathbf{r}}_i \Delta t + \frac{1}{2}\frac{\mathbf{F}_i}{m_i} \Delta t^2. \tag{0.4}$$

The distance between the positions of the point at the instant $t + \Delta t$ during the action of constraints and without them is equal to

$$\mathbf{B}^{(1)}\mathbf{B} = \Delta \mathbf{r}_i = \mathbf{r}_i(t + \Delta t) - \mathbf{r}_i^{(1)}(t + \Delta t) = \frac{1}{2}\left(\ddot{\mathbf{r}}_i - \frac{\mathbf{F}_i}{m_i}\right)\Delta t^2. \tag{0.5}$$

The vector $\Delta \mathbf{r}_i$ defines the deviation of a material point at its actual motion from the position in which it would find itself at free motion. For a measure of deviation of a point from its free motion Gauss takes a quantity z_i, proportional to the square of deviation $|\Delta \mathbf{r}_i|^2$, which is referred to as a "constraint"

$$z_i = \frac{1}{2} m_i \left(\ddot{\mathbf{r}}_i - \frac{\mathbf{F}_i}{m_i}\right)^2. \tag{0.6}$$

For all points of the system we have

$$z = \frac{1}{2}\sum_{i=1}^{n} m_i \left(\ddot{\mathbf{r}}_i - \frac{\mathbf{F}_i}{m_i}\right)^2 \tag{0.7}$$

or in scalar form

$$z = \frac{1}{2}\sum_{i=1}^{n}\left[\sum_{j=1}^{3} m_i \left(\ddot{x}_{ij} - \frac{F_{ij}}{m_i}\right)^2\right] = \sum_{i=1}^{n}\frac{m_i}{2} z_i^{2(1)}, \tag{0.8}$$

where

$$z_i^{2(1)} = \sum_{j=1}^{3}\left(\ddot{x}_{ij} - \frac{F_{ij}}{m_i}\right)^2.$$

The Gauss principle is that at any instant of time the constraint for the actual motion of a system as compared with a kinematically possible motion has the least value

$$z = \sum_{i=1}^{n} \frac{m_i}{2} z_i^{2(1)} = \min . \tag{0.9}$$

The idea of the measure of deviation of a system from a free motion in the form of the sum of quantities proportional to the squares of deviations of the material points of a system is closely connected with the works of Gauss on the theory of errors, in particular, with the method of the least squares allowing us to determine the unknown quantity with the least root-mean-square error. The method of the least squares relevant to the analysis of random phenomena results in relationships similar to relationship (0.9).

In the theory of the least squares the basic problem is formulated as follows: there are unknown functions y_i and their known values y_{i0}, then the measurement errors are

$$\delta_i = y_i - y_{i0} .$$

If y_{i0} have different variances σ_i, then the multipliers p_i, dependent on σ_i are entered. Assuming that the mistakes δ_i obey the normal law, the following theorem is proved. For the unknown y_i to have the most probable values, it is necessary that the expression

$$z = \sum_{i=1}^{k} p_i \delta_i^2$$

take the minimal value, and this expression at

$$p_i = \frac{m_i}{2}, \quad \delta_i^2 = z_i^{2(1)}$$

and $k = n$ transforms into expression (0.9).

Turning back to the Gauss principle, we can formulate (with due account of the result adopted from the theory of errors and cited above) in terms of the theory of probabilities, namely, the true motion of a system differs from its kinematically possible one by having the greatest probability. The relation between the method of least squares and the Gauss principle of least constraint represents something more than a simple analogy, i.e. the difference between the actual motion of a body and its possible motion is probabilistic in character. The Gauss principle has an essential advantage over that of d'Alambert: it enables us to obtain the equations of motion of a system at any nonholonomic constraints. This means that the Gauss principle is the most general principle of mechanics and allows a probabilistic interpretation! In modern physics we have come to realize with the utmost clarity the irrefutable fact that randomness cannot be completely excluded from analysis and should be taken into account as a component of any theory.

Not all researchers, however, have shared this point of view. Many physicists and mechanics considered taking randomness into account a temporary deviation from classical theories and believed that in the process of storing knowledge the role of randomness would be reduced to zero. According to them, a much better understanding of physical processes in the future will make it possible to explain the phenomena that seem random to us today as a manifestation of entirely deterministic processes now inaccessible to our mind.

The proponents of the opposite viewpoint on randomness maintain that only the probability methods of analyzing the processes can give us a correct answer. In the opinion of the supporters of the probability interpretation of the laws of nature, classical mechanics is a particular case of statistical mechanics impossible to achieve in practice because, in fact, the exact value of initial conditions and forces cannot be found out not only in the microworld, but in the macrouniverse as well.

There are also scientists who argue that determinism is a mathematical convention allowing us to simplify the analysis of many complex processes, in which we can limit our analysis to mere average values.

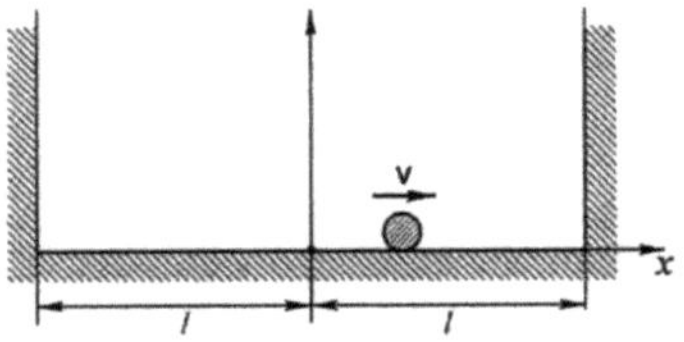

Fig. 0.4.

Consider a ball (Fig. 0.4) [11] moving in the direction x between two rigid walls positioned at a distance $\pm l$ from the origin of coordinates. Let us assume that at the initial instant of time the ball was in the origin of coordinates and obtained the velocity $\dot{x}_0$ under the action of an impulse of force. Theoretically, we can predict the subsequent motion of the ball and its exact position between the walls at any instant of time t (provided the sphere is producing a totally elastic impact on the walls). This would make the motion of the ball completely deterministic. But in a real rather than theoretical set up the motion acquires a different pattern because it is impossible to determine the initial velocity with absolute accuracy. We know this velocity with a certain error $\Delta\dot{x}_0$, no matter how small it may be. Therefore the position of the ball (coordinate x) will also be determined with an error equal to

$$\Delta x = \Delta\dot{x}_0 t,$$

which at sufficiently large time will exceed the distance $2l$ between the walls. Henceforward we can only tell that the ball is somewhere between the walls — the answer that we can hardly call deterministic.

The considered points of view are diametrically opposite and have a mere philosophical value. In the applied context, none of the physical processes may be regarded as either strictly deterministic or completely random. As to what category a given specific process falls into, it depends entirely on the accuracy of the available initial data, on what final results we are after and on how accurate they must be.

There are many applied problems that can be described by exact mathematical relationships. Practically, we may consider them deterministic (e.g., the motion of a satellite in an orbit). However, there is no less extensive class of physical processes that are obviously not deterministic in character, such as the vibrations of a vehicle at its motion on a road with random irregularities (Fig. 0.1 *a*), the vibrations of elastic systems under the action of a random wind load (Fig. 0.5), the launch of rockets at random misalignments of thrust (Fig 0.5), etc. In the presented examples the effect of the action of random excitations plays a rather fundamental and sometimes the determining role. That is why we must not neglect random excitations. The principal general properties of random processes are the uncertainty of expected behaviour for any single realization of a process and the obviously pronounced statistical characteristics of a set of a large number of realizations. Designers show less interest than philosophers and theoretical physicists for our world in the context of it being casual or cause-determined. The former are much more interested in finding reasonable solutions to applied problems they tackle. If a designer knows everything that is necessary for the solution of a problem, can estimate the influence of the initial data's scatter on the final result and

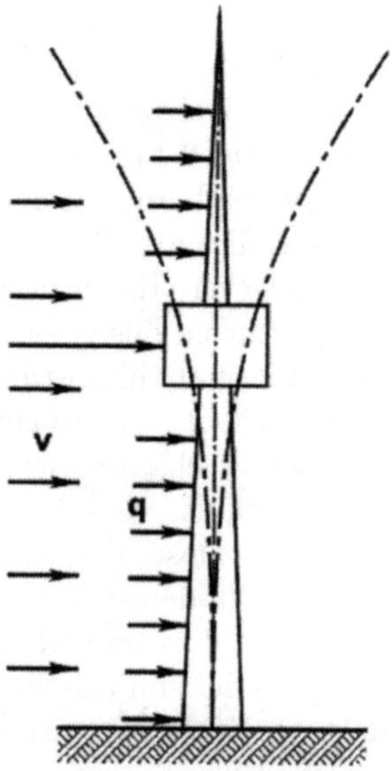

Fig. 0.5.

establishes that this influence may be neglected, he uses methods of classical mechanics. If the uncertainty of the initial data or an indeterminacy in specifying forces that act on the system are great, he has to address himself to methods allowing him to eliminate this uncertainty, in particular, to methods of probability theory and statistical dynamics. The idea that some time in the future this uncertainty will be eliminated is a poor consolation for the designer at the given moment.

One of the prejudices against the propagation of the probability methods of analyzing mechanical systems is connected with the impossibility of an unequivocal quantitative prediction of the system's behaviour as it can be done on the basis of the laws and methods of classical mechanics. Many researchers hold that the only kind of prediction having the right to be called scientific is an exact quantitative prediction of the future events. Some complain of the imperfection of statistical laws to the effect that they do not allow us to draw certain conclusions and make definite predictions concerning individual events. However, when it is required to predict results characterizing the behaviour of a large number of separate random events, statistical methods give more substantial information, and a prediction of the behaviour of a separate body based on these methods is as definite as that made using methods of classical mechanics.

The possibility of repeated realization of a random event in practically homogeneous conditions is indispensable for applying probability theory methods. The use of probabilistic methods of investigation becomes meaningful only at mass events. When using probabilistic methods in design practice, the principal difficulty is that the probability characteristics of stochastic functions can be obtained only given a large number of random process realizations. This can entail large technical difficulties in carrying out experiments or large economic expenses. In order to obtain the probability characteristics of random irregularities of a road or an airfield pavement (Fig. 0.1 *b*) we need a set of records of microprofiles of different sections of the road, and for each road (earth, Belgian pave, etc.) these records of microprofiles are different. Their obtainment is a very labour-consuming work. However, the process of accumulating statistical information about random perturbations goes on and therefore the role probabilistic methods becomes increasingly important.

Today the correlation theory is one of the principal methods of analyzing random processes. It allows us to obtain similar probability characteristics of the output at known probability characteristics of the input. We emphasize once again that these characteristics are meaningful as those of a set of processes and not of an individual one. If, for example, 1000 identical automobiles move at an equal speed on roads of the same type, it is possible to predict, on the average, how the given type of road (input) acts on this kind of automobile. For instance, we can determine the mathematical expectations and variances (output) in sections of the chassis frame. If only one automobile moves on the limited section of the road, it is impossible to obtain the probability characteristics of the output (without additional assumptions). A single launch of a rocket illustrates the point even more convincingly (Fig. 0.2).

During the motion along the guide random forces caused by the misalignments of thrust (input) act on the rocket, therefore at the moment of leaving the guide it obtains random scatters (output) of an angle $\Delta\varphi_k$ and an angular velocity $\Delta\dot{\varphi}_k$. Using the correlation theory we can formally determine the mathematical expectations (or rather mean values) of the scatters ($\Delta\varphi_k$ and $\Delta\dot{\varphi}_k$) and their variances, but the reliability of these results is very small.

In the rocket-launch example we deal with a more complex problem of taking into account the uncertainty of variation of thrust misalignments, because we do not know which of the possible random laws of thrust misalignment variation will be realized at the given single launch. If 1000 rockets were launched, we could use the probability characteristics of the input (of thrust misalignments) and obtain the probability characteristics of the output ($\Delta\varphi_k$ and $\Delta\dot{\varphi}_k$) for the whole series of launches. It cannot be done at a at a single launch, but we may assert that with an increase in the number of launches the obtained values of the output ($\Delta\varphi_k$ and $\Delta\dot{\varphi}_k$) approach, on the average, ever more closer their theoretical values. Usually, during the analysis of random processes in mechanical systems, in particular, of non-stationary processes it is tacitly implied (an assumption enabling us to use the mathematical apparatus of the theory of random processes) that the condition of making the process massive is carried out. As regards stationary random processes the acceptance of the ergodicity-of-the process hypothesis allows us to consider only one realization instead of a large number of realizations and to obtain information sufficient (within the framework of the correlation theory) to predict the behaviour of the system. This approach to the analysis of random processes appears to be quite sufficient for many applied problems. That is why the correlation theory has become so widespread.

Having an algorithm making it possible to determine the probability of no-failure operation, we can increase this probability by changing the structural parameters of a system, i.e. to design a more reliable product. It does not always happen, however, that the probability estimations of quality prove acceptable. Not infrequently it is required that the warranted performance criteria of the process be satisfied. For example, during the launch of one rocket the need arises for it assuredly hitting the given area. The 0.9 probability estimate of such a hit, for instance, does not guarantee a success.

Any estimation of a random process or of test outcomes by probability theory methods is not absolute and may have different interpretations depending on the requirements to final results. In one of his poems M.G. Kendall [4] aptly depicts how varied viewpoints on the same final result can be. The poem is given in Appendix 4.

So, a designer should independently choose methods of analysis on the basis of available information about random perturbations.

1. Fundamentals of the Probability Theory and the Theory of Random Processes

1.1 Brief Information on the Probability Theory

1.1.1 Basic Concepts of the Probability Theory

In many fields of technology we have to deal with special phenomena which are usually called random phenomena. Let us consider, for example, the process of manufacturing parts of the same type. We may establish, that the dimensions of the parts will vary about a certain predetermined value. Since these deviations are of random nature, the measurements of the finished parts do not allow us to predict the dimensions of the next part. For large batches of the parts, however, dimensional deviations begin to follow certain laws, which are studied by a special mathematical discipline – the probability theory, that reflects the laws inherent in random events (phenomena) of a mass character. There are many monographs on the probability theory containing a detailed discussion of the basic concepts and methods of that theory as well as of the random functions theory. This chapter, therefore, introduces only those concepts and results related to the probability theory which have been used in the subsequent chapters of the book. One of the principal virtues of the probability theory, that enables us to use the latter effectively, for example, in the mechanical structures design, is the possibility to estimating quantitatively such emotional concepts as "probably", "hardly probably", "highly probable" etc. We know, that in order to design a machine, an instrument or a flying vehicle, it is necessary to obtain the numerical values of its structures parameters and of its quality (serviceability) criteria including the probability quality criteria. To compare structures according to the probability criteria we must know the numerical values of probabilities (for example, the probability of no-failure operation). The probability theory and the sections devoted to the statistical mechanics of mechanical systems, based on this theory, allow us to accomplish all of this.

Phenomena that either may or may not occur, are referred to as random events. For example, an air gust, acting on a television tower (Fig. 1.1) is a random event. The event, that unavoidably occur, is called a certain event. An event, that definitely can not occur is called an impossible event. In order to establish whether some random event will happen or not, it is necessary to carry out an experiment or, as it usually said, to make a trial.

Investigations show, that events, random at a single trial, at a large number of trials (under invariable conditions of trial) begin to follow some non-random laws, which have come to be known as probability laws. The number of event occurrences at trials is characterized by the frequency of an event W occurrence. The ratio of the number of trials n, at which the event occurred, to the total number of performed trials N is referred to as the frequency of event W.

This implies, that the results of events in a series of trials are mutually independent (random events are called independent, if the occurrence or lack of one of them in no way affects the occurrence of another).

The frequency of events is to some extent an inherent characteristic of a phenomenon. It is a random quantity, however, depending on a particular series of trials. At a very large number of trials the frequency W almost ceases to vary, approaching some quantity P, which is referred to as probability.

It is worth noting, that the character of a frequency approaching the probability as the number of trials increases somewhat differs from the tend to a limit as it is understood in higher mathematics. When in higher mathematics we say that variable x_n tends to a constant limit a with an increase of n, we mean, that the difference $|x_n - a|$ becomes less than any positive number ε for all values n, beginning from some sufficiently large number. We cannot make such statement about the frequency of an event and its probability, because it is quite possible, that at a large number of trials the frequency of an event will deviate considerably from its probability. The more the number of trials, however, the less probable is such pronounced deviation. The concept of convergence in probability is introduced in the probability theory, i.e. the random quantity x_n converges in probability to the quantity a as n increases, when the probability of an inequality $|x_n - a| < \varepsilon$ indefinitely approaches unity as n increases. This statement is the content of the Bernoulli theorem – "when the number of homogeneous independent trials increases indefinitely we may safely say, that the frequency of an event will differ from its probability as little as one likes".

Therefore, if event A repeats a large number of times (N) and in so doing has an indication B in n cases, and the results of the events in this sequence are mutually independent, the probability of indication B occurrence is

$$P(B) = \lim_{N \to \infty} \frac{n}{N}.$$

Knowing the probability of an event, we may predict, without performing any trials, the frequency of its occurrence at a large number of trials. We may also state that the probability of an event is a measure of the possibility of its occurrence at one trial.

The statistical definition of a probability allows us to determine:

1) The probability of a certain event (equal to unity);

2) The probability of an impossible event (equal to zero);

3) The probability of an arbitrary random event A (equal to a positive number which is not exceeding a unity, i.e. $0 \leq P(A) \leq 1$).

If we have two events A and B, and the probability of event A does not depend on whether event B occurred or not, such two events are referred to as independent. Event A is termed dependent on event B, if the probability of event A varies depending on whether event B occurred or not. The probability of event A, calculated on condition that event B occurred, is called the conditional probability of event A and designated as $P(A|B)$

Basic Probability Theorems

1. **The Probability Addition Theorem** The probability of a sum of two incompatible events is equal to the sum of their probabilities, i.e.

$$P(A+B) = P(A) + P(B). \tag{1.1}$$

2. **The Probability Multiplication Theorem** The probability of a product (coincidence) of two events A and B is equal to the product of event A probability by the conditional probability of event B, i.e.

$$P(AB) = P(A)\,P(B\,|A). \tag{1.2}$$

The probability of a product of two events A and B can be expressed in terms of the event A conditional probability. In this case

$$P(AB) = P(B)\,P(A\,|B).$$

The probability of the joint occurrence of two independent events is equal to the product of their probabilities, i.e.

$$P(AB) = P(A)\,P(B). \tag{1.3}$$

3. **The Formula of Total Probability** The corollary of the probability addition theorem and the probability multiplication theorem taken together is a so-called formula of total probability. Let it be required to determine the probability of an event A, which can occur together with one of the events B_j $(j = 1, 2, \ldots, n)$ that form a complete group of incompatible events, termed hypotheses. Several events in the given trial form a group of events, if, as a result of trial, only one of them unavoidably occurs. As the hypotheses B_j constitute a complete group, event A can appear only in a combination with any of these hypotheses, i.e.

$$A = B_1A + B_2A + \ldots + B_nA.$$

Since the hypotheses B_j are incompatible, the combinationsB_jA are incompatible too. Consequently, applying the addition theorem to them, we get

$$P(A) = \sum_{j=1}^{n} P(B_j A).$$

Taking advantage of the multiplication theorem, we finally have for each of the terms in the right-hand side of the obtained relation

$$P(A) = \sum_{j=1}^{n} P(B_j) P(A|B_j). \tag{1.4}$$

The expression (1.4) is referred to as the formula of total probability.

4. **The Bayesian Formula (Theorem of Hypotheses)** This formula is a corollary of the multiplication theorem and the formula of total probability. Let there be a complete group of incompatible hypotheses $B_j (j = 1, 2, \ldots, n)$. We know the probabilities of these hypotheses before trial. They are equal to $P(B_j)$. A trial has been performed resulting in the occurrence of an event A. The question arises: how should we change the probabilities of the hypotheses in connection with this occurrence?

Let us apply the formula for conditional probability $P(B_j|A)$ to each hypothesis and take advantage of the multiplication theorem

$$P(AB_j) = P(A) P(B_j|A) = P(B_j) P(A|B_j).$$

It follows from the obtained relation that

$$P(B_j|A) = \frac{P(B_j) P(A|B_j)}{P(A)}$$

Using the formula for total probability, we finally get

$$P(B_j|A) = \frac{P(B_j) P(A|B_j)}{\sum_{j=1}^{n} P(B_j) P(A|B_j)} \tag{1.5}$$

1.2 The Distribution Function and the Probability Density of a Random Variable

The Distribution Function One of the principal concepts of the probability theory is that of the random variable. A quantity will be called random, if, as a result of trial, it takes this or that value, unknown in advance. A random quantity can be discrete or continuous. A random variable will be completely defined from the probability point of view, if we know the probability of occurrence of each of the values, taken by the random variable. Such correspondence is referred to as a distribution law of a discrete random variable. The distribution law of a discrete random variable X, which will take one of $x_j (j = 1, 2, \ldots, n)$ possible values as a result of trial, can be presented as a table (or a row of distribution):

Table 1.1.

x_1	x_2	x_3	...	x_n
P_1	P_2	P_3	...	P_n

Knowing the row of a random discrete quantity distribution, we may obtain the function defining this distribution, which takes the form

$$F(x) = \sum_{x_j < x} P(X = x_j). \tag{1.6}$$

The inequality $x_j < x$ under the summation sign indicates, that the summation covers all values of x_j, which are smaller than x. The function $F(x)$ is referred to as the distribution function of a random quantity x. When a variable x goes through the possible value X, the function $F(x)$ varies stepwise, and the magnitude of the jump is equal to the probability P_j of the value x_j occurrence. The incompatible events x_j form a complete group, therefore

$$F(x) = \sum_{j=1}^{n} P_j = 1,$$

i.e. the distribution function can not exceed unity.

For a continuous random quantity the probability of an event means the probability of an event $X < x$, where x is some current variable. In this case the probability $P(X < x)$ is some function of x, which by analogy with a discrete random quantity is referred to as a distribution function:

$$F(x) = P(X < x). \tag{1.7}$$

Sometimes the function $F(x)$ is called a distribution function or an integral distribution law. The distribution function is the most universal characteristic of random quantities, both discrete and continuous.

The distribution function should meet the condition

$$0 \leq F(x)) \leq 1. \tag{1.8}$$

It follows from the function $F(x)$ definition (see (1.7)), that at any arbitrary large x the function $F(x)$ can not exceed unity (as $F(x)$ is a probability and the probability can not be more than unity)

$$F(\infty) = \lim_{x \to \infty} F(x) = 1.$$

As the probability $P(X < x)$ can not be smaller than zero, it follows from (1.7), that

$$F(-\infty) = \lim_{x \to -\infty} F(x) = 0.$$

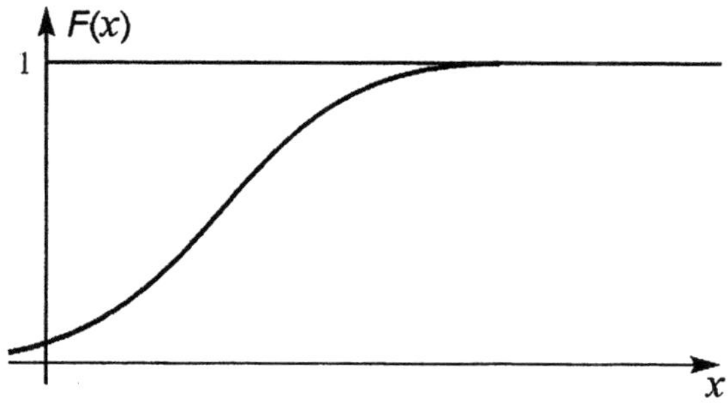

Fig. 1.1.

The approximate view of a distribution function for a continuous random quantity is shown in Fig. 1.1. When solving applied probability problems, we often find it necessary to determine the probabilities of a random variable X taking a value limited by a certain interval, for example, that of (a, b). For definiteness, let us agree to include the left end of the interval (a, b) in the latter, leaving its right end b outside it. Then, the fact that the random quantity X falls within the interval (a, b) is tantamount to the fulfillment of the inequality

$$a \leq X < b.$$

Let us express the probability of this event in terms of the random variable X distribution function. Let us consider three events: an event , consisting in that $X < b$; an event , consisting in that $X < a$; an event C, consisting in that $a \leq X < b$.

As

$$A = B + C,$$

then from the probabilities addition theorem (1.1) we have

$$P(X < b) = P(X < a) + P(a \leq X < b)$$

or

$$F(b) = F(a) + P(a \leq X < b).$$

Therefore we finally get

$$P(a \leq X < b) = F(b) - F(a). \tag{1.9}$$

Probability Density Let us consider a continuous random quantity X with a known continuous and differentiable distribution function $F(x)$. Let us determine the probability of a random quantity falling within an interval $(x,\, x + \Delta x)$

$$P(x < X < x + \Delta x) = F(x + \Delta x) - F(x), \tag{1.10}$$

i.e. the probability P is the increment of the distribution function in this section. It is clear, that the smaller Δx, the smaller the probability of X falling within the interval $(x,\, x + \Delta x)$. Therefore the left-hand side of equality (1.10) may be replaced by

$$\Delta P(x < X < x + \Delta x).$$

Let us consider a ratio of a probability ΔP to the length of an interval at Δx approaching zero

$$\lim_{\Delta x \to 0} \frac{\Delta P}{\Delta x} = \lim_{\Delta x \to 0} \frac{F(x + \Delta x) - F(x)}{\Delta x}.$$

Limiting ourselves to the linear part of the expansion $F(x + \Delta x)$ in series we get

$$\frac{\mathrm{d}P}{\mathrm{d}x} = F'(x). \tag{1.11}$$

Let us introduce the notations

$$F'(x) = f(x), \tag{1.12}$$

therefore

$$\mathrm{d}P = f(x)\,\mathrm{d}x.$$

The function $f(x)$ is referred to as the continuous random variable X distribution density. It follows from (1.12) that

$$F(x) = \int_{-\infty}^{x} f(x)\,\mathrm{d}x. \tag{1.13}$$

Letting the upper limit to approach infinity, we shall get

$$F(\infty) = \int_{-\infty}^{\infty} f(x)\,\mathrm{d}x = 1.$$

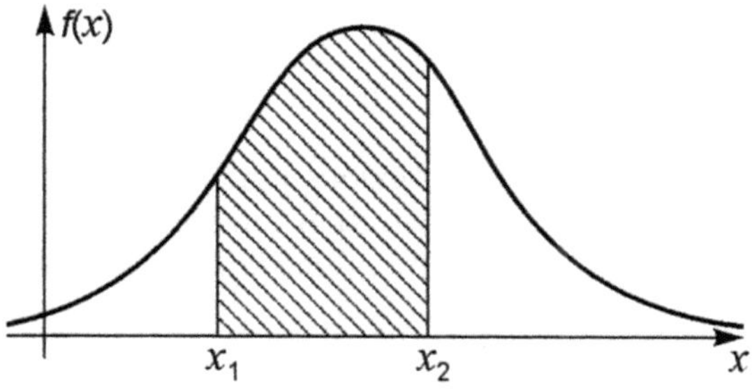

Fig. 1.2.

The probability of the random quantity X being in the interval (x_1, x_2) is equal to

$$P(x_1 \leq X \leq x_2) = \int_{x_1}^{x_2} f(x)\, dx. \tag{1.14}$$

An approximate form of the probability density function is shown in Fig. 1.2. The probability $P(x_1 \leq X \leq x_2)$ is numerically equal to the shaded area in Fig. 1.2. The distribution function of a random quantity and its probability density are different forms of the random quantity distribution law.

1.3 Numerical Characteristics of Random Quantities and Their Principal Properties

The distribution law given by the function $F(x)$ or by the density of distribution $f(x)$ is the exhaustive characteristic of a random variable (or a system of random variables). In practice, however, this exhaustive characteristic can not always be obtained owing to the limited nature of experimental results either because of the complexity of their acquisition or due to of their large cost. In such cases the approximate description of a random quantity obtained with the help of the minimum number of nonrandom characteristics reflecting the most essential features of distributions are used instead of distribution laws. It is often sufficient to point out only separate numerical parameters characterizing the essential properties of a random variable distribution, for example, a mean value, about which the possible values of a random quantity are clustered or a number describing the degree of the scatter of the random quantity about its mean value. Such nonrandom characteristics, which allow us to express the most essential features of a distribution in a condensed form, are referred to as the numerical characteristics of a random quantity. For example, such numerical (nonrandom) characteristics for a single random quantity X are its mathematical expectation and variance.

Expectation An arithmetical mean value is the simplest numerical characteristic of a discrete random quantity X in the given series of trials

$$M[X] = m'_x = \frac{1}{N} \sum_{i=1}^{N} x_i, \tag{1.15}$$

where m'_x is the mean value of a random quantity; N is the number of trials; x_i are the values of a random quantity, which it took at these trials; M is an averaging operation.

In technical literature two more notations for an averaging operation are also used: $\langle\ \rangle$, E. Later on the notation M is taken for an averaging operation.

If some values x_j, taken by a random variable X, repeat themselves n_j times, it is possible to divide all values x_j into R groups and to present the expression (1.15) as

$$m'_x = \sum_{j=1}^{R} \frac{n_j}{N} x_j = \sum_{j=1}^{k} W_j x_j, \qquad \left(\sum_{j=1}^{k} n_j = N \right), \tag{1.16}$$

where W_j is the frequency (statistical probability P_j) of x_j occurrence. Therefore the mean value of a discrete random variable is equal to

$$m_x^{(1)} = \sum_{j=1}^{k} x_j P_j.$$

The mathematical expectation for a continuous random variable X is equal to

$$M[x] = m_x = \int_{-\infty}^{\infty} x \, \mathrm{d}P = \int_{-\infty}^{\infty} x f(x) \, \mathrm{d}x. \tag{1.17}$$

Variance It is possible to take the mean value of any positive measure of a random quantity deviation from its mean value, for example, the square of a difference between the values of a random quantity and its mean value for characterizing a scatter of discrete random quantity values in the given series of trials

$$M\left[\overset{\circ}{X}{}^2\right] = D'_x = \sum_{i=1}^{k} \frac{n_i}{N} (x_i - m'_x)^2 = \sum_{i=1}^{k} \overset{\circ}{x}_i{}^2. \quad \left(\overset{\circ}{x}_i = x_i - m'_x\right), \tag{1.18}$$

where $\overset{\circ}{x}_i = x_i - m'_x$ is referred to as a centered random quantity.

The quantity D'_x is referred to as the statistical variance of a random quantity X. In practical calculations it is more convenient to use the quantity

$$\sigma'_x = \sqrt{D'_x}, \tag{1.19}$$

which is referred to as a statistical standard deviation.

For a continuous random quantity the variance and the standard deviation are equal to

$$D_x = M\left[\overset{\circ}{X}{}^2\right] = \int\limits_{-\infty}^{\infty} (x - m_x)^2 f(x)\, \mathrm{d}x$$
$$\sigma_x = \sqrt{D_x}. \tag{1.20}$$

It is possible to estimate the probability of a random quantity deviation from its mean value by some value α using the formula of P. Chebyshev

$$P(|X - m_x| \geq \alpha) \leq \frac{D_x}{\alpha^2}. \tag{1.21}$$

Principal Properties of Expectation and Variance

1. The expectation of a deterministic quantity is equal to this quantity

$$M[c] = \int\limits_{-\infty}^{\infty} cf(x)\, \mathrm{d}x = c\int\limits_{-\infty}^{\infty} f(x)\, \mathrm{d}x = c.$$

2. The expectation of a random quantity multiplied by a deterministic factor is

$$M[cX] = \int\limits_{-\infty}^{\infty} cxf(x)\, \mathrm{d}x = cm_x. \tag{1.22}$$

3. The variance of a deterministic quantity C is zero

$$M\left[\overset{\circ}{X}{}^2\right] = \int\limits_{-\infty}^{\infty} (x - m_x)^2 f(x)\, \mathrm{d}x = \int\limits_{-\infty}^{\infty} (c - c)^2 f(x)\, \mathrm{d}x = 0.$$

4. The variance of a random quantity multiplied by a deterministic factor is

$$M\left[\left(c\overset{\circ}{X}\right)^2\right] = c^2 \int\limits_{-\infty}^{\infty} (x - m_x)^2 f(x)\, \mathrm{d}x = c^2 D_x. \tag{1.23}$$

Example 1.1. A random force with known characteristics m_Q and D_Q is acting on the rod shown in Fig. 1.3. It is required to determine the expectations and variances of reactions, of the deflection in the point of a force Q application and of the maximum normal stress in the rod (in the section K). The flexural rigidity of a beam is EJ_x.

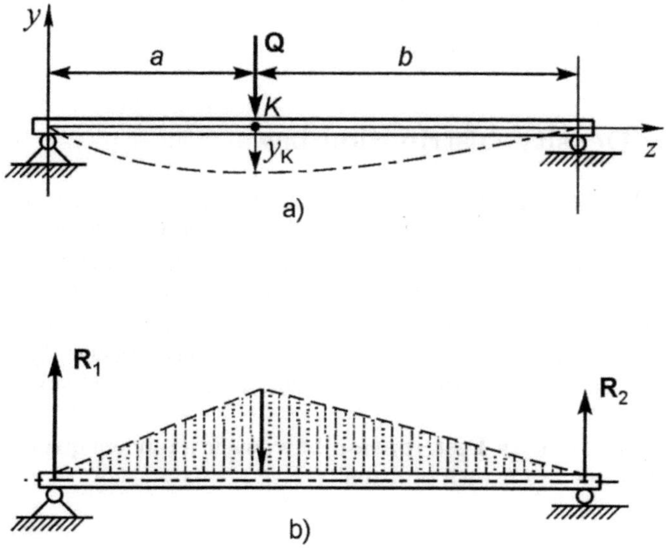

Fig. 1.3.

Let us determine reactions R_1 and R_2 and the deflection y_k in the point of force application and the maximum normal stress in the section according to the formulas obtained for deterministic forces:

$$R_1 = \frac{b}{a+b}Q; \qquad R_2 = \frac{a}{a+b}Q;$$

$$y_k = \frac{a^2b^2}{3EJ_x(a+b)}Q; \qquad \sigma_{\max} = \frac{M_{\max}}{W_x} = \frac{ab}{(a+b)W_x}Q.$$

Using the previous results, we shall obtain the probability characteristics of R_1, R_2, y_k and $\sigma_{\max}$:

$$m_{R_1} = \frac{b}{a+b}m_Q; \qquad m_{R_2} = \frac{a}{a+b}m_Q;$$

$$m_{y_k} = \frac{a^2b^2}{3EJ_x(a+b)}m_Q; \qquad m_{\sigma_{\max}} = \frac{ab}{(a+b)W_x}m_Q;$$

$$D_{R_1} = M\left[\overset{\circ}{R}{}_1^2\right] = \frac{b^2}{(a+b)^2}D_Q; \qquad D_{R_2} = \frac{a^2}{(a+b)^2}D_Q;$$

$$D_{y_k} = \left[\frac{a^2b^2}{3EJ_x(a+b)}\right]^2 D_Q; \qquad D_{\sigma_{\max}} = \frac{a^2b^2}{(a+b)^2W_x^2}D_Q.$$

Let us estimate a probability that the difference $(\sigma_{\max} - m_{\sigma_{\max}})$ will take a value exceeding the yield point of a beam material. Using the Chebyshev inequality (1.21), we shall get

$$P\left[\left(\sigma_{\max} - m_{\sigma_{\max}}\right) \geq \sigma_T\right] \leq \frac{a^2 b^2}{(a+b)^2 W_x^2} \frac{D_Q}{\sigma_T^2}.$$

1.4 Probability Density Distribution Laws

The laws of the probability theory represent a mathematical expression of real laws manifesting in mass random phenomena. The distribution laws of random quantities (and more general distribution laws of random functions) primarily apply to such mathematical laws. As a rule, random quantities distribution laws are determined on the basis of experimental investigations. Sometimes, however, the distribution laws can be obtained theoretically. An example of the theoretical determination of a distribution law is presented at the end of this paragraph.

1. **The Probability Density Distribution Law**, when a random quantity X takes a single value $x = a$ with probability $P = 1$.

In this case

$$f(x) = \delta(x - a), \tag{1.24}$$

where $\delta(x-a)$ is the delta function.

The principal properties of the delta function are given in Appendix 1.

2. **Probability Density Normal Distribution Law (Gaussian Law)** (Fig. 1.4). The normal distribution law is a most commonly encountered practical laws that is characterized by the probability density

$$f(x) = \frac{1}{\sigma_x \sqrt{2\pi}} \exp\left\{-\frac{(x - m_x)^2}{2\sigma_x^2}\right\}. \tag{1.25}$$

3. **Rayleigh Probability Density Distribution Law** In many applied problems random quantities can take only positive values (vibration amplitudes, the potential energy of an elastic system at random deformations, the

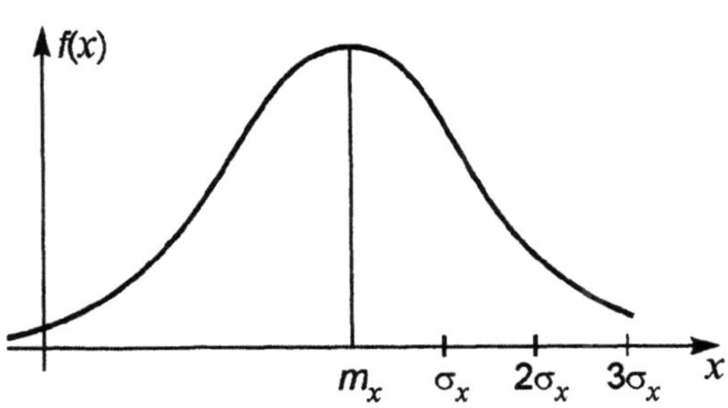

Fig. 1.4.

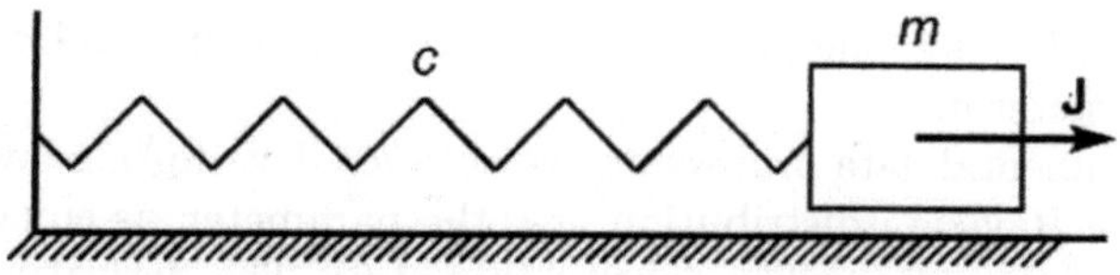

Fig. 1.5.

kinetic energy of a system at random velocities etc.). For example, an impulse of a random magnitude J is acting on a mass m (Fig. 1.5), imparting the latter a random velocity $\dot{x}$ equivalent to a random kinetic energy $\frac{m\dot{x}^2}{2}$, that is independent on the velocity $\dot{x}$ sign. Random variables taking only positive values frequently follow the Rayleigh distribution function

$$F(x) = \begin{cases} 1 - \exp\left\{-\dfrac{x^2}{2a^2}\right\}, & x \geq 0; \\ 0, \quad x < 0. \end{cases} \tag{1.26}$$

In this case, the probability density function (Fig. 1.6) is

$$f(x) = \begin{cases} \dfrac{x}{a^2} \exp\left\{-\dfrac{x^2}{2a^2}\right\}, & x \geq 0, \\ 0, \; x < 0. \end{cases} \tag{1.27}$$

The mathematical expectation and root-mean-square value of X are equal to

$$m_x = \int_0^\infty x f(x)\,\mathrm{d}x = \int_0^\infty \frac{x^2}{a^2} \exp\left\{-\frac{x^2}{2a^2}\right\} \mathrm{d}x = a\sqrt{\frac{\pi}{2}},$$

$$\sigma_x = 0{,}655a.$$

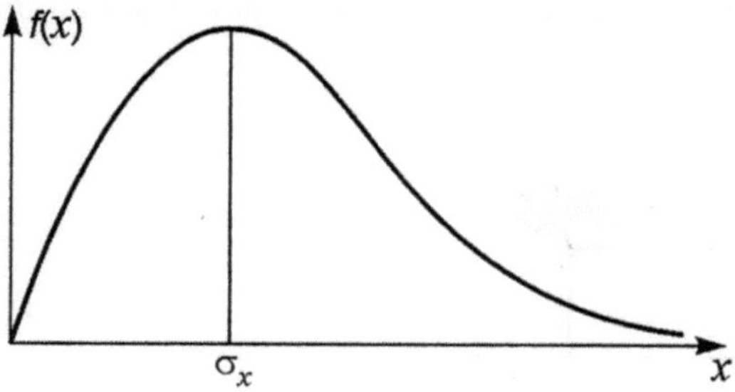

Fig. 1.6.

The Rayleigh distribution is a one-parameter distribution, as m_x and σ_x depend on a single parameter a.

If, as a result of experimental data processing, we obtain a distribution law which we may consider a Rayleigh distribution law, the parameter is equal to the maximum value of the probability density (Fig. 1.6). The derivative of $f(x)$ (1.27) is equal to

$$\frac{\mathrm{d}f}{\mathrm{d}x} = \frac{1}{a^2}\left(1 - \frac{x^2}{a^2}\right)\exp\left\{-\frac{x^2}{2a^2}\right\} = 0.$$

Therefore, the extremal value of $f(x)$ is reached at $x = a$.

4. **Poisson Distribution Law** A discrete random quantity X (non-dimensional) is regarded as distributed according to the Poisson law, if its possible values are equal to 0, 1, 2, ..., n and given the probability that $X = n$ is expressed by the relationship

$$P(X = n) = \frac{a}{n!}\mathrm{e}^{-a}, \tag{1.28}$$

where $a > 0$.

The expectation and variance of a random variable X distributed according to the Poisson law are equal to: $m_x = a$ and $D_x = a$.

This property of the given distribution is used to verify the hypothesis that a random quantity X is distributed according to the Poisson law. If the statistical characteristics $m_x^{(1)}$ and $D_x^{(1)}$ are close, it confirms the validity of the assumption, that the random quantity X is distributed according to the Poisson law.

5. **Law of the Rectangular Distribution of Probability Density** If the continuous random quantity possible values fall within some particular interval and within it all values of the random quantity are equally probable, the latter is said to have a rectangular distribution (Fig. 1.7).

In this case

$$f(x) = \begin{cases} a, & x_1 < x < x_2 \\ 0, & x < x_1 \text{ or } x > x_2 \end{cases} \quad \left(a = \frac{1}{x_2 - x_1}\right), \tag{1.29}$$

where a is the density of distribution; (x_1, x_2) is the interval of the random quantities possible values.

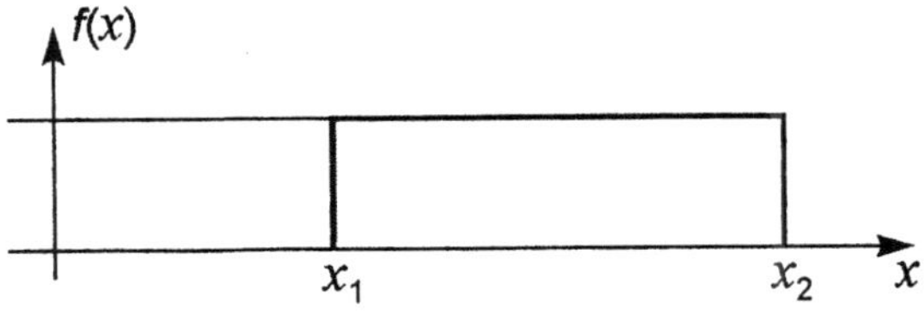

Fig. 1.7.

The mathematical expectation and variance of the random quantity X that is uniformly distributed within the interval (x_1, x_2) are equal to

$$m_x = \frac{x_1 + x_2}{2}; \quad D_x = \frac{(x_1 - x_2)^2}{12}.$$

respectively.

6. Distribution Law of the Probability Density of the Modulus of a Random Quantity Distributed According to the Normal Law A random quantity y is equal to the modulus of a random quantity x, i.e. $y = |x|$. The random quantity x has normal distribution. The distribution law of the probability density of Y takes the form:

$$f(y) = f(|x|) = \frac{1}{\sqrt{2\pi}\sigma_x}\left[\exp\left\{-\frac{(y - m_x)^2}{2\sigma_x^2}\right\} + \left\{-\frac{(y + m_x)^2}{2\sigma_x^2}\right\}\right], \quad (1.30)$$

where m_x and σ_x are the parameters of the random quantity x normal distribution law. The mathematical expectation and variance of the random quantity are equal to

$$m_y = 2\left[m_x \Phi_0\left(\frac{m_x}{\sigma_x}\right) + \sigma_x \varphi\left(\frac{m_x}{\sigma_x}\right)\right],$$

$$\sigma_y = \sigma_x^2 + m_x^2 - m_y^2,$$

where

$$\Phi_0\left(\frac{m_x}{\sigma_x}\right) = \frac{1}{\sqrt{2\pi}}\int_0^{\frac{m_x}{\sigma_x}} \exp\left\{-\frac{t^2}{2}\right\} dt,$$

$$\varphi\left(\frac{m_x}{\sigma_x}\right) = \frac{1}{\sqrt{2\pi}} \exp\left\{\frac{m_x^2}{2\sigma_x^2}\right\}.$$

The plots of $f(y)$ as a function of m_x are presented in Fig. 1.8.

In the particular case of $m_x = 0$,

$$f(y) = \frac{2}{\sqrt{2\pi}\sigma_x} \exp\left\{-\frac{y^2}{2\sigma_x^2}\right\} \quad (1.31)$$

7. The Law of the Logarithmically Normal Distribution of the Probability Density Let us consider random quantities related by the formula $x = \lg y$. If the random quantity x is distributed normally, then

$$f_1(y) = \frac{1}{\sqrt{2\pi}\sigma[\lg y]} \exp\left\{-\frac{(\lg y - \lg m)^2}{2\sigma^2[\lg y]}\right\}. \quad (1.32)$$

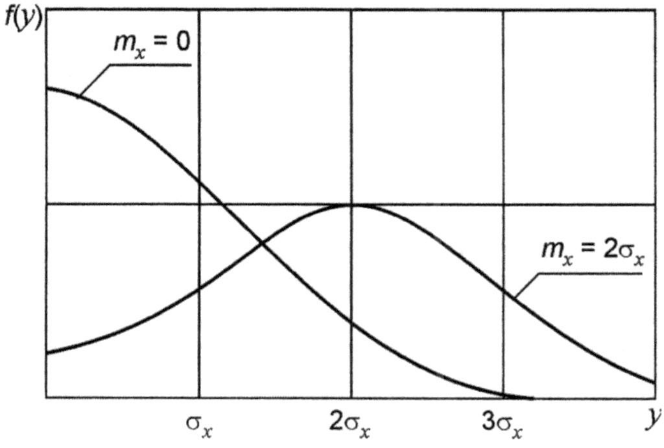

Fig. 1.8.

The behavior of the function $f(y)$ for a number of σ values at $\lg m = 1$ is shown in Fig. 1.9.

8. **Gamma Distribution Law** For this law we have

$$f(x) = \begin{cases} \dfrac{1}{\beta \Gamma(\alpha+1)} \left(\dfrac{x}{\beta}\right)^{\alpha} \exp\left\{-\dfrac{x}{\beta}\right\} \text{ at } x \geq 0, \\ 0 \text{ at } x < 0, \end{cases} \tag{1.33}$$

where $\alpha > -1$; $\beta > 0$; and $\Gamma(\alpha+1) = \int\limits_0^\infty e^{-t} t^\alpha \, dt$ is the gamma function.

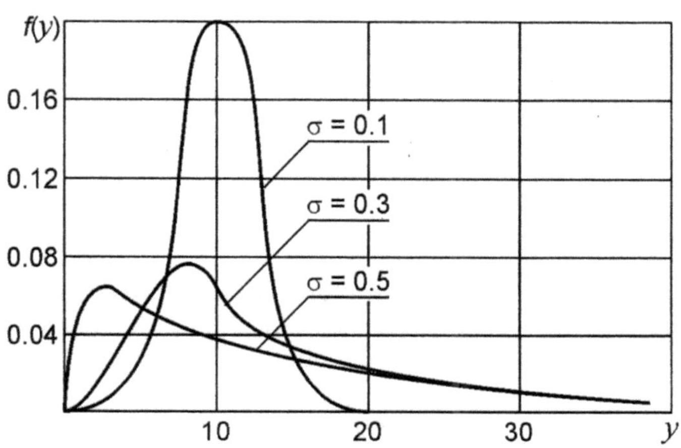

Fig. 1.9.

The parameters of the gamma distribution are

$$\alpha = \left(\frac{m_x}{\sigma_x}\right)^2 - 1; \quad \beta = \frac{\sigma_x^2}{m_x}.$$

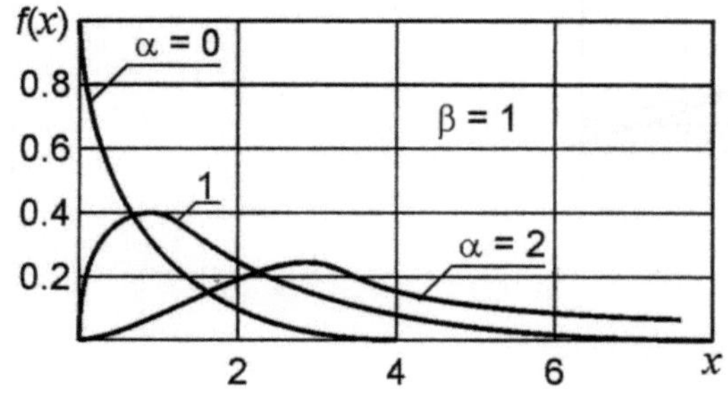

Fig. 1.10.

The probability density plots for $\alpha = 0; 1; 2$ at $\beta = 1$ are given in Fig. 1.10. For large values of α the gamma distribution goes into a normal one. If we introduce in (1.33) $\beta = \frac{1}{\lambda}$ and $\alpha = k$, where k is an integer positive number, we get the Erlangian probability density

$$f(x) = \frac{\lambda (\lambda x)^k}{k!} \mathrm{e}^{-\lambda x} \quad (x > 0), \tag{1.34}$$

which is used in the reliability theory.

9. **Weibull Probability Density Distribution Law** The Weibull distribution function and the Weibull density of distribution at $x \geq x_a$, $\gamma > 0$ take the form

$$F(x) = 1 - \exp\left\{-\frac{(x - x_a)^\gamma}{x_0}\right\}; \tag{1.35}$$

$$f(x) = \frac{\gamma}{x_0}(x - x_a)^{\gamma - 1} \exp\left\{-\frac{(x - x_a)^\gamma}{x_0}\right\}, \tag{1.36}$$

where x_a, x_0 and γ are free parameters.

In the particular case, where $x_a = 0$ and $x_0 = 1$, we get

$$f(x) = \gamma x^{\gamma - 1} \mathrm{e}^{-x\gamma} (x > 0, \ \gamma > 0). \tag{1.37}$$

The plot of the function $f(x)$ for a number of γ values is shown in Fig. 1.11.

The expectation and variance of the random quantity following the distribution (1.37) are:

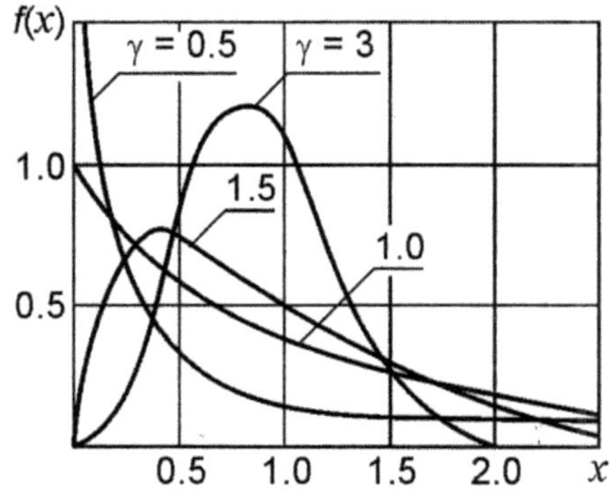

Fig. 1.11.

$$m_x = \Gamma\left(\frac{\gamma+1}{\gamma}\right);$$
$$\sigma_x^2 = \Gamma\left(\frac{\gamma+2}{\gamma}\right) - \Gamma^2\left(\frac{\gamma+1}{\gamma}\right).$$

Let us consider the following problem as an example of the probability density analytical determination.

A point mass m (Fig. 1.12) executes simple harmonic oscillations (steady-state oscillations under the action of a harmonic force).

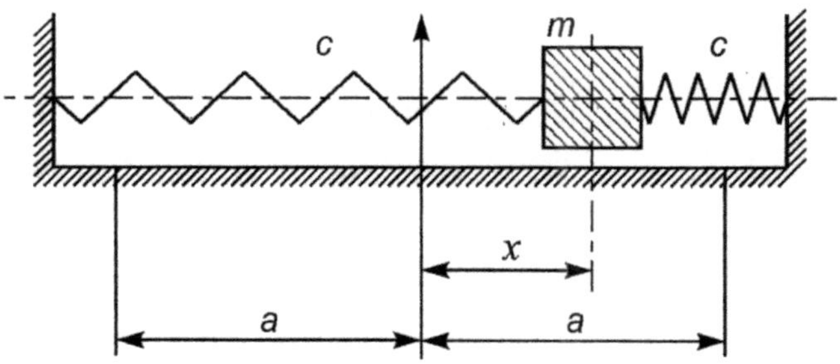

Fig. 1.12.

In this case, the mass m displacement at an arbitrary instant is

$$x = a \sin \omega t,$$

where a is the oscillation amplitude.

It is required to determine the probability density of an event, in which at a random instant the point m finds itself at some distance x from the equilibrium position. We may consider that the probability of the mass falling within the interval $(x,\, x + \mathrm{d}x)$ at an arbitrary instant is proportional to the length of this interval dx and inversely proportional to the point motion velocity, i.e.

$$\mathrm{d}P\,(x < X < x + \mathrm{d}x) = c_1 \frac{\mathrm{d}x}{\dot{x}} = c_1 \frac{\mathrm{d}x}{\dfrac{\mathrm{d}x}{\mathrm{d}t}} = c_1\,\mathrm{d}t,$$

where c_1 is a proportionality factor.

Since

$$\mathrm{d}P = f\,(x)\,\mathrm{d}x,$$

then by way of manipulation we get

$$f\,(x) = c_1 \frac{\mathrm{d}t}{\mathrm{d}x}.$$

Let us use the law of the mass m motion to eliminate time t from this expression for $f\,(x)$

$$\frac{\mathrm{d}x}{\mathrm{d}t} = a\omega\cos\omega t = a\omega\sqrt{1 - \sin^2\omega t} = a\omega\sqrt{1 - \left(\frac{x}{a}\right)^2} = \omega\sqrt{a^2 - x^2},$$

therefore

$$f\,(x) = \frac{c_1}{\omega\sqrt{a^2 - x^2}}.$$

Then we determine c_1 value from the normalization condition

$$\int\limits_{-\infty}^{\infty} f\,(x)\,\mathrm{d}x = \int\limits_{-a}^{a} f\,(x)\,\mathrm{d}x = 1.$$

Through calculations we get $c_1 = \omega/\pi$. The final expression for the probability density is

$$f\,(x) = \frac{1}{\pi\sqrt{a^2 - x^2}}.$$

The plot of the probability density function $f\,(x)$ is presented in Fig. 1.13. The distribution function is

$$F\,(x) = \int\limits_{-a}^{x} f\,(x)\,\mathrm{d}x = \frac{1}{\pi}\left(\arcsin\frac{x}{a} + \frac{\pi}{2}\right).$$

The probability of the mass m being in the interval $(x_1,\, x_2)$ at an arbitrary instant is

$$P\,(x_1 \le X \le x_2) = \int\limits_{x_1}^{x_2} f\,(x)\,\mathrm{d}x = \frac{1}{\pi}\left(\arcsin\frac{x_2}{a} - \arcsin\frac{x_1}{a}\right).$$

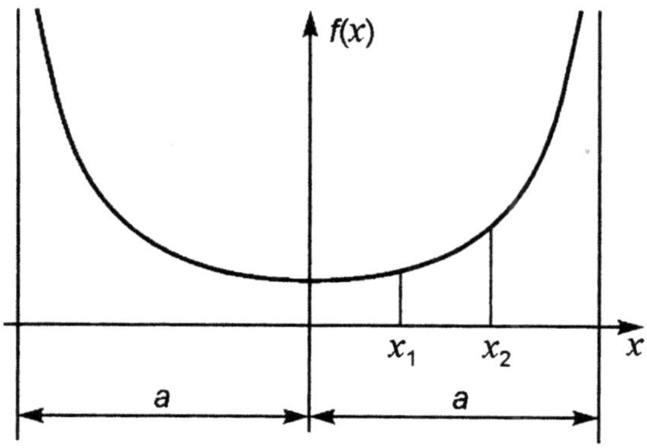

Fig. 1.13.

1.5 Determination of the Probability of a Normally Distributed Random Quantity Lying in the Given Range

Let us use the relation (1.9) to determine the probability that a random quantity lies in the range (x_1, x_2).

The distribution function is

$$F(x) = \int_{-\infty}^{x} f(x)\,\mathrm{d}x = \frac{1}{\sigma\sqrt{2\pi}} \int_{-\infty}^{x} \exp\left\{-\frac{(x-m_x)^2}{2\sigma^2}\right\} \mathrm{d}x. \tag{1.38}$$

Introducing the notation $\dfrac{x-m_x}{\sigma} = t$, we get the probability integral

$$F(t) = \frac{1}{\sqrt{2\pi}} \int_{t_1}^{t_2} \exp\left\{-\frac{t^2}{2}\right\} \mathrm{d}t = \Phi(t).$$

Then

$$P(x_1 X x_2) = \frac{1}{\sqrt{2\pi}} \int_{t_1}^{t_2} \exp\left\{-\frac{t^2}{2}\right\} \mathrm{d}t = \Phi(t_2) - \Phi(t_1),$$

where

$$t_1 = \frac{x_1 - m_x}{\sigma}; \quad t_2 = \frac{x_2 - m_x}{\sigma}.$$

The $\Phi(t_2)$ and $\Phi(t_1)$ numerical values may be found in handbooks on higher mathematics.

Let us determine the probability that a random quantity X falls within the σ length interval (Fig. 1.14).

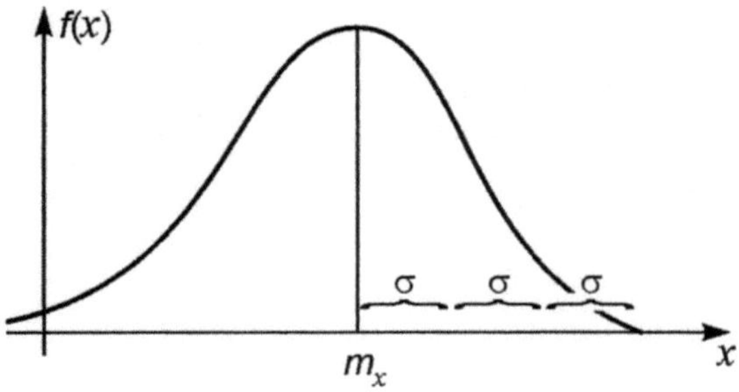

Fig. 1.14.

The probability that x falls within the interval $(m_x, m_x + \sigma)$ is

$$P = (m_x < x < m_x + \sigma) = \Phi(t_2) - \Phi(t_1),$$

where

$$t_2 = \frac{m_x + \sigma - m_x}{\sigma} = 1;$$

$$t_1 = \frac{m_x - m_x}{\sigma} = 0,$$

therefore

$$P(m_x < x < m_x + \sigma) = \Phi(1) - \Phi(0).$$

Substituting the values $\Phi(1)$ and $\Phi(0)$, we finally get

$$P(m_x < x < m_x + \sigma) = 0,341.$$

In the same way we determine the probability that $\dot{x}$ falls within the following σ length intervals:

$$P(m_x + \sigma < X < m_x + 2\sigma) = \Phi(2) - \Phi(1) = 0,136;$$
$$P(m_x + 2\sigma < X < m_x + 3\sigma) = \Phi(3) - \Phi(2) = 0,012.$$

The sum of the three above-mentioned probabilities is approximately equal to 0,5 (correct to 1 %). This means that for a normally distributed random quantity the spread is well within the interval $m_x \pm 3\sigma$. The obtained result allows us to determine the approximate range of random quantity possible values on the basis of the known values of their mathematical expectation and standard deviation. This estimation method is called the "three sigma rule" and can be used only in cases, where we can ignore the realization of an event with small probability.

Example 1.2. A periodic force $F = F_0 \cos\omega t$ with a random normally distributed amplitude F_0 is acting on an inertialess rod (Fig. 1.15). Its probability characteristics are $m_{F_0} = 6{,}5\,\mathrm{N}$, $\sigma_{F_0} = 2{,}4\,\mathrm{N}$. It is required to determine the probability that this force acting on the rod lies in the $100 \div 120\,\mathrm{N}$ range.

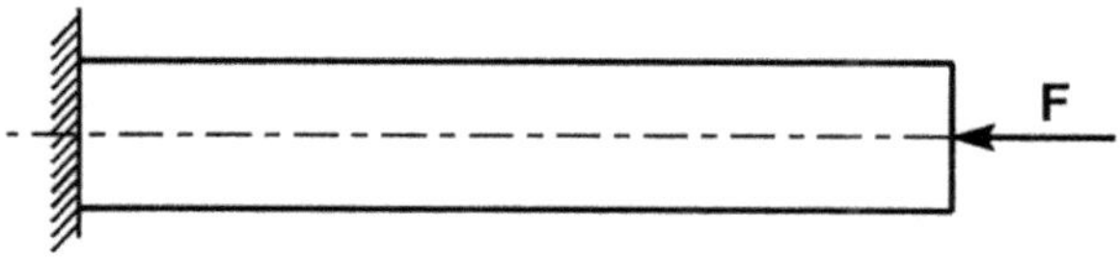

Fig. 1.15.

Let us determine the values t_2 and t_1 :

$$t_2 = \frac{120 - 65}{24} = 2,292;$$
$$t_1 = \frac{100 - 65}{24} = 1,458.$$

The probability that such amplitude of the force occurs within the interval $100 \div 120\,\mathrm{N}$ is

$$P\,(100 < F_0 < 120) = \Phi\,(t_2) - \Phi\,(t_1) = 0{,}0614.$$

1.6 Random Quantity Systems and Their Numerical Characteristics

Let us consider the simplest case of such systems, a system of two random quantities X, Y.

The probability of a simultaneous fulfillment of two inequalities $X < x$ and $Y < y$, i.e.

$$F\,(x,\, y) = P\,(X < x,\; Y < y)\,. \tag{1.39}$$

is referred to as joint distribution function of two random quantities X and Y.

Geometrically, the relation (1.39) is the probability that a random point falls within the shaded area (Fig. 1.16). A two-dimensional probability density may be introduced in the same way as that of one dimension:

$$f\,(x,\, y) = \lim_{\substack{\Delta x \to 0 \\ \Delta y \to 0}} \frac{P\,(x \le X \le x + \Delta x;\; y \le Y \le y + \Delta y)}{\Delta x \Delta y} = \frac{\partial^2 F\,(x,\, y)}{\partial x\, \partial y}. \tag{1.40}$$

It follows from the relation (1.40), that $f\,(x, y)\;\mathrm{d}x\,\mathrm{d}y$ is the probability of a point falling within the infinitely small rectangle (see Fig. 1.16). The probability that a point falls within some finite area B of the plane is equal to

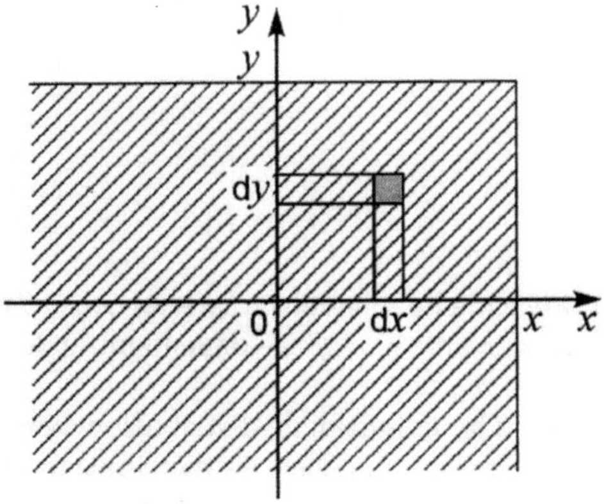

Fig. 1.16.

$$P = \iint_B f(x, y)\, dx\, dy. \tag{1.41}$$

The distribution function and the probability density are related by the following equation

$$F(x, y) = \int_{-\infty}^{x} \int_{-\infty}^{y} f(x, y)\, dx\, dy. \tag{1.42}$$

From this formula we may obtain the relation

$$\frac{\partial^2 F(x, y)}{\partial x\, \partial y} = f(x, y). \tag{1.43}$$

The function $F(x, y)$ meets the conditions:
1) $0 \leq F(x, y) \leq 1$;
2) $F(-\infty, y) = 0$ at any y;
3) $F(x, -\infty) = 0$ at any x;
4) $F(x, \infty) = P(X < x) = F_1(x)$;
5) $F(\infty, y) = P(Y < y) = F_2(y)$;
6) $F(\infty, \infty) = 1$.
The function $f(x, y)$ must satisfy the condition

$$\int_{-\infty}^{\infty} \int_{-\infty}^{\infty} f(x, y)\, dx\, dy = 1. \tag{1.44}$$

We may obtain the probability density of each random quantity in terms of joint probability density from the formulas:

$$f_1(x) = \int_{-\infty}^{\infty} f(x, y)\, \mathrm{d}y; \tag{1.45}$$

$$f_2(y) = \int_{-\infty}^{\infty} f(x, y)\, \mathrm{d}x. \tag{1.46}$$

The random quantities X and Y are called dependent, if events consisting in the fulfillment of the inequalities $X < x$ and $Y < y$ are dependent at least for one pair of x and y values.

Conditional Distribution Laws The conditional distribution law of one of the quantities (X, Y) entering a system, is called its distribution law deduced subject to the condition that the other random quantity has taken a specific value, i.e.

$$F(x, y) = F_1(x)\, F_2(y \,|X < x)$$

or

$$F(x, y) = F_2(y)\, F_1(x \,|Y < y). \tag{1.47}$$

Differentiating firstly with respect to x, and then with respect to y, we obtain with the use of the density-multiplication theorem

$$\frac{\partial^2 F(x, y)}{\partial x\, \partial y} = f(x, y) = f_1(x)\, f_2(y \,|x) \tag{1.48}$$

or

$$\frac{\partial^2 F(x, y)}{\partial x\, \partial y} = f(x, y) = f_2(y)\, f_1(x \,|y). \tag{1.49}$$

The relations (1.48) and (1.49) are referred to as the distribution laws multiplication theorem. From (1.48) and (1.49) we determine the conditional distribution laws in terms of unconditional laws:

$$f(y \,|x) = \frac{f(x, y)}{f_1(x)}; \quad f(x \,|y) = \frac{f(x, y)}{f_2(y)}$$

or

$$f(y \,|x) = \frac{f(x, y)}{\int_{-\infty}^{\infty} f(x, y)\, \mathrm{d}y}; \quad f(x \,|y) = \frac{f(x, y)}{\int_{-\infty}^{\infty} f(x, y)\, \mathrm{d}x}.$$

For conditional densities the following conditions

$$f(x \,|y) \geq 0; \qquad \int_{-\infty}^{\infty} f(x \,|y)\, \mathrm{d}x = 1.$$

$$f(y\,|x) \geq 0; \qquad \int_{-\infty}^{\infty} f(y\,|x)\,\mathrm{d}y = 1.$$

are fulfilled.

Random quantities X and Y are considered independent, if events consisting in the fulfillment of the inequalities $X < x$, $Y < y$ are independent at any values of x and y. For independent random quantities X, Y a joint distribution function (based on the rules of multiplying the probabilities of independent events) is

$$F(x,\,y) = P(X < x)\;P(Y < y) \tag{1.50}$$

or

$$F(x,\,y) = F_1(x)\;F_2(y)\,. \tag{1.51}$$

Differentiating the relation (1.51) initially with respect to x, and then with respect to y, we get

$$\frac{\partial^2 F(x,\,y)}{\partial x\,\partial y} = \frac{\partial F_1}{\partial x}\cdot\frac{\partial F_2}{\partial y} = f(x,\,y) = f_1(x)\;f_2(y)\,. \tag{1.52}$$

Numerical Characteristics of a System of Two Random Quantities In order to establish a "linkage" between two random quantities X and Y, a nonrandom numerical characteristic — the correlation moment K_{xy} — is introduced. For discrete random quantities it is equal to

$$K_{xy} = \sum_{i=1}^{n}\sum_{j=1}^{n}(x_i - m_x)\,(y_j - m_y)\,P_{ij},$$

where P_{ij} is the probability that the system $(X,\,Y)$ will take the value $(x_i,\,y_j)$. For continuous random quantities, the correlation (cross-correlation) moment is equal to

$$K_{xy} = M\left[\overset{\circ}{X}\,\overset{\circ}{Y}\right] = \iint_{-\infty}^{\infty}(x - m_x)\,(y - m_y)\,f(x,\,y)\,\mathrm{d}x\,\mathrm{d}y,$$

where $f(x,\,y)$ is the joint distribution law of the probability density of the random quantities $(X,\,Y)$ system. To illustrate the joint distribution law of two random quantities, we may cite the normal distribution law

$$f(x,\,y) = \frac{1}{2\pi\sigma_x\sigma_y\sqrt{1 - r_{xy}}}\exp\left\{-\frac{1}{2\,(1 - r_{xy})}\left[\frac{(x - m_x)^2}{\sigma_x^2} - \frac{2r_{xy}(x - m_x)(y - m_y)}{\sigma_x\sigma_y} + \frac{(y - m_y)^2}{\sigma_y^2}\right]\right\}. \tag{1.53}$$

We have the following probability characteristics for two random continuous quantities:

$$m_x = \iint\limits_{-\infty}^{\infty} xf(x,\, y)\, \mathrm{d}x\, \mathrm{d}y;$$
$$m_y = \iint\limits_{-\infty}^{\infty} yf(x,\, y)\, \mathrm{d}x\, \mathrm{d}y; \tag{1.54}$$

$$D_x = \iint\limits_{-\infty}^{\infty} (x - m_x)^2 f(x,\, y)\, \mathrm{d}x\, \mathrm{d}y;$$
$$D_y = \iint\limits_{-\infty}^{\infty} (y - m_y)^2 f(x,\, y)\, \mathrm{d}x\, \mathrm{d}y; \tag{1.55}$$

$$K_{xy} = \iint\limits_{-\infty}^{\infty} (x - m_x)(y - m_y) f(x,\, y)\, \mathrm{d}x\, \mathrm{d}y. \tag{1.56}$$

Since for independent random quantities $f(x,\, y) = f_1(x) f_2 t(y)$, we get from the formula (1.56)

$$K_{xy} = \int\limits_{-\infty}^{\infty} f_1(x)(x - m_x)\, \mathrm{d}x \int\limits_{-\infty}^{\infty} f_2(y)(y - m_y)\, \mathrm{d}y$$
$$= (m_x - m_x)(m_y - m_y) = 0.$$

Hence, if the correlation moment of two random quantities differs from zero, it points to the presence of a relationship between them. Random quantities with a zero correlation moment are referred to as uncorrelated random quantities. It follows from the formula (1.56) that a correlation moment characterizes not only the relationship of quantities, but also their scatter. If, for example, the quantity X or the quantity Y insignificantly deviates from its mathematical expectation, their correlation moment will be small irrespective of the kind of relationship between them. Therefore, to eliminate this shortcoming a non-dimensional characteristic is introduced, namely the correlation coefficient (or normalized correlation moments):

$$r_{xy} = \frac{K_{xy}}{\sigma_x \sigma_y}, \tag{1.57}$$

where σ_x, σ_y are the standard deviations of the random quantities X and Y. The correlation coefficient can exceed zero or be less than it. The factor r_{xy} varies within the limits

$$-1 < r_{xy} < 1. \tag{1.58}$$

To prove the validity of the relation (1.58) let us consider a random quantity z:

$$z = \sigma_y X \pm \sigma_x Y,$$

where σ_x, σ_y are the random centered quantities X and Y root-mean-square values. The random quantity Z variance is equal to

$$D_Z = M\left[Z^2\right] = \sigma_y^2 D_x + \sigma_x^2 D_y \pm 2\sigma_x\sigma_y K_{xy}.$$

The variance is more than zero. Therefore

$$\sigma_y^2 D_x + \sigma_x^2 D_y \pm 2\sigma_x\sigma_y K_{xy} \geq 0$$

or (as $D_x = \sigma_x^2$, $D_y = \sigma_y^2$)

$$\sigma_y\sigma_x \pm K_{xy} \geq 0.$$

It follows from the last relation that

$$|K_{xy}| \leq \sigma_x\sigma_y$$

or

$$|r_{xy}| \leq 1. \tag{1.59}$$

Let us determine, for example, r_{xy} for linearly related random centered quantities Y and X:

$$Y = aX.$$

The variance D_y is equal to $a^2 D_x$. The cross-correlation moment K_{xy} is equal to

$$K_{xy} = M\left[XY\right] = aM\left[X^2\right] = aD_x.$$

Therefore

$$r_{xy} = \frac{aD_x}{a\sigma_x\sigma_x} = 1.$$

The reverse is also true: if the correlation coefficient r_{xy} is close to unity, the relationship of the random quantities X and Y differs little from a linear one. The correlation coefficient r_{xy} (1.57) determining the degree of correlation between the two random functions, has no strict functional character. A correlation dependence and not the functional one is used when one of the quantities depends not only on the second quantity, but also on a series of random factors that cannot be taken into account when determining relationships between the quantities.

The dependence between random quantities X and Y manifests itself in the fact that a conditional probability, for example, occurrence of Y_j, during the realization of an event x_k differs from the unconditional probability. In other words, the influence of one random quantity on another is characterized by the conditional distribution of one of them at a fixed value of another. The practical use of the correlation coefficient during the quantitative estimation of a degree of relationship (degree of dependence) between two random quantities is usually satisfied when the law of distribution is normal. In this case, the independence of random quantities follows from the equality $K_{xy} = 0$. It is impossible to use K_{xy} for the estimation of a degree of interdependence of two arbitrary random quantities, because even with a functional relationship between the two quantities (one-to-one dependence) the correlation moment can be zero, i.e. the concepts of noncorrelatedness and independence are not being same.

Let us discuss in greater detail whether the concept of the noncorrelatedness of random quantities is equivalent to the concept of independence. It has already been shown, that two independent random quantities are uncorrelated. The question arises: is the opposite statement true? Let us consider an example. A system of two random quantities (X, Y) has a uniform probability density inside a circle of some radius R, see the area D in Fig. 1.17. The joint density function of the random quantities X and Y is expressed by the formula

$$f(x, y) = \begin{cases} a \text{ at } x^2 + y^2 < R^2 \\ 0 \text{ at } x^2 + y^2 > R^2 \end{cases}.$$

From the normalization condition of the function $f(x, y)$ we get

$$\iint\limits_{-\infty}^{\infty} f(x, y)\, dx\, dy = \iint\limits_{D} dx\, dy = 1$$

therefore

$$a = \frac{1}{\pi R^2}.$$

Let us show, that in this example the random quantities X and Y are dependent. If the random quantity Y has taken the value of 0, the random quantity X can with an equal probability take all values from $-R$ to $+R$.

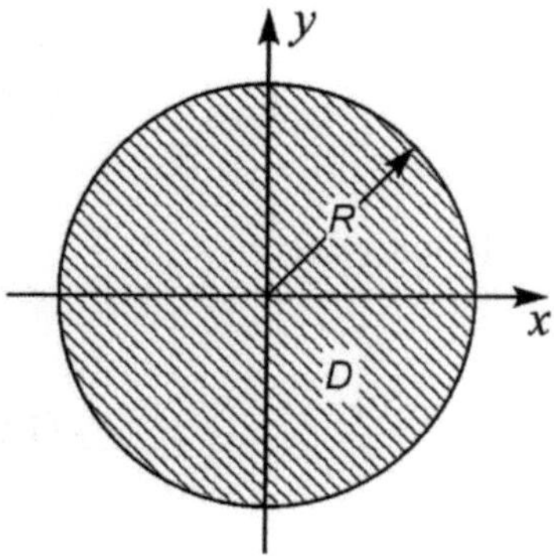

Fig. 1.17.

If the random quantity Y has taken the value of R, the random quantity X can only take a zero value. This means that the range of X possible values depends on the value taken by quantity Y. Therefore, the random quantities X and Y are dependent. Now we shall find out, whether they are correlated. Let us find their cross-correlation moment K_{xy} (random quantities X and Y are centered)

$$K_{xy} = \iint_D xy\, f(xy)\, \mathrm{d}x\, \mathrm{d}y = \frac{1}{\pi R^2} \iint_D xy\, \mathrm{d}x\, \mathrm{d}y .$$

As axes x, y (Fig. 1.17) are the axes of the area D symmetry, the integral $\iint_D xy\, \mathrm{d}x\, \mathrm{d}y$ is zero, that is $K_{xy} = 0$.

The obtained result indicates that the random quantities X and Y are not correlated. Therefore, the noncorrelated nature of the random quantities does not always mean their independence. During practical calculations, however, the correlation coefficient gives a qualitative information about the interdependence of two random quantities. For example, if $r_{xy} > 0$, the increase of one random quantity usually leads to the growth of the other and at $r_{xy} < 0$ at increase in one random quantity, as a rule, makes the other one decrease.

We may present the variances and correlation moments of two random quantities as the matrix

$$K = \begin{Vmatrix} K_{xx} & K_{xy} \\ K_{yx} & K_{yy} \end{Vmatrix}, \quad (K_{xx} = D_x, K_{yy} = D_y) . \tag{1.60}$$

For a system of n random quantities their variance and correlation moments may be presented in the way it is done with two random quantities systems, i.e. as the correlation matrix

$$K = \begin{Vmatrix} K_{x_1x_1} & K_{x_1x_2} & \dots & K_{x_1x_n} \\ \vdots & \vdots & \ddots & \vdots \\ K_{x_nx_1} & K_{x_nx_2} & \dots & K_{x_nx_n} \end{Vmatrix}, \tag{1.61}$$

where $K_{x_ix_i} = D_{x_i}$, or as the normalized correlation matrix

$$K = \begin{Vmatrix} 1 & r_{x_1x_2} & \dots & r_{x_1x_n} \\ r_{x_2x_1} & 1 & \dots & \dots \\ \vdots & \vdots & \ddots & \vdots \\ r_{x_nx_1} & \dots & \dots & 1 \end{Vmatrix} \tag{1.62}$$

Numerical Characteristics of a Sum and a Product of Two Random Quantities The mathematical expectation of a sum of two random quantities is equal to the sum of their mathematical expectations

$$M[X+Y] = M[X] + M[Y]. \tag{1.63}$$

The variance of a sum of two random quantities is equal to the sum of their variances plus the double correlation moment

$$D(X+Y) = D_x + D_y + 2K_{xy}. \tag{1.64}$$

The mathematical expectation of a product of two random quantities is equal to the product of their mathematical expectations plus the correlation moment

$$M[XY] = M[X]\,M[Y] + K_{xy}. \tag{1.65}$$

The variance of a product of independent random quantities is equal to

$$D(XY) = M\left[X^2\right] \cdot M\left[Y^2\right] - m_x^2 m_y^2, \tag{1.66}$$

where $M\left[X^2\right] = D_x + m_x^2$, $M\left[Y^2\right] = D_y + m_y^2$.

For centered random quantities:

$$D\left(\overset{\circ}{X}\overset{\circ}{Y}\right) = D_x D_y. \tag{1.67}$$

1.7 Complex Random Quantities

When solving applied problems we often have to consider not only real, but also complex random quantities. This makes us generalize the concepts of an expectation, a variance and a correlation function to complex quantities. A complex random quantity is defined as a random quantity whose possible values constitute complex numbers, i.e.

$$Z = X + iY, \tag{1.68}$$

where X, Y are real random numbers, i is the imaginary unit. The complex quantities play an essential role in the theory of random functions.

Considering the expression (1.68) as a sum of random quantities and in accordance with (1.63), we get the mathematical expectation of the random quantity Z

$$m_z = m_x + im_y. \tag{1.69}$$

The variance of a complex random quantity Z is referred as the expectation of the square of a centered random quantity $\overset{\circ}{Z}$ modulus:

$$D_z = M\left[\left|\overset{\circ}{Z}\right|^2\right], \tag{1.70}$$

where $\left|\overset{\circ}{Z}\right|$ is the modulus of a centered random quantity.

Under this definition the variance of a complex random quantity is always real and positive, i.e. the principal property of a variance is retained.

Since

$$\left|\overset{\circ}{Z}\right|^2 = \overset{\circ}{Z}\,\overset{\circ}{Z}{}^* = \overset{\circ}{X}{}^2 + \overset{\circ}{Y}{}^2$$

where $\overset{\circ}{Z}{}^*$ is a conjugate complex quantity, we get from (1.70)

$$D_z = M\left[\overset{\circ}{X}{}^2 + \overset{\circ}{Y}{}^2\right] = D_x + D_y. \tag{1.71}$$

The correlation moment of two complex random quantities $X = X_1 + iX_2$ and $Y = Y_1 + iY_2$ in a particular case, where $X = Y$, should be equal to the variance D_x. This occurs, if we assume that

$$K_{xy} = M\left[\overset{\circ}{X}\,\overset{\circ}{Y}{}^*\right], \tag{1.72}$$

where $\overset{\circ}{Y}{}^*$ is a centered conjugate random quantity.

Indeed, with this definition of a correlation moment at $X = Y$ we get

$$K_{xy} = M\left[\overset{\circ}{X}\,\overset{\circ}{X}{}^*\right] = D_x.$$

For arbitrary complex random quantities we have:

$$\begin{aligned} K_{xy} &= M\left[\left(\overset{\circ}{X}_1 + i\overset{\circ}{X}_2\right)\left(\overset{\circ}{Y}_1 - i\overset{\circ}{Y}_2\right)\right] \\ &= K_{x_1y_1} + K_{x_2y_2} + i\left(K_{x_2y_1} - K_{x_1y_2}\right). \end{aligned} \tag{1.73}$$

The correlation moment of complex random quantities depends on a sequence, in which they are taken, namely, when the sequence of the random quantities alters, their correlation moment changes into the complex conjugate quantity

$$K_{xy} = M\left[\overset{\circ}{X}\,\overset{\circ}{Y}^{*}\right] = M\left[\overset{\circ}{X}^{*}\,\overset{\circ}{Y}\right] = K_{yx}^{*}. \tag{1.74}$$

1.8 Numerical Characteristics of Functions of Random Arguments

When solving various engineering problems associated with the analysis of random phenomena, we often have to consider random functions depending on random quantities with known distribution laws. Knowing the distribution laws of the arguments of a composite function, we may determine the distribution law of this function. During the solution of applied problems, however, it is usually sufficient to have the numerical characteristics of the function of random arguments, and they are much easier to obtain, than the distribution law.

Let us consider the problem of determining the numerical characteristics of a function Y of random argument x at the known distribution law of argument x, i.e. it is required to determine m_y and D_y, if

$$Y = \varphi(x). \tag{1.75}$$

For a discrete random quantity X, the probabilities of discrete values x_j occurrences are known, i.e. the distribution law $(x_j \rightarrow p_j)$ is known. Therefore, from (1.75) we obtain the relation $(y_j \rightarrow p_j)$, where $Y_j = \varphi(x_j)$, i.e. we get the table:

Table 1.2.

$\varphi(x_1)$	$\varphi(x_2)$	$\varphi(x_3)$	...	$\varphi(x_n)$
p_1	p_2	p_3	...	p_n

It is similar to the table (see section 1.2) defining the distribution law of a random quantity X. In the presented table, $\varphi(x_i)$ do not necessarily go in ascending order. In addition to this, the coincidences of $\varphi(x_j)$ at different x_j are possible, but we always may arrange the quantities $\varphi(x_j)$ in ascending order, to combine columns with equal $\varphi(x_j)$ by adding their probabilities, i.e. we may obtain a table completely corresponding to the distribution function. Therefore, for the discrete random quantity Y the expectation is equal to

$$M\left[Y\right]=M\left[\varphi\left(x\right)\right]=\sum_{j=1}^{n}\varphi\left(x_j\right)p_j=m_y^{(1)}. \tag{1.76}$$

For the continuous random quantity Y:

$$M\left[Y\right]=\int_{-\infty}^{\infty}\varphi\left(x\right)f\left(x\right)\,\mathrm{d}x=m_y. \tag{1.77}$$

The variance of a function depending on one random argument, is equal to

$$D_y=D\left[\varphi\left(x\right)\right]=\int_{-\infty}^{\infty}\left[\varphi\left(x\right)-m_y\right]^2f\left(x\right)\,\mathrm{d}x. \tag{1.78}$$

For a function depending on two random arguments,

$$Z=\varphi\left(X,Y\right) \tag{1.79}$$

the expectation and variance are equal to

$$M\left[Z\right]=\iint_{-\infty}^{\infty}\varphi\left(x,\,y\right)\,f\left(x,y\right)\,\mathrm{d}x\,\mathrm{d}y=m_z, \tag{1.80}$$

$$D_z=D\left[\varphi\left(x,\,y\right)\right]=\iint_{-\infty}^{\infty}\left[\varphi\left(x,\,y\right)-m_z\right]^2f\left(x,\,y\right)\,\mathrm{d}x\,\mathrm{d}y. \tag{1.81}$$

2. Non-Stationary Random Functions (Processes)

2.1 Introduction

In engineering practice, we have to examine random quantities depending on continuously varying nonrandom arguments, such as the time t and the coordinates x, y, z. This kind of random quantities is referred to as random functions.

Figure 2.1 shows records $X_j(t)$ of the random function $X(t)$, depending on time alone. Each particular record is termed as a random function $X(t)$ realization. The set of all possible realizations, which a random phenomenon $X(t)$ can achieve, is referred to as a random or stochastic process. The pertinent branch of science, which studies random processes, may be called "random phenomena dynamics".

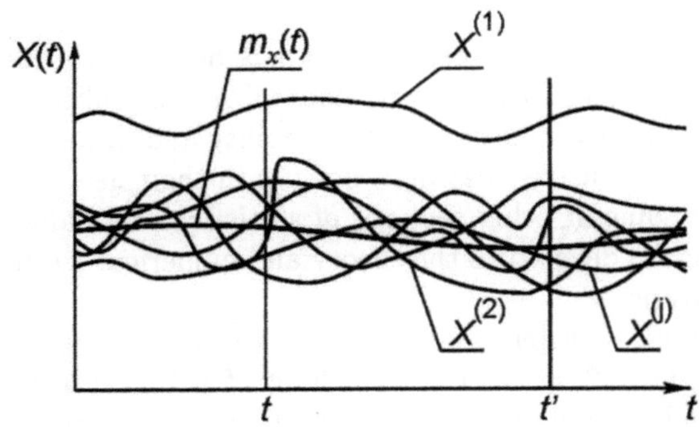

Fig. 2.1.

The random function $X(t)$, depending on time alone, will be referred to as a random process, by analogy with the dynamic deterministic process. It is common practice to term a random function depending on time and space, as a random space-time process, or, when it does not lead to confusion, simply as a random process.

Examples of dynamic systems, whose investigation involves the solution of differential equations of disturbed motion caused by random forces, that vary in time, have been given in the introduction. These problems refer to "dynamic random phenomena", i.e. to random processes. For example, scatter ΔR of engine thrust R , depending on time (Fig. 0.2), is a random function. It is worth noting that during the study of random processes it is not the properties of separate random functions $X_j(t)$, characterizing the process, that are examined, but the properties of the full set of functions in their entirety. This makes it possible to analyze the motion of a mechanical system under the action of random disturbances, to investigate its behavior relative not to any one action, but to a variety of possible random actions.

Let us consider a random function $X(t)$. Suppose that; n independent trials have been carried out, as a result of which; n realizations $X_j(t)$ have been obtained (Fig. 2.1). Each realization is a nonrandom function, but before a trial it is impossible to predict the way $X_j(t)$ would vary. The $X_j(t)$ variation from zero to t_1, which came to light after the trial, doesn't allow us to predict $X_j(t)$ behavior at $t > t_1$, that is $X_j(t)$ is not determined at $t > t_1$. If we fix argument $t = t_1$ the random function $X(t)$ will turn into random quantity X, which is the subject matter of the probability theory. Let us call this random quantity random function $X(t)$ section corresponding to a given moment t_1 (Fig. 2.1). At $t > t_1$ we obtain n values of the random function. It is possible to fix time and consider the process variation in a set of trials. Therefore each realization characterizing the process depends on the following two nonrandom arguments: a realization discrete number and a continuous time. For example, if we record vertical accelerations at any point of a car (Fig. 0.2, *a*), then during each travel along the same road at one and the same speed, well obtain one realization $X_j(t)$, the set of these realizations determining the random process.

In engineering practice, a particularly keen interest has been shown for the analysis of machines, devices and structures random vibrations substantially affecting their serviceability (fatigue strength, reliability and lifetime). Presently there is no branch of industry, that does not use in their new machinery designs and analyses some or other sections of statistical mechanics and, in particular, a main section devoted to the theory and numerical methods of random processes analysis.

Let us consider the mechanical system vibrations due to random disturbances in greater detail. Figure 0.2 shows the launching of a rocket from an inclined guide. Due to random combined technological and gasdynamical misalignments (linear e and angular α) varying with time, thrust R is not directed along the rocket axis, this bringing into existence two random disturbances — force $N_c = |\mathbf{R}|\,\alpha$ and moment $M_c = |\mathbf{R}|e$ (except for thrust scatter ΔR). Figure 0.1 *a* shows the car moving over the road with random irregularities ($h(x)$). Taking into consideration that at constant velocity $x = vt$, $h(vt)$ is a random function dependent on time.

Figure 0.5 shows a high-rise structure (for example, television tower or antenna mast) under the action of wind load, which is a distributed aerodynamic force $\mathbf{q}(t)$, having a random component due to a random component of flow velocity, that is

$$\mathbf{q}(t) = \mathbf{q}_0(t) + \Delta\mathbf{q}(t)$$

where $\mathbf{q}_0\,(t)$ is the deterministic component of the distributed load, $\Delta\mathbf{q}\,(t)$ is the random component of the distributed load depending on time.

The loads $\mathbf{q}_0$ and $\Delta\mathbf{q}$ depend on $\mathbf{v}_0(t)$ and $\Delta\mathbf{v}(t)$ respectively, where $\mathbf{v}_0(t)$ is the modulus of the wind velocity deterministic component and $\Delta\mathbf{v}(t)$ is the modulus of the projection of the flow velocity random component on vector $\mathbf{v}_0$ direction. If wind velocity is constant through the height (does not depend on z), random distributed load $\mathbf{q}(t)$ depends on time only. Generally, wind velocity $\mathbf{v}$ for high-rise structures may be dependable also on coordinate z. Therefore

$$\mathbf{v}\,(t,\,z) = \mathbf{v}_0\,(z,\,t) + \Delta\mathbf{v}\,(z,\,t)$$

where $\Delta\mathbf{v}(z,\,t)$ is a one-dimensional vector random function. Distributed aerodynamic load arising from the action of random wind load on such structure elements as plates and shells depends on two coordinates and time (Fig. 2.2)

$$\mathbf{q} = \mathbf{q}\,(t,\,x,\,y)\,.$$

In engineering practice, we have to deal with different random processes, description methods for these processes largely depending on their nature;

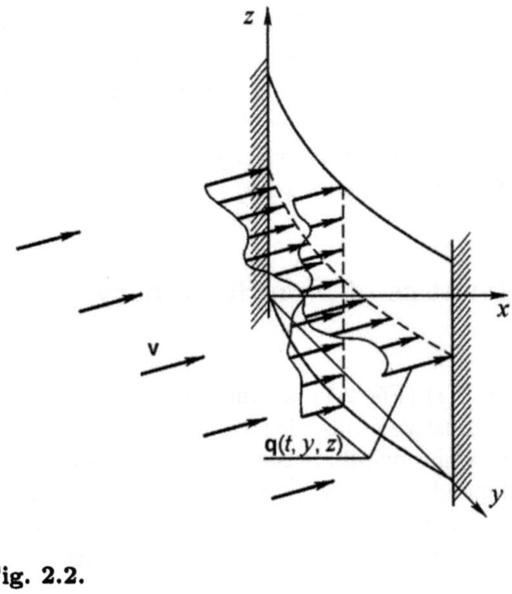

Fig. 2.2.

therefore it is appropriate to classify random processes in such a way as to simplify the selection of these methods in applied problems investigation. This requires the introduction of appropriate terms, making it possible to answer the question: what specific random process is under consideration?

It will be recalled, that classical analytical dynamics and the theory of vibrations study deterministic processes, regarding the dynamic process as the systems response to external deterministic excitation whose variation with time is known exactly. The main feature of deterministic processes lies in the fact that process behavior in the future can be predicted exactly, if we know its behavior in the past.

As stated above, the analog of these processes in statistical dynamics are stationary and non-stationary random processes similar to deterministic stationary and non-stationary processes. The classification of dynamic processes as deterministic or random processes can sometimes cause doubts. Deterministic and random processes are two limiting cases. Any real dynamic process always contains a random component. When its influence on the final result is insignificant, it may be disregarded and we can consider the process to be deterministic. Small random components, however, often may give rise to a great scatter in final results, which can not be neglected.

It is safe to say that all physical processes are not fully deterministic, because the appearance of a non-controllable random disturbance is always possible, and it will make the initially deterministic process a random one. Random dynamic processes that are brought about in mechanical systems, are the systems response to random external forces which in turn are random processes.

To investigate the random motion of mechanical system we need at least minimum of information about random external forces including, for example, random processes probability characteristics. That is why this chapter is devoted to the presentation of the general theory of random functions or random processes, in particular, random processes describing external loads ("input") acting on mechanical systems. It is impossible to obtain "output" probability characteristics necessary for structure "strength" evaluation without the knowledge of input probability characteristics.

2.2 Probability Characteristics of Non-Stationary Random Functions

Let us consider random function $X(t)$ (Fig. 2.1), a random quantity at each given argument t value; whose exhaustive probability characteristic is its distribution law. It is referred to as the one-dimensional distribution law of random function X. This law depends on parameter t, and can be given by one-dimensional probability density $f(x,\, t)$. The one-dimensional distribution law $f(x,\, t)$, however, does not constitute the exhaustive characteristic of random function $X(t)$. Function $f(x,\, t)$ only characterizes distribution law $X(t)$ for a given, even if arbitrary time t. Knowing $f(x,\, t)$ we cannot answer the question about the dependence of random quantities $X(t)$ at different t. Fuller characteristic of random function $X(t)$ is the two-dimensional distribution law:

$$f\left(x_1,\, x_2;\, t_1,\, t_2\right), \tag{2.1}$$

which is a two random quantities $X(t_1)$, $X(t_2)$ system distribution for two arbitrary sections of random function $X(t)$. This characteristic, however, also cannot be generally described as exhaustive, because three-dimensional distribution law may give even a fuller characteristic

$$f\left(x_1,\, x_2,\, x_3;\, t_1,\, t_2,\, t_3\right). \tag{2.2}$$

Theoretically, we may to increase the number of arguments beyond all bounds and get even more comprehensive information about the random function. It is, in fact, impossible, however, to use such cumbersome characteristics that depend on many arguments in practical analysis. Therefore, only one- and two-dimensional distributions are employed in the solution of applied problems of random dynamic processes analysis. The theory of random functions using one- and two-dimensional distribution laws is called the correlation theory.

It has been shown in the first chapter that such nonrandom numerical characteristics of random quantities as mathematical expectation and variance for one random quantity, mathematical expectations and correlation matrix for the system of random quantities, play a major role in the probability theory. The ability of using numerical characteristics is the heart of the applied probability theory. They constitute a rather versatile and powerful mathematical apparatus, which makes it possible to find rather simple solutions to many practical problems.

The simplest basic characteristics, similar to the numerical characteristics of random quantities, are also introduced for random functions, with the rules of procedure being established for them. This apparatus proves to be sufficient for the solution of many practical problems.

Unlike the numerical characteristics of random quantities representing certain numbers, the characteristics of random functions generally represent functions rather than numbers.

Let us consider the random function $X(t)$ section at a given t. We have an ordinary random quantity in this section; let us define its mathematical expectation. It is obvious that in the general case it depends on t, i.e. represents any function of t (Fig. 2.1)

$$m_x\left(t\right) = M\left[X\left(t\right)\right]. \tag{2.3}$$

This means that nonrandom function $m_x(t)$, which equals to the mathematical expectation of the random function appropriate section at each argument t, is referred to as the mathematical expectation of random function $X(t)$.

The random function variance is defined in a similar way.

Nonrandom function $D_x(t)$, whose value for each t equals to the variance of the appropriate random function section is referred to as the variance of the random function $X(t)$:

$$D_x(t) = D[X(t)]. \tag{2.4}$$

The variance of the random function at each t defines the scatter of the possible realizations of the random function about the mean, in other words, "the extent of randomness" of the random function.

As in the case of random quantities $D_x(t)$ is a non-negative function. Taking the root of this function, we obtain function $\sigma_x(t)$ or the standard deviation of the random function

$$\sigma_x(t) = \sqrt{D_x(t)}. \tag{2.5}$$

Mathematical expectation and variance are rather important characteristics of the random function; these characteristics, however, are not enough to describe the main features of random functions. To make sure that this is true, let us consider two random functions $X_1(t)$ and $X_2(t)$, vividly shown by the families of realizations in Fig. 2.3 *a* and 2.3 *b*.

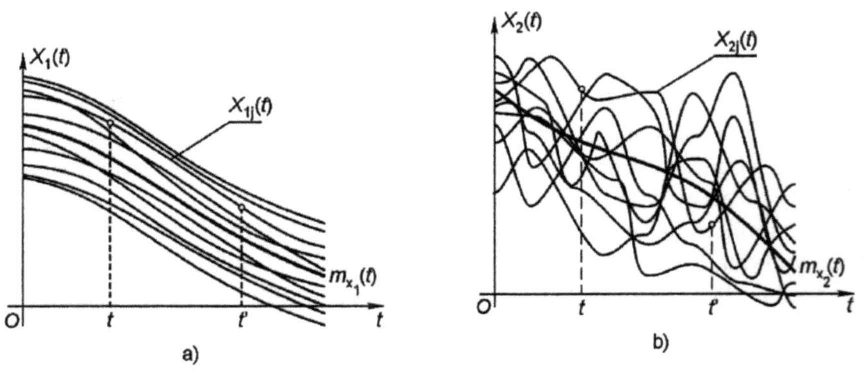

Fig. 2.3.

Random functions $X_1(t)$ and $X_2(t)$ have almost the same mathematical expectations and variances; however, the nature of the variation of these random functions realizations is absolutely different. The random function $X_1(t)$ is marked by a pronounced dependence between its values at different t. On the contrary, the random function $X_2(t)$ is of erratic oscillatory character. Such random function is marked by a rapid decline in relationship between its values at t' and t with an interval $t' - t$ increase.

The internal structure of these random processes is absolutely dissimilar, but this difference is recovered neither by mathematical expectation nor by variance. Therefore, another nonrandom characteristic referred to as the correlation function (or autocorrelation function) is being introduced. The correlation function defines the degree of relationship between random function sections at two different instants of time ($t'$and t).

Let us consider two random function $X(t)$ sections (Fig. 2.3 *a*) associated with different instants t and t'. It is apparent that at close t and t' values quantities $X(t)$ and $X(t')$ are closely related: if quantity $X(t)$ took any value, there would be the high probability of quantity $X(t')$ taking the value close to that of quantity $X(t)$. In general, interrelation between quantities $X(t)$ and $X(t')$ must weaken with an increase of the interval between sections t and t'.

The degree of relationship between quantities $X(t)$ and $X(t')$ may be largely defined by their correlation moment.

Thus, the non-random function of two arguments $K_x(t, t')$, which equals to the correlation moment of the appropriate random function sections at each pair of values t, t', is termed as the correlation function of the random function $X(t)$:

$$K_x(t, t')) = M\left[\overset{\circ}{X}(t)\,\overset{\circ}{X}(t')\right], \tag{2.6}$$

where

$$\overset{\circ}{X}(t) = X(t) - m_x(t), \quad \overset{\circ}{X}(t') = X(t') - m_x(t').$$

($\overset{\circ}{X}(t)$ and $\overset{\circ}{X}(t')$) are centered random functions.

The random functions $X_1(t)$ and $X_2(t)$, considered as an example, have the same mathematical expectations and variances, but totally different correlation functions. The correlation function of the random function $X_1(t)$ (Fig. 2.3 *a*) decreases slowly with an increase in the interval (t, t'); on the contrary, the correlation function of the random function $X_2(t)$ (Fig. 2.3 *b*) decreases rapidly with an increase in this interval.

When it arguments coincide the correlation function $K_x(t, t')$ equals to the variance of the random function.

$$K_x(t, t') = M\left[\left(\overset{\circ}{X}(t)\right)^2\right] = D_x(t). \tag{2.7}$$

Since the correlation moment of the two random quantities $X(t)$ and $X(t')$ does not depend on the sequence of considering these quantities, the correlation function is symmetrical about its arguments, i.e. remains the same when arguments interchange their positions:

$$K_x(t, t') = K_x(t', t). \tag{2.8}$$

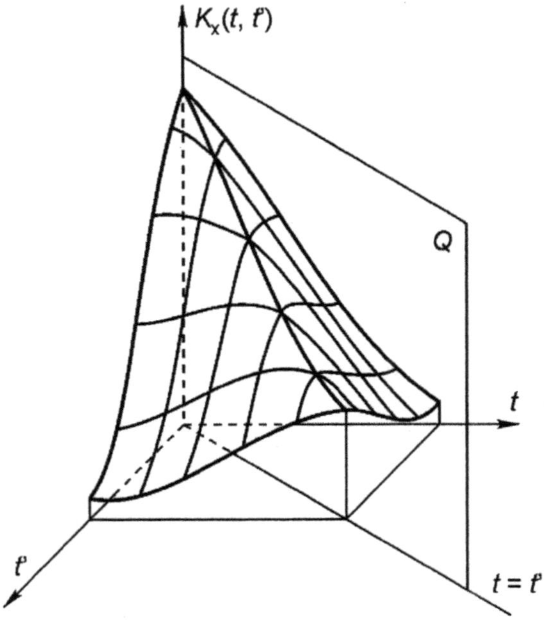

Fig. 2.4.

The qualitative variation of $K_x(t,\, t')$ as a function of t and t' is shown on Fig. 2.4.

Instead of correlation function $K_x(t,\, t')$, we may use the normalized correlation function:

$$r_x\,(t,\, t') = \frac{K_x\,(t,\, t')}{\sigma_x\,(t)\;\sigma_x\,(t')}, \tag{2.9}$$

which is the correlation coefficient of the quantities $X(t)$, $X(t')$. The normalized correlation function is similar to the normalized correlation matrix of the random quantities system. It satisfies the condition (similar to condition (1.59))

$$|r_x\,(t,\, t')| \le 1$$

because

$$|K_x\,(t,\, t')\,| \le \sigma_x\,(t)\,\sigma_x\,(t')\,.$$

At $t = t'$ the normalized correlation function is equal to unity:

$$r_x\,(t,\, t) = \frac{K_x\,(t,\, t)}{[\sigma_x\,(t)]^2} = \frac{D_x\,(t)}{[\sigma_x\,(t)]^{\,2}} = 1.$$

Let us consider how the principal random function characteristics vary when a nonrandom term is added to the function and when the latter is multiplied by a nonrandom factor. These nonrandom terms and multipliers can be either constant quantities or the functions of t (in general case).

Adding the nonrandom term $\varphi(t)$ to the random function $X(t)$, we obtain the following new random function:

$$Y(t) = X(t) + \varphi(t). \tag{2.10}$$

According to the theorem of adding mathematical expectations, we get

$$m_y(t) = m_x(t) + \varphi(t), \tag{2.11}$$

i.e., when adding a nonrandom term to a random function the same nonrandom term is added to its mathematical expectation.

Let us obtain the correlation function and variance of the random function $Y(t)$:

$$\begin{aligned}K_y(t, t') &= M\left[\overset{\circ}{Y}(t)\overset{\circ}{Y}(t')\right] = M\left[(Y(t) - m_y(t))(Y(t') - m_y(t'))\right]\\ &= M\left[(X(t) + \varphi(t) - m_x(t) - \varphi(t))(X(t') + \varphi(t') - m_x(t') - \varphi(t'))\right]\\ &= M\left[(X(t) - m_x(t))(X(t') - m_x(t'))\right] = K_x(t, t').\end{aligned}$$

$$D_y(t) = D_x(t),$$

i.e., when adding a nonrandom term, the correlation function and variance of the random function do not alter.

Let us multiply the random function $X(t)$ by the nonrandom function $\varphi(t)$:

$$Y(t) = \varphi(t) X(t). \tag{2.12}$$

By factoring the nonrandom function $\varphi(t)$ from the sign of mathematical expectation, we obtain

$$m_y(t) = M\left[\varphi(t) X(t)\right] = \varphi(t) m_x(t), \tag{2.13}$$

i.e., when multiplying a random function by a nonrandom factor, its mathematical expectation is multiplied by the same factor.

Let us derive the correlation function and variance

$$\begin{aligned}K_y(t, t') &= M\left[\overset{\circ}{Y}(t)\overset{\circ}{Y}(t')\right] = M\left[(Y(t) - m_y(t))(Y(t') - m_y(t'))\right]\\ &= M\left[\varphi(t)\varphi(t')(X(t) - m_x(t))(X(t') - m_x(t'))\right] = \varphi(t)\varphi(t')K_x(t, t'),\\ D_y(t) &= \varphi^2(t) D_x(t),\end{aligned} \tag{2.14}$$

i.e., when multiplying a random function by the nonrandom function $\varphi(t)$, its correlation function is multiplied by $\varphi(t)\varphi(t')$, and its variance is multiplied by $\varphi^2(t)$.

In particular, when $\varphi(t) = c$ (does not depend on t), the correlation function is multiplied by c^2.

Example 2.1. It is required to determine the normalized correlation function, if the random function is

$$X(t) = A\mathrm{e}^{-\alpha t},$$

where A is a random quantity with the known mathematical expectation m_A and the variance D_A.

The mathematical expectation of the random function is

$$m_x = M\left[\mathrm{e}^{-\alpha t}\right] = \mathrm{e}^{-\alpha t} m_A.$$

The correlation function of the random function $X(t)$ is

$$\begin{aligned} K_x(t,\, t') &= M\left[\left(\mathrm{e}^{-\alpha t} - m_A \mathrm{e}^{-\alpha t}\right)\left(A\mathrm{e}^{-\alpha t'} - m_A \mathrm{e}^{-\alpha t'}\right)\right] \\ &= \mathrm{e}^{-\alpha t}\,\mathrm{e}^{-\alpha t'} M\left[(A - m_A)^2\right] = \mathrm{e}^{-\alpha(t+t')} D_A. \end{aligned}$$

The variance of the random function $X(t)$ is equal to

$$D_x(t) = D_A \mathrm{e}^{-2\alpha t}.$$

The standard deviations $X(t)$ for the instants t' and t are equal to

$$\sigma_x(t) = \sigma_A \mathrm{e}^{-\alpha t};$$
$$\sigma_x(t') = \sigma_A \mathrm{e}^{-\alpha t'}.$$

The normalized correlation function is

$$r_x(t,\, t') = \frac{k_x(t,\, t')}{\sigma_x(t)\, \sigma_x(t')} = \frac{D_A \mathrm{e}^{-\alpha(t+t')}}{\sigma_A^2 \mathrm{e}^{-\alpha(t+t')}} = 1.$$

The result obtained indicates that there is a linear relationship between the values of the random function $X(t)$ at different instants t' and t.

Example 2.2. It is required to determine the correlation function of the random function $X(t)$, if

$$X(t) = A \sin\omega t = B\cos\omega t,$$

where A and B are random quantities, for which m_A, m_B, σ_A, σ_B and K_{AB} are known ($m_A = m_B = 0$). By definition

$$\begin{aligned} K_x(t,\, t') &= M\left[(A\sin\omega t + B\cos\omega t)(A\sin\omega t' + B\cos\omega t')\right] \\ &= M\left[A^2 \sin\omega t \sin\omega t' + AB\sin\omega t\cos\omega t'\right. \\ &\quad \left. + AB\cos\omega t \sin\omega t' + B^2 \cos\omega t\cos\omega t'\right], \end{aligned}$$

or

$$K_x(t,\, t') = \sin\omega t \sin\omega t' M\left[A^2\right] + \sin\omega\,(t+t')\, M\,[AB] \\ + \cos\omega t \cos\omega t' M\left[B^2\right].$$

Finally we obtain

$$K_x(t,\, t') = \sigma_A^2 \sin\omega t \sin\omega t' + K_{AB} \sin\omega\,(t+t') + \sigma_B^2 \cos\omega t \cos\omega t'.$$

Let us consider a particular case, where $K_{AB} = 0$, and $\sigma_A^2 = \sigma_B^2$,

$$K_x\,(t,\, t') = \sigma_A^2 \cos\omega\,(t - t').$$

2.3 Random Function Systems and Their Probability Characteristics

Let us consider the system of the following two random functions $X(t)$ and $Y(t)$, characterizing different random processes. For example, the random force N_l and the random moment M_l are acting on a rocket at launch (Fig. 0.2). Therefore it would be useful to be aware of the correlation relationship of this random functions when studying the rockets disturbed motion.

Let us consider the second product moment of the real centered random functions $\overset{\circ}{X}$ and $\overset{\circ}{Y}$ for different instants of time:

$$M\left[\overset{\circ}{X}(t)\,\overset{\circ}{Y}(t')\right] = K_{xy}\,(t,\, t'), \tag{2.15}$$

where $K_{xy}(t,\, t')$ is a cross-correlation function.

The non-random function of the two arguments t and t' which is equal to the correlation moment of the appropriate sections of the random function $X(t)$ and random function $Y(t)$ at each pair of two values t, t', is referred to as the cross-correlation function of two random functions.

The cross-correlation function does not satisfy the symmetry condition relative to its arguments, i.e.

$$K_{xy}\,(t,\, t') \neq K_{xy}\,(t',\, t),$$

but at a simultaneous permutation of instants of time and indexes we have

$$K_{xy}\,(t,\, t') = K_{yx}\,(t',\, t). \tag{2.16}$$

If the cross-correlation function K_{xy} is not identically equal to zero, the random functions X and Y are called correlated by analogy with random quantities; if, however, K_{xy} is identically equal to zero, these random functions are called non-correlated functions.

With respect to applied problems it is convenient to use the following normalized cross-correlation function

$$r_{xy}(t, t') = \frac{K_{xy}(t, t')}{\sigma_x(t)\,\sigma_y(t')}. \tag{2.17}$$

As in the case of two random quantities system we can introduce the correlation matrix

$$K = \begin{Vmatrix} K_x(t, t') & K_{xy}(t, t') \\ K_{yx}(t, t') & K_y(t, t') \end{Vmatrix} \tag{2.18}$$

For the system of n real random functions we have the correlation matrix

$$K = \begin{Vmatrix} K_{x_1x_1} & K_{x_1x_2} & \dots & K_{x_1x_n} \\ \dots & \dots & \dots & \dots \\ \vdots & \vdots & \ddots & \vdots \\ K_{x_nx_1} & K_{x_nx_2} & \dots & K_{x_nx_n} \end{Vmatrix} \tag{2.19}$$

Considering the random functions $X_j(t)$ as the components of the $n-$th vector, we can represent the matrix K (2.19) in a more compact form, using a two-vector dyadic (tensor) product (1.50):

$$K = M\left[\overset{\circ}{\bar{X}}(t) \otimes \overset{\circ}{\bar{X}}(t')\right].$$

For complex-valued random functions, by analogy with complex-valued random quantities, the cross-correlation function is

$$K_{xy}(t, t') = M\left[\overset{\circ}{X}(t)\,\overset{\circ}{Y}{}^*(t')\right], \tag{2.20}$$

where $\overset{\circ}{Y}{}^*$ is the centered conjugate random function. In a particular case where $t = t'$ and $X = Y$, the correlation function (2.20) is the random complex function X variance, which must be positive; this occurs, if we take the product of the complex function $X(t)$ into the conjugate function $X^*(t)$.

$$\begin{aligned} D_x(t) &= M\left[\overset{\circ}{X}(t)\,\overset{\circ}{X}{}^*(t)\right] = M\left[\left(\overset{\circ}{X}_1(t) + i\,\overset{\circ}{X}_2(t)\right)\left(\overset{\circ}{X}_1(t) - i\,\overset{\circ}{X}_2(t)\right)\right] \\ &= M\left[\left(\overset{\circ}{X}{}_1^2 + \overset{\circ}{X}{}_2^2\right)\right]. \end{aligned}$$

Let us show that for a complex random function the condition

$$K_{xy}(t, t') = K^*_{yx}(t', t). \tag{2.21}$$

is satisfied. The correlation functions $K_{xy}(t, t')$ and $K_{yx}(t', t)$ in their more comprehensive form are

$$K_{xy}(t, t') = M[(X_1(t) + iX_2(t))(Y_1(t') - iY_2(t'))]$$
$$= K_{x_1y_1} + K_{x_2y_2} + i(K_{x_1x_2} - K_{x_2y_1}).$$
$$K_{yx}(t', t) = M[(Y_1(t') + iY_2(t'))(X_1(t) - iX_2(t))]$$
$$= K_{x_1y_1} + K_{x_2y_2} - i(K_{x_1y_2} - K_{x_2y_1}).$$

Hence the conjugate function $K^*_{yx}(t', t)$ is equal to $K_{xy}(t, t')$.

In accordance with (2.21) the elements of the correlation matrix for complex random functions, symmetric about the principal diagonal, are complex conjugate functions.

Example 2.3. There are the following two random functions $X_1 = A\cos\omega_1 t$ and $X_2 = B\cos\omega_2 t$. The amplitudes A and B are random quantities with known probability characteristics m_A, m_B, D_A, D_B and K_{AB}. It is required to determine the cross-correlation function and its value at $t = t'$.

According to (2.15)

$$K_{x_1x_2} = M[(A - m_A)\cos\omega_1 t\,(B - m_B)\cos\omega_2 t'] = K_{AB}\cos\omega_1 t\cos\omega_2 t'.$$

At $t = t'$

$$K_{x_1x_2}(t, t) = K_{AB}\cos\omega_1 t\cos\omega_2 t.$$

Example 2.4. Figure 2.5 shows a beam loaded with concentrated random-magnitude moments M_{x_0} and M_{y_0} having known probability characteristics m_x, m_y, D_x, D_y and K_{xy}. It is required to determine: 1) the mathematical expectation and variance of the maximum normal stress in an arbitrary section; 2) the maximum value of the maximum normal stress in the dangerous section on the assumption that the maximum stress has normal distribution.

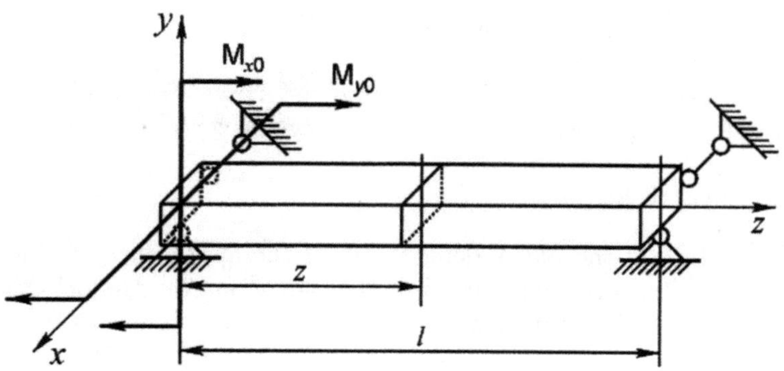

Fig. 2.5.

The bending moments in an arbitrary section are

$$M_x = M_{xo}\left(1-\frac{z}{l}\right);$$
$$M_y = M_{yo}\left(1-\frac{z}{l}\right).$$

The maximum normal stress in an arbitrary section is

$$\sigma_{\max} = \frac{M_x}{J_x} + \frac{M_y}{J_y} = a_1 M_{xo}\left(1-\frac{z}{l}\right) + a_2 M_{yo}\left(1-\frac{z}{l}\right),$$

where x_1, y_1 are the coordinates of the outermost from the neutral line point, $a_1 = y_1/J_x$, $a_2 = x_1/J_y$.

The mathematical expectation of the maximum normal stress is

$$m_{\sigma_m} = a_1\left(1-\frac{z}{l}\right)m_x + a_2\left(1-\frac{z}{l}\right)m_y.$$

The correlation function of the maximum normal stress is

$$\begin{aligned}
&K_{\sigma_m}(z,\, z')\\
&=M\left[(a_1 M_{xo} + a_2 M_{yo})^2\left(1-\frac{z}{l}\right)\left(1-\frac{z'}{l}\right)\right]\\
&=\left(1-\frac{z}{l}\right)\left(1-\frac{z'}{l}\right)\left[a_1^2 M\left[M_{xo}^2\right] + 2a_1a_2 M\left[M_{xo}M_{yo}\right] + a_2^2 M\left[M_{yo}^2\right]\right]\\
&=\left(1-\frac{z}{l}\right) = \left(1-\frac{z'}{l}\right)\left(a_1^2 D_x + 2a_1a_2K_{xy} + a_2^2 D_y\right).
\end{aligned}$$

The variance of the maximum normal stress in an arbitrary section is

$$D\sigma_m = \left(1-\frac{z}{l}\right)^2\left(a_1^2 D_x + 2a_1a_2K_{xy} + a_2^2 D_y\right).$$

With the respect to the example considered the dangerous section is the section at $z = 0$, therefore we obtain $\max\sigma_m$ using the three sigma rule

$$\max\sigma_m = m_{\sigma_m} + 3\sigma_{\sigma_m}.$$

If the limit state of the bar is determined by the appearance of plastic deformations, the structure parameters of the bar at known random loads must be selected from the condition

$$\max\sigma_m = \frac{\sigma_y}{n_y},$$

where σ_y is the yield stress of the bar material, n_y is the safety factor.

2.4 Random Functions Linear Transformations

Let us consider the linear transformations of random functions, when the relation between "input" and "output" is established using linear operators L:

$$Y(t) = L[X(t)]. \tag{2.22}$$

We keep in mind the statistical relation between input and output, when it is necessary to determine the probability characteristics of the $Y(t)$ from the known probability characteristics of the random function $X(t)$.

Let us consider the linear transformations of random functions most frequently used in the analysis of random vibrations.

1. If the input $\mathbf{X}$ and the output $\mathbf{Y}$ are random n-dimensional vectors and related by the linear relation of the form

$$\mathbf{Y}(t) = B(t)\mathbf{X}(t),$$

where $B(t)$ is the matrix n by n, whose elements $b_{ij}(t)$ are nonrandom functions, the mathematical expectations of the components y_j of the vector $\mathbf{Y}$, the correlation matrix $K(t, t')$ and the cross-correlation functions $K_{y_i y_j}$ are equal to

$$m_{y_j}(t) = M\left[\sum_{i=1}^{n} {}_{ji}x_i(t)\right] = \sum_{i=1}^{n} b_{ji}m_{x_i}(t) \quad (j = 1, 2, \dots, n) \tag{2.23}$$

$$K(t, t') = M\left[\overset{\circ}{\mathbf{Y}}(t) \otimes \overset{\circ}{\mathbf{Y}}(t')\right] = \begin{bmatrix} K_{y_1 y_1} & K_{y_1 y_2} & \dots & K_{y_1 y_n} \\ \dots & \dots & \dots & \dots \\ \vdots & \vdots & \ddots & \vdots \\ K_{y_1 y_1} & K_{y_n y_2} & \dots & K_{y_n y_n} \end{bmatrix}$$

where

$$K_{y_i y_j} = M\left[\overset{\circ}{Y}_j(t)\overset{\circ}{Y}_i(t')\right] = \sum_{\rho=1}^{n}\sum_{\nu=1}^{n} b_{i\rho}(t)\, b_{j\nu}(t')\, K_{x_\rho x_\nu}(t, t').$$

The variance of the vector $\mathbf{Y}$ components is

$$D_{y_i}(t) = \sum_{\rho=1}^{n}\sum_{\nu=1}^{n} {}_{i\rho}(t)_{j\nu}(t)\, K_{x_\rho x_\nu}(t, t'). \tag{2.24}$$

2. The integral of the random function $X(t)$

$$Y(t) = \int_0^t X(t_1)\,\mathrm{d}t_1. \tag{2.25}$$

Assuming that averaging and integration operations are interchangeable, the mathematical expectation is determined as

$$m_y(t) = M\left[\int_0^t X\,\mathrm{d}t_1\right] = \int_0^t M[X]\,\mathrm{d}t_1 = \int_0^t m_x(t_1)\,\mathrm{d}t_1. \tag{2.26}$$

The interchange of the mathematical expectation operation and other linear mathematical operations is used in subsequent transformations.

The correlation function is

$$\begin{aligned} K(t, t') &= M\left[(Y(t) - m_y(t))(Y(t') - m_y(t'))\right] \\ &= M\left\{\int_0^t [X(t_1) - m_x(t_1)]\,\mathrm{d}t_1 \int_0^{t_1} \left[X\left(t_1^{(1)}\right) - m_x\left(t_1^{(1)}\right)\right] \mathrm{d}t_1^{(1)}\right\}. \end{aligned} \tag{2.27}$$

The product of the two integrals under the sign of mathematical expectation in formula (2.27) is equal to the double integral

$$\int_0^t \int_0^{t'} (X(t_1) - m_x(t_1))\left(X\left(t_1^{(1)}\right) - m_x\left(t_1^{(1)}\right)\right) \mathrm{d}t_1\,\mathrm{d}t_1^{(1)},$$

therefore

$$\begin{aligned} &K_y(t, t') \\ &= M\left[\int_0^t \int_0^{t'} (X(t_1) - m_x(t_1))\left(X\left(t_1^{(1)}\right) - m_x\left(t_1^{(1)}\right)\right) \mathrm{d}t_1\,\mathrm{d}t_1^{(1)}\right] \\ &= \int_0^t \int_0^{t_1} M\left[(X(t_1) - m_x(t_1))\left(X\left(t_1^{(1)}\right) - m_x\left(t_1^{(1)}\right)\right)\right] \mathrm{d}t_1\,\mathrm{d}t_1^{(1)} \\ &= \int_0^t \int_0^{t_1} K_x\left(t_1, t_1^{(1)}\right) \mathrm{d}t_1\,\mathrm{d}t_1^{(1)}. \end{aligned} \tag{2.28}$$

Example 2.5. The beam shown in Fig. 2.6 is under the action of the random distributed load $q(z)$. The probabilistic characteristics of the load are known; in other words, it is the mathematical expectation $m_q(z)$ as a function of z and the correlation function $K_q(z, z')$ that are known. It is necessary to determine the probabilistic characteristics of the reaction forces R_1 and R_2.

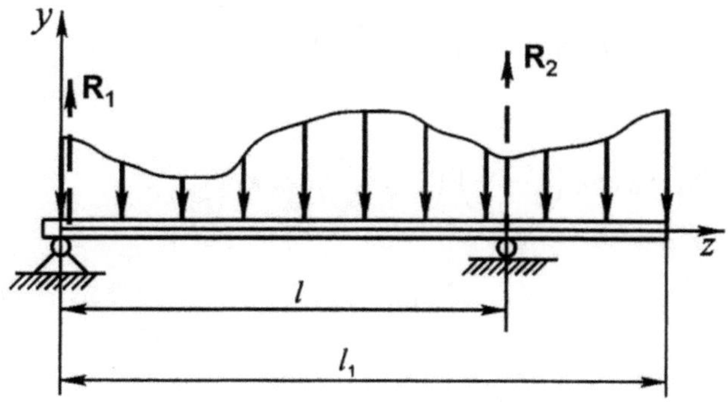

Fig. 2.6.

Considering the beam equilibrium, we obtain the following two equations

$$R_1 + R_2 = \int_0^{l_1} q(z)\,\mathrm{d}z; \qquad R_2 l = \int_0^{l_1} q(z)\,z\mathrm{d}z. \tag{2.29}$$

We determine the mathematical expectations of the reaction forces with the use of (2.29)

$$m_{R_1} + m_{R_2} = \int_0^{l_1} m_q(z)\,\mathrm{d}z;$$

$$l m_{R_2} = \int_0^{l_1} m_q(z)\,z\mathrm{d}z.$$

The centered random quantities $\overset{\circ}{R}_1$ and $\overset{\circ}{R}_2$ are equal to

$$\overset{\circ}{R}_1 = \int_0^{l_1} (q(z) - m_q(z))\left(1 - \frac{z}{l}\right)\mathrm{d}z,;$$

$$\overset{\circ}{R}_2 = \int_0^{l_1} (q(z) - m_q(z))\left(\frac{z}{l}\right)\mathrm{d}z.$$

The reaction variances are equal to

$$D_{R_1} = M\left[\left(\int_0^{l_1} q_0(z_1)\left(1 - \frac{z_1}{l}\right)\mathrm{d}z_1\right)\left(\int_0^{l_1} q_0(z)\left(1 - \frac{z}{l}\right)\mathrm{d}z\right)\right] =$$

$$= M\left[\int_0^{l_1}\int_0^{l_1} q_0(z_1)\, q_0(z)\left(1-\frac{z}{l}\right)\left(1-\frac{z_1}{l}\right)\mathrm{d}z\,\mathrm{d}z_1\right]$$

$$= \int_0^{l_1}\int_0^{l_1} M\left[q_0(z_1)\, q_0(z)\right]\left(1-\frac{z}{l}\right)\left(1-\frac{z_1}{l}\right)\mathrm{d}z\,\mathrm{d}z_1$$

$$= \int_0^{l_1}\int_0^{l_1} K_q(z,\, z_1)\left(1-\frac{z}{l}\right)\left(1-\frac{z_1}{l}\right)\mathrm{d}z\,\mathrm{d}z_1;$$

$$D_{R_2} = \int_0^{l_1}\int_0^{l_1} K_q(z,\, z_1)\left(\frac{z}{l}\right)\left(\frac{z_1}{l}\right)\mathrm{d}z\,\mathrm{d}z_1.$$

3. The probabilistic characteristics of the random function $Y(t)$

$$Y(t) = a(t)\,c + \int_0^t k(t,\, t_1)\, X(t_1)\,\mathrm{d}t_1, \tag{2.30}$$

where $a(t)$, $k(t,\, t_1)$ are nonrandom functions, c is a random quantity with known m_c and D_c; $X(t_1)$ is a random function with known $m_x(t)$ and $K_x(t,\, t')$.

The mathematical expectation of the $Y(t)$ is

$$m_y(t) = M\left[a(t)\,c + \int_0^t k(t,\, t_1)\, X\mathrm{d}t_1\right]$$

$$= a(t)\, M[c] + \int_0^t k(t,\, t_1)\, M[X]\,\mathrm{d}t_1$$

$$= a(t)\, m_c + \int_0^t k(t,\, t_1)\, m_x \mathrm{d}t_1. \tag{2.31}$$

The correlation function with independent c and X is

$$K_y(t,\, t') = M\left[\left(a(t)\overset{\circ}{c} + \int_0^t k(t,\, t_1)\,\overset{\circ}{X}_1(t_1)\,\mathrm{d}t_1\right)\right.$$

$$\left.\times\left(a(t')\overset{\circ}{c} + \int_0^{t'} k\left(t', t_1^{(1)}\right)\overset{\circ}{X}\left(t_1^{(1)}\right)\mathrm{d}t_1^{(1)}\right)\right] = a(t)\, a(t')\, M\left[\overset{\circ}{c}{}^2\right]$$

$$+ \int_0^t\int_0^{t_1} k(t,\, t_1)\, k\left(t', t_1^{(1)}\right) K_x\left(t_1,\, t_1^{(1)}\right)\mathrm{d}t_1\,\mathrm{d}t_1^{(1)}. \tag{2.32}$$

4. Let us consider the derivative of the random function

$$Y(t) = \frac{\mathrm{d}}{\mathrm{d}t} X(t). \tag{2.33}$$

The mathematical expectation of the $Y(t)$ is

$$M[Y(t)] = m_y(t) = M\left[\frac{\mathrm{d}X(t)}{\mathrm{d}t}\right] = \frac{\mathrm{d}}{\mathrm{d}t} M[X(t)] = \frac{\mathrm{d}m_x(t)}{\mathrm{d}t}. \tag{2.34}$$

The correlation function $K_y(t, t')$ is

$$\begin{aligned} K_y(t, t') &= M[(Y(t) - m_y(t))(Y(t') - m_y(t'))] \\ &= M\left[\left(\frac{\mathrm{d}\overset{\circ}{X}(t)}{\mathrm{d}t}\right)\left(\frac{\mathrm{d}\overset{\circ}{X}(t')}{\mathrm{d}t'}\right)\right]. \end{aligned} \tag{2.35}$$

The product of the derivatives under the sign of mathematical expectation can be represented as

$$\frac{\mathrm{d}}{\mathrm{d}t} \overset{\circ}{X}(t) \frac{\mathrm{d}}{\mathrm{d}t'} \overset{\circ}{X}(t') = \frac{\partial^2 \left[\overset{\circ}{X}(t) \overset{\circ}{X}(t')\right]}{\partial t\, \partial t'}. \tag{2.36}$$

As a result we obtain

$$\begin{aligned} K_y(t,t') &= M\left[\frac{\partial^2 \left[\overset{\circ}{X}(t) \overset{\circ}{X}(t')\right]}{\partial t\, \partial t'}\right] \\ &= \frac{\partial^2}{\partial t\, \partial t'} M\left[\overset{\circ}{X}(t) \overset{\circ}{X}(t')\right] = \frac{\partial^2 K_x(t, t')}{\partial t\, \partial t'}. \end{aligned} \tag{2.37}$$

If the random functions $X(t)$ and $Y(t)$ are connected by the relation $Y(t) = \dfrac{\mathrm{d}^2 X(t)}{\mathrm{d}t^2}$, the correlation function $K_y(t, t')$ is equal to

$$K_y(t, t') = \frac{\partial^4 K_x(t, t')}{\partial t^2\, \partial t'^2}. \tag{2.38}$$

For the general case when

$$Y = \frac{\mathrm{d}^n X}{\mathrm{d}t^n}, \tag{2.39}$$

the correlation function K_y is equal to

$$K_y = \frac{\partial^{2n} X}{\partial t^n\, \partial t'^n}. \tag{2.40}$$

5. The cross-correlation function of the random function X and its derivative is

$$K_{x\dot{x}} = M\left[\overset{\circ}{X}(t)\,\frac{\mathrm{d}\overset{\circ}{X}(t')}{\mathrm{d}t}\right] = M\frac{\partial\,\overset{\circ}{X}(t)\,\overset{\circ}{X}(t')}{\partial t'} = \frac{\partial K_x}{\partial t'}. \tag{2.41}$$

Similarly, we obtain an expression for the cross-correlation function of the different order derivatives of the random function $X(t)$

$$K_{x_p x_q} = \frac{\partial^{p+q} K_x(t,\, t')}{\partial t^p \partial t'^q}, \tag{2.42}$$

where p, q are orders of the derivatives ($p = 0,\, 1, \dots,\, n,\; q = 0,\, 1, \dots,\, n$).

Example 2.6. Determine a correlation function and a variance for the derivative of the random function $X(t)$, if $X(t) = A\sin t$ (m_A and σ_A are known).

The correlation function of the random function $X(t)$ is

$$K_x(t,\, t') = M\left[((A - m_A)\sin t)\,((A - m_A)\sin t')\right] = \sin t \sin t'\, \sigma_A^2.$$

Using the formula (2.37) we obtain

$$K_y(t,\, t') = K_{\dot{x}}(t,\, t') = \sigma_A^2 \cos t \cos t'.$$

The variance is

$$D_{\dot{x}}(t) = D_A \cos^2 t.$$

Example 2.7. Determine the correlation function of the random function

$$Y(t) = a(t)\,X(t) + b(t)\,\frac{\mathrm{d}X(t)}{\mathrm{d}t},$$

where $a(t)$, $b(t)$ are nonrandom functions; X is a random function with known characteristics $m_x = 0$ and $K_x(t,\, t')$.

The correlation function is

$$K(t,\, t') = M\left[\overset{\circ}{Y}(t)\,\overset{\circ}{Y}(t')\right],$$

or

$$\begin{aligned} K_y(t,\, t') &= a(t)\,a(t')\,M\left[X(t)\,X(t')\right] + a(t)\,b(t')\,M\left[X(t)\,\frac{\mathrm{d}X(t')}{\mathrm{d}t'}\right] \\ &+ a(t')\,b(t)\,M\left[\frac{\mathrm{d}X(t)}{\mathrm{d}t}X(t')\right] + b(t)\,b(t')\,M\left[\frac{\mathrm{d}X(t)}{\mathrm{d}t}\frac{\mathrm{d}X(t')}{\mathrm{d}t'}\right]. \end{aligned}$$

Since

$$M\left[X(t)\,\frac{\mathrm{d}X(t')}{\mathrm{d}t}\right] = \frac{\partial}{\partial t'}M\left[X(t)\,X(t')\right] = \frac{\partial}{\partial t'}K_x(t,\, t'),$$

after the necessary transformations we obtain

$$\begin{aligned} K_y(t,\, t') &= a(t)\,a(t')\,K_x + a(t)\,b(t')\,\frac{\partial K_x}{\partial t'} \\ &+ a(t')\,b(t)\,\frac{\partial K_x}{\partial t} + b(t)\,b(t')\,\frac{\partial^2 K_x}{\partial t\,\partial t'}. \end{aligned} \tag{2.43}$$

2.5 The Probabilistic Characteristics of the Linear Differential Equations at Non-stationary Random Disturbances

1. The linear non-homogeneous differential equation of the first order

$$\dot{y} + ky = X(t), \tag{2.44}$$

where $X(t)$ is a random function with known $m_x(t)$ and $K_x(t, t')$.

The equation (2.44) describes the physical process varying in time, such as the work of an engine, shown in Fig. 2.7. In steady-state conditions, the engine moment $M_q(t)$ is balanced by the moment of resistance $M_r(\omega)$ depending on angular velocity ω and load moment M_l. The random variation of the load moment by ΔM_l will alter the angular velocity by $\Delta\omega$ and lead to the occurrence of inertia moment equal to $J\dfrac{\mathrm{d}\Delta\omega}{\mathrm{d}t}$, where J is the rotating parts moment of inertia reduced to the axis of the shaft.

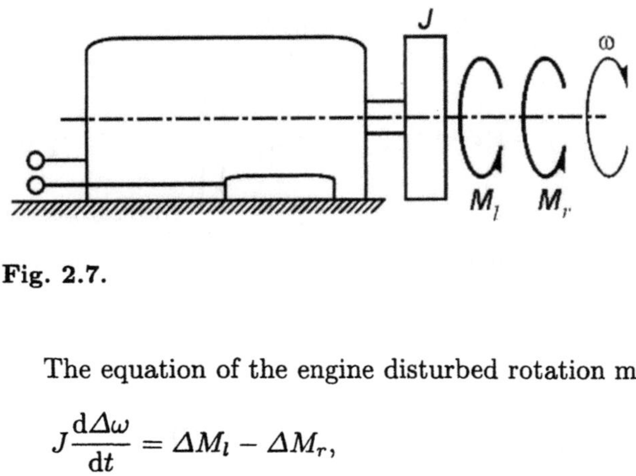

Fig. 2.7.

The equation of the engine disturbed rotation motion is

$$J\frac{\mathrm{d}\Delta\omega}{\mathrm{d}t} = \Delta M_l - \Delta M_r, \tag{2.45}$$

where ΔM_r is the moment of resistance variation. At small deviations from steady-state operating conditions we can assume that $\Delta M_r = k_1 \Delta\omega$.

Then the equation (2.45) will take the form similar to the equation (2.44):

$$\frac{\mathrm{d}\Delta\omega}{\mathrm{d}t} + \frac{k_1}{J}\Delta\omega = \frac{\Delta M_r}{J}. \tag{2.46}$$

General solution of the equation (2.46) with arbitrary right-hand side (putting $y = \Delta\omega$ and $\Delta M_l(t)/J = X(t)$ is

$$y = c\mathrm{e}^{-kt} + \int\limits_0^t \mathrm{e}^{-k(t-t_1)} X(t_1)\,\mathrm{d}t_1, \quad \left(k = \frac{k_1}{J}\right). \tag{2.47}$$

The probabilistic characteristics of the random moment ΔM_l ($m_{\Delta M_l}(t)$ and $K_{\Delta M_l}(t, t')$) and hence the $X(t)$ are considered to be known.

At $t = 0$, $y(0) = y_0$, therefore $c = y_0$. The initial value of Y may be both random and nonrandom.

Let us consider a case where the initial value y_0 is random, with random quantity y_0 and random function $X(t)$ being independent. We consider that the probabilistic characteristics of y_0 are known and equal to m_0 and D_0.

The mathematical expectation of the equation (2.47) solution and the correlation function will be:

$$m_y = M\left[y_0 e^{-kt}\right] + \int_0^t M\left[e^{-k(t-t_1)} X(t_1)\right] dt_1$$

$$= m_0 e^{-kt} + \int_0^t e^{-k(t-t_1)} m_x(t_1) dt_1; \qquad (2.48)$$

$$K_y(t, t') = M\left[\left\{\overset{\circ}{y}_0 e^{-kt} + \int_0^t e^{-k(t-t_1)} \overset{\circ}{X}(t_1) dt_1\right\}\right.$$

$$\left.\times \left\{\overset{\circ}{y}_0 e^{-kt^{(1)}} + \int_0^{t'} e^{-k\left(t'-t^{(1)}\right)} \overset{\circ}{X}\left(t_1^{(1)}\right) dt_1^{(1)}\right\}\right]. \qquad (2.49)$$

Since $\overset{\circ}{y}_0$ and $\overset{\circ}{X}(t)$ are independent, after the necessary transformations we get

$$K_y(t, t') = D_0 e^{-k(t+t')}$$

$$+ \int_0^t \int_0^{t'} e^{-k(t-t_1)} e^{-k\left(t'-t_1^{(1)}\right)} K_x\left(t_1, t_1^{(1)}\right) dt_1\, dt_1^{(1)}. \qquad (2.50)$$

In order to obtain the variance it is necessary to put $t' = t$ in the expression (2.50) after the integration.

Let us find the correlation function and the variance of the equation (2.47) solution, if ΔM_l is the random function with zero mathematical expectation and constant variance D_l (at $t = 0, y = 0$). After the necessary transformations we obtain from (2.50)

$$K_y(t, t') = \frac{D_l}{k^2}\left(e^{-kt} - 1\right)\left(e^{-kt'} - 1\right);$$

$$D_y(t) = \frac{D_l}{k^2}\left(e^{-kt} - 1\right)^2.$$

When solving applied problems, it is often necessary to know the probabilistic characteristics of the derivatives $\dot{y}$. The information that can be obtained by considering the cross-correlation function $K_{y\dot{y}}$ happens to be useful. From equation (2.44) we obtain an expression for the derivative

$$\dot{y}(t) = -ky(t) + X(t). \tag{2.51}$$

The same expression can be obtained by the differentiation of solution (2.47) with respect to t. When differentiating an integral in the right-hand side of the relationship (2.47), one must use the rule of integral differentiation with respect to the parameter

$$\begin{aligned} \frac{\partial J}{\partial t} &= \int\limits_{\alpha(t)}^{\beta(t)} f(t, t_1)\,\mathrm{d}t_1 \\ &= \int\limits_{\alpha(t)}^{\beta(t)} \frac{\partial f(t, t_1)}{\partial t} \alpha t_1 + \frac{\mathrm{d}\beta}{\mathrm{d}t} f(t, \beta) - \frac{\mathrm{d}\alpha}{\mathrm{d}t} f(t, \alpha(t)). \end{aligned} \tag{2.52}$$

When differentiating (2.46) with due account of (2.50), we obtain

$$\dot{y} = -k\left(c\mathrm{e}^{-kt} + \int\limits_0^t \mathrm{e}^{-k(t-t_1)} X(t_1)\,\mathrm{d}t_1\right) + X(t)$$

or

$$\dot{y} = -ky + X(t).$$

The mathematical expectation and the correlation function for $\dot{y}$ are

$$\begin{aligned} m_{\dot{y}}(t) &= -km_y(t) + m_x(t); \\ K_{\dot{y}} &= M\left[\frac{\mathrm{d}\overset{\circ}{Y}(t)}{\mathrm{d}t}\frac{\mathrm{d}\overset{\circ}{Y}(t')}{\mathrm{d}t'}\right] \\ &= M\left[\left(-k\overset{\circ}{y}(t) + \overset{\circ}{X}(t)\right)\left(-k\overset{\circ}{y}(t') + \overset{\circ}{X}(t')\right)\right] \\ &= k^2K_y - kM\left[\overset{\circ}{y}(t)\overset{\circ}{X}(t')\right] - kM\left[\overset{\circ}{y}(t')\overset{\circ}{X}(t)\right] + K_x. \end{aligned}$$

Since the random initial deviation y_0 and $X(t)$ are independent, after the necessary transformations the mathematical expectations will be

$$M\left[\overset{\circ}{y}(t)\overset{\circ}{X}(t')\right] = \int\limits_0^t \mathrm{e}^{-k(t-t_1)} K_x(t_1, t')\,\mathrm{d}t_1;$$

$$M\left[\overset{\circ}{y}(t')\,X(t)\right] = \int_0^{t'} \mathrm{e}^{-k\left(t'-t_1^{(1)}\right)} K_x\left(t_1^{(1)},\,t\right) \mathrm{d}t_1^{(1)}.$$

Finally we obtain

$$K_{\dot{y}}(t,\,t') = k^2 K_y(t,\,t')\,K_x(t,\,t') + \int_0^{t} \mathrm{e}^{-k(t-t_1)} K_x\left(t_1,\,t_1^{(1)}\right) \mathrm{d}t_1$$

$$+ \int_0^{t'} \mathrm{e}^{-k\left(t'-t_1^{(1)}\right)} K_x\left(t_1^{(1)},t\right) \mathrm{d}t_1^{(1)}. \tag{2.53}$$

The cross-correlation function is

$$K_{y\dot{y}}(t,\,t') = M\left[\overset{\circ}{Y}(t)\,\frac{\mathrm{d}\overset{\circ}{Y}(t'')}{\mathrm{d}t'}\right].$$

After the necessary transformations we obtain

$$K_{y\dot{y}}(t,\,t') = -kK_y(t,\,t') + \int_0^{t} \mathrm{e}^{-k(t-t_1)} K_x(t_1,\,t')\,\mathrm{d}t_1. \tag{2.54}$$

2. The linear equation of the second order with constant coefficients and the random right-hand side

$$\ddot{y} + 2ny + p_0^2 y = X(t)\,. \tag{2.55}$$

General solution of equation (2.55) can be represented in the form

$$y = c_1 f_1(t) + c_2 f_2(t) + \int_0^{t} g(t-t_1)\,X(t_1)\,\mathrm{d}t_1\,, \tag{2.56}$$

where

$$f_1 = \mathrm{e}^{-nt}\cos pt, \qquad f_2 = \mathrm{e}^{-nt}\sin pt;$$

$$g(t-t_1) = \mathrm{e}^{-n(t-t_1)}\sin p(t-t_1)\,, \quad \left(p = \sqrt{p_0^2 - n^2}\right).$$

The expression for the time derivative of y is

$$\dot{y} = c_1\dot{f}_1 + c_2\dot{f}_2 + \int_0^{t} \frac{\partial g(t-t_1)}{\partial t} X(t_1)\mathrm{d}t_1. \tag{2.57}$$

The arbitrary constants c_1 and c_2 can be determined from the initial conditions ($t = 0,\ y = y_0,\ \dot{y} = \dot{y}_0$)

$$c_1 = y_0,\ c_2 = (\dot{y}_0 + n y_0)/p.$$

Expression for the y takes the following form

$$y = y_0 f_{12} + \frac{\dot{y}_0}{p} f_2(t) + \int_0^t g(t - t_1) X(t_1)\,\mathrm{d}t_1, \tag{2.58}$$

where

$$f_{12} = f_1 + \frac{n}{p} f_2.$$

In determining the probabilistic characteristics one can use the results of sect. 2.4.

Considering the initial values $y_0, \dot{y}_0$ and the disturbance $X(t)$ as independent random quantities with known probabilistic characteristics, let us determine the mathematical expectation

$$m_y = m_{y_0} f_{12} + m_{\dot{y}_0} f_2 + \int_0^t g(t - t_1)\, m_x(t_1)\,\mathrm{d}t_1 \tag{2.59}$$

and the correlation function

$$\begin{aligned} K_y &= M\left[\overset{\circ}{Y}(t),\, \overset{\circ}{Y}(t')\right] \\ &= f_{12}(t) f_{12}(t') D_{y_0} + f_2(t) f_2(t') D_{\dot{y}_0} \\ &+ \int_0^t \int_0^t g(t - t_1)\, g\left(t' - t_1^{(1)}\right) K_x\left(t_1, t_1^{(1)}\right) \mathrm{d}t_1^{(1)} \mathrm{d}t_1 . \end{aligned} \tag{2.60}$$

3. Let us consider a linear non-homogeneous equation of the second order with variable coefficients

$$\ddot{y} + a_1(t)\,\dot{y} + a_2(t)\,y = x(t). \tag{2.61}$$

Assuming that $\dot{y} = z_1,\ y = z_2$, let us represent the equation as a system of two equations of the first order

$$\begin{cases} \dot{z}_1 + a_1 z_1 + a_2 z_2 = x, \\ \dot{z}_2 - z_1 = 0 \end{cases}$$

or

$$\dot{\mathbf{Z}} + \mathrm{A}(t)\,\mathbf{Z} = \mathbf{f}, \tag{2.62}$$

where

$$\mathbf{Z} = \begin{vmatrix} z_1 \\ z_2 \end{vmatrix}; \qquad \mathrm{A} = \begin{vmatrix} a_1 & a_2 \\ -1 & 0 \end{vmatrix}; \quad \mathbf{f} = \begin{vmatrix} x \\ 0 \end{vmatrix}.$$

The equation (2.62) solution takes the form

$$\mathbf{Z} = \mathrm{B}(t)\,\mathbf{C} + \int_0^t \mathrm{G}(t,\, t_1)\,\mathbf{f}(t_1)\,dt_1, \tag{2.63}$$

where $\mathrm{B}(t)$ is the fundamental matrix of homogeneous equation (2.61), satisfying the condition $\mathrm{B}(0) = \mathrm{E}$; E is the identity matrix and $\mathrm{G}(t,\, t_1)$ is the Green matrix

$$\mathrm{G}(t,\, t_1) = \mathrm{B}(t)\,\mathrm{B}^{-1}(t_1), \tag{2.64}$$

$\mathbf{C}$ is the initial values vector.

The components of the vector $\mathbf{C}$ may be both deterministic and random. Let us assume that their probabilistic characteristics (vector $\mathbf{C}$ is equal to $\mathbf{Z}(0)$) m_{c_j}, D_{c_j} and $K_{c_i c_j}$ are known. The mathematical expectation of the vector $\mathbf{Z}$ (at random initial values) is

$$\mathbf{m}_z = \mathrm{B}(t)\,\mathbf{m}_c + \int_0^t \mathrm{G}(t,\, t_1)\,\mathbf{m}_f(t_1)\,dt_1 \tag{2.65}$$

or in its scalar form

$$m_{z_1} = m_{\dot{y}} = b_{11}m_{c_1} + b_{12}m_{c_2} + \int_0^t g_{11}m_f dt_1;$$

$$m_{z_2} = m_y = b_{21}m_{c_1} + b_{22}m_{c_2} + \int_0^t g_{21}m_f dt_1.$$

The correlation functions matrix for the components of the vector $\mathbf{Z}$ is

$$K = M\left[\overset{\circ}{\mathbf{Z}}(t) \otimes \overset{\circ}{\mathbf{Z}}(t')\right] = \begin{bmatrix} K_{z_1 z_1} & K_{z_1 z_2} \\ K_{z_2 z_1} & K_{z_2 z_2} \end{bmatrix}.$$

According to the accepted condition that the components of the $\mathbf{C}$ and $\mathbf{f}$ vectors are independent, we have

$$K = M\left[\mathrm{B}(t)\,\overset{\circ}{\mathbf{C}} \otimes \mathrm{B}(t')\,\overset{\circ}{\mathbf{C}}\right]$$
$$+ \int_0^t \int_0^{t'} M\left[\mathrm{G}(t,\, t_1)\,\overset{\circ}{\mathbf{f}} \otimes \mathrm{G}(t',\, t_1')\,\overset{\circ}{\mathbf{f}}(t')\right] dt_1\, dt_1^{(1)}. \tag{2.66}$$

After the necessary transformations we obtain

$$K = K_c + \int_0^t \int_0^{t'} K_f \mathrm{d}t_1 \mathrm{d}t_1'. \tag{2.67}$$

The elements of the matrixes K_c and K_f are

$$\begin{aligned}K_{c_i c_j} &= b_{i1}(t)\, b_{j1}(t')\, K_{c_1 c_1} + b_{i1}(t)\, b_{j2}(t')\, K_{c_1 c_2} \\ &\quad + b_{i2}(t)\, b_{j1}(t)\, K_{c_2 c_1} + b_{i2}(t)\, b_{i2}(t')\, K_{c_2 c_2}, \\ K_{f_i f_j}(t_1, t_1') &= g_{i1}(t_1)\, g_{j1}(t_1)\, K_x(t_1, t_1')\,.\end{aligned}$$

The probabilistic characteristics of the equation (1.155) solution in the vector-matrix form, obtained by way of this example are readily extended to a case, where vector $\mathbf{Z}$ has n components.

3. Stationary Random Functions (Processes)

3.1 Probability Characteristics of Stationary Random Functions

Random processes that proceed in time with approximate homogeneity and have the form of continuous random oscillations about a certain mean value are widespread. Their probability characteristics do not depend on the choice of time reference point, i.e. are invariant relative to the shift of time. Accordingly, a random function $X(t)$ is defined as stationary, if the probability characteristics of a random function $X(t+t')$ at any t' coincide with the appropriate characteristics of $X(t)$. This occurs only when the mathematical expectation and the variance of a random function do not depend on time, and the correlation function depends only on the difference of arguments $(t'-t)$. The stationary process may be considered as a process, that proceeds in time without limit. In this context the stationary process is similar to the steady-state vibrations, whose parameters are independent of a time reference point.

In applied problems, where we have to deal only with the moments of the first two orders – mathematical expectations and correlation functions – it is enough to consider a random function as stationary given the constancy of its mathematical expectation and variance and the dependence of the correlation function solely on the difference of arguments. Such random functions will be referred to as *stationary functions in the broad sense of the term* [29].

A stationary random function $X(t)$, whose distribution laws of all possible orders $f_n(x_1, x_2, \ldots, x_n, t_1, t_2, \ldots t_n)$ depend only on intervals $t_2 - t_1$, $t_3 - t_1, \ldots, t_n - t_1$, and are independent of their position on the axis of time, will be referred to as *the stationary function in the narrow sense of the term.*

Below we shall interpret stationary functions in the broad context and discuss only those of them, which are most often used in applied problems. Let us dwell now on the principal properties of stationary functions.

As indicated above, a random function $X(t)$ is called stationary, if its mathematical expectation and variance do not depend on time, i.e.

$$m_x(t) = \text{const}, \quad D_x(t) = \text{const}. \tag{3.1}$$

Let us consider the correlation moment $K_x(t, t')$ of the random stationary function $X(t)$ (Fig. 3.1) for two instants of time separated by an interval τ.

For a stationary process the correlation moment is independent of the specific values t and t', and only depends on their difference $t' - t = \tau$, i.e. (Fig. 3.1)

$$K_x(t, t') = K_x(t, t + \tau) = K_x(\tau). \tag{3.2}$$

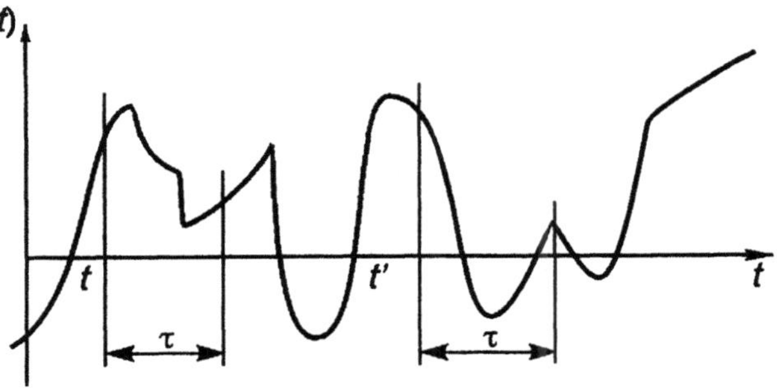

Fig. 3.1.

It follows from the relation (3.2) that the correlation function $K_x(\tau)$ of a stationary random process is a function of one argument τ.

As the correlation function is symmetric (2.8)

$$K_x(t, t') = K_x(t', t),$$

and assuming that $t' - t = \tau$, we obtain

$$K_x(\tau) = K_x(-\tau), \tag{3.3}$$

that is the correlation function is an even function.

The random function variance is equal to

$$D_x(t) = K_x(t, t).$$

Therefore, for the stationary random function we have

$$D_x = K_x(0). \tag{3.4}$$

The normalized correlation function

$$\rho_x(\tau) = \frac{K_x(\tau)}{D_x} \tag{3.5}$$

is often used instead of the correlation function $K_x(\tau)$.

As $|\rho_x(\tau)| \leq 1$, it follows from (3.5) that:

$$|K_x(\tau)| \leq D_x.$$

The plots of autocorrelation functions $K(\tau)$ most frequently used in applied problems are shown in Appendix 3.

Appendix 3 contains, among its other $K(\tau)$ functions, the following correlation function, proportional to the Dirac delta function and known as stationary white noise or delta-correlated random process:

$$K(\tau) = S_0 \delta(\tau). \tag{3.6}$$

where S_0 is a constant factor; $\delta(\tau)$ is the Dirac delta function (See Appendix 1).

The white noise is defined as a random process, with the values of the random function $X(t)$ being non-correlated for arbitrary close instants of time. A random process of the white noise type cannot be realized in real conditions, because, firstly, random functions at sufficiently close instants of time (at very small τ) are, in fact, always dependable and, secondly, as (3.6) shows, at $\tau = 0$ the process variance is equal to infinity Therefore the realization of such process requires an infinite power, that is impossible to obtain in real conditions. Nevertheless, random processes of the white noise type are widely used in many divisions of statistical dynamics.

Example 3.1. The task is to determine, whether the random function

$$X(t) = \sum_{j=1}^{n} (A_j \cos \omega_j t + B_j \sin \omega_j t) \tag{3.7}$$

is a stationary random function, if A_j and B_j are random mutually independent quantities with zero mathematical expectations and equal variances $\left(D_{A_j} = D_{B_j} = D_j\right)$.

The mathematical expectation of the random function $X(t)$ is zero.

The correlation function is

$$K_x(t, t') = M\left[\left(\sum_{j=1}^{n} (A_j \cos \omega_j t + B_j \sin \omega_j t)\right) \times \left(\sum_{k=1}^{n} (A_k \cos \omega_k t' + B_k \sin \omega_k t')\right)\right].$$

Some transformations yield

$$K_x(t, t') = \sum_{j=1}^{n} D_j \cos \omega_j (t - t'). \tag{3.8}$$

The obtained expression (3.8) depends only on the difference $t - t'$, i.e. $X(t)$ is a stationary function. Then the random function $X(t)$ variance will be

$$D_x = K_x(0) = \sum_{j=1}^{n} D_j. \tag{3.9}$$

Example 3.2. The task is to determine the correlation function K_x, if the random function X is

$$X = \sum_{j=1}^{n} A_j e^{i\omega_j t}, \tag{3.10}$$

where $A_j = A_{1j} + iA_{2j}$ is a complex amplitude; and j is the imaginary unit.

The non-correlated random quantities A_{1j} and A_{2j} have zero mathematical expectations and equal variances.

Let us take advantage of the formula (2.20), (assuming that $Y^*(t') = X^*(t')$)

$$K_x(t, t') = M[X(t)X^*(t')] = M\left[\left(\sum_{j=1}^{n} A_j e^{i\omega_j t}\right) \cdot \left(\sum_{k=1}^{n} A_k e^{-i\omega_k t'}\right)\right].$$

where $A_k^* = A_{1k} - iA_{2k}$.

By manipulation we shall get

$$K_x(t, t') = \sum_{j=1}^{n} M\left[A_j A_j^*\right] e^{i(t-t')\omega} = \sum_{j=1}^{n} 2D_j e^{i\tau\omega}. \tag{3.11}$$

It follows from (3.11), that the random function (3.10) at $m_{1j} = m_{2j} = 0$, $D_{1j} = D_{2j} = D_j$ is a stationary random function for the non-correlated A_{1j}, A_{2j}.

Let us consider a system of two random functions $X_1(t)$ and $X_2(t)$.

The two random functions $X_1(t)$, $X_2(t)$ of the same argument are called stationary connected, if their cross-correlation function depends on a difference of arguments, i.e.

$$K_{x_1x_2}(t, t') = K_{x_1x_2}(\tau), \qquad (\tau = t' - t). \tag{3.12}$$

If condition (3.12) is satisfied for the random stationary functions X_1 and X_2, they are considered stationary and stationary connected. It follows from the cross-correlation function property, that

$$K_{x_1x_2}(t, t') = K_{x_1x_2}(\tau) = K^*_{x_2x_1}(-\tau)$$

or for real-valued random functions

$$K_{x_1x_2}(\tau) = K_{x_2x_1}(-\tau). \tag{3.13}$$

In the general case, cross-correlation functions are not even functions of argument τ.

The system $X_j(t)$ $(j = 1, 2, \ldots, n)$ of stationary and stationary connected random functions is characterized by the correlation matrix

$$K(\tau) = \begin{bmatrix} K_{x_1x_1}(\tau) & \cdots\cdots\cdots & K_{x_1x_n}(\tau) \\ \cdots\cdots & \cdots\cdots & \cdots\cdots \\ K_{x_nx_1}(\tau) & \cdots\cdots\cdots & K_{x_nx_n}(\tau) \end{bmatrix}.$$

Example 3.3. The task is to determine the cross-correlation functions $K_{xy}(t, t')$ and $K_{yx}(t, t')$ of the random stationary functions

$$X(t) = A_1 \cos\omega t + B_1 \sin\omega t;$$
$$Y(t) = -A_1 \sin\omega t + B_1 \cos\omega,$$

where A_1, B_1 are non-correlated random variables with a zero mathematical expectation and equal variances D.

The cross-correlation functions (assuming that $t' - t = \tau$) are

$$K_{xy}(t, t') = M\left[\overset{\circ}{X}(t)\overset{\circ}{Y}(t')\right] = D\sin\omega\tau;$$
$$K_{yx}(t, t') = M\left[\overset{\circ}{Y}(t)\overset{\circ}{X}(t')\right] = -D\sin\omega\tau.$$

It follows from the relations obtained, that the random functions $X(t)$ and $Y(t)$ are stationary connected. If we replace τ by $-\tau$ in the expression for K_{yx}, the condition (3.13) fulfills

$$K_{xy}(\tau) = K_{yx}(-\tau). \tag{3.14}$$

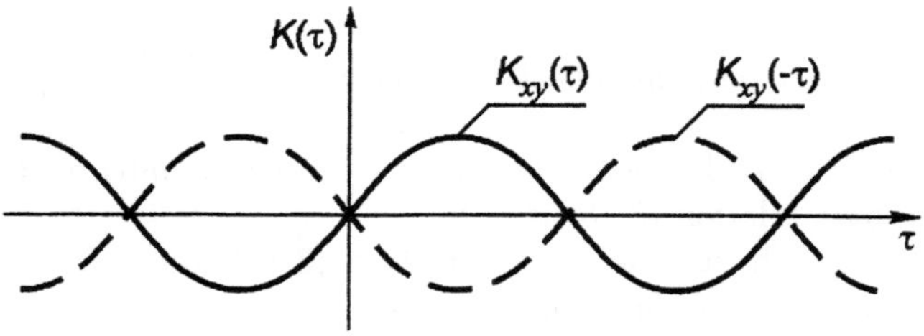

Fig. 3.2.

The plot of the function $K_{yx}(-\tau)$ is a mirror reflection of the plot of the function $K_{xy}(\tau)$ (Fig. 3.2).

3.2 The Ergodic Property of a Stationary Random Function

To obtain the stationary random function characteristics (the mathematical expectation m_x and the correlation function $K_x(\tau)$) we need a considerable number of random function $X(t)$ realizations. The approximate values of m_x and $K_x(\tau)$ may be determined on the basis of realization records. The question arises: is it possible to obtain the same characteristics using only one realization of the random function $X(t)$. As the random process is stationary and proceeds in time homogeneously (the mathematical expectation does not depend on time, and the correlation function is independent of the origin), we may well assume, that a single realization is sufficient for the random function characteristics determination. The possibility of obtaining the probability characteristics of a stationary random function on the basis of its single realization is very important in the practical context, as this allows us to decrease the extent of experimental investigations, and thus to diminish the expenditures.

Stationary random functions, whose probability characteristics can be determined through their single realization are referred to as random functions with an ergodic property, or simply ergodic stationary random functions. The ergodic property is a property making it possible to judge the statistical properties of all set of realizations of a random function on the basis of its single realization. The time-mean value (on rather large observation interval) of a random function $X(t)$ possessing the ergodic property is approximately equal to the number-of-observations mean value, i.e.

$$m_x \approx \frac{1}{T}\int_0^T X(t)\,\mathrm{d}t, \tag{3.15}$$

where m_x is the value averaged over the collection of realizations; $\frac{1}{T}\int_0^T X(t)\,\mathrm{d}t$ is time-mean value for a single realization. The task is to find out, under what conditions the approximate relation (3.15) is fulfilled.

The integral in the right-hand side of the relation (3.15) has different random values for different realizations, therefore

$$Y_T = \frac{1}{T}\int_0^T X(t)\,\mathrm{d}t, \tag{3.16}$$

where Y_T is a random quantity.

The necessary and sufficient condition of the approximate equality (3.16) fulfillment is a small value of the random variable Y_T variance. The random variable Y_T variance is equal to

$$D\left[Y_t\right] = M\left[\left(Y_T - m_x\right)^2\right]$$
$$= M\left[\left\{\frac{1}{T}\int\limits_0^T \left(X - m_x\right) dt\right\}^2\right] = \frac{1}{T^2}\int\limits_0^T\int\limits_0^T K_x\left(t - t'\right) \mathrm{d}t\,\mathrm{d}t'. \tag{3.17}$$

It follows from (3.17), that the expression (3.15) for the mathematical expectation of a stationary random function will be an exact equality at sufficiently large T

$$m_x = \lim \frac{1}{T}\int\limits_0^T X\left(t\right) \mathrm{d}t, \tag{3.18}$$

only provided that

$$\lim_{T\to\infty} D\left[Y_T\right] = \lim_{T\to\infty} \frac{1}{T^2}\int\limits_0^T\int\limits_0^T K_x\left(t - t'\right) \mathrm{d}t\,\mathrm{d}t' = 0. \tag{3.19}$$

Let us transform the expression (3.19) by introducing a new variable $\tau = t - t'$. When t and t' vary within the interval $(0, T)$, τ varies in the interval $(-T, T)$. Then

$$I = \frac{1}{T^2}\int\limits_0^T\int\limits_0^T K_x\left(t - t'\right) \mathrm{d}t\,\mathrm{d}t' = \frac{1}{T^2}\int\limits_0^T \left(\int\limits_{-t'}^{T-t'} K_x\left(\tau\right) \mathrm{d}\tau\right) \mathrm{d}t'. \tag{3.20}$$

Since $K_x\left(\tau\right) = K_x\left(-\tau\right)$, it follows, that $K_x\left(\tau\right)$ is a function of $|\tau|$. Changing the order of integration, we shall get

$$I = \frac{1}{T^2}\int\limits_{-T}^T \left[\int\limits_{|\tau|}^T K_x\left(\tau\right) \mathrm{d}t'\right] \mathrm{d}\tau = \frac{1}{T}\int\limits_{-T}^T \left(1 - \frac{|\tau|}{T}\right) K_x\left(\tau\right) \mathrm{d}\tau. \tag{3.21}$$

As $\left(1 - \frac{|\tau|}{T}\right)$ and $K_x\left(\tau\right)$ are even functions,

$$I = \frac{2}{T}\int\limits_0^T \left(1 - \frac{\tau}{T}\right) K_x\left(\tau\right) \mathrm{d}\tau. \tag{3.22}$$

It results from (3.22) that for the random stationary function $X\left(t\right)$ to become ergodic (fulfillment of the condition (3.19)) its correlation function must satisfy the following necessary and sufficient condition:

$$\lim_{T\to\infty}\frac{1}{T}\int\limits_0^T\left(1-\frac{\tau}{T}\right)K_x(\tau)\,d\tau=0. \tag{3.23}$$

The sufficient condition of a stationary random function ergodicity is an unbounded decrease of its correlation function $K_x(\tau)$ absolute value at $|\tau|\to\infty$.

Since the correlation function $K_x(\tau)$ of a stationary random function can be represented as

$$K_x(\tau)=M\left[\overset{\circ}{X}(t)\,\overset{\circ}{X}(t+\tau)\right] \tag{3.24}$$

and does not depend on time t, it can also be determined by a single realization through averaging the right-hand side of the expression (3.24) with respect to time t i.e.

$$K_x(\tau)\approx\frac{1}{T}\int\limits_0^T\overset{\circ}{X}(t)\,\overset{\circ}{X}(t+\tau)\,\mathrm{d}t,$$

or, through proceeding to the limit

$$K_x(\tau)=\lim_{T\to\infty}\frac{1}{T}\int\limits_0^T\overset{\circ}{X}(t)\,\overset{\circ}{X}(t+\tau)\,\mathrm{d}t. \tag{3.25}$$

The relation (3.25) is true, provided the fulfillment of the sufficient condition [29]: at $|\tau|\to\infty\quad K_x(\tau)\to 0$.

Example 3.4. Determine, whether the stationary stochastic function $X(t)$ with a correlation function

$$K_x(\tau)=D_x\mathrm{e}^{-\alpha|\tau|},$$

is ergodic.

By substituting $K_x(\tau)$ in the relation (3.23), we'll get

$$\lim_{T\to\infty}\frac{1}{T}\int\limits_0^T\left(1-\frac{\tau}{T}\right)D_x\mathrm{e}^{-\alpha\tau}\mathrm{d}\tau$$

$$=\lim_{T\to\infty}\frac{1}{T}\left[-\frac{1}{\alpha}\left(\mathrm{e}^{-\alpha T}-1\right)-\frac{D_x}{T}\left(-\frac{T}{\alpha}\mathrm{e}^{-\alpha T}-\frac{1}{\alpha^2}\mathrm{e}^{-\alpha T}+\frac{1}{\alpha^2}\right)\right]=0,$$

which shows that the stationary random function $X(t)$ is ergodic.

3.3 Derivatives and Integrals of Stationary Functions

3.3.1 Probability Characteristics of Stationary Random Function Derivatives

Let us use the expressions for probability characteristics of non-stationary function derivatives obtained in section 2.4.

If

$$Y(t) = \frac{\mathrm{d}}{\mathrm{d}t} X(t), \tag{3.26}$$

then the mathematical expectation m_y of the stationary function $X(t)$ is zero

$$m_y = M\left[\frac{\mathrm{d}}{\mathrm{d}t} X(t)\right] = \frac{\mathrm{d}}{\mathrm{d}t} m_x = 0.$$

The correlation function (2.37) for the stationary function $X(t)$ is

$$K_y(t,\, t') = \frac{\partial^2 K_x(t,\, t')}{\partial t\, \partial t'} = \frac{\partial^2 K_x(\tau)}{\partial t\, \partial t'}.$$

Since $\tau = t' - t$, then proceeding to a derivative with respect to τ, we get

$$K_y(t,\, t') = -\frac{\mathrm{d}^2 K_x(\tau)}{\mathrm{d}\tau^2}. \tag{3.27}$$

Similarly, it is possible to determine the correlation function of the second derivative and derivatives of a higher-order

$$K_{\ddot{x}}(t,\, t') = \frac{\mathrm{d}^4 K_x(\tau)}{\mathrm{d}\tau^4}; \qquad K_{\dddot{x}} = -\frac{\mathrm{d}^6 K_x(\tau)}{\mathrm{d}\tau^6}. \tag{3.28}$$

It follows from (3.28), that stationary function derivatives are stationary functions, as their correlation functions solely depend on τ.

Let us consider a number of examples of differentiating the correlation function of a stationary random function.

The task is to determine the correlation functions K_y:

$$Y = \frac{\mathrm{d}}{\mathrm{d}t} X(t).$$

If

1) $K_x = D_x \mathrm{e}^{-\alpha|\tau|}$;

2) $K_x(\tau) = D_x \mathrm{e}^{-\alpha|\tau|}(1 + \alpha|\tau|)$;

3) $K_x = D_x \mathrm{e}^{-\alpha|\tau|} \cos \beta\tau$.

As

$$K_y = -\frac{d^2 K_x}{d\tau^2},$$

Then after the twofold successive differentiating of $K_x$with respect to τ, we'll get (See Appendix 1):

$$1)\ K_y = -\frac{d}{d\tau}\left(-D_x \alpha\, e^{-\alpha|\tau|}\frac{d|\tau|}{d\tau}\right) = D_x \alpha \frac{d}{d\tau}\left(e^{-\alpha|\tau|}\operatorname{sign}\tau\right)$$

$$= D_x \alpha\left[-\alpha e^{-\alpha|\tau|}(\operatorname{sign}\tau)^2 + e^{-\alpha|\tau|}\frac{d\operatorname{sign}\tau}{d\tau}\right]$$

$$= D_x \alpha\left[2\delta(\tau) - \alpha(\operatorname{sign}\tau)^2\right]e^{-\alpha|\tau|};$$

$$2)\ K_y = -D_x \frac{d}{d\tau}\left[-\alpha e^{-\alpha|\tau|}\operatorname{sign}\tau\,(1+\alpha|\tau|) + \alpha e^{-\alpha|\tau|}\operatorname{sign}\tau\right]$$

$$= D_x \alpha^2 \frac{d}{d\tau}\left[e^{-\alpha|\tau|}\operatorname{sign}\tau\,|\tau|\right] = D_x \alpha^2 \frac{d}{d\tau}\left(e^{-\alpha|\tau|}\tau\right)$$

$$= D_x \alpha^2\left[-\alpha e^{-\alpha|\tau|}\operatorname{sign}\tau\cdot\tau + e^{-\alpha|\tau|}\right] = D_x \alpha^2 e^{-\alpha|\tau|}(1-\alpha|\tau|);$$

$$3)\ K_y = -D_x \frac{d}{d\tau}\left(-\alpha e^{-\alpha|\tau|}\frac{d|\tau|}{d\tau}\cos\beta\tau - \beta e^{-\alpha|\tau|}\sin\beta\tau\right)$$

$$= -D_x\left[\alpha^2 e^{-\alpha|\tau|}\left(\frac{d|\tau|}{d\tau}\right)^2\cos\beta\tau - \alpha e^{-\alpha|\tau|}\frac{d^2|\tau|}{d\tau^2}\cos\beta\tau\right.$$

$$\left.+2\alpha\beta e^{-\alpha|\tau|}\frac{d|\tau|}{d\tau}\sin\beta\tau - \beta^2 e^{-\alpha|\tau|}\cos\beta\tau\right]$$

$$= D_x e^{-\alpha|\tau|}\left[\left(\beta^2 - \alpha^2\operatorname{sign}^2\tau\right)\cos\beta\tau 2\alpha\delta(\tau)\cos\beta\tau - 2\alpha\beta\sin\beta|\tau|\right];$$

$$\left(\sin\beta|\tau| = \sin\beta\tau\operatorname{sign}\tau,\quad \operatorname{sign}^2\tau = \begin{cases}1 \text{ at } \tau\neq 0,\\ 0 \text{ at } \tau = 0.\end{cases}\right)$$

As another example, let us obtain an expression for the cross-correlation function of the stationary random function $X(t)$ and its derivative $\dot{\mathring{X}}(t)$:

$$K_{x\dot{x}}(t, t') = I\left[\mathring{X}(t)\frac{\partial}{\partial t'}\mathring{X}(t')\right]$$

$$= \frac{\partial}{\partial t'} I\left[\mathring{X}(t)\mathring{X}(t')\right] = \frac{\partial}{\partial t'}K_x(\tau) = \frac{dK_x(\tau)}{d\tau}. \qquad (3.29)$$

It follows from (3.29) that the stationary function $X(t)$ and its first derivative are stationary-connected random functions.

Since $K_x(\tau)$ is an even function of τ, then

$$\left.\frac{\mathrm{d}K_x(\tau)}{\mathrm{d}\tau}\right|_{\tau=0} = 0,$$

i.e. the stationary function $X(t)$ and its derivative for the same instant of time are non-correlated random quantities. The correlation functions of the first two derivatives of the stationary random function $X(t)$ are equal to:

$$K_{\dot{x}\dot{x}} = -\ddot{K}_x(\tau), \quad K_{x\ddot{x}} = \ddot{K}_x(\tau),$$
$$K_{\dot{x}\ddot{x}} = -\dddot{K}_x(\tau), \quad K_{\ddot{x}\ddot{x}} = \ddddot{K}_x(\tau),$$

which shows that they are stationary connected.

3.3.2 Probability Characteristics of the Integral of Stationary Random Functions

Let us consider the integral (2.25) of the stationary random function $X(t)$, whose correlation function is equal to

$$K_x(\tau) = D_x \cos\omega\tau.$$

The correlation functions $K_y(t, t')$ and $K_x(t, t')$ are related by the equation (2.28). In the case considered we have ($\tau = t_1' - t_1$)

$$K_y(t, t') = \int_0^t \int_0^{t'} K_x(t_1' - t_1)\, \mathrm{d}t_1\, \mathrm{d}t_1',$$

or

$$K_y(t, t') = \int_0^t \int_0^{t'} D_x \cos\omega(t_1' - t_1)\, \mathrm{d}t_1\, \mathrm{d}t_1'.$$

By integrating we get:

$$K_y(t, t') = \frac{1}{\omega^2}\cos\omega\tau + \frac{1}{\omega^2}(1 - \cos\omega t - \cos\omega t').$$

It follows from the last formula that the random process $Y(t)$ is not stationary that is (and it is also true for the general case) the integral of the stationary function does not have the stationarity property.

3.4 The Spectral Representation of Stationary Random Processes

The plots of realizations of the random functions $X_1(t)$ and $X_2(t)$ (see Fig. 2.3 *a*, *b*) shows great differences in the variations of this functions in time.

A nonrandom function – the correlation function $K_x(t, t')$, or $K_x(\tau)$ for stationary process, — that reflects the internal structure of a corresponding random process, for example, the process shown in Fig. 2.3 *b*, has been introduced to estimate random functions behavior in time. All realizations of this process have an obvious oscillatory nature. This suggests that we should characterize this process by its spectral properties, the way we did with the deterministic oscillatory process. We know, that, if any oscillatory process is represented as a sum of harmonic functions with different frequencies, the function describing the distribution of amplitudes in frequency is referred to as the spectrum of this process. It shows what components dominate in the given oscillatory process. Let us consider the random function

$$X(t) = \sum_{j=1}^{n} (A_j \cos \omega_j t + B_j \sin \omega_j t)\,, \tag{3.30}$$

where A_j, B_j are independent random quantities with zero mathematical expectations and equal variances ($D_{Aj} = D_{Bj} = D_j$). It has already been shown (example 3.1), that $X(t)$ is a stationary function. Its correlation function $K_x(\tau)$ equals

$$K_x(\tau) = \sum_{j=1}^{n} D_j \omega_j \cos \omega_j \tau. \tag{3.31}$$

At $\tau = 0$ we get the variance of the random function $X(t)$

$$D_x = \sum_{j=1}^{n} D_j(\omega_j)\,. \tag{3.32}$$

The formula (3.32) shows, that the variance D_x is distributed in different frequencies. The qualitative nature of D_x distribution in frequencies is shown in Fig. 3.3. The spectrum shown in Fig. 3.3, is referred to as the discrete spectrum. The variance D_x is finite; therefore as the number of terms n approaches infinity, we may consider that amplitude variances D_j are small quantities ΔD_j, i.e.

$$D_x = \sum_{j=1}^{\infty} \Delta D_j(\omega_j)\,. \tag{3.33}$$

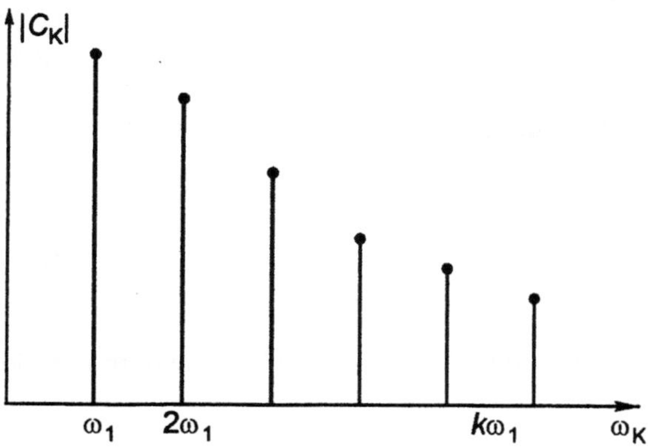

Fig. 3.3.

As the number of terms increases, the discrete spectrum will approach the continuous spectrum, where the elementary variance $\Delta D_j(\omega_j)$ corresponds to the small interval of frequencies $\Delta\omega_j$. As ΔD_j is small, let us plot the ratio $\dfrac{\Delta D_j}{\Delta\omega_j}$, characterizing the average (finite) variance density on the axis of ordinates.

Let us introduce the notation

$$S_x(\omega_j) = \frac{\Delta D_j}{\Delta\omega_j}. \tag{3.34}$$

for the average density of variance. The average density of variance is a finite number. The plot of the discrete average density of variance is similar to that shown in Fig. 3.3.

The correlation function (3.31) at large n with due account of (3.33) and (3.34) is equal to

$$K_x(\tau) = \sum_{j=1}^{n} S_x(\omega_j)\,\Delta\omega_j \cos\omega_j\tau.$$

In the limit at $n \to \infty$ we have $\omega_j \to \omega$; $\Delta\omega_j \to \mathrm{d}\omega$; $\sum_{j=1}^{\infty} \to \int_0^{\infty}$; $S(\omega_j) \to S(\omega)$ and get

$$K_x(\tau) = \int_0^{\infty} S_x(\omega)\cos\omega\tau\,\mathrm{d}\omega. \tag{3.35}$$

The function $S_x(\omega)$ is referred to as spectral density. Since the spectral density characterizes the variance distribution, it is always positive

$(S_x(\omega > 0))$ and an even function of ω

$$S(\omega) = S(-\omega). \tag{3.36}$$

Taking the inverse Fourier transform of the above formula, we obtain another useful relation

$$S_x(\omega) = \frac{1}{\pi}\int_0^\infty K_x(\tau)\cos\omega\tau \mathrm{d}\tau. \tag{3.37}$$

We can obtain a more general relation (3.37) if we consider a random function of the form

$$X(t) = \sum_{j=1}^{n} A_j \mathrm{e}^{i\omega_j t},$$

where A_j are independent complex random quantities with zero expectations and equal variances. Under these conditions the correlation function K_x is

$$K_x(\tau) = \sum_{j=1}^{n} D_j \mathrm{e}^{i\omega_j \tau}, \tag{3.38}$$

or at $n \to \infty$

$$K_x(\tau) = \int_0^\infty S_x(\omega)\,\mathrm{e}^{i\omega\tau}\mathrm{d}\omega. \tag{3.39}$$

Using the formula of the inverse Fourier transform, we get

$$S_x(\omega) = \frac{1}{\pi}\int_0^\infty K_x(\tau)\,\mathrm{e}^{-i\omega\tau}\mathrm{d}\tau. \tag{3.40}$$

If we carry out the summation in the expression (3.38) from $-n$ to $+n$, we'll get (at $|n| \to \infty$)

$$K_x(\tau) = \int_{-\infty}^{\infty} S_x(\omega)\,\mathrm{e}^{i\omega\tau}\mathrm{d}\omega, \tag{3.41}$$

$$S_x(\omega) = \frac{1}{2\pi}\int_{-\infty}^{\infty} K_x(\tau)\,\mathrm{e}^{-i\omega\tau}\mathrm{d}\tau. \tag{3.42}$$

The relationships (3.41) and (3.42) are referred to as Wiener–Khintchin formulas. The relationships (3.35) and (3.37) represent particular cases of the formulas (3.41)-(3.42), with the imaginary parts of their integrals being zero, which occurs, when the functions $S_x(\omega)$ and $K_x(\tau)$ are even.

The variance of a stationary function is connected with spectral density by the relationship

$$D_x = K_x(0) = \int_{-\infty}^{\infty} S_x(\omega)\, d\omega \tag{3.43}$$

or

$$D_x = K_x(0) = 2\int_{0}^{\infty} S_x(\omega)\, d\omega.$$

Let us consider another version of deriving the relationship (3.41). For this purpose we shall use the representation of a centered random function $X(t)$ in a frequency domain (Fourier transform)

$$X(t) = \int_{-\infty}^{\infty} X_0(i\omega)\, e^{i\omega t} d\omega.$$

We determine the correlation function of the complex function

$$\begin{aligned} K_x(t, t') &= M\left[X(t)\, X^*(t')\right] \\ &= M\left[\iint X_0(i\omega)\, X_0^*(i\omega')\, e^{i\omega t} e^{-i\omega' t'} d\omega\, d\omega'\right] \end{aligned}$$

or

$$K_x(t, t') = M \int_{-\infty}^{\infty}\int_{-\infty}^{\infty} M\left[X_0(i\omega)\, X_0^*(i\omega')\right] e^{i(\omega t - \omega' t')} d\omega\, d\omega'. \tag{3.44}$$

For a stationary function the correlation function depends on a difference of time instants. Therefore, if $X(t)$ is a stationary function, then the correlation function K_x will depend on the difference of time instants, if we assume that

$$M\left[X_0(i\omega)\, X_0^*(i\omega')\right] = S_x(i\omega')\, \delta(\omega' - \omega). \tag{3.45}$$

Substituting (3.45) in the right-hand side of the relationship (3.44) and integrating over ω', we get

$$K_x(t, t') = \int_{-\infty}^{\infty} S_x(\omega)\, e^{i\omega(t-t')} d\omega = K_x(\tau), \qquad (t - t' = \tau).$$

3.4.1 Spectral Densities of Stationary Function Derivatives

The following correlation functions of stationary function derivatives have been obtained in Sect. 3.3:

$$K_{\dot{x}} = -\frac{\mathrm{d}^2 K_x}{\mathrm{d}\tau^2}, \qquad K_{\ddot{x}} = \frac{\mathrm{d}^4 K_x}{\mathrm{d}\tau^4}.$$

Since

$$K_x = \int_{-\infty}^{\infty} S_x(\omega)\, \mathrm{e}^{i\omega\tau} d\omega,$$

we have

$$K_{\dot{x}} = \int_{-\infty}^{\infty} S_x(\omega)\, \omega^2 \mathrm{e}^{i\omega\tau} d\omega. \tag{3.46}$$

It follows from (3.46) that the spectral density of the derivative of the stationary function $X(t)$ is

$$S_{\dot{x}}(\omega) = S_x(\omega)\, \omega^2. \tag{3.47}$$

Similarly, we obtain the spectral density of the second derivative (and so forth)

$$S_{\ddot{x}}(\omega) = S_x(\omega)\, \omega^4. \tag{3.48}$$

3.4.2 Determination of Spectral Density (Examples)

Example 3.5. The correlation function of the random stationary function $X(t)$ (Fig. 3.4) is

$$K_x(\tau) = D_x \mathrm{e}^{-\alpha|\tau|}.$$

Using formula (3.42), we get

$$\begin{aligned} S_x(\omega) &= \frac{1}{2\pi} \int_{-\infty}^{\infty} D_x \mathrm{e}^{-\alpha|\tau|} \mathrm{e}^{-i\omega\tau} \mathrm{d}\tau = \frac{D_x}{2\pi} \int_{-\infty}^{\infty} \mathrm{e}^{-\alpha|\tau| - i\omega\tau} \mathrm{d}\tau \\ &= \frac{D_x}{2\pi} \left[\int_{-\infty}^{0} \mathrm{e}^{(\alpha - i\omega)\tau} \mathrm{d}\tau + \int_{0}^{\infty} \mathrm{e}^{-(\alpha + i\omega)\tau} \mathrm{d}\tau \right] \\ &= \frac{D_x}{\pi} \cdot \frac{\alpha}{(\alpha^2 + \omega^2)}. \end{aligned} \tag{3.49}$$

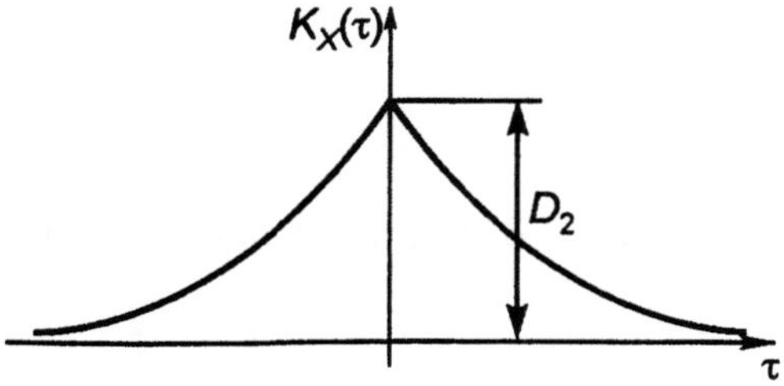

Fig. 3.4.

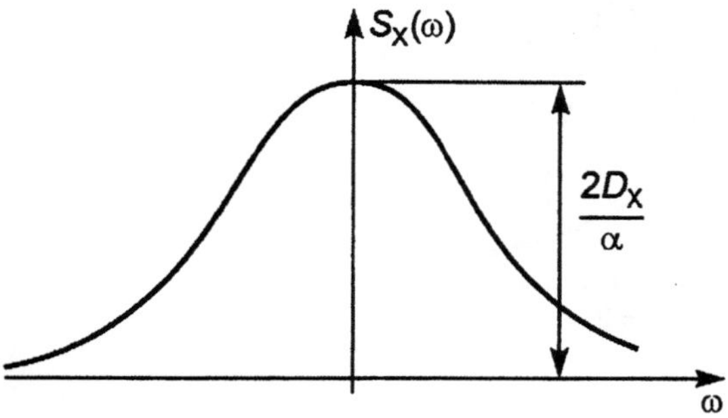

Fig. 3.5.

The plot of $S_x(\omega)$ versus ω is shown in Fig. 3.5.

Example 3.6. The correlation function $K_x(\tau)$ is equal to

$$K_x(\tau) = D_x e^{-\alpha|\tau|} \cos\beta\tau \qquad (\alpha > 0).$$

Since

$$\cos\beta = \frac{1}{2}\left(e^{i\beta\tau} + e^{-i\beta\tau}\right),$$

we get

$$S_x(\omega) = \frac{D_x}{4\pi}\int_{-\infty}^{\infty} e^{-\alpha|\tau|}\left(e^{i\beta\tau} + e^{-i\beta\tau}\right)e^{-i\omega\tau}d\tau$$

or

$$S_x(\omega) = \frac{D_x}{4\pi}\int_{-\infty}^{0} e^{\alpha\tau}\left(e^{i\beta\tau} + e^{-i\beta\tau}\right)e^{-i\omega\tau}d\tau$$

$$+ \frac{D_x}{4\pi}\int_{0}^{\infty} e^{\alpha\tau}\left(e^{i\beta\tau} + e^{-i\beta\tau}\right)e^{i\omega\tau}d\tau.$$

By manipulations we get

$$S_x(\omega) = \frac{D_x}{2\pi}\left[\frac{\alpha}{\alpha^2+(\omega+\beta)^2} + \frac{\alpha}{\alpha^2+(\omega-\beta)^2}\right]. \tag{3.50}$$

The behaviour of $S_x(\omega)$ as a function of ω for: 1) $\alpha = 1$, $\beta = 2$ and 2) $\alpha = 3$, $\beta = 2$ is shown in Fig. 3.6 (at $D_x = 1$).

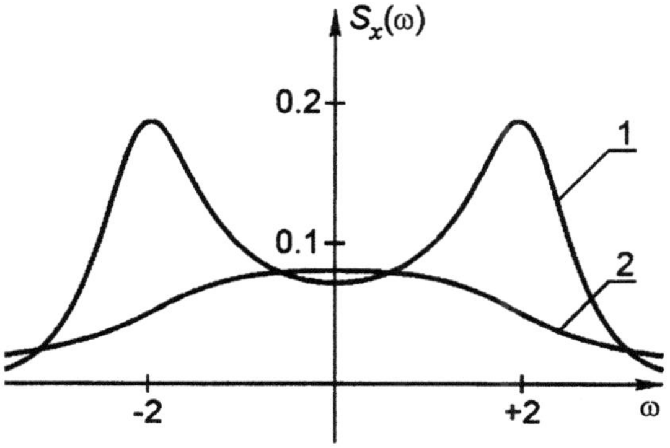

Fig. 3.6.

It follows from the plots, that at $\alpha = 1$ (curve *1*) the random function spectrum exhibits a pronounced maximum in the frequency area $\omega = \pm\beta$. At $\alpha = 3$ (curve *2*) the spectral density remains almost constant in a considerable range.

Example 3.7. The correlation function has the form

$$K_x(\tau) = D_x e^{-\alpha|\tau|}\left(\cos\beta\tau + \frac{\alpha}{\beta}\sin\beta|\tau|\right)$$
$$= K_x^{(1)}(\tau) + D_x\frac{\alpha}{\beta}e^{-\alpha|\tau|}\frac{e^{i\beta|\tau|} - e^{-i\beta|\tau|}}{2i}.$$

The spectral density is

$$S_x(\omega) = S_x^{(1)}(\omega) + \frac{D_x\alpha}{4\pi i\beta}\left[\int_{-\infty}^{\infty} e^{-\alpha|\tau|+i\beta|\tau|-i\omega\tau}d\tau - \int_{-\infty}^{\infty} e^{-\alpha|\tau|-i\beta|\tau|-i\omega\tau}d\tau\right].$$

The first term $S_x^{(1)}(\omega)$ is equal to the spectral density of the previous example. Let us consider the terms depending on integrals, that can be transformed to

$$\int_{-\infty}^{\infty} e^{-\alpha_1|\tau|-i\omega\tau}d\tau - \int_{-\infty}^{\infty} e^{-\alpha_2|\tau|-i\omega\tau}d\tau,$$

where $\alpha_1 = \alpha - i\beta$, $\alpha_2 = \alpha + i\beta$.

Thereupon each of these terms is represented as

$$\int_{-\infty}^{\infty} \mathrm{e}^{-\alpha_1|\tau| - i\omega\tau} \mathrm{d}\tau = \int_{-\infty}^{0} \mathrm{e}^{(\alpha_1 - i\omega)\tau} \mathrm{d}\tau + \int_{0}^{\infty} \mathrm{e}^{-(\alpha_1 + i\omega)\tau} \mathrm{d}\tau,$$

$$\int_{-\infty}^{\infty} \mathrm{e}^{-\alpha_2|\tau| - i\omega\tau} \mathrm{d}\tau = \int_{-\infty}^{0} \mathrm{e}^{(\alpha_2 - i\omega)\tau} \mathrm{d}\tau + \int_{0}^{\infty} \mathrm{e}^{-(\alpha_2 + i\omega)\tau} \mathrm{d}\tau.$$

The terms obtained are easily integrated and by manipulations we get the following expression for a spectral density

$$S_x(\omega) = \frac{2D_x\alpha}{\pi} \cdot \frac{\alpha^2 + \beta^2}{\left(\omega^2 - \beta^2 - \alpha^2\right)^2 + 4\alpha^2\omega^2}. \tag{3.51}$$

Example 3.8. The correlation function is

$$K_x(\tau) = D_x \mathrm{e}^{-\alpha^2\tau^2}. \tag{3.52}$$

The spectral density is

$$S_x(\omega) = D_x \frac{1}{2\pi} \int_{-\infty}^{\infty} \mathrm{e}^{-\alpha^2\tau^2 - i\omega\tau} \mathrm{d}\tau.$$

Let us take advantage of the well-known formula

$$\int_{-\infty}^{\infty} \mathrm{e}^{-Ax^2 \pm 2Bx - C} \mathrm{d}x = \sqrt{\frac{\pi}{A}} \exp\left\{\frac{AC - B^2}{A}\right\} \qquad (A > 0). \tag{3.53}$$

In the case considered $A = \alpha^2$; $2B = i\omega$; $C = 0$. As a result we get

$$S_x(\omega) = \frac{D_x}{2\pi}\sqrt{\frac{\pi}{\alpha^2}} \exp\left\{-\frac{\omega^2}{4\alpha^2}\right\} = \frac{D_x}{2\alpha\sqrt{\pi}} \exp\left\{-\frac{\omega^2}{4\alpha^2}\right\}. \tag{3.54}$$

Example 3.9. The correlation function is equal to

$$K_x(\tau) = D_x \mathrm{e}^{-\alpha^2\tau^2} \cos\beta\tau. \tag{3.55}$$

The spectral density is

$$S_x(\omega) = D_x \frac{1}{2\pi} \left[\int_{-\infty}^{\infty} \mathrm{e}^{-\alpha^2\tau^2 - i\omega\tau + i\beta\tau} \mathrm{d}\tau + \int_{-\infty}^{\infty} \mathrm{e}^{-\alpha^2\tau^2 - i\omega\tau - i\beta\tau} \mathrm{d}\tau\right].$$

Using the formula (3.53), we carry out the necessary transformations and get

$$S_x(\omega) = \frac{D_x}{4\alpha\sqrt{\pi}}\left[\exp\left\{-\frac{(\omega+\beta)^2}{4\alpha^2}\right\} + \exp\left\{-\frac{(\omega-\beta)^2}{4\alpha^2}\right\}\right]. \tag{3.56}$$

Let us consider as the last example the stationary white noise, whose correlation function depends on the Dirac delta function

$$K_x(\tau) = S_0\delta(\tau).$$

Let us determine the white noise spectral density

$$S_x(\omega) = \frac{1}{2\pi}\int_{-\infty}^{\infty} S_0\delta(\tau)\,\mathrm{e}^{-i\omega\tau}\mathrm{d}\tau$$

or

$$S_x(\omega) = \frac{S_0}{2\pi}, \tag{3.57}$$

(as $\int_{-\infty}^{\infty}\delta(\tau)\,\mathrm{e}^{-i\omega\tau}\mathrm{d}\tau = 1$).

Let us recall a transformwise formula relating the Dirac delta function with the Fourier integral transform, namely

$$\delta(\tau) = \frac{1}{2\pi}\int_{-\infty}^{\infty}\mathrm{e}^{i\omega\tau}\mathrm{d}\omega \quad \text{or} \quad \delta(\tau) = \frac{1}{2\pi}\int_{-\infty}^{\infty}\cos\omega\tau d\omega. \tag{3.58}$$

It follows from the result obtained, that the spectral density of the stationary white noise is constant and equal to S_0. As stated above, the white noise variance is equal to infinity. It takes an infinite power to produce such random process, involving, for example, a random force that would continuously obtain random increments with an infinite variance. The concept of the white noise is therefore a mathematical abstraction. It is highly instrumental, however, in the solution of many applied problems dealing with generalized functions.

Appendix 3 contains plots of correlation functions and their corresponding spectral densities.

3.5 Cross-Spectral Densities and Their Properties

If the cross-correlation functions $K_{x_j x_\nu}(t, t')$ of the stationary random functions $X_j(t)$, $X_\nu(t)$ depend on a difference of time instants $(t' - t = \tau)$, such stationary random functions are referred to as stationary-connected functions (Sect. 3.1).

Using a Wiener–Khintchin formula (3.42), we get a cross-spectral density

$$S_{x_j x_\nu}(i\omega) = \frac{1}{2\pi}\int_{-\infty}^{\infty} K_{x_j x_\nu}(\tau)\,\mathrm{e}^{-i\omega\tau}\mathrm{d}\tau. \tag{3.59}$$

It follows from (3.59) that

$$\begin{aligned} S_{x_j x_\nu}(i\omega) &= \frac{1}{2\pi}\int_0^{\infty}\left[K_{x_j x_\nu}(\tau)\,\mathrm{e}^{-i\omega\tau} + K_{x_j x_\nu}(-\tau)\,\mathrm{e}^{i\omega\tau}\right]\mathrm{d}\tau \\ &= \int_{-\infty}^{0} K_{x_j x_\nu}(\tau)\,\mathrm{e}^{-i\omega\tau}\mathrm{d}\tau = \int_0^{\infty} K_{x_j x_\nu}(-\tau)\,\mathrm{e}^{i\omega\tau}\mathrm{d}\tau. \end{aligned}$$

Since

$$K_{x_\nu x_j}(\tau) = K_{x_j x_\nu}(-\tau),$$

then

$$S_{x_j x_\nu}(i\omega) = \frac{1}{2\pi}\int_0^{\infty}\left[K_{x_j x_\nu}(\tau)\,\mathrm{e}^{-i\omega\tau} + K_{x_\nu x_j}(\tau)\,\mathrm{e}^{i\omega\tau}\right]\mathrm{d}\tau. \tag{3.60}$$

A similar expression will be obtained for the cross-spectral density $S_{x_\nu x_j}(i\omega)$

$$S_{x_\nu x_j}(i\omega) = \frac{1}{2\pi}\int_0^{\infty}\left[K_{x_\nu x_j}(\tau)\,\mathrm{e}^{-i\omega\tau} + K_{x_j x_\nu}(\tau)\,\mathrm{e}^{i\omega\tau}\right]\mathrm{d}\tau. \tag{3.61}$$

By adding spectral densities (3.60) and (3.61) we get

$$S_{x_j x_\nu}(i\omega) + S_{x_\nu x_j}(i\omega) = \frac{1}{2\pi}\int_0^{\infty}\left(K_{x_j x_\nu} + K_{x_\nu x_j}\right)\left(\mathrm{e}^{i\omega\tau} + \mathrm{e}^{-i\omega\tau}\right)\mathrm{d}\tau$$

or

$$S_{x_j x_\nu}(\omega) + S_{x_\nu x_j}(i\omega) = \frac{1}{2\pi}\int_0^{\infty}\left(K_{x_j x_\nu} + K_{x_\nu x_j}\right)\cos\omega\tau\mathrm{d}\tau. \tag{3.62}$$

The right-hand side of the relationship (3.62) is a real quantity; therefore, the sum of spectral densities is a real function of ω. Since these spectral densities can be presented as a sum (just as any function depending on an imaginary argument)

$$S_{x_j x_\nu}(i\omega) = S^{(1)}_{x_j x_\nu}(\omega) + iS^{(2)}_{x_j x_\nu}(\omega),$$
$$S_{x_\nu x_j}(i\omega) = S^{(1)}_{x_\nu x_j}(\omega) + iS_{x_\nu x_j}(\omega),$$

it results from the relation (3.62) that

$$S^{(2)}_{x_j x_\nu}(\omega) = -S^{(2)}_{x_\nu x_j}(\omega) \tag{3.63}$$

therefore

$$S^{(1)}_{x_j x_\nu}(\omega) + S^{(1)}_{x_\nu x_j}(\omega) = \frac{1}{\pi}\int_0^\infty \left(K_{x_j x_\nu} + K_{x_\nu x_j}\right)\cos\omega\tau \mathrm{d}\tau. \tag{3.64}$$

The right-hand side of the relationship (3.64) is an even function of ω, therefore $S_{x_j x_\nu}$ and $S_{x_\nu x_j}$ are even functions, i.e.

$$\begin{aligned} S^{(1)}_{x_j x_\nu}(\omega) &= S^{(1)}_{x_j x_\nu}(-\omega), \\ S^{(1)}_{x_\nu x_j}(\omega) &= S^{(1)}_{x_\nu x_j}(-\omega). \end{aligned} \tag{3.65}$$

This prompts us the following general conclusion: real parts of cross-spectral densities are even functions of ω. Let us show now, that imaginary parts of cross- spectral densities are odd functions of ω. It follows from (3.59), that

$$S^{(1)}_{x_j x_\nu}(\omega) + iS^{(2)}_{x_j x_\nu}(\omega) = \frac{1}{2\pi}\int_{-\infty}^{\infty} K_{x_j x_\nu}\cos\omega\tau \mathrm{d}\tau - \frac{i}{2\pi}\int_{-\infty}^{\infty} K_{x_j x_\nu}\sin\omega\tau \mathrm{d}\tau. \tag{3.66}$$

It results from (3.66)

$$S^{(2)}_{x_j x_\nu}(\omega) = -\frac{1}{2\pi}\int_{-\infty}^{\infty} K_{x_j x_\nu}\sin\omega\tau \mathrm{d}\tau. \tag{3.67}$$

The right-hand side of the relation (3.67) is an odd function of ω. Therefore $S^{(2)}_{x_j x_\nu}(\omega)$ is an odd function

$$S^{(2)}_{x_j x_\nu}(\omega) = S^{(2)}_{x_j x_\nu}(-\omega), \tag{3.68}$$

that is, imaginary parts of cross-spectral densities are odd functions of ω.

For a system of n random stationary and stationary-connected functions $X_j(t)$ we may obtain a matrix (similar to that of cross-correlation functions)

$$S(i\omega) = \begin{bmatrix} S_{x_1 x_1}, S_{x_1 x_2}, \ldots\ldots\ldots\ldots\ldots\ldots & S_{x_1 x_n} \\ S_{x_2 x_1} \ldots\ldots\ldots\ldots\ldots\ldots\ldots\ldots & S_{x_2 x_n} \\ \ldots\ldots\ldots\ldots\ldots\ldots\ldots\ldots\ldots\ldots & \\ S_{x_n x_1}, \ldots\ldots\ldots\ldots\ldots\ldots\ldots\ldots & S_{x_n x_n} \end{bmatrix}. \tag{3.69}$$

The diagonal terms of the matrix $S(\omega)$ are real non-negative even functions of ω:

$$S_{x_i x_i}(\omega) = S_{x_i x_i}(-\omega).$$

Example 3.10. Given a stationary random function $X(t)$ with probability characteristics $m_x = 0$, $K_x(\tau) = D_x e^{-\alpha|\tau|}$. The task is to find the cross-spectral density of the stationary function and of its first derivative. The expression for cross-correlation function (3.29) is

$$K_{x\dot{x}} = \frac{\mathrm{d}}{\mathrm{d}\tau} K_x(\tau) = -D_x \alpha e^{-\alpha|\tau|} \sin \tau \mathrm{d}\tau.$$

Let us use relationship (3.59) to determine the cross-spectral density

$$\begin{aligned} S_{x\dot{x}}(i\omega) &= -\frac{D_x\alpha}{2\pi} \int_{-\infty}^{\infty} e^{-\alpha|\tau| - i\omega\tau} \mathrm{sign}\tau \mathrm{d}\tau \\ &= -\frac{D_x\alpha}{2\pi} \left[\int_{0}^{-\infty} e^{-(\alpha+i\omega)\tau} (1) \mathrm{d}\tau + \int_{-\infty}^{0} e^{\alpha\tau - i\omega\tau} (-1) \mathrm{d}\tau \right] \\ &= -\frac{D_x\alpha}{2\pi} \left[\int_{0}^{\infty} e^{-(\alpha+i\omega)\tau} \mathrm{d}\tau - \int_{0}^{\infty} e^{-(\alpha-i\omega)\tau} \mathrm{d}\tau \right] = -\frac{D_x \alpha i\omega}{\pi(\alpha^2 + \omega^2)}. \end{aligned}$$

3.6 Determination of the Spectral Densities of the Linear Differential Equations with Constant Coefficients Solutions

Solving linear differential equations at steady-state excitations may involve the following cases.

1. An equation (or a set of equations) has coefficients variable in time, as for example, the equation (2.61). In this case a solution of an equation with a stationary right-hand side will be non-stationary at any duration of the process. The launch of a flying vehicle may be an example of the non-stationary motion of a system, described by the differential equations with variable coefficients (See Fig. 0.2).

2. An equation has constant coefficients, for example, equation (2.55). In this case, two conditions of the systems motion at stationary excitations are possible: 1) non-stationary motion and 2) stationary motion. The non-stationary motion (for example, oscillations) of a system under the action of stationary random excitations occurs, when the system is at rest before the action of these random excitations.

Stationary oscillations occur at stationary excitations after a certain period of time has passed since the beginning of oscillations.

Stationary random oscillations are possible in stable systems. Let us consider the algorithm of determining the solution spectral density assuming, that at stationary excitation the system is under the conditions of stationary motion. We shall consider only equations of the first and second order (the case of determining the spectral density of the solutions of n–th order system of equations is discussed in a broader context in chapter 6).

Let us recall the essentials of operational calculus pertaining to the Laplace transform. Given a certain function $f(t)$ of the independent real variable t, the Laplace transform (a for direct transformation and b for inverse transformation) are defined by the formulas

$$\begin{aligned} &a) \quad f_0(p) = \int_0^\infty f(t)\, \mathrm{e}^{-pt} \mathrm{d}t, \\ &b) \quad f(t) = \int_0^\infty f_0(p)\, \mathrm{e}^{pt} \mathrm{d}p. \end{aligned} \tag{3.70}$$

where p is a certain complex quantity. The function $f(t)$ is referred to as the original, while the function $f_0(p)$ is defined as a transform. The originals and transforms of the function $f(t)$ and its derivatives as well as of the function with delay $f(t-t_0)$ are given in Table 3.1 at zero initial values. The Fourier transforms of the same functions are presented in the third row of Table 3.1.

Table 3.1.

Original	$\frac{\mathrm{d}f}{\mathrm{d}t}$	$\frac{\mathrm{d}^2 f}{\mathrm{d}t^2}$	$\frac{\mathrm{d}^n f}{\mathrm{d}t^n}$
Transform	$pf_0(p)$	$p^2 f_0(p)$	$p^n f_0(p)$
Frequency domain	$i\omega f_0(i\omega)$	$(i\omega)^2 f_0(i\omega)$	$(i\omega)^n f_0(i\omega)$

Table 3.1 (continuation)

Original	$f(t-t_0)$	$\int f(t)\mathrm{d}t$	$f_1 + f_2$
Transform	$\mathrm{e}^{-pt_0} f_0(p)$	$\frac{1}{p} f_0(p)$	$f_1(p) + f_2(p)$
Frequency domain	$\mathrm{e}^{-i\omega t_0} f_0(i\omega)$	$\frac{1}{i\omega} f_0(i\omega)$	$f_{01}(i\omega) + f_{02}(i\omega)$

The direct and inverse one-sided Fourier transformations are

$$f_0(i\omega) = \frac{1}{\pi}\int_0^\infty f(t)\,\mathrm{e}^{-i\omega t}\mathrm{d}t,$$
$$f(t) = \int_0^\infty f_0(i\omega)\,\mathrm{e}^{i\omega t}\mathrm{d}\omega. \tag{3.71}$$

or bilaterial transformations

$$f_0(i\omega) = \frac{1}{\pi}\int_{-\infty}^\infty f(t)\,\mathrm{e}^{-i\omega t}\mathrm{d}t,$$
$$f(t) = \int_{-\infty}^\infty f_0(i\omega)\,\mathrm{e}^{i\omega t}\mathrm{d}\omega. \tag{3.72}$$

Let us consider the equation of the first order

$$L(t) = \dot{Y}(t) + kY(t) - f(t) = 0 \tag{3.73}$$

where $f(t)$ is a stationary function with a zero mathematical expectation and a known spectral density. If the function $f(t)$ has a non-zero mathematical expectation it may be presented in the form

$$f(t) = m_f + \overset{\circ}{f}(t),$$

where $\overset{\circ}{f}(t)$ is a centered random function. Since m_f is a constant quantity at stationary oscillations, the mathematical expectation m_y will be constant and equal to

$$m_y = \frac{1}{k}m_f.$$

Therefore by manipulations we get the equation (3.73), where $f(t)$ and $Y(t)$ are centered stationary functions. Using the Laplace transform (at zero initial values) we get

$$L(p) = \int_0^\infty \left[\dot{Y}(t) + kY(t) - f(t)\right]\mathrm{e}^{-ipt}\mathrm{d}t = pY(p) + kY(p) - f_0(p) = 0.$$

The transform of the function $Y(t)$ is

$$Y(p) = \tfrac{1}{(p+k)} f_0(p) = W(p)\,f_0(p), \tag{3.74}$$

where $W(p)$ is a transfer function.

In frequency domain we get

$$Y\left(i\omega\right)=\frac{1}{\left(i\omega+k\right)}f_0\left(i\omega\right)=W\left(i\omega\right)f_0\left(i\omega\right). \tag{3.75}$$

Using the inverse bilateral Fourier transform, we get

$$Y\left(t\right)=\int\limits_{-\infty}^{\infty}Y_0\left(i\omega\right)\mathrm{e}^{i\omega t}d\omega. \tag{3.76}$$

Therefore, it is possible to represent the correlation function $K_y\left(t,\,t'\right)$ as

$$\begin{aligned}K_y\left(t,\,t'\right)&=M\left[Y\left(t\right)Y^*\left(t'\right)\right]\\&=M\left[\left(\int\limits_{-\infty}^{\infty}Y_0\left(i\omega\right)\mathrm{e}^{i\omega t}d\omega\right)\left(\int\limits_{-\infty}^{\infty}Y_0^*\left(i\omega'\right)\mathrm{e}^{-i\omega't'}d\omega'\right)\right]\end{aligned}$$

or

$$K_y=\int\limits_{-\infty}^{\infty}\int\limits_{-\infty}^{\infty}M\left[Y\left(i\omega\right)Y^*\left(i\omega'\right)\right]\mathrm{e}^{i\left(\omega t-\omega't'\right)}\mathrm{d}\omega\,\mathrm{d}\omega'. \tag{3.77}$$

For the function K_y to depend on a difference of time instants $(t-t'=\tau)$, which is a must at a stationary process, we cannot do without the fulfillment of the condition

$$M\left[Y\left(i\omega\right)Y^*\left(i\omega'\right)\right]=S_y\left(\omega'\right)\delta\left(\omega'-\omega\right).$$

Substituting this relationship in the left-hand side of (3.77) and integrating over ω', we get

$$K_y\left(\tau\right)=\int\limits_{-\infty}^{\infty}S_y\left(\omega\right)\mathrm{e}^{i\omega\tau}\mathrm{d}\omega. \tag{3.78}$$

We may obtain the correlation function $K_y\left(t,\,t'\right)$ using equation (3.75)

$$\begin{aligned}K_y\left(t,\,t'\right)&=M\left[Y\left(t\right)Y^*\left(t'\right)\right]\\&=M\left[\int\limits_{-\infty}^{\infty}\left(W\left(i\omega\right)f_0\left(i\omega\right)\mathrm{e}^{i\omega t}\mathrm{d}\omega\right)\cdot\int\limits_{-\infty}^{\infty}\left(W^*\left(i\omega'\right)f_0^*\left(i\omega'\right)\mathrm{e}^{-i\omega't'}\mathrm{d}\omega'\right)\right]\end{aligned}$$

or

$$K_y\left(t,\,t'\right)=\int\limits_{-\infty}^{\infty}\int\limits_{-\infty}^{\infty}W\left(i\omega\right)W^*\left(i\omega'\right)M\left[f_0\left(i\omega\right)f_0^*\left(i\omega'\right)\right]\mathrm{e}^{i\left(\omega t-\omega't'\right)}\mathrm{d}\omega\,\mathrm{d}\omega'.$$

(3.79)

The function $K_y(t, t')$ will depend on a difference of time instants given

$$M[f_0(i\omega) f_0^*(i\omega')] = S_f(\omega')\,\delta(\omega' - \omega)\,. \tag{3.80}$$

Excluding the mathematical expectation of the product of functions f_0 and f_0^* (3.80) from the right-hand side of (3.79), we get

$$K_y(\tau) = \int_{-\infty}^{\infty} W(i\omega)\, W^*(i\omega)\, S_f(\omega)\, \mathrm{e}^{i\omega\tau} \mathrm{d}\omega. \tag{3.81}$$

Since (3.78) and (3.81) are two representations of the same function $K_y(\tau)$, we should satisfy the identity

$$\int_{-\infty}^{\infty} [S_y(\omega) - W(i\omega)\, W^*(i\omega)\, S_f(\omega)]\, \mathrm{e}^{i\omega\tau} d\omega \equiv 0. \tag{3.82}$$

Therefore, we get from (3.82)

$$S_y(\omega) = W(i\omega)\, W^*(i\omega)\, S_f(\omega) = |W(i\omega)|^2\, S_f(\omega)\,, \tag{3.83}$$

where $|W_1(i\omega)|^2$ is the square of the frequency function modulus. The relationship (3.83) relates the "input" spectral density $S_y(\omega)$ to the "output" spectral density $S_y(\omega)$.

For the second-order equation

$$\ddot{Y} + a_1\dot{Y} + a_2 Y = kf$$

We get in frequency domain

$$Y_0(i\omega) = \frac{k}{\left[(i\omega)^2 + (i\omega)\, a_1 + a_2\right]} f_0(i\omega) = W(i\omega)\, f_0(i\omega)\,, \tag{3.84}$$

where

$$W(i\omega) = \frac{k}{\left[(i\omega)^2 + (i\omega)\, a_1 + a_2\right]}.$$

The relation (3.84) is similar to that of (3.75), therefore the spectral density $S_y(\omega)$ is equal to

$$S_y(\omega) = |W(i\omega)|^2\, S_f(\omega)\,. \tag{3.85}$$

This result is true for any linear equation with constant coefficients (here an equation of n–th order is meant), for which the frequency function $W(i\omega)$ is equal to

$$W(i\omega) = \frac{k}{\left[(i\omega)^n a_0 + (i\omega)^{n-1} a_1 + (i\omega)^{n-2} a_2 + \cdots + a_n\right]}.$$

Knowing the "output" spectral density, we may determine the variance D_y

$$K_y(0) = D_y = \int_{-\infty}^{\infty} S_y(\omega)\, d\omega.$$

In the general case, we may present the spectral density $S_y(\omega)$ as

$$S_y(\omega) = |W(i\omega)|^2 S_f(\omega) = \frac{1}{2\pi} \cdot \frac{G(i\omega)}{|A(i\omega)|^2}, \tag{3.86}$$

where

$$G(i\omega) = b_0 (i\omega)^{2n-2} + b_1 (i\omega)^{2n-4} + \cdots + b_{n-1};$$
$$A(i\omega) = a_0 (i\omega)^n + a_1 (i\omega)^{n-1} + \cdots + a_n,$$

therefore

$$D_y = \frac{1}{2\pi} \int_{-\infty}^{\infty} \frac{G(i\omega)}{|A(i\omega)|^2} d\omega. \tag{3.87}$$

The values of integrals I_n are presented in Appendix 2.

4. Fundamentals of the Markov Processes Theory

4.1 Continuous One-Dimensional Markov Processes

The previous sections dealt with the correlation theory of random functions and only the first two moments of a random function — expectation and the correlation function has been considered. Unfortunately, far from all encountered applied problems can be solved by correlation theory methods. A case in point is the problem of determining the probability that the ordinate of a random function will exceed a particular given value, which often arises during the dynamic systems analysis. These problems become solvable if we restrict their treatment to processes not only possessing some special properties, but also interesting in the practical plane. Up to this point we have used correlation theory methods to analyze systems with a linear input–output relation. In this case the correlation theory enables us to obtain the probability characteristics of the differential equations solution, knowing the probability characteristics of perturbations. It is impossible to find a solution of nonlinear equations by correlation theory methods. However, if we confine ourselves to the processes possessing some special properties, we can obtain a solution of nonlinear problems of statistical dynamics. Markov processes, for the exhaustive characterization of which it is sufficient to know only two-dimensional distribution laws, are classified among such processes.

The theory of Markov processes assumes that the distribution law of the ordinate of a process at any future time instant t_i depends only on the value of the ordinate at a given instant t_{i-1} and is independent of the function's past ordinates. In other words, additional knowledge of the random function's values at $t < t_{i-1}$ does not alter the distribution character of its ordinates at $t \geq t_i$. Physically, this singularity of a random process is equivalent to the processes without aftereffect (processes independent on previous history).

For a Markov process any multi-dimensional distribution laws can be expressed in terms of two-dimensional laws. As an example let us consider a three-dimensional probability density $f(x_2, t_2, x_1, t_1, x_0, t_0)$, a probability density of the three ordinates of a random process taken at three sequential instants t_0, t_1, t_2 $(t_2 > t_1 > t_0)$. In accordance with the general formula for the conditional distribution laws (1.48)

$$f(x_2, t_2, x_1, t_1, x_0, t_0) = f(x_2, t_2 | x_1, t_1, x_0, t_0)\, f(x_1, t_1, x_0, t_0)\,. \quad (4.1)$$

In its turn,

$$f(x_1, t_1, x_0, t_0) = f(x_1, t_1 | x_0, t_0) f(x_0, t_0). \tag{4.2}$$

However, since

$$f(x_2, t_2 | x_1, t_1, x_0, t_0) = f(x_2, t_2 | x_1, t_1)$$

(this is the property only of Markov processes), we finally obtain

$$f(x_2, t_2, x_1, t_1, x_0, t_0) = f(x_2, t_2 | x_1, t_1) f(x_1, t_1 | x_0, t_0) f(x_0, t_0). \tag{4.3}$$

In the same way we may obtain an expression for the n-dimensional probability density

$$\begin{aligned} &f(x_n, t_n, \ldots, x_0, t_0) = \\ &= f(x_n, t_n | x_{n-1}, t_{n-1}) f(x_{n-1}, t_{n-1} | x_{n-2}, t_{n-2}) \ldots f(x_0, t_0), \end{aligned} \tag{4.4}$$

i.e. the probability density of the Markov process of any order ordinates can be expressed in terms of the conditional two-dimensional probability densities of the process ordinate at a time instant t_1. The relations (true for any probability density)

$$f(x_i, t_i | x_{i-1}) \geq 0; \tag{4.5}$$

$$\int_{-\infty}^{\infty} f(x_i, t_i | x_{i-1}, t_{i-1}) \, \mathrm{d}x_i = 1. \tag{4.6}$$

should be valid for functions $f(x_i, t| x_{i-1}, t_{i-1})$.

Besides, the two-dimensional probability density $f(x_i, t_i, x_{i-1}, t_{i-1})$ is related to the conditional probability density by the relation

$$f(x_1, t_1, x_{i-1}, t_{i-1}) = f(x_i, t_i | x_{i-1}, t_{i-1}) f(x_{i-1}, t_{i-1}). \tag{4.7}$$

It follows from (4.7) that

$$\begin{aligned} f(x_i, t_i) &= \int_{-\infty}^{\infty} f(x_i, t_i, x_{i-1}, t_{i-1}) \, \mathrm{d}x_{i-1} \\ &= \int_{-\infty}^{\infty} f(x, t | x_{i-1}, t_{i-1}) f(x_{i-1}, t_{i-1}) \, \mathrm{d}x_{i-1}. \end{aligned} \tag{4.8}$$

Let us consider three sequential instants: t_0, τ and $t\,(t_0 < \tau < t)$. At the instant t_0 (Fig. 4.1) the process had an ordinate x_0. Then the elementary probabilities of transition from the state x_0 to a state in the interval $(z, z + \mathrm{d}z)$ (where $z = x(\tau)$) at the instant τ and to a state in the

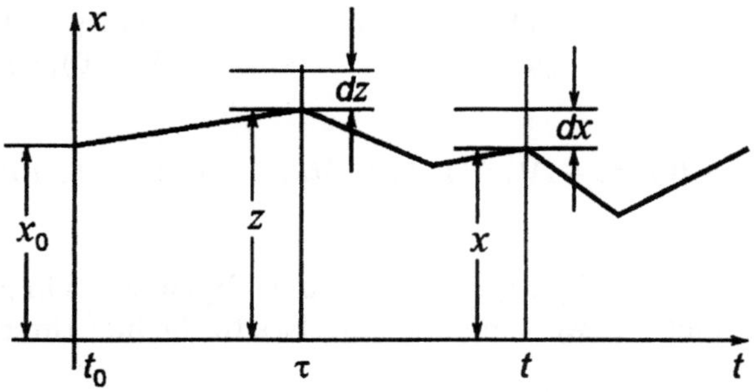

Fig. 4.1.

interval $(x,\ x+\mathrm{d}x)$ at the instant t are equal to $f(z,\ \tau|\,x_0,\ t_0)\,\mathrm{d}z$ and $f(x,\ t|\,x_0,\ t_0)\,\mathrm{d}x$ respectively.

The probability of transition from the state $(z,\ \tau)$ to the interval $(x,\ x+$ $+\mathrm{d}x)$ at the instant t is equal to $f(x,\ t|\,z,\ \tau)\,\mathrm{d}x$. The total probability of transition from the state $(x,\ x+\mathrm{d}x)$ at the instant t is obtained by integration of the probabilities product $(x,\ f(x,\ t|\,z,\ \tau)\,\mathrm{d}x\cdot\ f(z,\ \tau|\,x_0,\ \tau_0)\,\mathrm{d}z)$ with respect to all intermediate values z, i.e.

$$f(x,\ t|\,x_0,\ t_0)\,\mathrm{d}x=\left(\int\limits_{-\infty}^{\infty} f(x,\ t|\,z,\ \tau)\,f(z,\ \tau|\,x_0,\ t_0)\,\mathrm{d}z\right)\mathrm{d}x$$

or

$$f(x,\ t|\,x_0,\ t_0)=\int\limits_{-\infty}^{\infty} f(x,\ t|\,z,\ \tau)\,f(z,\ \tau|\,x_0,\ t_0)\,\mathrm{d}z. \tag{4.9}$$

In mathematical literature the equation (4.9) is referred to as the Chapman-Kolmogorov equation, in the literature on physics it is called the Smolukhowski equation, who obtained it when studying the Brownian motion of a particle. The equation (4.9) sets a rather strict limits on the form of a transition probability conditional density, namely, the integration over z should lead to the elimination of z with the form of the function f remaining unchanged.

The limits of integration should not necessarily be equal to infinity.

4.2 The Fokker–Planck–Kolmogorov Equation

Let us obtain equations satisfied by a conditional probability density $f(x,\ t|\,x_0,\ t_0)$ (Kolmogorov equations) for a continuous random process. The process $X(t)$ is considered continuous, if within small time intervals $X(t)$ can obtain increments of noticeable magnitude with only small probability. Let us assume for this purpose that in the equation (4.9) $\tau = t_0 + \Delta t \, (\Delta t > 0)$, i.e.

$$f(x, t | x_0, t_0) = \int_{-\infty}^{\infty} f(x, t | z, t_0 + \Delta t) \, f(z, t_0 + \Delta t | x_0, t_0) \, dz. \tag{4.10}$$

Let us then expand the function $f(x, t | z, t_0 + \Delta t)$ in a Taylor series in z in the neighbourhood of a point $z = x_0$ (confining ourselves to the first three expansion terms):

$$\begin{aligned} f(x, t | z, t_0 + \Delta t) &= f(x, t | x_0, t_0 + \Delta t) + \frac{\partial f}{\partial x_0}(z - x_0) \\ &+ \frac{1}{2}\frac{\partial^2 f}{\partial x_0^2}(z - x_0)^2 + \frac{1}{6}\frac{\partial^3 f}{\partial x_0^3}(z - x_0)^3 + \dots . \end{aligned} \tag{4.11}$$

Substituting (4.11) in the equation (4.10), we obtain

$$\begin{aligned} f(x, t | x_0, t_0) = \int_{-\infty}^{\infty} \Big[& f(x, t | x_0, t_0 + \Delta t) + \frac{\partial f}{\partial x_0}(z - x_0) \\ & + \frac{1}{2}\frac{\partial^2 f}{\partial x_0^2}(z - x_0)^2 + \frac{1}{6}\frac{\partial^3 f}{\partial x_0^3}(z - x_0)^3 \Big] \\ & \times f(z_0, t_0 + \Delta t | x_0, t_0) \, dz. \end{aligned} \tag{4.12}$$

Since

$$\int_{-\infty}^{\infty} f(z, t_0 + \Delta t | x_0, t_0) \, dz = 1, \tag{4.13}$$

then by manipulations we obtain

$$\begin{aligned} & f(x, t | x_0, t_0) - f(x, t | x_0, t_0 + \Delta t) \\ & = \frac{\partial f}{\partial x_0} \int_{-\infty}^{\infty} (z - x_0) \, f(z, t_0 + \Delta t | x_0, t_0) \, dz \\ & + \frac{1}{2}\frac{\partial^2 f}{\partial x_0^2} \int_{-\infty}^{\infty} (z - x_0)^2 f(z, t_0 + \Delta t | x_0, t_0) \, dz \\ & + \frac{1}{6}\frac{\partial^3 f}{\partial x_0^3} \int_{-\infty}^{\infty} (z - x_0)^3 f(z, t_0 + \Delta t | x_0, t_0) \, dz. \end{aligned} \tag{4.14}$$

Dividing both sides of the equation (4.14) by Δt and proceeding to the limit, we obtain

$$-\frac{\partial f\left(x, t \mid x_0, t_0\right)}{\partial t_0}=a_1 \frac{\partial f}{\partial x_0}+\frac{b_1}{2} \frac{\partial^2 f}{\partial x_0^2}+\frac{c_1}{6} \frac{\partial^3 f}{\partial x_0^3}, \tag{4.15}$$

where

$$a_1=a_1\left(x_0, t_0\right)=\lim_{\Delta t \rightarrow 0} \frac{1}{\Delta t} \int_{-\infty}^{\infty}\left(z-x_0\right) f\left(z, t_0+\Delta t \mid x_0, t_0\right) \mathrm{d} z;$$

$$b_1=b_1\left(x_0, t_0\right)=\lim_{\Delta t \rightarrow 0} \int_{-\infty}^{\infty}\left(z-x_0\right)^2 f\left(z, t_0+\Delta t \mid x_0, t_0\right) \mathrm{d} z; \tag{4.16}$$

$$c_1=c_1\left(x_0, t_0\right)=\lim_{\Delta t \rightarrow 0} \frac{1}{\Delta t} \int_{-\infty}^{\infty}\left(z-x_0\right)^3 f\left(z, t_0+\Delta t \mid x_0, t_0\right) \mathrm{d} z.$$

Two of the introduced coefficients have special names: the drift coefficient for a_1 and the diffusion coefficient for b_1.

We may record these expressions for coefficients a_1, b_1 and c_1 in a more compact fashion:

$$\begin{aligned} a_1 &= \lim_{\Delta t \rightarrow 0} \frac{M\left[z-x_0\right]}{\Delta t}=\lim_{\Delta t \rightarrow 0} \frac{M[\Delta z]}{\Delta t}; \\ b_1 &= \lim_{\Delta t \rightarrow 0} \frac{M\left[\left(z-x_0\right)^2\right]}{\Delta t}=\lim_{\Delta t \rightarrow 0} \frac{M\left[\Delta z^2\right]}{\Delta t}; \\ c_1 &= \lim_{\Delta t \rightarrow 0} \frac{M\left[\left(z-x_0\right)^3\right]}{\Delta t}=\lim_{\Delta t \rightarrow 0} \frac{M\left[\Delta z^3\right]}{\Delta t}. \end{aligned} \tag{4.17}$$

The $M[\,]$ operation in the expressions (4.17) is an operation of conditional mathematical expectation. The coefficient $a_1\left(x_0, t_0\right)$ constitutes the limit of the ratio of the random process difference mathematical expectation to a time interval Δt, for which the ordinates have been taken when the interval tended to zero, i.e. $a_1\left(x_0, t_0\right)$ characterizes the velocity of random function variation. The coefficient $b_1\left(x_0, t_0\right)$ determines the limit of the ratio of the ordinates difference square mathematical expectation to a time interval Δt as Δt tends to zero. In other words, it specifies the random function conditional variance variation velocity.

The theory of Markov processes usually assumes that the moment of the third order $M\left[\left(z-x_0\right)^3\right]$ tends to zero faster than Δt, therefore

$$c_1=\lim_{\Delta t \rightarrow 0} \frac{M\left[\left(z-x_0\right)^3\right]}{\Delta t}=0. \tag{4.18}$$

The adopted condition $c_1\left(x_0, t_0\right)=0$ is not obvious and should be considered as the assumption that the probability of large deviations $z-x_0$

must decrease with the reduction of Δt rapid enough for all moments of this difference, beginning from the third one, tend to zero faster than Δt. This requirement allows us to regard a random function $X(t)$ in a system subjected to the action of random impulses as a continuously varying quantity. Hence, the solution of the equation (4.15) is true (at $c_1 = 0$) only for time intervals Δt, that far exceed time intervals between impulses.

Finally we obtain the following equation (Kolmogorov's first equation):

$$-\frac{\partial f}{\partial t_0} = a_1 \frac{\partial f}{\partial x_0} + b_1 \frac{\partial^2 f}{\partial x_0^2}. \tag{4.19}$$

The equation (4.19) enables us to determine the conditional probability density $f(x, t|\, x_0, t_0)$ as a function of the initial state, since the "past" time t_0 and the "past" state x_0 are independent variables.

Let us obtain an equation which allows us to determine the variation of the conditional probability density in the future, i.e. an equation relating the derivatives $f(x, t|\, x_0, t_0)$ with respect to t and x, rather than with respect to t_0 and x_0, as is the case in Kolmogorov's first equation. An equation, in which t and x are independent variables, is called Kolmogorov's second equation.

Let us take advantage of the conclusion given in work [17] and use in our consideration some arbitrary function $R(x)$ that in the whole variation range of the random function $X(t)$ is continuously differentiable up to the second order and becomes zero on the boundaries of the x variation interval (a, b) together with the first two derivatives

$$R(a) = R(b) = R'(a) = R'(b) = R''(a) = R''(b) = 0. \tag{4.20}$$

It is generally considered (by analogy with a normal distribution law) that the area of possible x values is equal to $(-\infty, \infty)$, i.e. $a = -\infty$, $b = \infty$. Let us consider the definition of a partial derivative

$$\frac{\partial f(x, t|\, x_0, t_0)}{\partial t} = \lim_{\Delta t \to 0} \frac{f(x, t + \Delta t|\, x_0, t_0) - f(x, t|\, x_0, t_0)}{\Delta t}. \tag{4.21}$$

As distinct from the derivation of Kolmogorov's first equation where the instant τ is taken as the one closely spaced with respect to the instant t_0, when deriving Kolmogorov's second equation, the instant τ is taken close to the instant t.

Let us multiply the relation (4.21) by $R(x)$ and integrate it between $-\infty$ and ∞, then

$$\int_{-\infty}^{\infty} \frac{\partial f(x, t|\, x_0, t_0)}{\partial t} R(x)\, \mathrm{d}x$$

$$= \lim_{\Delta t \to 0} \int_{-\infty}^{\infty} \frac{f(x, t + \Delta t|\, x_0, t_0) - f(x, t|\, x_0, t_0)}{\Delta t} R(x)\, \mathrm{d}x. \tag{4.22}$$

Let us use the equation (4.9), in which we shall replace t by $t + \Delta t$ and τ by t. As a result we obtain

$$f(x, t + \Delta t | x_0, t_0) = \int_{-\infty}^{\infty} f(x, t + \Delta t | z, t)\, f(z, t | x_0, t_0)\, dz. \tag{4.23}$$

Let us eliminate $f(x, t| x_0, t_0)$ that enters the expression (4.22), using the equation (4.23):

$$\int_{-\infty}^{\infty} \frac{\partial f(x, t| x_0, t_0)}{\partial t} R(x)\, dx$$
$$= \lim_{\Delta t \to 0} \frac{1}{\Delta t} \int_{-\infty}^{\infty} \left[\int_{-\infty}^{\infty} f(x, t + \Delta t | z, t)\, f(z, t| x_0, t_0)\, R(x)\, dz \right.$$
$$\left. - f(x, t| x_0, t_0)\, R(x) \right] dx$$
$$= \lim_{\Delta t \to 0} \frac{1}{\Delta t} \left[\int_{-\infty}^{\infty} f(z, t| x_0, t_0) \int_{-\infty}^{\infty} f(x, t + \Delta t | z, t)\, R(x)\, dx\, dz \right.$$
$$\left. - \int_{-\infty}^{\infty} f(x, t| x_0, t_0)\, R(x)\, dx \right]. \tag{4.24}$$

Replacing the variable of integration x by the variable of integration z in the expression (the addend in the formula (4.24))

$$\int_{-\infty}^{\infty} f(z, t| x_0 t_0) \int_{-\infty}^{\infty} f(x, t + \Delta t| z, t)\, R(x)\, dx\, dz$$

we obtain

$$\int_{-\infty}^{\infty} \frac{\partial f(x, t| x_0, t_0)}{\partial t} R(x) dx$$
$$= \lim_{\Delta t \to 0} \frac{1}{\Delta t} \int_{-\infty}^{\infty} f(x), t| x_0, t_0) \left[\int_{-\infty}^{\infty} f(z, t + \Delta t| x, t)\, R(z)\, dz - R(x) \right] dx.$$

Let us expand the function $R(z)$ in a Taylor series in powers of $(x-z)$ near the point $z=x$:

$$R(z) = R(x) + R'(x)(x-z) + \frac{1}{2}R''(x)(x-z)^2 + O(x-z),$$

where $O(x-z)$ are terms of a higher order of smallness.

Let us consider an expression

$$\int_{-\infty}^{\infty} f(z,\, t+\Delta t|\, x,\, t)\Big[R(x) + R'(x-z)$$

$$+\frac{1}{2}R''(x-z)^2 + O(x-z)\Big]\mathrm{d}z$$

$$= \int_{-\infty}^{\infty} f(z,\, t+\Delta t|\, x,\, t)\, R(x)\,\mathrm{d}z$$

$$+ R'(x)\int_{-\infty}^{\infty}(x-z)\, f(z,\, t+\Delta t|\, x_0,\, t_0)\,\mathrm{d}z$$

$$+\frac{1}{2}R''(x)\int_{-\infty}^{\infty}(x-z)^2 f(z,\, t+\Delta t|\, x,\, t)\,\mathrm{d}z. \tag{4.25}$$

Since

$$\int_{-\infty}^{\infty} f(z, t+\Delta\, t|\, x, t)\, dz = 1,$$

then, omitting the series terms $0(x-z)$, we obtain

$$\int_{-\infty}^{\infty}\frac{\partial f(x,\, t|\, x_0, t_0)}{\partial t}R(x)\,\mathrm{d}x$$

$$= \int_{-\infty}^{\infty}\Big[f(x,\, t|\, x_0, t_0)\, R'(x)\, a_1(x,t) + \frac{1}{2}R''(x)\, b_1(x,t)\Big]\,\mathrm{d}x, \tag{4.26}$$

where

$$
\begin{aligned}
a_1 = a_1(x,t) &= \lim_{\Delta t \to 0} \frac{1}{\Delta t} \int_{-\infty}^{\infty} (x-z)\, f(z, t+\Delta t | x, t)\, \mathrm{d}z \\
&= \lim_{\Delta t \to 0} \frac{M[x-z]}{\Delta t}; \\
b_1 = b_1(x,t) &= \lim_{\Delta t \to 0} \frac{1}{\Delta t} \int_{-\infty}^{\infty} (x-z)^2 f(z,\, t+\Delta t | x_0,\, t_0)\, \mathrm{d}z \\
&= \lim_{\Delta t \to 0} \frac{M\left[(x-z)^2\right]}{\Delta t}.
\end{aligned}
\tag{4.27}
$$

It is assumed in expression (4.25) that

$$
\lim_{\Delta t \to 0} \frac{1}{\Delta t} \int_{-\infty}^{\infty} O(x-z)\, f(z, t+\Delta t | x, t)\, dz = 0.
$$

The expressions (4.27) for the coefficients a_1 and b_1 are similar to those of (4.16) and (4.17), but now they depend on x and t.

Integrating the expressions entered in the right-hand side of (4.26)

$$
\int_{-\infty}^{\infty} fR' a_1 \mathrm{d}x, \qquad \int_{-\infty}^{\infty} fR'' b_1 \mathrm{d}x
$$

by parts, we obtain (at $R(-\infty) = R(\infty) = 0$)

$$
\begin{aligned}
\int_{-\infty}^{\infty} f a_1 R' \mathrm{d}x &= - \int_{-\infty}^{\infty} \frac{\partial}{\partial x}(f a_1)\, R \mathrm{d}x + (f a_1 R) \Big|_{-\infty}^{\infty} \\
&= - \int_{-\infty}^{\infty} \frac{\partial (f a_1)}{\partial x} R \mathrm{d}x; \\
\int_{-\infty}^{\infty} f b_1 R'' \mathrm{d}x &= \int_{-\infty}^{\infty} R \frac{\partial^2 (f b_1)}{\partial x^2} \mathrm{d}x.
\end{aligned}
\tag{4.28}
$$

Let us substitute (4.28) in (4.26), then

$$
\int_{-\infty}^{\infty} \left[\frac{\partial f}{\partial t} + \frac{\partial (f a_1)}{\partial x} - \frac{1}{2} \frac{\partial^2 (f b_1)}{\partial x^2} \right] R(x)\, \mathrm{d}x = 0. \tag{4.29}
$$

Since the function $R(x)$ is arbitrary, the identical vanishing of the left-hand side of the relationship (4.29) is possible only when

$$\frac{\partial f}{\partial t}+\frac{\partial (fa_1)}{\partial x}-\frac{1}{2}\frac{\partial^2 (fb_1)}{\partial x^2}=0. \tag{4.30}$$

The obtained equation (4.30) is Kolmogorov's second equation for one-dimensional processes that allows us to determine the variation of the conditional probability density in the future with a known probability density at an initial instant.

Kolmogorov's second equation enjoys the broadest use in problems of statistical dynamics and the theory of random vibrations. According to the current classification of partial differential equations, Kolmogorov's equations (4.19) and (4.30) are of the parabolic type. For the solution of an equation to be unambiguous, we must to know the initial and boundary conditions for the desired function (for the probability density $f(x, t|\, x_0, t_0)$). The function f must satisfy not only the initial and boundary conditions, but also conditions true for any probability density

$$f \geq 0, \quad \int_{-\infty}^{\infty} f(x, t|\, x_0, t_0)\, \mathrm{d}x = 1.$$

Two versions of the initial conditions are possible at $t = t_0$: 1) The values of the random function $x = x_0$ ordinate are random quantities with a known probability density $f(x_0,\, t_0)$; 2) The values of the random function x_0 ordinate are given (a nonrandom quantity).

In the first version the condition

$$f(x,\; t|\, x_0, t_0)\,|_{t=t_0} = f(x_0,\; t_0)\,. \tag{4.31}$$

must be fulfilled.

In the second version, where x_0 is given,

$$f(x,\; t|\, x_0, t_0)|_{t=t_0} = \delta(x - x_0)\,. \tag{4.32}$$

This condition can be obtained as follows. At a given x_0 the probability that possible values x are more or less than x_0, is equal to zero. Therefore, we may present the probability density at $t = t_0$, which should meet the condition

$$\int_{-\infty}^{\infty} f(x, t_0)\, \mathrm{d}x = 1$$

as

$$\int_{-\infty}^{x_0-\varepsilon} f(x, t_0)\, \mathrm{d}x + \int_{x_0-\varepsilon}^{x_0+\varepsilon} f(x, t_0)\, \mathrm{d}x + \int_{x_0+\varepsilon}^{\infty} f(x, t_0)\, \mathrm{d}x = 1, \tag{4.33}$$

where ε is a small positive quantity.

The first and the third integrals in the left-hand side of the expression (4.33) are zero, since they are equal to the probabilities of the occurrence of x values smaller and larger than x_0, therefore

$$\int_{x_0-\varepsilon}^{x_0+\varepsilon} f(x,t_0)\,\mathrm{d}x = 1. \tag{4.34}$$

The relation (4.34) at $\varepsilon \to 0$ is a condition defining the delta function, i.e.

$$f(x,\,t_0) = \delta(x - x_0)\,. \tag{4.35}$$

If a random function $x(t)$ can take any values in the interval $(-\infty,\,\infty)$, then the conditions at $\pm\infty$ must be taken as boundary conditions.

When determining the distributions that are homogeneous in time, the function f and the coefficients a and b do not depend on t, therefore the equation (4.30) takes the form

$$\frac{1}{2}\frac{\mathrm{d}^2}{\mathrm{d}x^2}\left[b_1(x)\,f\right] - \frac{\mathrm{d}}{\mathrm{d}x}\left[a_1(x)\,f\right] = 0. \tag{4.36}$$

Or, integrating over x

$$\frac{\mathrm{d}}{\mathrm{d}x}\left[b_1(x)\,f\right] - 2a_1(x)\,f = c. \tag{4.37}$$

If we suppose that at $x \to \infty$ f and $\dfrac{\mathrm{d}f}{\mathrm{d}x}$ tend to zero, then $c = 0$, therefore the solution of equation (4.36) takes the form

$$f(x) = c_1 \exp\left\{-\int_0^x \left(\frac{b_1 - 2a_1}{b_1}\right)\mathrm{d}h\right\}. \tag{4.38}$$

The integration constant c_1 will be determined from the condition

$$\int_{-\infty}^{\infty} f(x)\,\mathrm{d}x = 1.$$

Let us cite some examples of the equation (4.30) solution.

Example 4.1. Let us consider a particular case at $a_1(x,t) = 0$ and $b_1(x,t) = b_{10} = \text{const}:$

$$\frac{\partial f}{\partial t} = \frac{1}{2}b_{10}\frac{\partial^2 f}{\partial x^2}. \tag{4.39}$$

It is required to find the solution of equation (4.39) for the x variation interval $x(-\infty,\infty)$ meeting the conditions

$$1)\ t = 0, f(x, t|\, x_0, 0) = \delta(x - x_0),$$

$$2)\ \int_{-\infty}^{\infty} f(x, t|\, x_0, 0)\, dx = 1, \tag{4.40}$$

$$3)\ f \geq 0.$$

The solution of the equation (4.39) that meets the conditions (4.40) has the form

$$f(x, t|\, x_0, 0) = \frac{1}{\sqrt{2\pi b_{10} t}} \exp\left\{-\frac{(x - x_0)^2}{2b_{10}t}\right\}. \tag{4.41}$$

We may transform the right-hand side of (4.41) to the form

$$f = \frac{1}{\sqrt{\pi}} m \exp\left\{-m^2 (x - x_0)^2\right\}, \quad \left(m = \frac{1}{\sqrt{2b_{10}t}}\right).$$

At $t \to 0$, $m \to \infty$, therefore the right-hand side of (4.41) at $t \to 0$ is the delta function (Appendix 1).

The expression (4.41) obtained for a conditional probability density is a Gaussian normal distribution law that varies in time and has a mathematical expectation $m_x = x_0$ and a variance $D_x = b_{10}t$.

Let us consider a random process, when it is homogeneous in coordinate x, i.e. the probability of passage from the state x_0 to the state x depends only on a difference $x - x_0$.

In this case, the coefficients a_1 and b_1 do not depend on x.

Therefore, from the equation (4.30) we obtain

$$\frac{\partial f}{\partial t} = a_1(t) \frac{\partial f}{\partial x} + \frac{1}{2} b_1(t) \frac{\partial^2 f}{\partial x^2}. \tag{4.42}$$

Let us introduce a new variable

$$x_1 = x - \int_0^t A(t_1)\, dt_1; \qquad t_2 = \int_0^t B(t_1)\, dt_1.$$

Then, the equation (4.42) will change into the form

$$\frac{\partial f}{\partial t_2} = \frac{1}{2} \frac{\partial^2 f}{\partial x_1^2}. \tag{4.43}$$

The equation (4.43) solution is similar to that of (4.41) (at the same initial conditions):

$$f(x, t|\, x_0, 0) = \frac{1}{\sqrt{2\pi a_1(t)}} \exp\left\{-\frac{[x - x_0 - b_1(t)]^2}{2a_1^2(t)}\right\},$$

where

$$a_1(t) = \int_0^t B(t_1)\,\mathrm{d}t_1; \qquad b_1(t) = \int_0^t A(t_1)\,\mathrm{d}t_1.$$

The obtained equations (4.19) and (4.30) enable us to investigate the variation in time of the distribution conditional density. For a full solution of these equations, however, we must generally have an explicit dependence of the coefficients a_1 and b_1 upon the variables x_0, t_0 for the first equation, and upon x, t for the second equation. As the conditional probability densities, defined by the equations (4.19), (4.30), describe (in a probabilistic sense) the state of any object, for example, that of a mechanical system, there should be a relation between Kolmogorov's equations and the equations of this systems motion. In order to establish this relationship, let us consider the equation of motion of the first order system

$$\dot{x} = F(x) + \varepsilon(t), \tag{4.44}$$

where $\varepsilon(t)$ is a random disturbance (the white noise) with known probability characteristics $m_\varepsilon = 0, k_\varepsilon = S_0\delta(\tau)$. The white noise variance and power are equal to infinity which contradicts the mechanical and physical notions of real processes, i.e. the delta-correlated random function is a rather crude approximation. Nevertheless, the white noise and its properties are widely used in the theory of random processes, and, in particular, in the theory of Markov processes. The use of the white noise in theoretical investigations has allowed us to obtain classical results in the nonlinear systems statistical dynamics.

Let us integrate the equation (4.44) between t and $t + \Delta t$, then

$$x(t+\Delta t) - x(t) = \Delta x = x - z = \int_t^{t+\Delta t} F(x)\,\mathrm{d}t_1 + \int_t^{t+\Delta t} \varepsilon\,\mathrm{d}t_1,$$

or, using the mean-value theorem, we obtain

$$\Delta x = x - z = F(x)\,\Delta t + \int_t^{t+\Delta t} \varepsilon \mathrm{d}t_1. \tag{4.45}$$

The conditional mathematical expectation of both sides of equality (4.45) is

$$M[x - z] = \int_{-\infty}^{\infty} (x - z)\, f(z, t + \Delta\, t \,|\, x, t)\,\mathrm{d}x,$$

or

$$M[x-z] = M[F(x)\Delta t] + M\left[\int_t^{t+\Delta t} \varepsilon(t_1)\,dt_1\right]. \tag{4.46}$$

As $F(x)$ is taken at the given x,

$$M[x-z] = F(x)\Delta t + \int_t^{t+\Delta t} m_\varepsilon(t_1)\,dt_1 = F(x)\Delta t. \tag{4.47}$$

Dividing both parts of the relation (4.47) by Δt and proceeding to the limit in (4.27), we obtain the coefficient a_1

$$a_1 = \lim_{\Delta t \to 0} \frac{1}{\Delta t} \frac{M[x-z]}{\Delta t} = F(x). \tag{4.48}$$

Let us determine the coefficient $b_1(x,t)$, considering the conditional mathematical expectation of the square of a difference $(x-z)$:

$$\begin{aligned} b_1 &= \lim_{\Delta t \to 0} \frac{M\left[(x-z)^2\right]}{\Delta t} \\ &= \lim_{\Delta t \to 0} \frac{1}{\Delta t} M\left[\left(F(x\Delta t) + \int_t^{t+\Delta t} \varepsilon(t_1)\,dt_1\right)^2\right] \\ &= \lim_{\Delta t \to 0} \left[F^2(x)\Delta t^2 + \int_t^{t+\Delta t}\int_t^{t+\Delta t} K_\varepsilon(t_1,t_2)\,dt_1\,dt_2\right]; \end{aligned}$$

since

$$K_\varepsilon(t_1,t_2) = K_\varepsilon(t_2-t_1) = S_0\delta(t_2-t_1),$$

we obtain

$$b_1 = \lim_{\Delta t \to 0} \frac{S_0}{\Delta t} \int_t^{t+\Delta t}\int_t^{t+\Delta t} \delta(t_2-t_1)\,dt_1\,dt_2 = S_0. \tag{4.49}$$

Example 4.2. A liquid damper is shown in Fig. 4.2. A random force $\varepsilon(t)$ is acting on the piston rod. Considering the rod and the piston inertialess, we obtain the equation of the pistons motion $\alpha\dot{x} = \varepsilon(t)$, where α is a coefficient of viscous friction; $\varepsilon(t)$ is a random stationary excitation (force) with known probability characteristics $m_\varepsilon = 0$, $K_\varepsilon = S_0\delta_\varepsilon(\tau)$.

It is required to determine the conditional probability density $f(x,t\,|\,x_0,0)$ of the rod displacement x at an arbitrary instant t that at $t=0$ must meet the condition

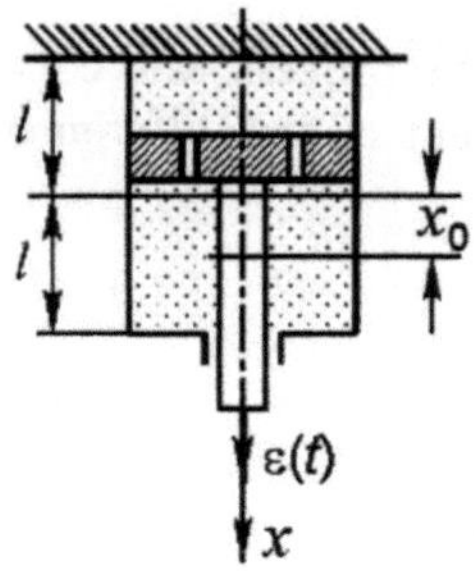

Fig. 4.2.

$$f(x,\ 0|\, x_0,\ 0) = \delta(x - x_0)\,, \tag{4.50}$$

and at an arbitrary t must satisfy the normalization condition

$$\int\limits_{-\infty}^{\infty} f(x,\ t|\, x_0,\ 0) = 1. \tag{4.51}$$

For this problem the condition (4.51) is approximate, since the displacements of the piston are limited ($-l \leq x \leq l$) and the coefficients a_1 and b_1 are equal to

$$\begin{aligned} a_1 &= 0, \\ b_1 &= \lim_{\Delta t \to 0} \frac{1}{\Delta t} \int\limits_{t}^{t+\Delta t} \int\limits_{t}^{t+\Delta t} \frac{1}{\alpha^2} K_\varepsilon (t_2 - t_1)\, \mathrm{d}t_1\, \mathrm{d}t_2 = \frac{S_0}{\alpha^2}. \end{aligned} \tag{4.52}$$

Therefore, the equation for determining the conditional probability density is of the form

$$\frac{\partial f}{\partial t} = \frac{S_0}{2\alpha^2} \frac{\partial^2 f}{\partial x^2}. \tag{4.53}$$

The equation (4.53) coincides with the equation (4.39).

Taking advantage of the expression (4.41), we obtain the following solution of the equation (4.53):

$$f(x,\, t\,|\, x_0, 0) = \frac{\alpha}{\sqrt{S_0}} \sqrt{\frac{1}{2\pi t}} \exp\left\{ -\frac{\alpha^2 (x - x_0)^2}{2 S_0 t} \right\}. \tag{4.54}$$

Let us consider the steady-state (stationary) condition, when the conditional probability density $f(x,\, t\,|\, x_0, t_0)$ and the coefficients of the equation (4.30) do not depend on time. This is tantamount to the assumption that at large time intervals, elapsed since the beginning of the process, the conditional probability density can be considered independent of the difference $t - t_0$.

In this case, we obtain from equation (4.30)

$$\frac{1}{2}\frac{\mathrm{d}^2 (b_1 f)}{\mathrm{d}x^2} - \frac{\mathrm{d}}{\mathrm{d}x}(a_1 f) = 0. \tag{4.55}$$

Let us integrate equation (4.55) over x

$$\frac{\mathrm{d}}{\mathrm{d}x}(b_1 f) - 2a_1 f = c. \tag{4.56}$$

It has already been shown (equation (4.37)) that $c = 0$.

It follows from (4.56) at $c = 0$:

$$\frac{\mathrm{d}f}{f} = \left(\frac{2a_1}{b_1}\right)\mathrm{d}x - \frac{\mathrm{d}b_1}{b_1}. \tag{4.57}$$

Integrating (4.57), we obtain

$$f(x) = \frac{c_1}{b_1(x)} \exp\left\{2\int \frac{a_1(x)}{b_1(x)}\mathrm{d}x\right\}. \tag{4.58}$$

We may determine this arbitrary constant c_1 from the condition

$$\int_{-\infty}^{\infty} f(x)\,dx = 1.$$

Example 4.3. Figure 4.3 shows a body of a mass m moving horizontally under the action of a force $R_0 + \Delta R$, where R_0 is the thrust force nominal value, and ΔR is a random component. The equation of the body motion has the form

$$m\dot{v} + F(v) = R_0 + \Delta R, \tag{4.59}$$

where v is a velocity of motion; $F(v)$ is a resistance force. Due to a scatter of thrust ΔR the body motion velocity v is equal to $v = v_0 + v_1$, where v_0

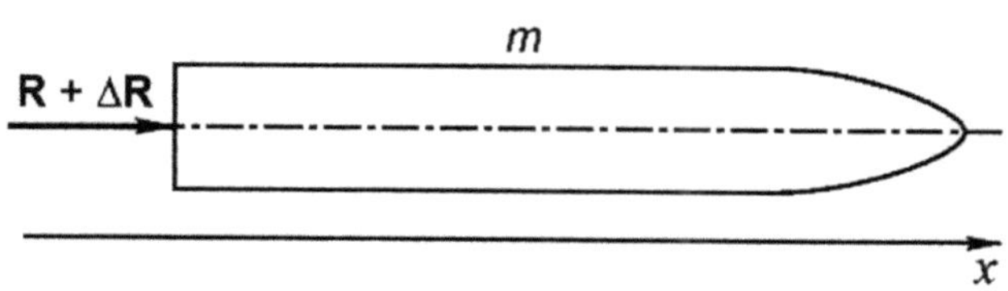

Fig. 4.3.

is the velocity of the body at $\Delta R = 0, v_1$ is a scatter of a velocity resulting from the scatter of thrust.

Let us assume, that the random function ΔR presents the process of the white noise type with a zero mathematical expectation and the correlation function $K_{\Delta R} = \delta(t - t_1)$. It is required to determine the steady-state probability distribution $f(v_1)$ and the variance of a velocity v_1.

Let us expand $F(v)$ in a series near $v = v_0$, considering that the resistance force ΔF is the odd function of v_1:

$$F(v) = F(v_0) + \left.\frac{\partial F}{\partial v}\right|_{v=v_0} v_1 + \frac{1}{6}\left.\frac{\partial^3 F}{\partial v^3}\right|_{v=v_0} v_1^3 + \dots . \tag{4.60}$$

As a result, we obtain the following equation

$$\dot{v}_1 + F_1(v_1) = \varepsilon(t), \qquad \left(\varepsilon(t) = \frac{\Delta R}{m}\right), \tag{4.61}$$

where

$$F(v_1) = \frac{1}{m}\left(\left.\frac{\partial F}{\partial v}\right|_{v=v_0} v_1 + \frac{1}{6}\frac{\partial^3 F}{\partial v^3}\Big|_{v=v_0} v_1^3\right) = \alpha_1 v_1 + \alpha_2 v_1^3.$$

In order to determine the coefficients a_1 and b_1 let us integrate the equation of motion (4.61) between $t, t + \Delta t$

$$\Delta v_1 = -F_1 \Delta t + \int\limits_t^{t+\Delta t} \varepsilon(t)\,\mathrm{d}t.$$

According to (4.48) and (4.49), the coefficients a_1 and b_1 are:

$$a_1 = \lim_{\Delta t \to 0} \frac{1}{\Delta t} M[\Delta v_1] = -F_1(v_1);$$
$$b_1 = \lim_{\Delta t \to 0} \frac{1}{\Delta t} M[\Delta v_1(t)\,\Delta v_1(t_1)] = \frac{S_0}{m^2}.$$

Kolmogorov's second equation for the above problem is of the form

$$\frac{\sigma^2}{2}\frac{\mathrm{d}^2 f}{\mathrm{d}v_1^2} + \frac{\mathrm{d}}{\mathrm{d}v_1}[F_1(v_1) f] = 0, \quad \left(\sigma^2 = \frac{S_0}{m^2}\right),$$

or

$$\frac{\sigma^2}{2}\frac{\mathrm{d}f}{\mathrm{d}v_1} + F_1(v_1) f = c. \tag{4.62}$$

We may consider that the probability of large values v_1 occurrence is small, therefore the probability density should meet the condition $f(v_1)|_{v_1 \to \infty} = 0$. If, in addition to this, we assume, that the derivative $f(v_1)$ at $|v_1| \to \infty$ is also zero, then it follows from the equation (4.62), that $c = 0$.

The obtained solution substantiates the adopted assumptions about the behavior of the probability density and its derivative at large values of v_1. If this solution meets the adopted assumptions, they may be consider true.

At $c = 0$, the solution of the equation (4.62) has the following form

$$f = c_1 \exp\left\{-\frac{2}{\sigma^2}\int_0^{v_1} F_1(v_1)\,dv_1\right\}. \tag{4.63}$$

Let us determine the arbitrary constant c_1 from the normalization condition

$$\int_{-\infty}^{\infty} f dv_1 = 1.$$

For example, in the case on hand at

$$F_1 = \alpha_1 v_1 + \alpha_2 v_1^3$$

we have

$$f = c_1 \exp\left\{-\frac{2}{\sigma^2}\left(\alpha_1\frac{v_1^2}{2} + \alpha_2\frac{v_1^4}{4}\right)\right\}. \tag{4.64}$$

It follows from (4.64) that f and $\frac{df}{dv_1}$ at $|v_1| \to \infty$ become zero. As the mathematical expectation $M[v_1]$ is zero, the variance will be

$$D_{v_1} = c_1 \int_{-\infty}^{\infty} v_1^2 \exp\left\{-\frac{2}{\sigma^2}\left(\frac{\alpha_1}{2}v_1^2 + \frac{\alpha_2}{4}v_1^4\right)\right\} dv_1. \tag{4.65}$$

At a linear dependence of a resistance force on v_1, i.e. at $F_1 = \alpha_1 v_1$, the probability density distribution follows the normal law.

Let us determine an approximate value of the integration constant c_1, assuming that

$$\exp\left\{-\frac{2\alpha_2}{4\sigma^2}v_1^4\right\} \approx 1 - \frac{\alpha_2}{2\sigma^2}v_1^4, \tag{4.66}$$

then

$$1 = c_1\left[\int_{-\infty}^{\infty} \exp\left\{-\frac{\alpha_1}{\sigma^2}v_1^2\right\} dv_1 - \frac{\alpha_2}{2\sigma^2}\int_{-\infty}^{\infty} v_1^4 \exp\left\{-\frac{\alpha_1}{\sigma^2}v_1^2\right\} dv_1\right]. \tag{4.67}$$

By integration we obtain

$$1 = c_1 \left[\frac{\sqrt{\pi}\sigma}{\sqrt{\alpha_1}} - \frac{\alpha_2}{2\sigma^2} \frac{3\sqrt{\pi}\sigma^5}{2\alpha_1^{5/2}} \right],$$

whence

$$c_1 = \frac{\sqrt{\alpha_1}}{\sigma\sqrt{\pi}\left(1 - \frac{3}{4}\frac{\alpha_2}{\alpha_1^2}\sigma^2\right)}. \tag{4.68}$$

The approximate value of the variance is

$$D_{v_1} = c_1 \int\limits_{-\infty}^{\infty} v_1^2 \exp\left\{-\frac{\alpha_1}{\sigma^2} v_1^2\right\} \left(1 - \frac{\alpha_2}{2\sigma^2} v_1^4\right) \mathrm{d}v_1 \tag{4.69}$$

integrating, we obtain

$$D_{v_1} = \frac{c_1}{2} \frac{\sqrt{\pi}\sigma^3}{\left(\sqrt{\alpha_1}\right)^3} \left(1 - \frac{15\alpha_2}{\alpha_1^2}\sigma^2\right). \tag{4.70}$$

By substituting the expression (4.68) in (4.70), we obtain

$$D_{v_1} = \sigma_{v_1}^2 = \frac{\sigma^2 \left(1 - \frac{15\alpha_2}{\alpha_1^2}\sigma^2\right)}{2\alpha_1 \left(1 - \frac{3\alpha_2}{4\alpha_1^2}\sigma^2\right)}. \tag{4.71}$$

4.3 Multidimensional Markov Processes

The basic concepts of a one-dimensional Markov process may be generalized for a case of multidimensional processes. A multidimensional process is considered a Markov process, if the distribution law of a system of random quantities $X_1, X_2, \ldots, X_n$ taken at an instant t and calculated given that the values $x_{10}, x_{20}, \ldots, x_{n0}$ of the random quantities $X_1, X_2, \ldots, X_n$ at an instant t_0 are known and do not depend on random functions $X_i(t)$ values at the preceding instants. Here, as in the case of one-dimensional Markov process, a two-dimensional conditional probability density

$$f(x_1, x_2, \ldots, x_n, t \,|\, x_{10}, x_{20}, \ldots, x_{n0}, t) = f(\bar{x}, t \,|\, \bar{x}_0, t_0),$$

is an exhaustive characteristic of the process, where $\bar{x}, \bar{x}_0$ are vectors with the components x_i and x_{i0}.

For multidimensional processes the Smolukhowski equation has the form

$$f\left(x_1, x_2, \ldots, x_n, t \middle| x_{10}, x_{20}, \ldots, x_{n0}, t_0\right) = \int\limits_{-\infty}^{\infty} \cdots \int\limits_{-\infty}^{\infty} f\left(x_1, \ldots, x_n, t \middle| z_1, \ldots, z_n, \tau\right) \times f\left(z_1, \ldots, z_n, \tau \middle| x_{10}, \ldots, x_{n0}, t_0\right) dz_1, \ldots, \mathrm{d}z_n .$$

Ignoring computations, we present Kolmogorov's second equation for multidimensional random processes [17, 40]

$$\frac{\partial f}{\partial t} + \sum_{i=1}^{n} \frac{\partial}{\partial x_i} \left[a_i f\right] - \frac{1}{2} \sum_{i=1}^{n} \sum_{j=1}^{n} \frac{\partial^2}{\partial x_i \partial x_j} \left[b_{ij} f\right] = 0. \tag{4.72}$$

In the specific case of two-dimensional random process, we have

$$\frac{\partial f}{\partial t} + \frac{\partial \left(a_1 f\right)}{\partial x_1} + \frac{\partial \left(a_2 f\right)}{\partial x_2} - \frac{1}{2} \frac{\partial^2 \left(b_{11} f\right)}{\partial x_1^2} - \frac{1}{2} \frac{\partial^2 \left(b_{22} f\right)}{\partial x_2^2} - \frac{1}{2} \frac{\partial^2 \left(b_{12} f\right)}{\partial x_1 \partial x_2} - \frac{1}{2} \frac{\partial^2 \left(b_{21} f\right)}{\partial x_1 \partial x_2} = 0. \tag{4.73}$$

The coefficients are

$$\begin{aligned} a_i &= \lim_{\Delta t \to 0} \frac{1}{\Delta t} M\left[x_i - z_i\right] = \lim_{\Delta t \to 0} \frac{M\left[\Delta x_i\right]}{\Delta t}; \\ b_{ij} = b_{ji} &= \lim_{\Delta t \to 0} \frac{1}{\Delta t} M\left[\left(x_i - z_i\right)\left(x_j - z_j\right)\right] = \lim_{\Delta t \to 0} \frac{M\left[\Delta x_i \Delta x_j\right]}{\Delta t}. \end{aligned} \tag{4.74}$$

Any solution of Kolmogorov's equations involves great difficulties (with the exception of elementary particular cases). These partial differential equations are classified among parabolic equations. Therefore, in order to obtain unambiguos solutions we must know the initial and boundary conditions, which the function f (the probability density distribution law) should meet.

Example 4.4. Let us consider a mechanical system, in cases where the force of inertia cannot be ignored (Fig. 4.4). The motion of a mass m is described by the following second-order equation:

$$\ddot{x} + 2n\dot{x} + p_0^2 x = \varepsilon\left(t\right). \tag{4.75}$$

For the complete description of this motion we must know not only the x value at $t = t_0$, but also the value of the first derivative $\dot{x}$, and, hence, the x value at $t < t_0$ (a process with aftereffect). Therefore, it is impossible to use a one-dimensional Markov process for the investigation of such system. However, if we present the equation (4.75) in the form of two equations of the first order (taking $\dot{x} = x_1, x = x_2$)

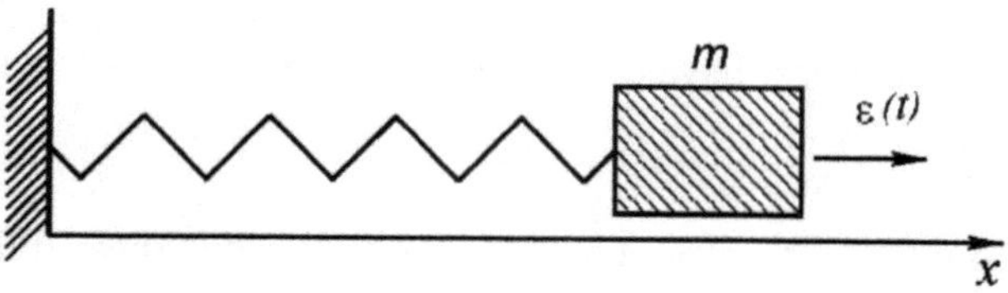

Fig. 4.4.

$$\dot{x}_1 = -2nx_1 - p_0^2 x_2 + \varepsilon(t);$$
$$\dot{x}_2 = x_1 \tag{4.76}$$

and consider x_1 and x_2 as the coordinates of a point on a phase plane, methods of Markov processes can be used for obtaining a solution to the system of equations (4.76). We may reduce a process with an aftereffect to a process without an aftereffect by increasing the dimensionality of a phase space. If we specify the state of a system that is described, for example, by a differential equation of the second order, by two coordinates x_1 and x_2 in two-dimensional phase space, we get a process without an aftereffect. In order to determine the joint probability density $f(x_1, x_2, t) = f(\dot{x}, x, t)$ for the equation (4.75) we must solve the equation (4.73) with due regard for appropriate initial and boundary conditions. In the general case we may always present the following set of mechanical system motion equations

$$\ddot{\mathbf{y}} = F_1(\mathbf{y}, \dot{\mathbf{y}}, t) + \boldsymbol{\varepsilon}_1(t) \tag{4.77}$$

as a system of G_1 equations of the first order

$$\dot{\mathbf{x}}_1 = F_1(\mathbf{x}_1, \mathbf{x}_2, t) + \boldsymbol{\varepsilon}_1;$$
$$\dot{\mathbf{x}}_2 = \mathbf{x}_1, \tag{4.78}$$

where $\dot{\mathbf{y}} = \mathbf{x}_1$; $\mathbf{y} = \mathbf{x}_2$.

Introducing a vector $\mathbf{x} = \begin{vmatrix} \mathbf{x}_1 \\ \mathbf{x}_2 \end{vmatrix}$, we obtain

$$\mathbf{x} = F(\mathbf{x}, t) + \boldsymbol{\varepsilon}, \qquad \left(\boldsymbol{\varepsilon} = (\varepsilon_1, 0)^{\mathrm{T}}\right), \tag{4.79}$$

or in a scalar form

$$\dot{x}_i = F_i(x_1, x_2, \ldots, x_n, t) + \varepsilon_i(t). \tag{4.80}$$

It should be emphasized, that the components $\varepsilon_j(t)$ of the vector $\boldsymbol{\varepsilon}_1(t)$ are stationary random functions of the normal white noise type ($m_{\varepsilon_j} = 0$, $K_{\varepsilon_j \varepsilon_i} = S_{0ji}\delta(\tau)$). Only in this case the set of equations (4.78) can be investigated with the use of Markov processes.

If the components of the vector $\boldsymbol{\varepsilon}$ are not a random white noise, for example, $\varepsilon_i(t)$ are independent and have correlation functions

$$K_{\varepsilon_i}(\tau) = \sigma_i^2 \mathrm{e}^{-\alpha_i|\tau|}, \tag{4.81}$$

the functions $\varepsilon_i(t)$ should be represented as processes following linear differential equations of the first order

$$\frac{\mathrm{d}\varepsilon_i}{\mathrm{d}t} + \alpha_i\varepsilon = a_i\varepsilon(t), \tag{4.82}$$

where $\varepsilon(t)$ is a normal white noise with the correlation function $K_\varepsilon(\tau) = S_0\delta(\tau)$, a_i are unknown coefficients. The spectral densities $S_{\varepsilon_i}(\omega)$ are determined from the equations (4.82)

$$S_{\varepsilon_i}(\omega) = \frac{a_i^2 S_0}{(\omega^2 + \alpha_i^2)}. \tag{4.83}$$

At known correlation functions $K_{\varepsilon_i}(t)$ (4.81) we may obtain the spectral densities $S_{\varepsilon_i}(\omega)$ from the Wiener–Khinchin relation (3.42)

$$S_{\varepsilon_i}(\omega) = \frac{\sigma_i^2\alpha_i}{\pi(\omega^2 + \alpha_i^2)}. \tag{4.84}$$

By equating the right-hand sides of expressions (4.83) and (4.84), we determine the unknown coefficients a_i

$$a_i^2 = \frac{\sigma_i^2\alpha_i}{\pi S_0}. \tag{4.85}$$

As a result, we obtain a set of simultaneous equations of the first order (4.78) and (4.82) involving a random function $\varepsilon(t)$ (a white noise) that can be investigated by methods of Markov processes.

In order to determine the coefficients of the equation (4.72) let us take advantage of the equations of motion (4.80). Let us integrate these equations between t and $t + \Delta t$:

$$\Delta x_i = \int\limits_t^{t+\Delta t} F_i(x_1, x_2, \ldots, x_n, t)\,\mathrm{d}t + \int\limits_t^{t+\Delta t} \varepsilon_i(t)\,\mathrm{d}t. \tag{4.86}$$

Applying the mean-value theorem, we obtain (similarly to (4.45))

$$\Delta x_i = F_j(x_1, x_2, \ldots, x_n, t)\,\Delta t + \int\limits_t^{t+\Delta t} \varepsilon_i(t)\,\mathrm{d}t. \tag{4.87}$$

Let us determine the coefficients a_i, b_{ij} of the equation (4.72) using (4.74) and (4.87):

$$a_i = \lim_{\Delta t \to 0} \frac{1}{\Delta t} \left[F_i \left(x_1, x_2, \ldots, x_n, t \right) \Delta t \right] = F_i;$$

$$b_{ij} = \lim_{\Delta t \to 0} \frac{1}{\Delta t} \left[F_i F_j \Delta t^2 + \int_t^{t+\Delta t} \int_t^{t+\Delta t} K_{\varepsilon_j \varepsilon_i} (t_2 - t_1) \mathrm{d}t_2 \, \mathrm{d}t_1 \right] = S_{0ji}.$$

For stationary conditions, when the conditional probability density f and the coefficients a_i, b_{ij} do not depend on time, we obtain from (4.72)

$$\sum_{i=1}^{n} \frac{\partial \left(a_i f \right)}{\partial x_i} - \frac{1}{2} \sum_{i=1}^{n} \sum_{j=1}^{n} \frac{\partial^2 \left(b_{ij} f \right)}{\partial x_i \, \partial x_j} = 0, \tag{4.88}$$

or for a two-dimensional process

$$\frac{\partial \left(a_1 f \right)}{\partial x_1} + \frac{\partial \left(a_2 f \right)}{\partial x_2} - \frac{1}{2} \frac{\partial^2 \left(b_{11} f \right)}{\partial x_1^2} - \frac{1}{2} \frac{\partial^2 \left(b_{22} f \right)}{\partial x_2^2} - \frac{1}{2} \frac{\partial^2 \left(b_{12} f \right)}{\partial x_1 \, \partial x_2} - \frac{1}{2} \frac{\partial^2 \left(b_{21} f \right)}{\partial x_1 \, \partial x_2} = 0. \tag{4.89}$$

4.4 Determination of the Probability of Attaining a Random Function Possible Values Area Boundaries

The theory of Markov processes allows us to investigate the problems pertaining to the analysis of transients in mechanical systems, to which we cannot obtain a solution by methods of the correlation theory. The problems of determining the probability of attaining a random function possible values area boundaries are problems that can be solved by the methods of Markov processes.

Let us consider a one-dimensional random process. It is required to determine the probability that the condition

$$x_1 \leq x \left(t \right) \leq x_2. \tag{4.90}$$

is satisfied for a random function $X(t)$ during a time interval $(t_0,\ t_0 + t_1)$.

Let us assume that the initial distribution

$$f \left(x,\, t_0 \right) = f_0 \left(x \right) \tag{4.91}$$

is given.

Then the required probability is

$$P \left(x_1 \leq X \leq x_2 \right) = \int_{x_1}^{x_2} f \left(x, t \right) \mathrm{d}x. \tag{4.92}$$

The average time of satisfying the condition (4.90) constitutes $t_{av} = t - t_0$. Let us confine ourselves to a case where the coefficients a_1 and b_1 do not depend on time. This occurs, for example, when a stationary random disturbance $\varepsilon(t)$ with a zero expectation and $K_\varepsilon = S_0\delta(\tau)$ is acting on the input of a system. We may assume that at the initial instant none of the process realizations had time to attain the boundary, therefore

$$P(t_0) = 1. \tag{4.93}$$

At a large value of time t any possible realizations of the process will attain the boundaries, i.e. at $t \to \infty$ $P(\infty) \to 0$. In the independent variables range $(x_1 \le x \le x_2)$; $(t_0 \le t \le \infty)$ the conditional probability density f varies according to the equation (4.30):

$$\frac{\partial f}{\partial t} = -\frac{\partial(a_1 f)}{\partial x} + \frac{1}{2}\frac{\partial^2(b_1 f)}{\partial x^2}. \tag{4.94}$$

One of the singularities of the given problem consists in a fact that the range of a random function x ordinates variation is limited, and we must take it into account when formulating the boundary conditions.

The probability that the trajectory has crossed the segment x, $x + \Delta x$, never reaching the boundary for the time $t - t_0$, is equal to

$$\Delta P = f(x, t)\,\Delta x. \tag{4.95}$$

Such trajectories are practically absent on the boundary $x = x_{1,2}$ at $t > t_0$, therefore $f(x, t)$ must satisfy the boundary conditions

$$f(x_1, t) = f(x_2, t) = 0. \tag{4.96}$$

The boundary conditions (4.96) and the initial condition (4.93) define the function f in a uniquely fashion. Let us suppose, that the solution of the equation (4.94) satisfying the boundary and initial conditions, has been obtained. Then, the function $P(t)$ is known and we may find the probability P_1 that $x(t)$ will reach the boundary within the time period $t - t_0$:

$$P_1 = P(t_0) - P(t) = 1 - P(t). \tag{4.97}$$

By differentiating the expression (4.97) with respect to t, we determine the probability density

$$f_1(t) = -\frac{\mathrm{d}P(t)}{\mathrm{d}t}. \tag{4.98}$$

Setting $t = t_0 + t_1$, where t_1 is a random variable, we obtain

$$f(t_1) = -\frac{\mathrm{d}P(t_0 + t_1)}{\mathrm{d}t_1}. \tag{4.99}$$

The relation (4.99) allows us to determine an average time (m_{t_1}) of the systems motion before reaching the boundary ($t_1 > 0$):

$$m_{t_1} = \int_0^\infty t_1 f_1(t_1)\,\mathrm{d}t_1 = -\int_0^\infty t_1\,\mathrm{d}P(t_0 + t_1). \tag{4.100}$$

Integrating the right-hand side of the expression (4.100) by parts, we obtain

$$m_{t_1} = -t_1 P\Big|_{t_1=0}^{\infty} + \int_0^\infty P(t)\mathrm{d}t$$

or (as $P(\infty) = 0$)

$$m_{t_1} = \int_0^\infty P(t)\,\mathrm{d}t. \tag{4.101}$$

The solution of the equation (4.94) at the given boundary and initial conditions enables us to determine the probability that the random function $X(t)$ will reach the boundaries by the fixed instant T. We determine this probability from the relation (4.95)

$$P(T) = \int_{x_1}^{x_2} f(x,T)\,\mathrm{d}x. \tag{4.102}$$

Example 4.5. An equation of motion of an inertialess piston (see Fig. 4.2) acted by a constant force a and a random force $\varepsilon(t)$, is of the form

$$\alpha\dot{x} = a + \varepsilon(t), \tag{4.103}$$

where $\alpha\dot{x}$ is a resistance force. The random force $\varepsilon(t)$ is a random process of the white noise type with a zero expectation and the correlation function $K_\varepsilon = S_0\delta(\tau)$. It is required to determine the probability that the random function $x(t)$ (the displacement of the piston) will not go beyond the boundaries of the interval $(-x_0, x_0)$ within the time period , if at $t = 0$ $x(0) = 0$.

In this case, the coefficients are $a_1 = a/\alpha$; $b_1 = S_0/\alpha^2$. The equation (4.94) takes the form

$$\frac{\partial f}{\partial t} + \frac{a}{\alpha}\frac{\partial f}{\partial x} - \frac{S_0}{2\alpha^2}\frac{\partial^2 f}{\partial x^2} = 0. \tag{4.104}$$

The desired probability density $f(x, t)$ should satisfy the initial condition

$$f(x,0) = \delta(x). \tag{4.105}$$

Let us use the Fourier method to solve the equations (4.104):

$$f = X(x)\,T(t). \tag{4.106}$$

As a result, we obtain the following two equations

$$\frac{\mathrm{d}T}{\mathrm{d}t} + \lambda^2 T = 0; \qquad \frac{\mathrm{d}^2X}{\mathrm{d}x^2} - \frac{2a\alpha}{S_0}\frac{\mathrm{d}X}{\mathrm{d}x} + \frac{2\alpha^2\lambda^2}{S_0}X = 0,$$

whose solutions are of the form

$$\begin{aligned} &T = c_1 \mathrm{e}^{-\lambda^2 t};\\ &X = \exp\left\{\frac{a\alpha}{S_0}x\right\}(c_1\cos\lambda_1 x + c_2\sin\lambda_1 x), \end{aligned} \tag{4.107}$$

where

$$\lambda_1 = \sqrt{\frac{2\alpha^2\lambda^2}{S_0} - \frac{a^2\alpha^2}{S_0^2}}. \tag{4.108}$$

The probability density $f(x, t)$ should satisfy the boundary conditions (4.96), therefore we obtain two equations to determine c_1 and c_2:

$$\begin{aligned} &X(x_0) = c_1\cos\lambda_1 x_0 + c_2\sin\lambda_1 x_0 = 0;\\ &X(-x_0) = c_1\cos\lambda_1 x_0 - c_2\sin\lambda_1 x_0 = 0. \end{aligned} \tag{4.109}$$

The system (4.109) has a nontrivial solution ($c_1 \neq 0$) at $c_2 = 0$ and $\cos\lambda_1 x_0 = 0$, from which it follows:

$$\lambda_{1k} = \frac{(2k+1)}{2}\frac{\pi}{x_0}. \tag{4.110}$$

We determine from (4.108)

$$\lambda_k^2 = \frac{S_0}{2\alpha^2}\left(\frac{(2k+1)^2}{4}\frac{\pi^2}{x_0^2} + \frac{a^2\alpha^2}{\alpha^4}\right).$$

Finally, we obtain the following expression for the solution of the equation (4.104):

$$f(x, t) = \exp\left\{\frac{a\alpha}{S_0}x\right\}\sum_{k=0}^{\infty}\left(c_k\cos\frac{(2k+1)}{2}\frac{\pi x}{x_0}\mathrm{e}^{-\lambda_k^2 t}\right). \tag{4.111}$$

At an initial instant ($t = t_0 = 0$) the obtained expression (4.111) must satisfy the initial condition (4.105):

$$\exp\left\{\frac{\alpha a}{S_0}x\right\}\sum_{k=0}^{\infty}\left(c_k\cos\frac{(2k+1)}{2}\frac{\pi x}{x_0}\right) = \delta(x),$$

from which we obtain

$$c_R \int\limits_{-x_0}^{x_0} \cos^2 \frac{(2k+1)\,\pi x}{2x_0} \mathrm{d}x = \int\limits_{-x_0}^{x_0} \exp\left\{-\frac{a\alpha}{S_0}x\right\} \cos \frac{(2k+1)\,\pi x}{2x_0} \delta\,(x)\,\mathrm{d}x. \tag{4.112}$$

It follows from (4.112) that $c_k = 1/x_0$.
The expression (4.111) takes the form

$$f\,(x,t) = \frac{1}{x_0} \exp\left\{\frac{\alpha a}{S_0}x\right\} \left(\sum_{k=1}^{\infty} \cos \frac{(2k+1)\,\pi x}{2x_0} \mathrm{e}^{-\lambda_k^2 t}\right). \tag{4.113}$$

The required probability is equal to

$$P\,(t) = \int\limits_{-x_0}^{x_0} f\,(x,\,t)\,\mathrm{d}x = \frac{1}{x_0} \int\limits_{-x_0}^{x_0} \exp\left\{\frac{\alpha a}{S_0}x\right\} \sum_{k=0}^{\infty} \left(\mathrm{e}^{-\lambda_k^2 t} \cos \frac{(2k+1)\,\pi x}{2x_0}\right) \mathrm{d}x. \tag{4.114}$$

By integration we obtain a relationship that is true for any fixed instant $t = T$:

$$P\,(t) = \frac{2\mathrm{ch}\dfrac{a\alpha}{S_0}x_0}{x_0} \sum_{k=0}^{\infty} \frac{\mathrm{e}^{-\lambda_k^2 T}\,(-1)^{2+k}\,b_k}{\left(\dfrac{\alpha^2 a^2}{S_0^2} + b_k^2\right)}, \tag{4.115}$$

where $b_k = \dfrac{(2k+1)}{2}\dfrac{\pi}{x_0}$.

As one would expect, the value obtained for the probability $P\,(T)$ does not depend on the sign of the force a. The example considered assumes the generalized form, when the area boundaries of the possible values of the random function $x\,(t)$ at an initial instant are random (x_0 are random quantities with known distribution law), e.g.

$$f\,(x_0) = \frac{\pi}{4x_0} \cos \frac{\pi x}{2x_0}. \tag{4.116}$$

In this case, the solution (4.111) at $t = 0$ must satisfy the condition

$$\exp\left\{\frac{a\alpha}{S_0}x\right\} \left(\sum_{k=0}^{\infty} c_k \cos \frac{(2k+1)\,\pi x}{2x_0}\right) = \frac{\pi}{4x_0} \cos \frac{\pi x}{2x_0}\pi,$$

which allows us to determine c_k

$$c_k = \frac{\pi}{8} \int_{-x_0}^{x} \exp\left\{-\frac{a\alpha}{S_0}x\right\} \left(\cos\frac{k\pi x}{x_0} + \cos\frac{(k+1)\,\pi x}{x_0}\right) dx. \tag{4.117}$$

Having determined the coefficients c_k, we find $f(x,t)$, and the probability $P(t)$, as in the preceding case. The obtained expression (4.115) enables us to determine the average time of the motion of the piston without contacting the end faces of the cylinder through the formula (4.100):

$$m_{t_1} = \frac{2\mathrm{ch}\frac{a\alpha}{S_0}x_0}{x_0} \sum_{k=0}^{\infty} \frac{(-1)^{2+k}\, b_k}{\lambda_k^2\left(\frac{\alpha^2 a^2}{S_0^2} + b_k^2\right)}. \tag{4.118}$$

Let us transform the expression (4.118), introducing a non-dimensional parameter $\gamma_1 = \frac{a\alpha}{\sigma^2}x_0$ and a characteristic time t_k. The time t_k, during which the piston covers a distance equal to x_0 under the action of the force a, is taken for the characteristic time. Since the nonrandom motion of the inertialess piston is described by the equation

$$\alpha\dot{x} = a,$$

the time t_k is equal to

$$t_k = \frac{\alpha}{a}x_0.$$

Dividing the expression (4.118) by t_k, we obtain the non-dimensional average time of the pistons motion without contacting the ends

$$\tilde{m}_{t_1} = \frac{m_{t_1}}{t_k} = 2\mathrm{ch}\gamma_1 \cdot \pi\gamma_1 \sum_{k=0}^{\infty} \frac{(-1)^{2+k}\,(2k+1)}{\left(\gamma_1^2 + b_{k_0}^2\right)^2}, \tag{4.119}$$

where

$$b_{k_0} = \frac{(2k+1)}{2}.$$

It has been assumed in the example 4.5 that the pistons force of inertia is less than the resistance force $\alpha\dot{x}$ and for this reason can be ignored.

Let us consider a mechanical system, whose motion is described by a differential equation of the first order similar to that of (4.44). Figure 4.5 shows a rotating shaft with a disk. A known moment M_0 and a random moment $\varepsilon(t)$ are acting on the shaft. Neglecting any moment of resistance, we have the following equation of the disks rotation

$$J\dot{\omega} = M_0 + \varepsilon(t). \tag{4.120}$$

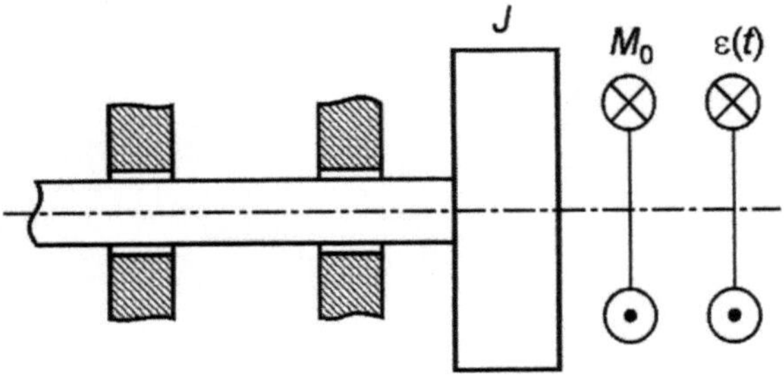

Fig. 4.5.

At an initial instant the angular velocity is zero. It is required to determine an average time m_{t_1}, at which the angular velocity will reach the value $|\omega_0|$. To determine the average time m_{t_1} we obtain an expression similar to that of (4.120), where we should replace x_0 with ω_0 and α with J.

5. Random Vibrations of Systems with One Degree of Freedom

The theory of random vibrations of mechanical systems finds an increasing application in the designing practice of almost all industries. The problems tackled in this field include analysis of object protection systems under the action of random disturbances, analysis of vibrations of airborne vehicle structures elements caused, for example, by the action of atmospheric turbulence, launch of airborne vehicles, the vehicles motion on roads with random irregularities, etc. The theory of random vibrations allows us to solve problems, requiring an estimate of the reliability and lifetime of structures. The theory of random vibrations plays an important role in vibroacoustical diagnostics.

5.1 Free Random Vibrations of Linear Systems

Let us consider the motion of a deterministic system (a system which does not contain random parameters), caused by random initial deflections from the equilibrium position.

In real conditions it is impossible to implement the motion of a mechanical system with absolutely exact values of initial conditions because of an inevitable scatter of initial data. Therefore, the real motion always differs from the designed one, which lead to the necessity of evaluating possible divergences between the two. The simplest problem here is that of determining the probabilistic characteristics of the motion the generalized coordinates and their first derivatives – at free vibrations caused by random deviations of initial data, and to solve it, it is sufficient to know the linear transformations of random functions set forth in Sect. 2.4.

Let us consider the free vibrations of a linear system with one degree of freedom (Fig. 5.1) with due regard for the resistance force proportional to the velocity. They are described by the equation

$$\ddot{y} + 2n\dot{y} + p_0^2 y = 0, \tag{5.1}$$

where $2n = \alpha/m$; $p_0^2 = c/m$.

The solution of the equation (5.1) takes the form

$$y = \mathrm{e}^{-nt}(c_1 \cos pt + c_2 \sin pt). \tag{5.2}$$

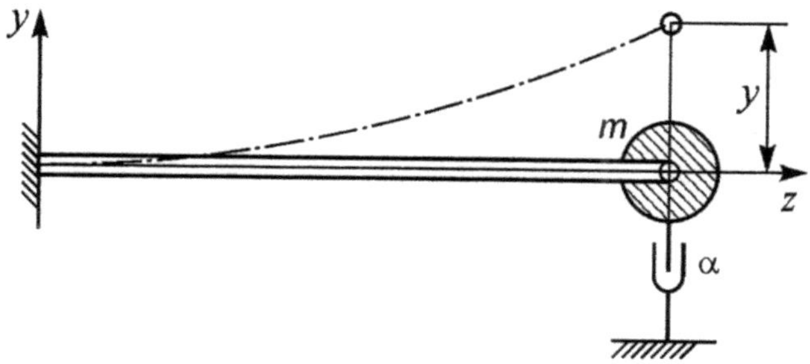

Fig. 5.1.

At $t = 0$ $y = y_0$ and $\dot{y} = \dot{y}_0$ (the most general case). Having determined arbitrary constants c_1 and c_2 we obtain

$$y = \mathrm{e}^{-nt}\left[y_0(\cos pt + \frac{n}{p}\sin pt) + \frac{\dot{y}_0}{p}\sin pt\right]. \tag{5.3}$$

The initial data y_0 and $\dot{y}_0$ are random quantities with known probability characteristics (we know their expectations m_{y_0}, $m_{\dot{y}_0}$, variances D_{y_0}, $D_{\dot{y}_0}$ and correlation moment $K_{y_0\dot{y}_0}$). It is required to determine the probability characteristics of y and of the first two derivatives $\dot{y}$ and $\ddot{y}$.

By differentiating (5.3) we find the following expression for the mass m velocity and acceleration:

$$\dot{y} = y_0\dot{f}_1 + \dot{y}_0\dot{f}_2; \qquad \ddot{y} = y_0\ddot{f}_1 + \dot{y}_0\ddot{f}_2, \tag{5.4}$$

where

$$f_1 = \mathrm{e}^{-nt}(\cos pt + \frac{n}{p}\sin pt), \qquad f_2 = \frac{1}{p}\mathrm{e}^{-nt}\sin pt. \tag{5.5}$$

In order to solve the formulated problem we must find the probability characteristics of the solution, i.e. $m_y(t)$, $D_y(t)$ and the autocorrelation function $K_y(t,\, t_1)$. Using the formula for the expectation of a sum of random functions, we obtain

$$m_y(t) = M\left[Y\right] = m_{y_0}\mathrm{e}^{-nt}(\cos pt + \frac{n}{p}\sin pt) + \frac{1}{p}m_{\dot{y}_0}\mathrm{e}^{-nt}\sin pt. \tag{5.6}$$

The autocorrelation function of the solution is

$$\begin{aligned} K_y &= M\left[\mathring{Y}(t)\,\mathring{Y}(t_1)\right] \\ &= D_{y_0}\mathrm{e}^{-n(t-t_1)}\left(\cos pt + \frac{n}{p}\sin pt\right) \times \left(\cos pt_1 + \frac{n}{p}\sin pt_1\right) \\ &+ \frac{1}{p}K_{y_0\dot{y}_0}\left[\mathrm{e}^{-n(t+t_1)}\left(\cos pt + \frac{n}{p}\sin pt\right)\sin pt_1\right. \end{aligned}$$

$$+\mathrm{e}^{-n(t_1+t)}\left(\cos pt_1+\frac{n}{p}\sin pt_1\right)\sin pt_1\Bigg]$$

$$+\frac{D_{\dot{y}_0}}{p^2}\sin pt\sin pt_1\mathrm{e}^{-n(t_1+t)}. \tag{5.7}$$

From the expression (5.7) at $t=t_1$ we obtain the deflection variance

$$D_y(t)=D_{y_0}f_1^2+2K_{y_0\dot{y}_0}f_1f_2+D_{\dot{y}_0}f_2^2. \tag{5.8}$$

For non-correlating initial data ($K_{y_0\dot{y}_0}=0$) we have

$$D_y(t)=D_{y_0}f_1^2+D_{\dot{y}_0}f_2^2. \tag{5.9}$$

The expressions for the mathematical expectations and autocorrelation functions of the first two solution derivatives take a similar form:

$$\begin{aligned} m_{\dot{y}}&=m_{y_0}\dot{f}_1+m_{\dot{y}_0}\dot{f}_2;\\ m_{\ddot{y}}&=m_{y_0}\ddot{f}_1+m_{\dot{y}_0}\ddot{f}_2; \end{aligned} \tag{5.10}$$

$$\begin{aligned} K_{\dot{y}}(t,t_1)&=D_{y_0}\dot{f}_1(t)\dot{f}_1(t_1)\\ &\quad+K_{y_0\dot{y}_0}\left[\dot{f}_1(t)\dot{f}_2(t_1)+\dot{f}_1(t_1)\dot{f}_2(t)\right]+D_{\dot{y}_0}\dot{f}_2(t)\dot{f}_2(t_1);\\ K_{\ddot{y}}(t,t_1)&=D_{y_0}\ddot{f}_1(t)\ddot{f}_1(t_1)\\ &\quad+K_{y_0\dot{y}_0}\left[\ddot{f}_1(t)\ddot{f}_2(t_1)+\ddot{f}_1(t_1)\ddot{f}_2(t)\right]+D_{\dot{y}_0}\ddot{f}_2(t)\ddot{f}_2(t_1). \end{aligned} \tag{5.11}$$

When investigating the random vibrations of systems with one degree of freedom, we must have two nonrandom functions that characterize the random vibrations (both free and forced), namely $m_y(t)$ and $K_y(t,t_1)$, which differs from the classical theory of vibrations. The later, therefore, may be considered as the particular case of the theory of random vibrations, where the conditions

$$D_{y_0}=D_{\dot{y}_0}=K_{y_0\dot{y}_0}=0.$$

are fulfilled.

What practical result can be obtained from the presented probabilistic characteristics of the solution (5.6–5.9)?

Let us consider the design scheme (see Fig. 5.1). A device can operate normally provided that the maximum normal stresses occurring in the built-in end at random vibrations do not exceed the yield point of the elastic element material, i.e.

$$\max(\sigma)\leq\frac{\sigma_y}{n_y}, \tag{5.12}$$

where σ_y is the yield point; n_y is the factor of safety with respect to yielding. The maximum normal stress $\sigma_{\max}$ (in clamped section) is related to the deflection y by relationship

$$\sigma_{\max} = \frac{3EJ_x}{l^2 W_x} y, \tag{5.13}$$

where is the Young modulus of a material; J_x, W_x are respectively the second moment of inertia of the section and the section modulus of the rod. The expectation of the stress $\sigma_{\max}$ and the variance will be respectively:

$$\begin{aligned} m_{\sigma_{\max}} &= am_y(t); \\ D_{\sigma_{\max}} &= a^2 D_y(t), \quad (a = 3EJ_x/l^2 w_x). \end{aligned} \tag{5.14}$$

Considering that there is a normal distribution for $\sigma_{\max}(t)$ at each instant, let us use the three sigma rule to determine the maximum stress $\max(\sigma_{\max})$:

$$\max(\sigma_{\max}) = am_y(t) + 3a\sigma_y(t), \qquad (\sigma_y = \sqrt{D_y}). \tag{5.15}$$

As the maximum stress depends on time, we should determine such value t_* at which $\max(\sigma_{\max})$ takes a maximum value in time. Having determined t_* we obtain the following final condition, the fulfillment of which ensures the normal operation of the elastic element:

$$\max(\sigma_m) = am_y(t_*) + 3a\sigma_y(t_*) < \frac{\sigma_y}{n_y}. \tag{5.16}$$

The motion of a system after the termination of the short-term random impulse action (whose time of action is far less than the period of natural vibrations) can be classed as free vibrations.

Example 5.1. Figure 5.2 shows a single degree-of-freedom mechanical system consisting of an absolutely rigid body, part of which is subjected to the action of a shock wave. In this case, the impulse of a moment J_M is acting on a rod, imparting it some initial angular velocity (at zero initial angular displacement), i.e. at $t = 0$, $\varphi_0 = 0$, $\dot{\varphi} = \dot{\varphi}_0$. The impulse of a moment is related to the initial velocity by the relationship

$$J_M = J_0 \dot{\varphi}_0, \tag{5.17}$$

where J_0 is the moment of the bodys inertia with respect to an axis perpendicular to the drawing and going through the point 0.

It is required to determine the parameters of the shock-absorption system (c and α) from the conditions: 1) the maximum angular displacement of the body under the action of the impulse of a moment should not exceed the admissible φ_D; 2) in a given time t_k the amplitude of the angular vibrations of the body should decrease by k times. If the probability characteristics of the random impulse J_M are known (m_J and D_J), then also known are the

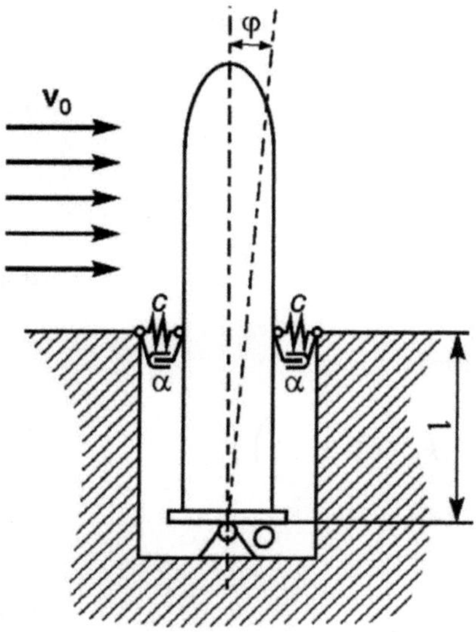

Fig. 5.2.

probability characteristics of the initial angular velocity $\dot{\varphi}_0$ $\left(m_{\dot{\varphi}_0} = \frac{m_J}{J_0},\right.$ $\left.D_{\dot{\varphi}_0} = \frac{D_J}{J_0^2}\right)$.

The body motion equation takes the form

$$\ddot{\varphi} + \frac{\alpha l^2}{J_0}\dot{\varphi} + \frac{cl^2}{J_0}\varphi = 0. \tag{5.18}$$

As $\varphi_0 = 0$, the probability characteristics of the solution are:

$$\begin{aligned} m_\varphi &= \frac{m_{\varphi_0}}{p}\mathrm{e}^{-nt}\sin pt;\\ K_\varphi &= \frac{D_{\varphi_0}}{p^2}\mathrm{e}^{-n(t-t_1)}\sin pt \sin pt_1;\\ \sigma_\varphi &= \frac{\sigma_{\varphi_0}}{p}\mathrm{e}^{-nt}\sin pt; \qquad \left(p = \sqrt{p_0^2 - n^2};\ p_0^2 = \frac{cl^2}{J_0};\ n = \frac{\alpha l^2}{2J_0}\right). \end{aligned} \tag{5.19}$$

The maximum value of the angle φ for any instant (with the use of three sigma rule) is

$$\varphi_{\max} = m_\varphi + 3\sigma_\varphi = \frac{(m_J + 3\sigma_J)}{J_0 p}\mathrm{e}^{-nt}\sin pt. \tag{5.20}$$

The function $\varphi_{\max}$ attains the first maximum value at an instant $t_1 = \frac{\pi}{2}\frac{1}{p}$, therefore, neglecting the influence of friction forces ($\mathrm{e}^{-nt} \approx 1,\quad p \approx p_0$) at a time interval $(0,\ t_1)$ and using the first condition, we obtain

$$\varphi_{\max} = \varphi_D = \frac{m_J + 3\sigma_J}{J_0 p_0}. \tag{5.21}$$

From (5.21) we determine

$$c = \frac{(m_J + 3\sigma_J)^2}{J_0 \varphi_D^2 l^2}. \tag{5.22}$$

The full time (see the second condition)

$$t_k = t_1 + t_{k_1}, \quad \left(t_{k_1} = \frac{(2k_1 - 1)}{2p}\pi, \quad (k_1 = 1, 2, \ldots)\right).$$

From expression (5.20)

$$\varphi_{mfx} = \varphi_D k = \frac{(m_J + 3\sigma_J)}{J_0 p} e^{-nt_{k_1}} \sin pt_{k_1}. \tag{5.23}$$

By manipulations we obtain from (5.23)

$$k\sqrt{1 - n_1^2} = \exp\left\{-\frac{(2k_1 - 1)\pi n_1}{2\sqrt{1 - n_1^2}}\right\}, \qquad \left(n_1 = \frac{n}{p_0}\right). \tag{5.24}$$

Depending on specific values of k and k_1, we find n_{1*} – the root of the equation (5.24). Knowing n_{1*}, we then determine the optimum coefficient of the viscous friction force in the shock-absorption system

$$\alpha_* = \frac{2p_0 J_0 n_{1*}}{l^2}. \tag{5.25}$$

5.2 Forced Random Vibrations of Linear Systems

5.2.1 Non-Stationary Vibrations

An equation of forced vibrations of a linear one degree-of-freedom system with constant parameters takes the form

$$\ddot{y} + 2n\dot{y} + p_0^2 y = \frac{1}{m} f(t), \tag{5.26}$$

where $f(t)$ is a random force.

In order to determine the probability characteristics of the solution of the equation (5.26) we must have the probability characteristics of the input and the probability characteristics of its initial data, i.e. m_f, K_f, m_{y_0}, $m_{\dot{y}_0}$, $K_{y_0\dot{y}_0}$,

D_{y_0}, $D_{\dot{y}_0}$, K_{fy_0}, $K_{f\dot{y}_0}$. Let us confine ourselves to the case where the initial data and the random force f are independent, i.e. $K_{y_0\dot{y}_0} = K_{fy_0} = K_{f\dot{y}_0} = 0$.

We may present the solution of the equation (5.26) for an arbitrary right-hand side as

$$y = \mathrm{e}^{-nt}\left[y_0\left(\cos pt + \frac{n}{p}\sin pt\right) + \frac{\dot{y}_0}{p}\sin pt\right] + \frac{1}{pm}\int_0^t \mathrm{e}^{-n(t-\tau)}\sin p(t-\tau)f(\tau)d\tau. \tag{5.27}$$

The expectation and autocorrelation function of the solution are

$$m_y = \mathrm{e}^{-nt}\left[m_{y_0}\left(\cos pt + \frac{n}{p}\sin pt\right) + \frac{m_{\dot{y}_0}}{p}\sin pt\right] + \frac{1}{mp}\int_0^t \mathrm{e}^{-n(t-\tau)}\sin p(t-\tau)m_f d\tau. \tag{5.28}$$

$$K_y(t,t_1) = \mathrm{e}^{-n(t+t_1)}\left[D_{y_0}\left(\cos pt + \frac{n}{p}\sin pt\right)\cdot\left(\cos pt_1 + \frac{n}{p}\sin pt_1\right) + \frac{1}{p^2}D_{\dot{y}_0}\sin pt\sin pt_1\right] + \frac{1}{m^2p^2}\int_0^t\int_0^{t_1} \mathrm{e}^{-n(t-\tau)}\sin p(t-\tau)\mathrm{e}^{-n(t_1-\tau_1)}\sin p(t_1-\tau_1)K_f(\tau,\tau_1)\mathrm{d}\tau_1\,\mathrm{d}\tau. \tag{5.29}$$

The variance of the solution is

$$D_y(t) = \mathrm{e}^{-2nt}\left[D_{y_0}\left(\cos pt + \frac{n}{p}\sin pt\right)^2 + \frac{1}{p^2}D_{\dot{y}_0}\sin^2 pt\right] + \frac{1}{m^2p^2}\int_0^t\int_0^{t_1} \mathrm{e}^{-n(t-\tau)}\mathrm{e}^{-n(t_1-\tau)}\sin p(t-\tau)\sin p(t_1-\tau_1)K_f\mathrm{d}\tau_1\,\mathrm{d}\tau. \tag{5.30}$$

In the particular case of $y_0 = \dot{y}_0 = 0$,

$$m_y = \frac{1}{mp}\int_0^t \mathrm{e}^{-n(t-\tau)}\sin p(t-\tau)m_f(\tau)\mathrm{d}\tau; \tag{5.31}$$

$$K_y(t, t_1) = \frac{1}{m^2p^2} \int_0^t \int_0^{t_1} \mathrm{e}^{-n(t-\tau)} \mathrm{e}^{-n(t_1-\tau_1)} \times \sin(t-\tau) \sin p(t_1-\tau_1) K_f(\tau, \tau_1) \mathrm{d}\tau \, \mathrm{d}\tau_1. \tag{5.32}$$

Let us consider the particular case of a random constant force with a hereafter retained constant value being applied to a system, i.e. $f = aH(t)$, where a is a random quantity (a force) that has known probability characteristics m_a and D_a; and $H(t)$ is the Heaviside function.

In this case, the probability characteristics of the solution are:

$$m_y = \frac{m_a}{mp} \int_0^t \mathrm{e}^{-n(t-\tau)} \sin p(t-\tau) d\tau; \tag{5.33}$$

$$K_y = \frac{D_a}{m^2p^2} \int_0^t \int_0^{t_1} \mathrm{e}^{-n(t-\tau)} \mathrm{e}^{-n(t_1-\tau_1)} \sin p(t-\tau) \times \sin p(t_1-\tau_1) \mathrm{d}\tau \, \mathrm{d}\tau_1. \tag{5.34}$$

By integrating we obtain

$$m_y = \frac{m_a}{mp} \frac{1}{(p^2+n^2)} \left[p(1-\cos pt \, \mathrm{e}^{-nt}) - n \sin pt \, \mathrm{e}^{-nt}\right]; \tag{5.35}$$

$$K_y = \frac{D_a}{m^2p^2} \frac{1}{(p^2+n^2)^2} \left[-n \sin pt \, \mathrm{e}^{-nt} + p(1-\cos pt \, \mathrm{e}^{-nt})\right] \times \left[-n \sin pt_1 \, \mathrm{e}^{-nt_1} + p(1-\cos pt_1 \, \mathrm{e}^{-nt_1})\right]. \tag{5.36}$$

The variance of the solution is

$$D_y = \frac{D_a}{m^2p^2(p^2+n^2)^2} \left[-n \sin pt \, \mathrm{e}^{-nt} + p(1-\cos pt \, \mathrm{e}^{-nt})\right]^2. \tag{5.37}$$

If an equation of small vibrations contains coefficients that vary in time, it is generally impossible to obtain a solution in the analytical form. For example, the approximate equation of the small vibrations of a rocket engine (Fig. 5.3), caused by a random scatter of the thrust ΔR due to non-uniform combustion of a charge, takes the form

$$m(t)\Delta\ddot{x} + \alpha\Delta\dot{x} + c\Delta x = -\Delta R \tag{5.38}$$

or

$$\Delta\ddot{x} + a_1(t)\Delta\dot{x} + a_2(t)\Delta x = f(t), \tag{5.39}$$

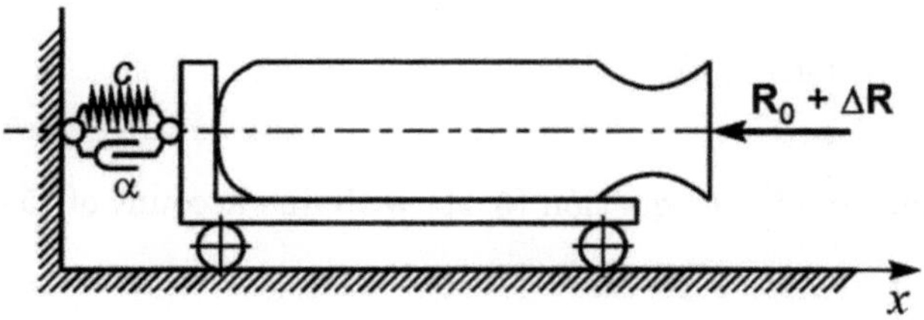

Fig. 5.3.

where $a_1(t) = \dfrac{\alpha}{m}$; $a_2(t) = \dfrac{c}{m}$; $f(t) = -\dfrac{\Delta R}{m}$.

Let us present the equation (5.39) as a system of two equations of the first order, setting $\Delta\dot{x} = y_1$; $\Delta x = y_2$:

$$\dot{y}_1 + a_1 y_1 + a_2 y_2 = j; \quad \dot{y}_2 - y_1 = 0, \tag{5.40}$$

or in the vector form

$$\dot{\mathbf{y}} + A(t)\mathbf{y} = \mathbf{f}, \tag{5.41}$$

where

$$\mathbf{y} = \begin{bmatrix} y_1 \\ y_2 \end{bmatrix}; \quad A(t) = \begin{bmatrix} a_1 & a_2 \\ -1 & 0 \end{bmatrix}; \quad \mathbf{f} = \begin{bmatrix} f \\ 0 \end{bmatrix}.$$

The solution of the homogeneous equation (5.41) is

$$\mathbf{y}_0 = K(t)\mathbf{y}_{00}, \tag{5.42}$$

where $K(t)$ is a fundamental matrix of solutions satisfying to the condition $K(0) = E$ (E is the identity matrix); and $\mathbf{y}_{00}$ is the initial values vector. We cannot always obtain the elements of the matrix $K(t)$ in the analytical form, but there is no need to do so when it comes to producing numerical solutions to this kind of equations.

Let us find the general solution of the equation (5.41), using the Lagrange method of variations of arbitrary constants. Considering $\mathbf{y}_{00}$ a function of time, let us substitute (5.42) in the equation (5.41):

$$\dot{K}\mathbf{y}_{00} + K\dot{\mathbf{y}}_{00} + AK\mathbf{y}_{00} = \mathbf{f}. \tag{5.43}$$

As K satisfies the corresponding homogeneous equation, we obtain from (5.43)

$$K\dot{\mathbf{y}}_{00} = \mathbf{f}, \tag{5.44}$$

whence

$$\mathbf{y}_{00} = \int_0^1 K^{-1}(\tau)\mathbf{f}(\tau)\mathrm{d}\tau + \mathbf{C}. \tag{5.45}$$

The general solution of the equation (5.41) with due account of (5.45) is

$$\mathbf{y} = K(t)\mathbf{C} + \int_0^t G(t,\tau)\mathbf{f}(\tau)\mathrm{d}\tau, \tag{5.46}$$

where $G(t,\tau) = K(t)K^{-1}(\tau)$ is a Green matrix.

The principal difficulty of numerical determination of a solution in the form (5.46) is the determination of the matrix $K(t,\tau)$ that depends on the inverse matrix $K^{-1}(\tau)$ necessarily obtained on each step of the numerical solution process. For equations whose solution can be obtained in special functions (Bessel equation, Legendre equation, Hermite equation, etc.), we may represent the matrix $K(t,\tau)$ in an analytical form in terms of special functions.

For equations with constant coefficients, the Green matrix depends on the difference of arguments

$$G(t,\tau) = K(t-\tau).$$

When it is necessary to obtain a solution for the fixed instant ($t = t_k$), we can obtain the Green matrix $G(t,\tau)$ without the determination of the matrix $K^{-1}(\tau)$.

Let us differentiate the identity $K(\tau)K^{-1}(\tau) = E$:

$$\frac{\mathrm{d}}{\mathrm{d}\tau}K(\tau)K^{-1}(\tau) + K(\tau)\frac{\mathrm{d}K^{-1}(\tau)}{\mathrm{d}\tau} = 0. \tag{5.47}$$

Let us multiply (5.47) from the right by the matrix $K(\tau)$, then

$$\dot{K} + K\dot{K}^{-1}K = 0. \tag{5.48}$$

The matrix K satisfies the equation

$$\dot{K}(\tau) + A(\tau)K(\tau) = 0,$$

therefore, excluding $\dot{K}$, we'll obtain

$$-AK + K\dot{K}^{-1}K = 0, \tag{5.49}$$

whence

$$K\dot{K}^{-1} - A = 0. \tag{5.50}$$

Let us multiply equation (5.50) from the left by K^{-1} :

$$\dot{K}^{-1}(\tau) - K^{-1}(\tau)A(\tau) = 0. \tag{5.51}$$

Let us multiply the equation (5.51) from the left by the matrix $K(t_k)$ and transform it to the form

$$\frac{\mathrm{d}}{\mathrm{d}\tau}G(t_k,\tau) - G(t_k,\tau)A(\tau) = 0, \tag{5.52}$$

where $G(t_k,\tau) = K(t_k)K^{-1}(\tau)$.

Hence it follows, that the Green matrix satisfies the matrix equation (5.52). Let us apply the conjugation operation (that of transposition for matrixes with real elements), to the equation (5.52), then

$$\frac{\mathrm{d}G^*}{\mathrm{d}\tau} - (GA)^* = 0, \tag{5.53}$$

or (since $(GA)^* = A^*G^*$)

$$\frac{\mathrm{d}G^*}{\mathrm{d}\tau} - A^*G^* = 0. \tag{5.54}$$

The value of the Green matrix is known at $\tau = t_k$ (since $G = K(t_k)K^{-1}(\tau)$, $G = E$ at $\tau = t_k$), therefore it is necessary to introduce a new independent variable $\tau_1 = t_k - \tau$. As a result the equation (5.54) takes the form

$$\frac{\mathrm{d}G^*}{\mathrm{d}\tau_1} + A^*(\tau_1)G^* = 0. \tag{5.55}$$

As the value of the integral in the expression (5.46) at the fixed limits does not depend on the direction of integration ($\tau = t_k - \tau_1$), a matrix $G^*(t_k,\tau_1)$, obtained when solving the equation (5.55) after a conjugation operation is used in the integral evaluation (the second term in (5.46)).

$$\int_0^{t_k} G(t_k,\tau)\mathbf{f}(\tau)\mathrm{d}\tau = -\int_{t_k}^0 G(t_k,\tau_1)\mathbf{f}(\tau_1)\mathrm{d}\tau_1 = \int_0^{t_k} G(t_k,\tau_1)\mathbf{f}(\tau_1)\mathrm{d}\tau_1.$$

The outlined methods of determining the solution of the equation (5.41) and the Green matrix are true not only for the system of the second order (5.40) (that was used as an example in performing all necessary transformations), but also for systems of any order. These methods will be applied later during the investigation of random vibrations of systems with n degrees of freedom and systems with distributed parameters.

In the process of numerical calculations we may obtain the matrix $K^*(t_k,\ \tau)$ in the following way. Let us consider a vector equation of the form

$$\frac{\mathrm{d}\mathbf{b}}{\mathrm{d}\tau_1} + A^*(\tau_1)\mathbf{b} = 0. \tag{5.56}$$

By solving the equation (5.56) two times (or $2n$ times for a system with n degrees of freedom) at the following initial conditions:
1) $\tau_1 = 0$; $b_1 = 1$; $b_2 = 0$;
2) $\tau_1 = 0$; $b_1 = 0$; $b_2 = 1$,
we obtain the columns of the matrix G^* or the rows of the Green matrix.

In a scalar form, the solution (5.46) takes the form

$$\begin{aligned} y_1 &= k_{11}c_1 + k_{12}c_2 + \int_0^t g_{11}(t,\tau) f \mathrm{d}\tau; \\ y_2 &= k_{21}c_1 + k_{22}c_2 + \int_0^t g_{21}(t,\tau) f d\tau. \end{aligned} \tag{5.57}$$

As at $t = 0$ $y(0) = y_{10}$, $y_2(0) = y_{20}$, and $K(0) = E$, the arbitrary constants are equal to $c_1 = y_{10}$, $c_2 = y_{20}$.

Let us find the probability characteristics of the solution (5.57), considering y_{10}, y_{20} and f to be independent:

$$\begin{aligned} m_{y_1} &= k_{11}m_{y_{10}} + k_{12}m_{y_{20}} + \int_0^t g_{11}m_f \mathrm{d}\tau; \\ m_{y_2} &= k_{21}m_{y_{10}} + k_{22}m_{y_{20}} + \int_0^t g_{21}m_f \mathrm{d}\tau. \end{aligned} \tag{5.58}$$

$$\begin{aligned} K_{y_1} &= D_{y_{10}}k_{11}(t)k_{11}(t_1) + D_{y_{20}}k_{12}(t)k_{12}(t_1) \\ &\quad + \int_0^t \int_0^{t_1} g_{11}(t,\tau)g_{11}(t_1,\tau_1)K_f \mathrm{d}\tau\, \mathrm{d}\tau_1; \\ K_{y_2} &= D_{y_{20}}k_{21}(t)k_{22}(t_1) + D_{y_{20}}k_{22}(t)k_{22}(t_1) \\ &\quad + \int_0^t \int_0^{t_1} g_{21}(t,\tau)g_{21}(t_1,\tau_1)K_f \mathrm{d}\tau \mathrm{d}\tau_1. \end{aligned} \tag{5.59}$$

In the general case, the calculation of the integrals involved in the expressions (5.59) offers some difficulties. It is most easy to calculate them when the autocorrelation function of a force can be represented as a product of two functions depending only on one argument, i.e.

$$K_f(\tau,\tau_1) = a\varphi(\tau_1)\varphi(\tau). \tag{5.60}$$

If the autocorrelation function of excitation takes the form of (5.60), we may present the integrals entering (5.59) as

$$\int_0^t \int_0^{t_1} g_{ij}(t,\tau) g_{ij}(t_1,\tau_1) K_f(\tau,\tau_1) \mathrm{d}\tau \mathrm{d}\tau_1$$

$$= a \int_0^t g_{ij}(t,\tau)\varphi(\tau)\mathrm{d}\tau \int_0^{t_1} g_{ij}(t_1,\tau_1)\varphi(\tau_1)\mathrm{d}\tau_1. \tag{5.61}$$

Let us consider several examples of the random non-stationary vibrations of a system with one degree of freedom.

Example 5.2. Figure 5.4 shows a mast with an antenna placed in a spherical shell transparent for radiowaves. The mass of the antenna is considered to be point. In order to increase the stiffness of the mast in a plane ZOY two cables with a tension N_0 are attached to it. At an instant $t_0 = 0$ a homogeneous air stream whose velocity v is a random variable, has suddenly acted on the mast with the antenna, resulting in the appearance of a random aerodynamic force F constant in time (Fig. 5.4 *b*). Let us suppose that the probability characteristics of the force F (m_F, D_F)are known. We neglect the aerodynamic forces acting on the mast. It is required to determine the greatest possible angle γ of antenna beam deflection from the given direction assuming, that the angle γ obeys a normal distribution law.

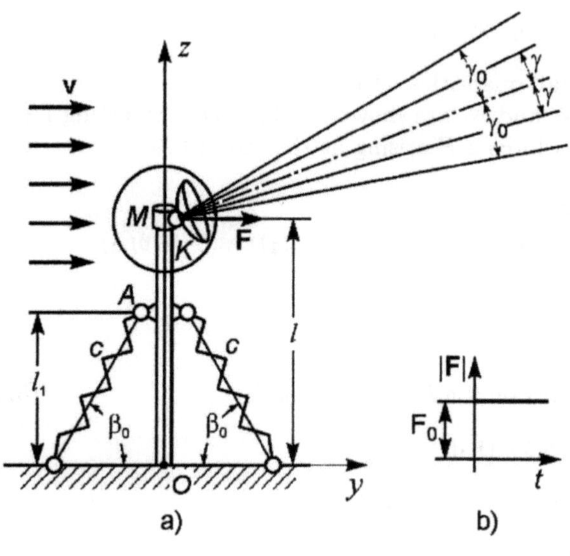

Fig. 5.4.

Vibrations caused by the variation of cable forces lead to the appearance of a force ΔP depending on displacement y_A (Fig. 5.5 *a*).

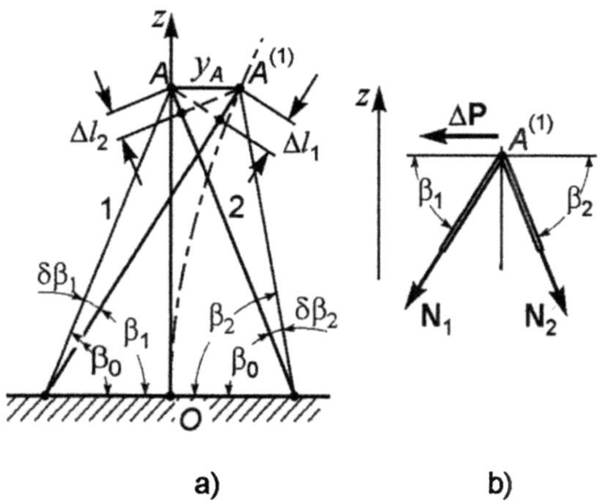

Fig. 5.5.

Let us consider the deflected position of the mast (Fig. 5.5 *a*). The variations of forces in cables ΔN_1 and ΔN_2 (Fig. 5.5 *b*) depend on "elongations" Δl_1 and Δl_2, i.e. $\Delta N_1 = c\Delta l_1$, $\Delta N_2 = c\Delta l_2$.

The variations in the length of cables are equal to

$$\Delta l_1 = y_A \cos\beta_1, \ \Delta l_2 = y_A \cos\beta_0. \tag{5.62}$$

As y_A and $\delta\beta_1$ are small quantities, $\Delta l_1 = y_A \cos\beta_0$.

The forces in cables N_1 and N_2 are shown in Fig. 5.5 *b*. Projecting N_1 N_2 on the axis y, we obtain a force ΔP arising at the vibrations

$$\begin{aligned} \Delta P &= N_1 \cos\beta_1 - N_2 \cos\beta_2 = \\ &= (N_0 + c\Delta l_1)\cos(\beta_0 - \delta\beta_1) - (N_0 - c\Delta l_2)\cos(\beta_0 + \delta\beta_2) \end{aligned}$$

or

$$\Delta P = 2y_A c\cos^2\beta_0 + N_0 \sin\beta_0 \delta\beta_1 + N_0 \sin\beta_0 \delta\beta_2. \tag{5.63}$$

It follows from Figure 5.5 *a* that

$$l\delta\beta_1 = y_A \sin\beta_1, \qquad l\delta\beta_2 = y_A \sin\beta_0.$$

Excluding $\delta\beta_1$ and $\delta\beta_2$ from (5.63), we obtain

$$\Delta P = 2\left(c\cos^2\beta_0 + \frac{N_0}{l}\sin^2\beta_0\right) y_A = c_1 y_A. \tag{5.64}$$

Using the force method, let us obtain the equations of small vibrations of the mast with the antenna (Fig. 5.4 *a*) (ignoring the dissipative forces)

$$y = \delta_{11}(-M\ddot{y} + F) + \delta_{12}\Delta P_1, \quad (\Delta P_1 = -\Delta P)$$
$$y_A = \delta_{21}(-M\ddot{y} + F) + \delta_{22}\Delta P_1. \tag{5.65}$$

Excluding y_A ($\Delta P_1 = -c_1 y_A$) from the system (5.65), we obtain by transformations the equation

$$\ddot{y} + p_0^2 y = \frac{F}{M}, \tag{5.66}$$

where

$$p_0^2 = \frac{(1 + c_1\delta_{22})}{M\left[\delta_{22} + c_1(\delta_{11}\delta_{22} - \delta_{12}\delta_{21})\right]}.$$

The solution of the equation (5.66) at zero initial data takes the form

$$y = \frac{1}{p_0}\int_0^t \sin p_0(t-\tau)\frac{F}{M}\mathrm{d}\tau = \frac{1}{p_0^2}\left(1 - \cos p_0 t\right)\frac{F}{M}. \tag{5.67}$$

Using the force method we can obtain the following expression for the angle of rotation of the mass M (the angle γ)

$$\gamma = \frac{1}{EJ}\left[\frac{\Delta P^{(1)} l_1^2}{2} + \frac{(J+F)l^2}{2}\right],$$

or

$$\gamma = \frac{1}{EJ}\left[-\frac{c_1 y_A l_1^2}{2} + \frac{(-M\ddot{y} + F)l^2}{2}\right]. \tag{5.68}$$

Excluding y_A and $\ddot{y}$ from (5.68), we obtain after transformations

$$\gamma = \frac{a}{EJ}(-\cos p_0 t + 1)F, \quad \left(a = -\frac{c_1 l_1^2 \delta_{21}}{2(1 + c_1\delta_{22})} + \frac{l^2}{2}\right),$$

where EJ is the flexural rigidity of the mast.

Let us determine the expectation and variance of the angle γ

$$m_\gamma = \frac{a}{EJ}\left(1 - \cos p_0 t\right) m_F,$$
$$D_\gamma = \frac{a^2}{EJ^2}\left(1 - \cos p_0 t\right)^2 D_F. \tag{5.69}$$

The greatest possible deflection of the angle γ is equal to

$$\max\gamma = m_\gamma + 3\sigma_\gamma = \frac{a}{EJ}\left(1 - \cos p_0 t\right)\left(M_F + 3\sigma_F\right). \tag{5.70}$$

Let us determine the maximum value $\max\gamma$ in time. Differentiating (5.70) with respect to t, we obtain

$$\frac{\mathrm{d}(\max\gamma)}{\mathrm{d}t} = \frac{a(m_F + 3\sigma_T)}{EJ} p_0 \sin p_0 t = 0.$$

The maximum value of the angle γ is reached at an instant $t_* = \dfrac{\pi}{p_0}$. At this instant the maximum value of the angle $(\max\gamma)$ is equal to

$$\max(\max\gamma) = 2\frac{a}{EJ}(m_F + 3\sigma_F). \tag{5.71}$$

Having determined the maximum value of the angle γ at the arising random vibrations of the mast with the antenna, we can check up the fulfillment of the condition of the antennas faultless operation

$$\max(\max\gamma) < \gamma_0, \tag{5.72}$$

where γ_0 is the admissible value of the antenna beam deflection angle.

Example 5.3. Figure 5.6 *a* shows the elevation mechanism with a mass m being clamped on the end of an extensible cable. At elevation (the drum rotates with constant angular velocity ω_0) the longitudinal vibrations of the mass m arise because of the random variation of a friction force between the mass m and the guides. These vibrations give rise to the occurrence of a cable dynamic tension ΔN, which fact must be taken into account at the analysis of cables strength and life. Assuming, that the random dynamic tension is normally distributed, let us determine the greatest possible tension according to the formula (at $m_{\Delta N} = 0$)

$$N_{\max} = mg + 3\sigma_{\Delta N}, \tag{5.73}$$

where $\sigma_{\Delta N}$ is the root-mean-square value of the random dynamic tension. It may be considered that the friction force between the mass m and the guides

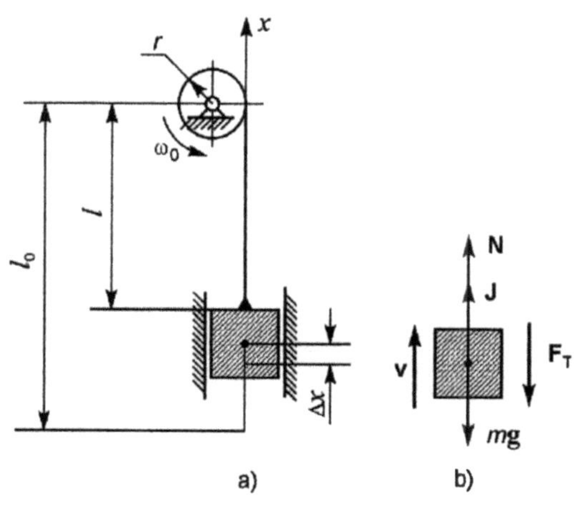

Fig. 5.6.

is a force of dry friction. With due regard for random scatter the force of dry friction is equal to

$$F_T = F_{T0}\mathrm{sign} v - \Delta F_T, \tag{5.74}$$

where ΔF_T is the random component of the friction force; v is the velocity of the weights motion. The velocity is

$$v = v_0 + \Delta\dot{x}, \qquad (v_0 = \omega_0 r = \mathrm{const}),$$

where Δx is the elongation of the cable at weight vibrations.

Let us assume, that the velocity v_0 exceeds the greatest possible random velocity $\Delta\dot{x}$, therefore $\mathrm{sign}\,\omega = 1$:

$$F_T = F_{T0} - \Delta F_T.$$

Using the d'Alembert principle, we obtain (Fig. 5.6 *b*) the equation

$$J - mg + N - F_T = 0. \tag{5.75}$$

Since

$$J = -m\dot{v}; \qquad N = N_0 + \Delta N;$$
$$N_0 = mg + F_{T0}; \quad \Delta N = -\Delta x EF,$$

from (5.75) we obtain the equation of the small vibrations of the mass

$$m\Delta\ddot{x} + \frac{EF}{l}\Delta x = \Delta F_T. \tag{5.76}$$

When determining the length of the cable l, we may neglect the elongation caused by the weight vibrations. In other words, it is possible to assume that $l = l_0 - v_0 t$.

The equation of the small vibrations of the mass (5.76) takes the form

$$\Delta\ddot{x} + \frac{EF}{l_0\left(1 - \frac{v_0}{l_0}t\right)m}\Delta x = \frac{\Delta F_T}{m}. \tag{5.77}$$

By replacing the argument t with

$$t_1 = 2\sqrt{\frac{EFl_0\left(1 - \frac{v_0}{l_0}t\right)}{mv_0^2}} = a\sqrt{1 - \frac{v_0}{l_0}t}$$

the equation (5.77) is reduced to Bessels equation.

The solution of Bessels homogeneous equation can be presented in an analytical form in terms of Bessel functions:

$$\begin{aligned} \Delta x &= c_1 t_1 J_1(t_1) + c_2 t_1 Y_1(t_1); \\ \Delta \dot{x} &= -\left[c_1 J_0(t_1) + c_2 Y_0(t_1)\right] \frac{2EF}{v_0 m}, \end{aligned} \tag{5.78}$$

or, turning back to the argument t, we obtain

$$\begin{aligned} \Delta \dot{x} &= c_1 f_{11}(t) + c_2 f_{12}(t); \\ \Delta x &= c_1 f_{21}(t) + c_2 f_{22}(t). \end{aligned} \tag{5.79}$$

The Green matrix for the given case is

$$K(t,\tau) = \begin{bmatrix} \dfrac{f_{11}(t)f_{22}(\tau) - f_{12}(t)f_{21}(\tau)}{D} & \dfrac{f_{12}(t)f_{11}(\tau) - f_{11}(t)f_{12}(\tau)}{D} \\ \dfrac{f_{21}(t)f_{22}(\tau) - f_{22}(t)f_{21}(\tau)}{D} & \dfrac{f_{22}(t)f_{11}(\tau) - f_{21}(t)f_{12}(\tau)}{D} \end{bmatrix}, \tag{5.80}$$

where $D = f_{11}(\tau)f_{22}(\tau) - f_{12}(\tau)f_{21}(\tau)$.

The solution of the equation (5.77) at zero initial data takes the form

$$\Delta \dot{x} = \int_0^t \frac{\left[f_{11}(t)f_{22}(\tau) - f_{12}(t)f_{21}(\tau)\right] \Delta F_T}{Dm} d\tau; \tag{5.81}$$

$$\Delta x = \int_0^t \frac{\left[f_{21}(t)f_{22}(\tau) - f_{22}(t)f_{21}(\tau)\right] \Delta F_T}{Dm} d\tau. \tag{5.82}$$

Experimental investigations show that the random scatter of a dry friction force can be presented as a process shown in Fig. 5.7. The random component of the friction force ΔF_T may be approximately regarded as a stationary random function limited in absolute value and having random instants of sign reversal. A realization of such process is shown in Fig. 5.7. To generate an exhaustive characteristic of the process we must also know the distribution of points of passage through zero (the distribution of zeros), i.e. the probability $P(n,\tau)$, where n is the number of zeros on an interval of time τ (n is a random quantity).

The plot of $\Delta F_T(t)$ variation in time, presented in Fig. 5.7, is an idealized one, as the finite quantity of an instant change in a friction force (in a random component) is impossible. This idealization, however, allows us to use the Poisson distribution law (see Sect. 1.4)

$$P(n,\,\tau) = \frac{(\mu\tau)^n}{n!} e^{-\mu\tau}, \tag{5.83}$$

where μ is the average frequency of zeros falling at a time interval unit.

We may express the expectation (parameter) in terms of μ and time t, assuming that $a = \mu t$. This has been accomplished in the expression (5.72).

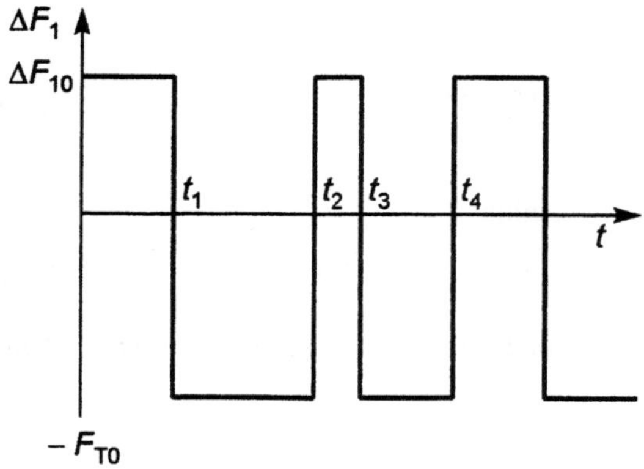

Fig. 5.7.

The section of a random function $\Delta F_T(t)$ has a distribution law presented by the series

Table 5.1.

$\Delta F_i(t)$	$-\Delta F_{TO}$	$+\Delta F_{TO}$
$P_i(t)$	$\frac{1}{2}$	$\frac{1}{2}$

As the sign reversal instants of the function ΔF_T are in no way related to the value of the random function, there is no reason for considering any of the values $+\Delta F_{TO}$, $-\Delta F_{TO}$ to be most probable, therefore

$$\begin{aligned} m_{\Delta F} &= +\frac{1}{2}\Delta F_{TO} - \frac{1}{2}\Delta F_{TO} = 0; \\ D_{\Delta F} &= (-\Delta F_{TO})^2\frac{1}{2} + \Delta F_{TO}^2\frac{1}{2} = \Delta F_{TO}^2. \end{aligned} \tag{5.84}$$

Let us find the correlation function, using its definition (see Sect. 2.2)

$$K_{\Delta F_T}(t, t_1) = M\left[\Delta F_T(t)\Delta F_T(t_1)\right]. \tag{5.85}$$

The product $\Delta F_T(t)\Delta F_T(t_1)$ is equal to -1, if an odd number of sign reversals occurs between the instants and changes its value to $+1$ when the number of sign reversals within this period becomes even. The probability that an even number of sign reversals will take place in time $\tau = t_1 - t$ is

$$P_e = \sum_{m=0}^{\infty} \frac{(\mu\tau)^{2m}}{(2m)!} e^{-\mu\tau} = e^{-\mu\tau}\frac{e^{\mu\tau} + e^{-\mu\tau}}{2}. \tag{5.86}$$

The probability that an odd number of sign reversals will take place in time is

$$P_o = \sum_{m=0}^{\infty} \frac{(\mu\tau)^{2m+1}}{(2m+1)!} e^{-\mu\tau} = e^{-\mu\tau} \frac{e^{\mu\tau} - e^{-\mu\tau}}{2}. \tag{5.87}$$

Knowing P_e and P_o, let us determine

$$K_{\Delta F_T}(t,\, t_1) = \Delta F_{TO}^2 P_e - \Delta F_{TO}^2 P_o = e^{-2\mu\tau} \Delta F_{TO}^2. \tag{5.88}$$

The expression obtained is true for $t_1 > t$. Similarly, at $t_1 < t$, we'll obtain

$$K_{\Delta F_T}(t,\, t_1) = \Delta F_{TO}^2 e^{-2\mu(-\tau)}. \tag{5.89}$$

Let us combine (5.88) and (5.89), then

$$K_{\Delta F_T}(t,\, t_1) = \Delta F_{TO}^2 e^{-2\mu|\tau|}. \tag{5.90}$$

The correlation function of the solution is

$$K_{\Delta x} = \int_0^t \int_0^{t_1} k_{21}(t,\tau) k_{21}(t_1,\tau_1) \Delta F_{TO}^2 e^{-2\mu|\varepsilon|} d\tau\, d\tau_1 \quad (\varepsilon = \tau - \tau_1). \tag{5.91}$$

The variance of the random tension is

$$D_{\Delta N} = \left(\frac{FF}{l}\right)^2 \int_0^t \int_0^t k_{21}(t,\tau) k_{21}(t,\tau_1) \Delta F_{TO}^2 e^{-2\mu|\varepsilon|} d\tau d\tau_1. \tag{5.92}$$

We can only integrate the right-hand side of the expression (5.92) numerically, but to do it we must know the value of the factor μ (the average frequency of zeros), which can only be determined experimentally. Therefore, let us determine the maximum value $\sigma_{\Delta N}$ (or $D_{\Delta N^*}$). It follows from the structure of expression for $D_{\Delta N}$ that the maximum value of $D_{\Delta N*}$ will be at $e^{-2\mu|\varepsilon|} = 1$, which corresponds to the limit case of $\mu = 0$. For this limit case

$$D_{\Delta N} < D_{\Delta N^*} = \left(\frac{EN}{l} \Delta F_{TO}\right)^2 \left(\int_0^t k_{21}(t,\tau) d\tau\right)^2. \tag{5.93}$$

As a result of calculations, we obtain an upper bound of the possible maximum values of the cables tension, assuming that ΔN^* has the following normal distribution

$$N_{\max_*} = mg + 3\sigma_{\Delta N^*}.$$

Let us consider the non-stationary vibrations of a mass m (the equation (5.26)), at a stationary random force of the white noise type

$$K_f(t - t_1) = s_0\delta(t - t_1), \tag{5.94}$$

where s_0 is the intensity (spectral density) of the stationary white noise.

The white noise (5.94) is a random function, whose values at $t \neq t_1$ are not correlated. In the physical context of view the condition of the noncorrelatedness of two values of a random function for an arbitrary small interval of time is equivalent to the condition of absolute inertialess process, which, of course, is incorrect. Any real physical process has an inertia, therefore the values of a random function describing the process, at a given instant partially determine its values at adjacent instants.

Any physical quantities, including time, are measured with a certain error; all values of the considered quantity, the difference of which is smaller than the error, are considered to be coinciding. Therefore, in practical plane we may regard a random function as a white noise, if the correlation of its values is only extended to the intervals of the arguments variation that is less than the minimum discernible one at the assumed accuracy of measurements. We can use the integration step $h = \Delta t$ as the characteristic interval of time. The correlation function of the stationary random function can be presented in the following form ($\tau = t - t_1$)

$$K_f(\tau) = D_f\varphi(\tau), \tag{5.95}$$

where $\varphi(\tau)$ is the decreasing function of the argument τ (at $\tau = 0$ $\varphi(0) = 1$). The relative variation of the correlation function within the integration step is

$$\Delta = \frac{K_f(h)}{D_f} = \varphi(h). \tag{5.96}$$

If at the assumed accuracy of the solution it is possible to put $\Delta \approx 0$, then the random function $f(t)$ can be considered a white noise.

In approximate calculations, when the process time is much greater than the interval of correlation, a random function can be approximately considered a white noise, the intensity of which (a special case of the Wiener–Khinchin formula at $\omega = 0$) is

$$s_0 = \int_{-\infty}^{\infty} K_f(\tau)\mathrm{d}\tau,$$

where $K_f(\tau)$ is a real correlation function, differring from the delta function. For example, if

$$K_f = D_f\mathrm{e}^{-\alpha|\tau|}, \tag{5.97}$$

then the intensity of the "real" white noise is

$$s_0 = D_f \int\limits_{-\infty}^{\infty} \mathrm{e}^{-\alpha|\tau|} \mathrm{d}\tau = \frac{2D_f}{\alpha}.$$

Let us determine the correlation function K_y for the case when K_f is defined by the expression (5.94)

$$K_y = \frac{1}{m^2 p^2} \int\limits_0^t \int\limits_0^{t_1} k(t-\tau) k(t_1 - \tau_1) s_0 \delta(\tau - \tau_1) \mathrm{d}\tau \, \mathrm{d}\tau_1 \tag{5.98}$$

or

$$K_y = \frac{s_0}{m^2 p^2} \int\limits_0^t k(t-\tau) \left[\int\limits_0^{t_1} k(t_1 - \tau_1) \delta(\tau - \tau_1) d\tau_1 \right] \mathrm{d}\tau$$

$$= \frac{s_0}{m^2 p^2} \int\limits_0^t k(t-\tau) k(t_1 - \tau) \mathrm{d}\tau, \tag{5.99}$$

where

$$k(t-\tau) = \mathrm{e}^{-n(t-\tau)} \sin p(t-\tau),$$
$$k(t_1-\tau) = \mathrm{e}^{-n(t_1-\tau)} \sin p(t_1-\tau).$$

The variance of a solution is

$$D_y(t) = \frac{s_0}{m^2 p^2} \int\limits_0^t k^2(t-\tau) \mathrm{d}\tau. \tag{5.100}$$

5.2.2 Stationary Forced Vibrations

If the linear equation of small vibrations of a system with one degree of freedom has constant coefficients and the solution of the homogeneous equation is asymptotically stable, random stationary vibrations (at a random stationary right-hand side) are possible in such a system.

Let us consider the stationary random vibrations of systems with one degree of freedom. If the motion of a system is described by linear equations with constant coefficients and the homogeneous part of these equations has asymptotically stable solutions, the conditions of stationary vibrations are possible (at a stationary right-hand side).

It is possible to find the spectral density of the stationary random function f (the equation (5.26)), knowing its correlation function $K_f(\tau)$

$$S_f(\omega) = \frac{1}{2\pi} \int\limits_{-\infty}^{\infty} K_f(\tau) \cos \omega\tau \mathrm{d}\tau. \tag{5.101}$$

It has been shown in Sect. 3.6, that the spectral density of the input is related to the spectral density of the output by relationship

$$S_y(\omega) = |W(i\omega)|^2 S_f(\omega), \tag{5.102}$$

where

$$|W(i\omega)| = \left| \frac{1}{m\left[(i\omega)^2 + 2ni\omega + p_0^2\right]} \right|.$$

From the spectral density of the output $S_y(\omega)$ we can find the correlation function of the output and the variance:

$$\begin{aligned} K_y(\tau) &= \int_{-\infty}^{\infty} S_y(\omega) \cos \omega\tau \mathrm{d}\omega \\ &= \frac{1}{m^2} \int_{-\infty}^{\infty} \left| \frac{1}{(i\omega)^2 + 2ni\omega + p_0^2} \right|^2 S_f(\omega) \cos \omega\tau \mathrm{d}\tau; \end{aligned} \tag{5.103}$$

$$D_y = \frac{1}{m^2} \int_{-\infty}^{\infty} \left| \frac{1}{(i\omega)^2 + 2ni\omega + p_0^2} \right|^2 S_f(\omega) \mathrm{d}\omega. \tag{5.104}$$

In a particular case, where we may approximately consider the random disturbance to be a white noise ($S_f(\omega) = S_0 = \text{const}$), we obtain

$$K_y(\tau) = \frac{S_0}{m^2} \int_{-\infty}^{\infty} \left| \frac{1}{(i\omega)^2 + 2ni\omega + p_0^2} \right|^2 \cos \omega\tau \mathrm{d}\tau; \tag{5.105}$$

$$D_y = \frac{S_0}{m^2} \int_{-\infty}^{\infty} \left| \frac{1}{(i\omega)^2 + 2ni\omega + p_0^2} \right|^2 \mathrm{d}\omega. \tag{5.106}$$

We may present the integrals entering the expressions (5.104) and (5.106) as (3.87)

$$J_n = \frac{1}{2\pi} \int_{-\infty}^{\infty} \frac{G(i\omega)}{|A(i\omega)|^2} \mathrm{d}\omega. \tag{5.107}$$

The values of integrals J_n are presented in Appendix 2.

If, for example,

$$K_f(\tau) = D_f e^{-\alpha|\tau|},$$

then the spectral density is equal to

$$S_f(\omega) = D_f \frac{2\alpha}{(\alpha^2 + \omega^2)} = \frac{2\alpha D_f}{|\alpha + i\omega|^2}.$$

Using (5.104), we obtain

$$D_y = \frac{2\alpha D_f}{m^2} \int_{-\infty}^{\infty} \left| \frac{1}{(i\omega)^2 + 2i\omega n + p_0^2} \right|^2 \frac{d\omega}{|\alpha + i\omega|^2},$$

or

$$D_y = \frac{2D_f\alpha}{m^2} \int_{-\infty}^{\infty} \frac{d\omega}{|(i\omega)^3 + (2n+\alpha)(i\omega)^2 + (p_0^2 + 2n\alpha)i\omega + p_0^2\alpha|^2}. \quad (5.108)$$

The integral in the expression (5.108) is a special case of the integral (5.107) at $G(i\omega) = 1$. Using the Appendix 2, we obtain

$$D_y = \frac{D_f}{m^2} \frac{\pi(2n+\alpha)}{n(p_0^2 + 2n\alpha + \alpha^2)p_0^2}. \quad (5.109)$$

The variances of the first and second derivatives of the solution are

$$D_{\dot{y}} = \int_{-\infty}^{\infty} |W(i\omega)|^2 \omega^2 S_f(\omega) d\omega$$

or

$$D_{\dot{y}} = \frac{2D_f\alpha}{m^2} \int_{-\infty}^{\infty} \frac{-(i\omega)^2 d\omega}{|(i\omega)^3 + (2n+\alpha)(i\omega)^2 + (p_0^2 + 2n\alpha)i\omega + p_0^2\alpha|^2}; \quad (5.110)$$

$$D_{\ddot{y}} = \frac{2D_f\alpha}{m^2} \int_{-\infty}^{\infty} \frac{(i\omega)^4 d\omega}{|(i\omega)^3 + (2n+\alpha)(i\omega)^2 + (p_0^2 + 2n\alpha)i\omega + p_0^2\alpha|^2}. \quad (5.111)$$

Using the value of J_3 (Appendix 2), we'll obtain

$$D_{\dot{y}} = \frac{D_f\alpha}{m^2} \frac{\pi}{n(p_0^2 + 2n\alpha + \alpha^2)};$$
$$D_{\ddot{y}} = \frac{D_f\alpha}{m^2} \frac{\pi(p_0^2 + 2n\alpha)}{n(p_0^2 + 2n\alpha + \alpha^2)}. \quad (5.112)$$

Example 5.4. In order to determine the horizontal acceleration of a body there is an accelerometer of a mass m_1 attached to the body of a mass m (Fig. 5.8) moving with an acceleration. The body moves under the action of a force R having a stationary random component $\Delta R(t)$ with a known correlation function ($m_{\Delta R} = 0$)

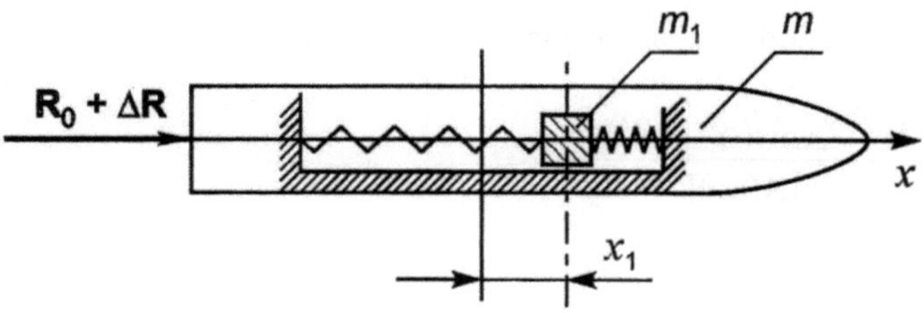

Fig. 5.8.

$$K_{\Delta R}(\tau) = D_{\Delta R}\mathrm{e}^{-\alpha|\tau|}\left(\cos\beta\tau + \frac{\alpha}{\beta}\sin\beta\,|\tau|\right). \tag{5.113}$$

It is required to determine the displacement variance of the mass m_1, caused by the random component of the force R. The equation of the random vibrations of the mass m_1 (neglecting the acceleration of a rocket caused by ΔR, as $m_1 \ll m$) is

$$\Delta\ddot{x}_1 + 2n\Delta x_1 + p_0^2\Delta x_1 = k\Delta R, \qquad \left(k = \frac{1}{m_1}\right), \tag{5.114}$$

where Δx_1 is the random relative displacement of the mass m_1, caused by the action of ΔR.

The spectral density of the thrust scatter

$$S_{\Delta R}(\omega) = D_{\Delta R}\int\limits_{-\infty}^{\infty} \mathrm{e}^{-\alpha|\tau|}\left(\cos\beta\tau + \frac{\alpha}{\beta}\sin\beta\,|\tau|\right)\cos\omega\tau\mathrm{d}\tau. \tag{5.115}$$

By integrating we obtain

$$S_{\Delta R}(\omega) = 4D_{\Delta R}\frac{\alpha(\alpha^2+\beta^2)}{(\omega^2-\alpha^2-\beta^2)+4\alpha^2\omega^2}. \tag{5.116}$$

We may present the expression (5.116) as

$$S_{\Delta R}(\omega) = \frac{4D_{\Delta R}\alpha(\alpha^2+\beta^2)}{\left|(i\omega)^2+2\alpha(i\omega)+\alpha^2+\beta^2\right|^2}. \tag{5.117}$$

The spectral density of a displacement Δx_1 is equal to

$$S_{\Delta x_1}(\omega) = \frac{4k^2D_{\Delta R}\alpha(\alpha^2+\beta^2)}{\left|(i\omega)^2+2ni\omega+p_0^2\right|^2\left|(i\omega)^2+2\alpha(i\omega)+\alpha^2+\beta^2\right|^2}. \tag{5.118}$$

The variance of a random displacement takes the form

$$D_{\Delta x_1} = \int_{-\infty}^{\infty} \frac{4k^2 D_{\Delta R} \alpha(\alpha^2+\beta^2) d\omega}{|A(i\omega)|^2},$$

where

$$\begin{aligned} |A(i\omega)|^2 &= \left|(i\omega)^4 + 2(\alpha+n)(i\omega)^3 + (\alpha^2+\beta^2+4n\alpha+p_0^2(i\omega)^2 \right. \\ &\quad + \left.\left[2n(\alpha^2+\beta^2)+2\alpha p_0^2\right](i\omega) + p_0^2(\alpha^2+\beta^2)\right|^2 \\ &= \left|\alpha_0(i\omega)^4 + \alpha_1(i\omega)^3 + \alpha_2(i\omega)^2 + \alpha_3(i\omega) + \alpha_4\right|^2. \end{aligned}$$

Using the Appendix 2, we obtain

$$D_{\Delta x_1} = \frac{4k^2 D_{\Delta R} \alpha(\alpha^2+\beta^2)\pi}{p_0^2(\alpha^2+\beta^2)(a_3^2+a_1^2 a_4 - a_1 a_2 a_3)}.$$

Assuming that Δx_1 has a normal distribution, we can estimate the greatest possible error of the accelerometer indications ($m_{\Delta x_1} = 0$)

$$\Delta = 3\sigma_{\Delta x}.$$

Let us consider the stationary vibrations of a mass m, caused by a stationary force $f(t)$ with a known spectral density S_f ($m_f = 0$) (Fig. 5.9). Let us determine the greatest possible value of a normal stress σ_m that arises in the section of a rod at $z = 0$,, assuming, that σ_m has a normal distribution. The simplest way to obtain the equation of the small vibrations of the mass m with due account taken of the damping device is to apply the force method based on the principle of superposition.

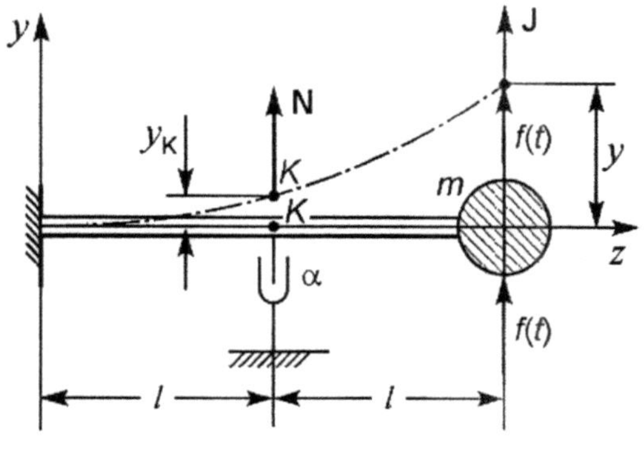

Fig. 5.9.

By the force method the displacements of masses G and m are equal to the sum of displacements due to each of the forces applied to a system, i.e.

$$\begin{aligned} y &= \delta_{11}(J+f)+\delta_{12}N, \qquad (J=-m\ddot{y}) \\ y_k &= \delta_{21}(J+f)+\delta_{22}N, \end{aligned} \tag{5.119}$$

where δ_{ij} are the compliances of the system, N is the force of viscous resistance that is equal to

$$N=-\alpha\dot{y}_k.$$

The force method is one of the most effective methods of deriving equations of small vibrations (when the stresses arising in elastic elements obey the Hooke law). For example, it is very difficult to obtain equations of small vibrations for the mechanical system presented in Fig. 5.9, using the Lagrange equations of the second kind.

An interesting feature of the given problem is that two differential equations have been obtained for a system with one degree of freedom. The point is that the mechanical system shown in Fig. 5.9 belongs to the systems with a fractional number of degrees of freedoms. The system of equations (5.119) is a third-order system.

Let us obtain the following images of the equations (5.119) in the frequency area

$$\begin{aligned} Y(i\omega) &= -\delta_{11}m(i\omega)^2Y(i\omega)-\delta_{12}\alpha(i\omega)Y_k(i\omega)+\delta_{11}f_0(i\omega); \\ Y_k(i\omega) &= -\delta_{21}m(i\omega)^2Y(i\omega)-\delta_{22}\alpha(i\omega)Y_k(i\omega)+\delta_{21}f_0(i\omega). \end{aligned} \tag{5.120}$$

From the system of algebraic equations (5.120) we determine the images Y and Y_k in the frequency area

$$\begin{aligned} Y(i\omega) &= W_1(i\omega)f_0(i\omega), \\ Y(i\omega) &= W_2(i\omega)f_0(i\omega), \end{aligned} \tag{5.121}$$

where

$$W_1(i\omega)=\frac{\delta_{11}+\alpha(i\omega)(\delta_{11}\delta_{22}-\delta_{12}\delta_{21})}{\Delta},$$

$$W_2(i\omega)=\frac{\delta_{21}}{\Delta},$$

$$\Delta=(1+\delta_{11}m(i\omega)^2)(1+\delta_{22}\alpha(i\omega))-\delta_{12}\delta_{21}\alpha^2(i\omega)^3.$$

The maximum normal stress in the rod at $z=0$ is equal to

$$\sigma_m(t)=\frac{M(t)}{W}, \tag{5.122}$$

where M is the bending moment, W is the section modulus.

The bending moment is equal to

$$M(t) = 2l(J + f) + Nl$$

or

$$M = -2lm\ddot{y} + 2lf - \alpha\dot{y}_k l. \tag{5.123}$$

Having passed to the image (5.123) in the frequency area, we obtain

$$M_0(i\omega) = -2lm(i\omega)^2 Y(i\omega) - \alpha l(i\omega)Y_k(i\omega) + 2lf_0(i\omega). \tag{5.124}$$

Excluding $Y(i\omega)$ and $Y_k(i\omega)$ from (5.124) we obtain

$$M_0(i\omega) = W_3(i\omega)f_0(i\omega), \tag{5.125}$$

where

$$W_3(i\omega) = \frac{\left[2\Delta - \alpha(i\omega)\delta_{22} - 2m(i\omega)^2(\delta_{11} + \alpha(i\omega)(\delta_{11}\delta_{22} - \delta_{12}\delta_{21})\right]}{\Delta} l.$$

As

$$\sigma_{m_0}(i\omega) = \frac{1}{W} M_0(i\omega) = \frac{1}{W} W_3(i\omega) f_0(i\omega),$$

the spectral density and variance of the maximum stress are equal to

$$\begin{aligned} S_{\sigma_m}(i\omega) &= \frac{1}{W^2} \left|W_3(i\omega)\right|^2 S_R(\omega), \\ D_{\sigma_m} &= \frac{1}{W^2} \int\limits_{-\infty}^{\infty} \left|W_3(i\omega)\right|^2 S_f(\omega) d\omega. \end{aligned} \tag{5.126}$$

The greatest possible value of σ_m is

$$\max \sigma_m = 3\sqrt{D_{\sigma_m}}.$$

5.3 Vibrations Caused by Random Kinematic Excitation

Let us consider vibrations of one-mass mechanical systems caused by the forced random displacement (or rotation) of elastic elements discrete points (Fig. 5.10 *a*) (or sections (Fig. 5.10 *b*)).

A special feature of the systems presented in Fig. 5.10, is that kinematic excitations are attached to zero-mass points, which complicates the derivation of equations of lumped masses m vibrations.

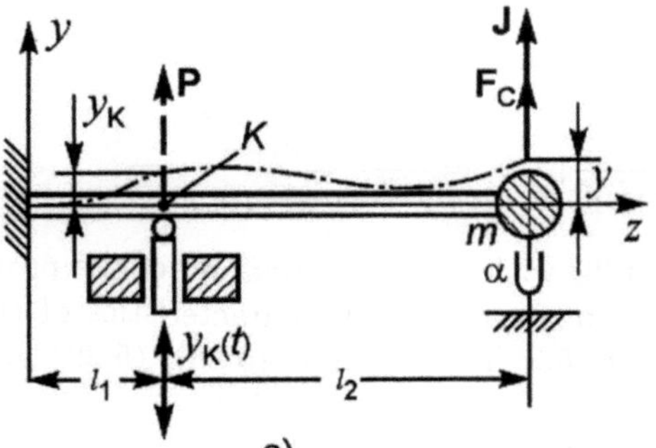

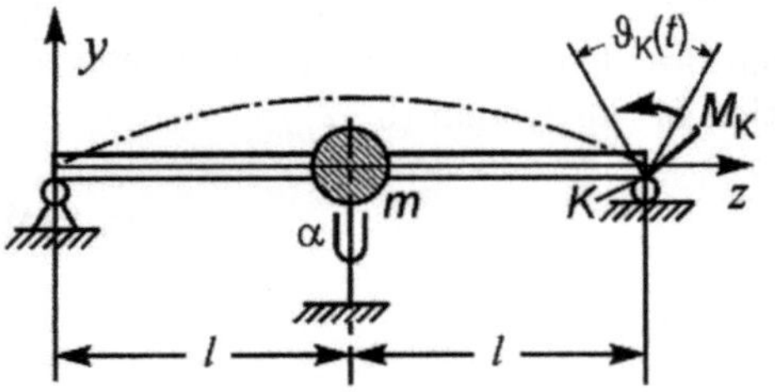

Fig. 5.10.

5.3.1 Non-Stationary Random Vibrations at Kinematic Excitation

Let us obtain the equations of small vibrations of a mass m, using the force method and introducing an unknown force at linear kinematic displacements applied to a point k (in Fig. 5.10 a the force is shown by the dotted line) or an unknown moment M_k at angular kinematic displacements (Fig. 5.10 b) as well as forces of inertia $J = -m\ddot{y}$ and forces of resistance $F_c = -\alpha\dot{y}$. For example, for the system shown in Fig. 5.10 a, we obtain two equations

$$\begin{aligned} y &= \delta_{11}(-m\ddot{y} - \alpha\dot{y}) + \delta_{12}P(t), \\ y_k(t) &= \delta_{21}(-m\ddot{y} - \alpha\dot{y}) + \delta_{22}P(t), \end{aligned} \tag{5.127}$$

where δ_{ij} are the compliances of the system and $y_k(t)$ is the given function.

Excluding the force from the first equation of the system (5.127) and carrying out the necessary transformations we obtain the equation

$$y = \left(\delta_{11} - \frac{\delta_{12}\delta_{21}}{\delta_{22}}\right)(-m\ddot{y} - \alpha\dot{y}) + \frac{\delta_{12}}{\delta_{22}}y_k \tag{5.128}$$

or

$$\ddot{y} + 2n\dot{y} + p_0^2 y = b_1 y_k, \tag{5.129}$$

where

$$2n = \frac{\alpha}{m}, \quad p_0^2 = \frac{\delta_{22}}{m(\delta_{11}\delta_{22} - \delta_{12}\delta_{21})}, \quad b_1 = \frac{\delta_{11}}{m(\delta_{11}\delta_{22} - \delta_{12}\delta_{21})}.$$

The solution of the equation (5.129) with the determination of the probability characteristics of y (at the known probability characteristics of the right-hand side) is presented in Sect. 5.2 (relationships (5.28), (5.29) and (5.30)).

Let us consider a problem where it is required to determine the variation in time of the greatest possible force arising in a section at the non-stationary vibrations of a mass m at zero initial data. When determining the greatest possible force , let us assume that the distribution of the probability density of the force $P(t)$ will obey the normal law. Let us obtain an expression for the force $P(t)$ from the second equation of the system (5.127)

$$P(t) = \frac{1}{\delta_{22}} y_k - \frac{\delta_{21}}{\delta_{22}}(-m\ddot{y} - \alpha\dot{y}). \tag{5.130}$$

We determine from the equation (5.128)

$$(-m\ddot{y} - \alpha\dot{y}) = (y - \frac{\delta_{12}}{\delta_{22}} y_k)\frac{\delta_{22}}{(\delta_{11}\delta_{22} - \delta_{12}\delta_{21})}.$$

Excluding $(-m\ddot{y} - \alpha\dot{y})$ from the equation (5.130), we obtain

$$P(t) = \gamma_1 y_k - \gamma_2 y, \tag{5.131}$$

where

$$\gamma_1 = \frac{\delta_{11}}{(\delta_{11}\delta_{22} - \delta_{12}\delta_{21})}; \qquad \gamma_2 = \frac{\delta_{21}}{(\delta_{11}\delta_{22} - \delta_{12}\delta_{21})}.$$

At zero initial data the solution of the equation (5.129) takes the form

$$y = \frac{b_1}{p}\int_0^t \mathrm{e}^{-n(t-\tau)} \sin p(t-\tau) y_k(\tau) \mathrm{d}\tau = \frac{b_1}{p}\int_0^t g(t-\tau) y_k(\tau) \mathrm{d}\tau.$$

The expectation $m_P(\tau)$ is equal to

$$m_P(\tau) = \gamma_1 m_{y_k} - \gamma_2 m_y.$$

Let us obtain the correlation function and the variance of the random force

$$K_P(t, t') = M\left[(\gamma_1 \overset{\circ}{y}(t)_k - \gamma_2 \overset{\circ}{y}(t))(\gamma_1 \overset{\circ}{y}_k(t') - \gamma_2 \overset{\circ}{y}(t'))\right]$$

or (omitting the intermediate transformations)

$$K_P(t,t') = \gamma_1^2 K_{y_k}(t,\, t') + \int\limits_0^t \int\limits_0^{t'} g(t-\tau)g(t'-\tau')K_{y_k}(\tau,\, \tau')\,\mathrm{d}\tau\,\mathrm{d}\tau'$$

$$-\gamma_1\gamma_2 \int\limits_0^{t'} g(t'-\tau')K_{y_k}(t,\tau)\,\mathrm{d}\tau' - \gamma_1\gamma_2 \int\limits_0^{t} g(t-\tau)K_{y_k}(t',\tau)\,\mathrm{d}\tau,$$

$$D_p(t) = K_p(t,t')\,|_{t'=t}.$$

The greatest possible value of the random force $P(t)$ (with the use of the three sigma rule) is equal to

$$\max P(t) = m_p(t) + 3\sigma_p(t).$$

The mass m connected with the randomly moving foundation is shown in Fig. 5.11. For greater definiteness we shall assume that the motion of the foundation began at an instant $t_0 = 0$.

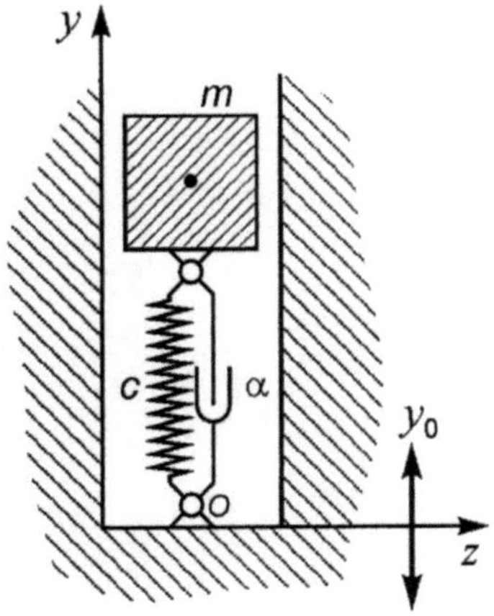

Fig. 5.11.

The equation of the motion of the mass m is of the form

$$\ddot{y} + 2n\dot{y} + p_0^2 y = 2n\dot{y}_0 + p_0^2 y_0. \tag{5.132}$$

The solution of the equation (5.132) at zero initial data is

$$y = \frac{1}{pm}\int\limits_0^t k(t-\tau)(2n\dot{y}_0 + p_0^2 y_0)\mathrm{d}\tau, \tag{5.133}$$

where

$$k(t-\tau) = \mathrm{e}^{-n(t-\tau)} \sin p(t-\tau).$$

We consider the displacement y_0 a non-stationary random function with known probability characteristics m_{y_0} and $K_{y_0}(t, t_1)$.

In order to obtain the probability characteristics of the solution we must know the probability characteristic of the random displacement derivative $\dot{y}_0$ and that of the product $y_0\dot{y}_0$. The expressions for the probability characteristics of the product of the random function by its derivative were obtained in Sect. 2.4

$$\begin{aligned} m_{\dot{y}_0} &= \frac{\mathrm{d}m_{y_0}}{\mathrm{d}t}; \quad K_{\dot{y}_0 y_0} = M\left[\overset{\circ}{\dot{y}}_0(t)\,\overset{\circ}{y}_0(t_1)\right] = \frac{\partial K_{y_0}}{\partial t}; \\ K_{y_0\dot{y}_0} &= \frac{\partial K_{y_0}}{\partial t_1}; \quad K_{\dot{y}_0\dot{y}_0} = \frac{\partial^2 K_{y_0}}{\partial t\,\partial t_1}. \end{aligned} \tag{5.134}$$

The probability characteristics of the solution of the equation (5.132) are

$$m_y = \frac{p_0^2}{p}\int_0^t k(t-\tau)m_{y_0}\mathrm{d}\tau + \frac{2n}{p}\int_0^t k(t-\tau)\dot{m}_{y_0}\mathrm{d}\tau; \tag{5.135}$$

$$\begin{aligned} K_y &= \frac{p_0^4}{p^2}\int_0^t\int_0^{t_1} k(t-\tau)k(t_1-\tau_1)K_{y_0}\mathrm{d}\tau\,\mathrm{d}\tau_1 \\ &+ \frac{2p_0^2 n}{p^2}\int_0^t\int_0^{t_1} k(t-\tau)k(t_1-\tau_1)\frac{\partial K_{y_0}}{\partial \tau_1}\,\mathrm{d}\tau\,\mathrm{d}\tau_1 \\ &+ \frac{2np_0^2}{p^2}\int_0^t\int_0^{t_1} k(t-\tau)k(t_1-\tau_1)\frac{\partial K_{y_0}}{\partial \tau}\,\mathrm{d}\tau\,\mathrm{d}\tau_1 \\ &+ \frac{4n^2}{p^2}\int_0^t\int_0^{t_1} k(t-\tau)k(t_1-\tau_1)\frac{\partial^2 K_{y_0}}{\partial\tau\partial\tau_1}\,d\tau\,d\tau_1. \end{aligned} \tag{5.136}$$

As a result of solving the equation (5.132) we have obtained the probability characteristics of y at non-stationary kinematic excitation.

Under the action of a stationary kinematic excitation both non-stationary and stationary vibrations are possible. The first can occur when the time of process is less than the time t_n necessary for the practical attenuation of transients, while the second may take place when the time of the process by far exceeds the time t_n.

Let us consider the non-stationary vibrations of a system under the action of a stationary kinematic excitation (Fig. 5.12). There is a point mass m on the end of the elastic element of a measuring instrument rigidly fixed to a foundation randomly displacing in a vertical direction. The vertical displacement y_0 is a stationary random function with known probability characteristics

$$m_{y_0} = 0; \qquad K_{y_0}(\tau) = D_{y_0}\mathrm{e}^{-\alpha|\tau|} \tag{5.137}$$

Before the instrument has been turned on the mass m cannot shift from the foundation because of the rigid tie A. At an instant $t = t_0 = 0$ the tie A disappears and the mass begins moving. The position of the mass at an arbitrary instant t is shown in Fig. 5.12 by the dashed line. It is required to determine the standard deviation of a difference $y_0 - y_k$ at a given instant t_k. The difference $y_0 - y$ represents a random displacement of the mass m relative to the foundation. The equation of small vibrations of the mass is

$$\ddot{y} + 2n\dot{y} + p_0^2 y = p_0^2 y_0 + 2n\dot{y}_0. \tag{5.138}$$

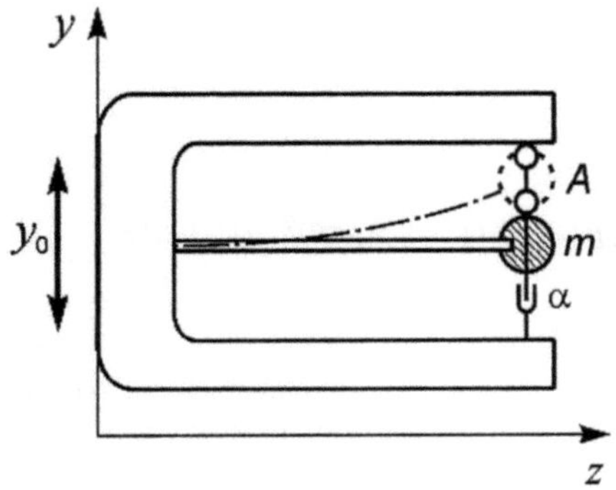

Fig. 5.12.

In order to determine the correlation function of the solution we can use the expression (5.136). In the considered case the correlation function K_{y_0}, entering underneath the sign of integrals (5.136), takes the form

$$K_{y_0} = D_{y_0}e^{-\alpha|\varepsilon|}, \quad (\varepsilon = \tau - \tau_1).$$

The derivatives of K_y with respect to τ and τ_1 are

$$\frac{\partial K_{y_0}}{\partial \tau} = D_{y_0}\frac{\partial \mathrm{e}^{-\alpha|\varepsilon|}}{\partial \tau} = D_{y_0}\frac{\partial \mathrm{e}^{-\alpha|\varepsilon|}}{\partial|\varepsilon|}\frac{\mathrm{d}|\varepsilon|}{\mathrm{d}\varepsilon}\frac{\mathrm{d}\varepsilon}{\mathrm{d}t} = -D_{y_0}\alpha\mathrm{e}^{-\alpha|\varepsilon|}\mathrm{sign}\varepsilon;$$

$$\frac{\partial K_{y_0}}{\partial \tau_1} = D_{y_0}\frac{\partial \mathrm{e}^{-\alpha|\varepsilon|}}{\partial \tau_1} = D_{y_0}\alpha\mathrm{e}^{-\alpha|\varepsilon|}\mathrm{sign}\varepsilon.$$

The second derivative of K_{y_0} with respect to τ and τ_1 is

$$\frac{\partial^2 K_{y_0}}{\partial\tau\,\partial\tau_1} = D_{y_0}\alpha e^{-\alpha|\varepsilon|}\left[2\delta(\varepsilon) - \alpha\right].$$

Through transformations the expression for the correlation function of the solution takes the form

$$K_y = \frac{D_y}{p^2}\int_0^t\int_0^{t_1} k(t-\tau)k(t_1-\tau_1)e^{-\alpha|\tau-\tau_1|} \times \left[p_0^4 + 8n^2\alpha\delta(\tau-\tau_1) - 4n^2\alpha^2\right]\,d\tau\,d\tau_1$$

or

$$K_y(t,t_1) = \frac{D_{y_0}(p_0^4 - 4n^2\alpha^2)}{p^2}\int_0^t\int_0^{t_1} k(t-\tau)k(t_1-\tau_1)e^{-\alpha|\tau-\tau_1|}\,d\tau\,d\tau_1 + \frac{8D_{y_0}n^2\alpha}{p^2}\int_0^{t_1} k(t_1-\tau_1)k(t-\tau_1)d\tau_1. \tag{5.139}$$

The variance is

$$D_y(t) = \frac{D_{y_0}(p_0^2 - 4n^2\alpha^2)}{p^2}\int_0^t\int_0^{t_1} k(t-\tau)k(t-\tau_1)e^{-\alpha|\tau-\tau_1|}\,d\tau\,d\tau_1 + \frac{8D_{y_0}n^2\alpha}{p^2}\int_0^{t} k^2(t-\tau)\,d\tau. \tag{5.140}$$

In order to determine the standard deviation of the difference $\Delta y = y_0 - y_1$ we must find the correlation function $K_{\Delta y}(t,\, t_1)$:

$$\begin{aligned} K_{\Delta y}(t,\, t_1) &= M\left[\Delta y(t)\Delta y(t_1)\right] = M\left[y_0(t) - y(t)\right]\left[y_0(t_1) - y(t_1)\right] \\ &= K_{y_0}(t,t_1) + K_y(t,t_1) - M_1 - M_2, \end{aligned}$$

where

$$M_1 = M\left[y_0(t)y(t_1)\right] \text{ and } M_2 = M\left[y_0(t_1)y(t)\right].$$

Substituting the expression of the solution $y(t_1)$ (5.133), we find

$$M_1 = M\left[y_0(t)\frac{1}{mp}\int_0^{t_1} k(t_1-\tau_1)\left[p_0^2 y_0(\tau_1)+2n\dot{y}_0(\tau_1)\right]\mathrm{d}\tau_1\right]$$
$$= \frac{1}{mp}\left\{p_0^2\int_0^{t_1} k(t_1-\tau_1)M\left[y_0(t)y_0(\tau_1)\right]\mathrm{d}\tau_1\right.$$
$$\left.+2n\int_0^{t_1} k(t_1-\tau_1)M\left[y_0(t)\dot{y}_0(\tau_1)\right]\mathrm{d}\tau\right\}. \qquad (5.141)$$

We may present the expectations entering under the sign of integrals as

$$M\left[y_0(t)y_0(\tau_1)\right] = K_{y_0}(t,\tau_1) = D_{y_0}\mathrm{e}^{-\alpha|t-\tau_1|};$$

$$M\left[y_0(t)\frac{\partial y_0}{\partial \tau_1}\right] = \frac{\partial}{\partial \tau_1}M\left[y_0(t)y_0(\tau_1)\right]$$
$$= \frac{\partial}{\partial \tau_1}D_{y_0}\mathrm{e}^{-\alpha|t-\tau_1|} = \alpha D_{y_0}\mathrm{e}^{-\alpha|t-\tau_1|}\mathrm{sign}(t-\tau_1).$$

Through transformations the expression for M_1 takes the form

$$M_1 = \frac{D_{y_0}}{p}\int_0^{t_1} k(t_1-\tau_1)\left[p_0^2+2n\alpha\mathrm{sign}(t-\tau_1)\right]\mathrm{e}^{-\alpha|t-\tau_1|}\mathrm{d}\tau_1. \qquad (5.142)$$

Carrying out similar transformations with M_2, we obtain

$$M_2 = \frac{D_{y_0}}{p}\int_0^{t} k(t-\tau)\left[p_0^2+2n\alpha\,\mathrm{sign}(t_1-\tau)\right]\mathrm{e}^{-\alpha|t_1-\tau|}\mathrm{d}\tau. \qquad (5.143)$$

The variance of Δy at an instant t_k is

$$D_{\Delta y}(t_k)=D_{y_0}+D_y(t_k)-2\frac{D_y}{p}\int_0^{t_r} k(t_k-\tau)\left[p_0^2+2n\alpha\,\mathrm{sign}(t_k-\tau)\right]\mathrm{e}^{-\alpha|t_k-\tau|}\mathrm{d}\tau.$$

As $t_k > \tau$, we have

$$D_{\Delta y}(t_k) = D_{y_0}+D_y(t_k)-\frac{2D_{y_0}}{p}(p_0^2+2n\alpha)\int_0^{t_k} k(t_k-\tau)\mathrm{e}^{-\alpha|t_k-\tau|}\,\mathrm{d}\tau. \quad (5.144)$$

The standard deviation of Δy is

$$\sigma_{\Delta y}(t_k) = \sqrt{D_{\Delta y}(t_k)}.$$

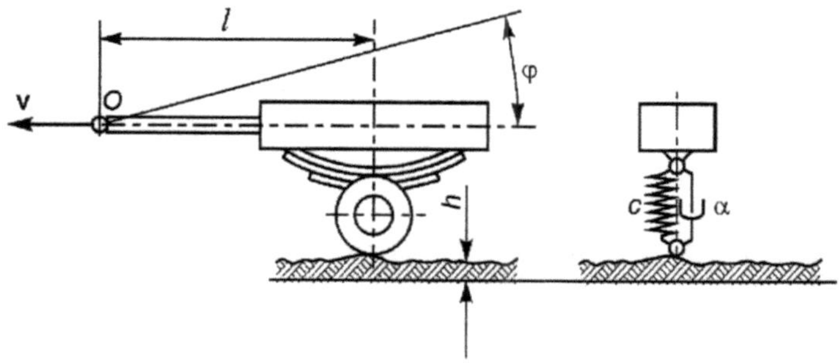

Fig. 5.13.

Let us consider the stationary vibrations of a system with one degree of freedom at a stationary kinematic excitation. Figure 5.13 *a* shows a trailer travelling along a road with random irregularities at a constant speed v. At the steady-state conditions of the motion the action of the road on the trailer can be regarded as a random stationary process. Let us assume that the point O of the trailers attachment to a motor vehicle has practically no vertical displacements. It is required to determine the probability characteristics of the stationary vibrations of the trailer (of its angular vibrations about the point O). In order to solve the problem, we must know the influence of the road on the trailer at different travel speeds, i.e. we need to know the spectrum of perturbations coming from the road, which depends on the road microprofile and the speed of travel. As an example a road section (the profile of a road section) with random irregularities is presented in Fig. 5.14. Many of roads sections have irregularities with a wide range of shape and length. The sequence of peaks and valleys of the road profile is random, therefore the magnitude and duration of the action of force impulses during the motion of a wheel when it passes these irregularities are random. To analyze the random vibrations of a motor vehicle during its movement along a road with random irregularities we must find out the dependence of the function h on time. To do this, it is sufficient to divide the abscissa of the plot shown in Fig. 5.14 on the travel speed v. As a result, we obtain the realization of the random function h in time. If we then divide the coordinate x by the

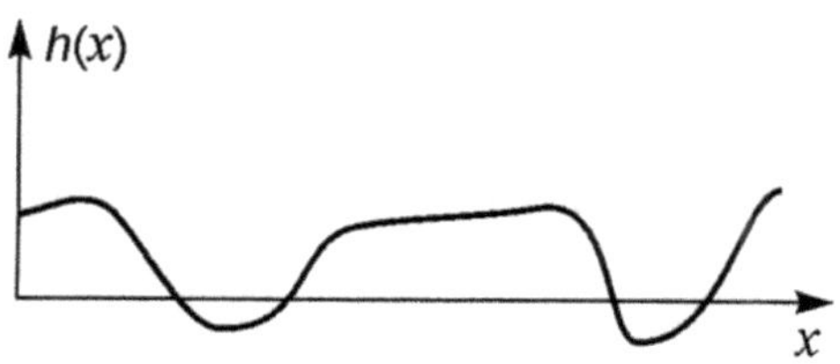

Fig. 5.14.

unit speed ($v_0 = 1$ m/sec), the values of the road profile function $h(x)$ will coincide with the values of the action function $h(t)$. With all other things being equal, the magnitudes and alternations of the force impulses acting on the motor vehicle during its movement along a certain section of the road at a constant speed, do not depend on when the vehicle traverses this section, hence, the action of the latter on the road will be the same at any instant of time, i.e. it will represent a stationary process.

The statistical processing of the road microprofiles measurement results allows us to obtain the probability characteristics of the random stationary function $h(t)$, i.e. m_h and $K_h(\tau)$ and, what is especially important, the spectral density $S_h(\omega)$. We cannot obtain the probability characteristics of the roads action on the motor vehicle true for all types of roads, therefore the latter are conventionally divided into a number of classes according to the root-mean-square height of their irregularities. More detailed information on statistical characteristics of the roads and the methods of obtaining them will be found in the special literature on the subject. In particular, the correlation functions of the action of roads on the motor vehicle $K_h(\tau)$ can be approximated by the following function

$$K_h(\tau) = D_h e^{-\alpha|\tau|} \cos \beta\tau, \tag{5.145}$$

where α, β are parameters depending on the road type and on the travel speed. We may present the parameters α and β as those bearing an explicit relationship to the travel speed v, m/sec:

$$\alpha = \alpha_1 v; \quad \beta = \beta_1 v,$$

where α_1, β_1 are the values of parameters at the unit travel speed that only depend on the type of a road.

The spectral density of the roads action on the motor vehicle is determined from the relation (for the assumed function $K_h(\tau)$)

$$S_h(\omega) = \frac{1}{2\pi} \int_{-\infty}^{\infty} K_h(\tau) \cos \omega\tau \, d\tau = \frac{4D_h \alpha(\alpha^2 + \beta^2 + \omega^2)}{\pi \left[(\alpha^2 + \beta^2)^2 + 2(\alpha^2 - \beta^2)\omega^2 + \omega^4\right]}$$

or with due account taken of the dependence of the factors α and β on the travel speed of the motor vehicle v :

$$S_h(\omega) = \frac{4D_h v \alpha_1 (v^2\gamma_1 + \omega^2)}{\pi \left[v^4\gamma_1^2 + 2v^2\gamma_2\omega^2 + \omega^4\right]}, \tag{5.146}$$

where $\gamma_1 = \alpha_1^2 + \beta_1^2$; $\gamma_2 = \alpha_1^2 - \beta_1^2$.

The following correlation function describing the action of a derno-podzolic soil country road on the motor vehicle in dry weather (at $v = 1$ m/sec) has been obtained by way of processing the results of experimental investigations

$$K_h(\tau) = 61.40\mathrm{e}^{-0.0627|\tau|}\cos 0.196\tau.$$

The spectral density corresponding to the function $K_h(\tau)$ and depending on the motor vehicles travel speed is

$$S_h(\omega) = 61.40\left[\frac{0.0627v}{0.0039v^2 + (\omega + 0.196v)^2} + \frac{0.0627v}{0.0039v^2 + (\omega - 0.196v)^2}\right]. \tag{5.147}$$

The plot of $S_h(\omega)$ versus ω at $v = 1\,\mathrm{m/sec}$ is presented in Fig. 5.15.

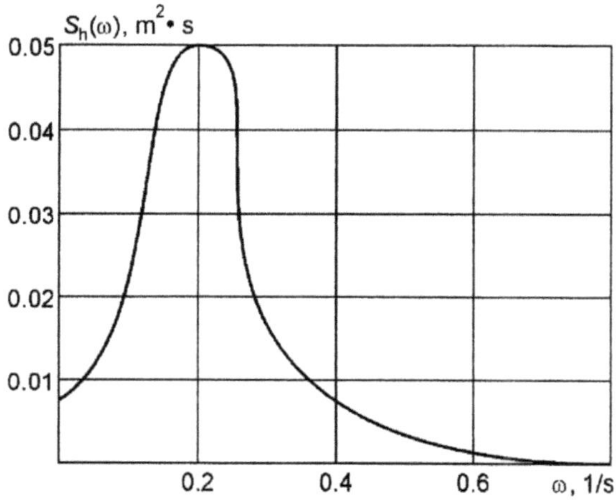

Fig. 5.15.

We can transform the expression for $S_h(\omega)$ (5.147) to the form

$$S_h(\omega) = 7.8v\frac{-(i\omega)^2 + 4.28\cdot 10^{-2}v^2}{|(i\omega)^2 + 0.125vi\omega + 4.28v^2\cdot 10^{-2}|^2}. \tag{5.148}$$

Neglecting the mass of the chassis, we may consider the trailer as a system with one degree of freedom. The equation of the small angular vibrations of the trailer about the point O is (Fig. 5.13 *a*, *b*)

$$\ddot{\varphi} + 2n\dot{\varphi} + p_0^2\varphi = p_0^2\frac{1}{l}h + 2n\frac{\dot{h}}{l}, \tag{5.149}$$

where $2n = \alpha l^2/J_O$; $p_0^2 = cl^2/J_O$.

In the frequency area we obtain

$$\left[(i\omega)^2 + 2ni\omega + p_0^2\right]\varphi(i\omega) = \frac{1}{l}(p_0^2 + 2n(i\omega))H(i\omega).$$

Therefore, the frequency function for the equation (5.149) is equal to

$$W(i\omega) = \frac{p_0^2 + 2ni\omega}{l((i\omega)^2 + 2ni\omega + p_0^2)}. \tag{5.150}$$

And so, the spectral density of a solution is

$$S_\varphi(\omega) = \left| \frac{p_0^2 + 2ni\omega}{l\left[(i\omega)^2 + 2ni\omega + p_0^2\right]} \right|^2 S_h(\omega). \tag{5.151}$$

Substituting the expression for

$$S_\varphi(\omega) = \frac{7.8v}{l^2} \frac{\left[b_1(i\omega)^4 + b_2(i\omega)^2 + b_3\right]}{\left|(i\omega)^4 + a_1(i\omega)^3 + a_2(i\omega)^2 + a_3(i\omega) + a_4\right|^2},$$

where

$$b_2 = -(p_0^2 + 2n^2 \cdot 4.3 \cdot 10^{-2}v^2); \quad b_3 = 4.3 \cdot 10^{-2}v^2 p_0^4; \quad b_1 = 4n^2;$$
$$a_1 = (0.125v + 2n); \quad a_2 = 2n \cdot 0.125v + 4.3 \cdot 10^{-2}v^2 + p_0^2;$$
$$a_3 = p_0^2 \cdot 0.125v + 2n \cdot 4.3 \cdot 10^{-2}v; \quad a_4 = 4.3 \cdot 10^{-2}p_0^2 v^2.$$

The plot of spectral density S_φ variation against ω is presented in Fig. 5.16 at $v = 20$ m/sec, $n = 0.3$.

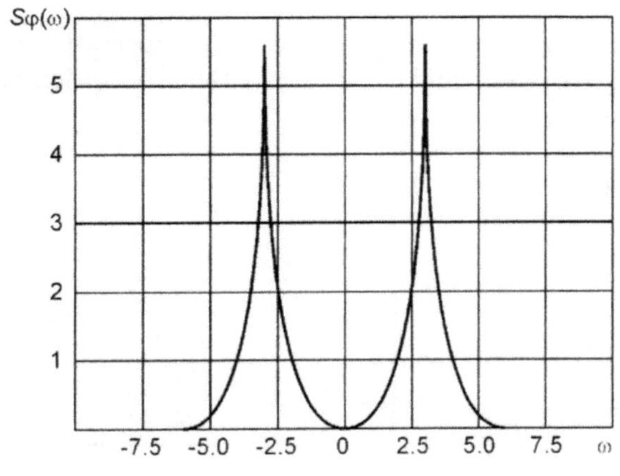

Fig. 5.16.

The variance of the angular deflection is

$$D_\varphi = \int_{-\infty}^{\infty} S_\varphi(\omega)\mathrm{d}\omega = D_\varphi(p_0, n, v). \tag{5.152}$$

The values of integrals of the form (5.152) are given in Appendix 2. Using the Appendix 2 ($D_\varphi = J_4$), we find

$$D_\varphi = 2\pi \frac{-a_0 a_1 b_2 + \dfrac{a_0 b_3}{a_4}(a_0 a_3 - a_1 a_2)}{2a_0(a_0 a_3^2 + a_1^2 a_4 - a_1 a_2 a_3)}.$$

The expression obtained for the variance depending on the parameters of the system and the travel speed enables us to investigate their influence on the variance of an angular deflection. The plot of angle φ root-mean-square value variation as a function of v at is shown in Fig. 5.17 *a*. The plot of σ_φ variation as a function of the viscous resistance force coefficient n at $v = 10$ m/sec is shown in Fig. 5.17 *b*. To make an analysis of the trailers suspension we must know the force acting on it during the travel of the trailer along a road with random irregularities. For the considered simplified model this force is equal to

$$N = c(h - \varphi l) + \alpha(\dot{h} - \dot{\varphi} l) = ml\ddot{\varphi}. \tag{5.153}$$

The spectral density of the force N (the relations (5.153) and (2.101)) are

$$S_N(\omega) = |W_1(\omega)|^2 S_h(\omega), \tag{5.154}$$

where

$$|W_1(\omega)|^2 = \left| \frac{c(i\omega)^2 + \alpha(i\omega)^3}{[(i\omega)^2 + 2ni\omega + p_0^2]} \right|^2.$$

As the expectation is constant at a stationary process, the expectation of the random function h derivative is zero and consequently the expectation of the angle φ is constant and equal to

$$m_\varphi = \frac{1}{l} m_h, \qquad (m_N = 0).$$

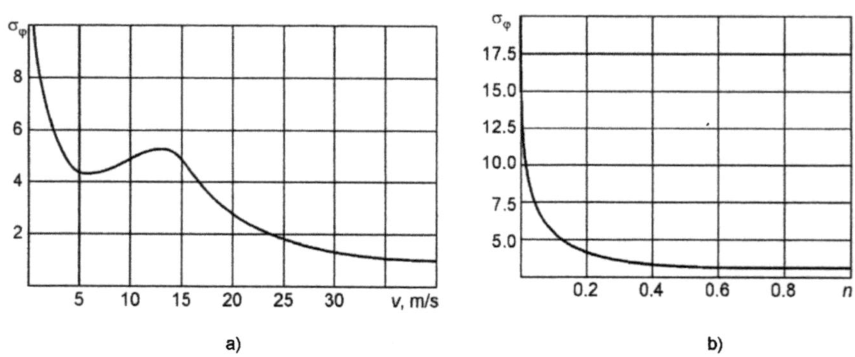

Fig. 5.17.

The variance of the force N is

$$D_N = \int_{-\infty}^{\infty} S_N(\omega)\,\mathrm{d}\omega.$$

Stresses occurring in the elastic suspension, are proportional to the force N, i.e. $\sigma = kN$.

With the knowledge of the probability characteristics m_N and D_N of the force N we can determine the probability characteristics of the stress σ : $m_\sigma = 0$, $\sigma_\sigma = k^2\sigma_N$.

Expression (5.154) with due account of (5.150) can be transformed to the form that is convenient for integration

$$S_N(\omega) = \frac{12.5v\left[(-i\omega)^6 + 4.3\cdot 10^{-2}v^2(i\omega)^4\right]}{\left|(i\omega)^4 + a_1(i\omega)^3 + a_2(i\omega)^2 + a_3(i\omega) + a_4\right|^2}.$$

Figure 5.18 *a*, *b* shows the plots of variation of the root-mean-square values σ_N as a function of the trailer motion velocity v at $n = 0.3$ (Fig. 5.18 *a*) and of the coefficient n at $v = 10$ m/sec (Fig. 5.18 *b*).

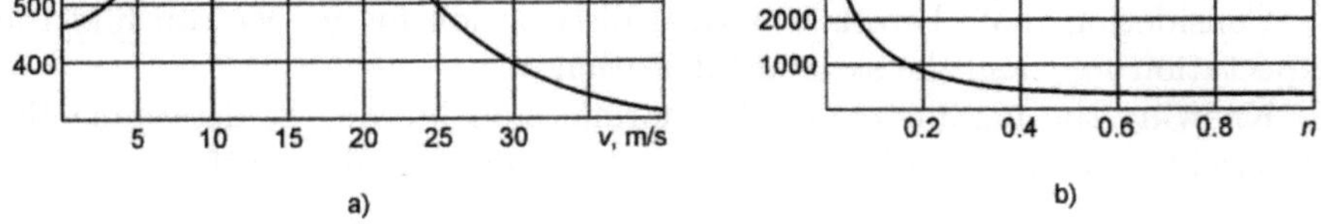

Fig. 5.18.

5.4 The Problem of Overshoots at Random Vibrations

When solving applied problems, we often have to determine the probability that a random function will overshoot the given level. A motor vehicle travel along a road with random irregularities is a case in point. The problem of determining the probability of a break-down in a suspension has become a

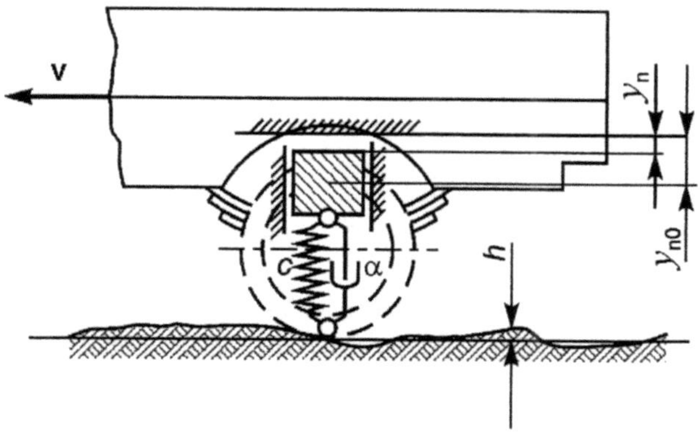

Fig. 5.19.

matter of practical interest. Any break-down of this kind involves large impact loads, which is undesirable. If we consider the relative vertical displacement of the suspension y_n to be a random function, the problem of the probability of such a break-down is equivalent to that of the probability of exceedance $y_n(t) \geq y_{n0}$, where y_{n0} is a free motion of the suspension (Fig. 5.19). This problem represents a special case of the general problem of overshoots. At the stationary vibrations of a motor vehicle we can obtain a relationship connecting the spectral density $S_{y_n}(\omega)$ of the relative displacement $y_n(t)$ with the spectral density of the road action $S_h(\omega)$:

$$S_{y_n}(\omega) = |W_{y_n}|^2 \, S_h(\omega), \tag{5.155}$$

where $|W_{y_n}|$ is the transfer function modulus relating the input $h(t)$ with the relative displacement of the suspension.

Considering that there is a normal distribution for y_n, we can find the expectation m_{y_n} and the standard deviation σ_{y_n}.

Knowing the parameters of this normal distribution, we determine the probability that the relative displacement y_n will exceed the value y_{n_0} :

$$P(y_n(t) \geq y_{n_0}) = \frac{1}{\sqrt{2\pi}} \int\limits_{t_1}^{\infty} \exp\left\{-\frac{t^2}{2}\right\} \mathrm{d}t = \Phi(\infty) - \Phi(t_1), \tag{5.156}$$

where $t_1 = \dfrac{y_{n_0} - m_{y_n}}{\sigma_{y_n}}$.

Example 5.5. Let us consider a simplified model (Fig. 5.19) ignoring the displacement of the motor vehicles body in comparison with the displacements of the suspension. In this case, we may consider the vertical displacement of the suspension separately, which will lead us to a system with one degree of freedom (the design scheme is shown in Fig. 5.19). At the designing stage, the problem of the exceedance of the suspensions structural dynamic motion y_{n_0} is traditionally considered without regard for the influence of the motion limiting device on the vibrations of the suspension. The dynamic motion y_{n_0} is chosen depending upon the permissible probability of a breakdown in the suspension at a design travel speed in the given road conditions. Let us determine the probabilities of y_n exceeding the value y_{n_0} as a function of the speed of the motor vehicle travelling along a Belgian pave road.

The numerical values of the system parameters are:

$$m = 3 \text{ kg}; \quad c = 300 \text{ N/cm}; \quad \alpha = 1.5 \text{ N sec/cm};$$

$$n = 0.5 \text{ 1/sec}; \quad p = \sqrt{\frac{c}{m}} = \sqrt{\frac{30}{0.3}} = 10 \text{ 1/sec}; \quad y_{n_0} = 8 \text{ cm}.$$

The equation of the motion of the mass m was presented in Sect. 5.3 and is of the form (the equation (5.138), where y_0 should be replaced with h)

$$\ddot{y}_n + 2n\dot{y}_n + p_0^2 y_n = 2n\dot{h} + p_0^2 h, \tag{5.157}$$

where $2n = \alpha/m$; $p_0^2 = c/m$.

Let us assume that the probability characteristics of the road irregularities are

$$\begin{aligned} &m_h = 0; \quad K_h(\tau) = 20\mathrm{e}^{-0,05|\tau|} \text{ cm}^2; \\ &S_h = 20\left(\frac{0.1v}{0.0025v^2 + \omega^2}\right), \text{ cm}^2 \cdot \text{sec}, \end{aligned} \tag{5.158}$$

where v is measured in m/sec.

The variance is

$$D_{y_n} = 2v \int_{-\infty}^{\infty} \left| \frac{2ni\omega + p_0^2}{(i\omega)^2 + 2ni\omega + p_0^2} \right|^2 \frac{\mathrm{d}\omega}{|i\omega + 0.05v|^2} \tag{5.159}$$

or

$$D_{y_n} = 2v \int_{-\infty}^{\infty} \frac{\left(-4n^2(i\omega)^2 + p_0^4\right) \mathrm{d}\omega}{|(i\omega)^3 + (2n + 0.05v)(i\omega)^2 + (p_0^2 + 0.1nv)i\omega + 0.05vp_0^2|^2}.$$

Taking the value of the integral J_3 from Appendix 2, we obtain

$$D_{y_n} = \frac{v\pi \left[4n^2 + \dfrac{(2n + 0.05v)p_0^4}{0.05vp_0^2}\right]}{[-0.05vp_0^2 + (2n + 0.05v)(p_0^2 + 0.1nv)]}. \tag{5.160}$$

Figure 5.20 shows the plot of σ_{y_n} (at $n = 0.05p_0$) variation as a function of travel speed of a motor vehicle and the probability p of y_n exceedance the value y_{n_0} as a function of motor vehicles travel speed is presented in Fig. 5.21.

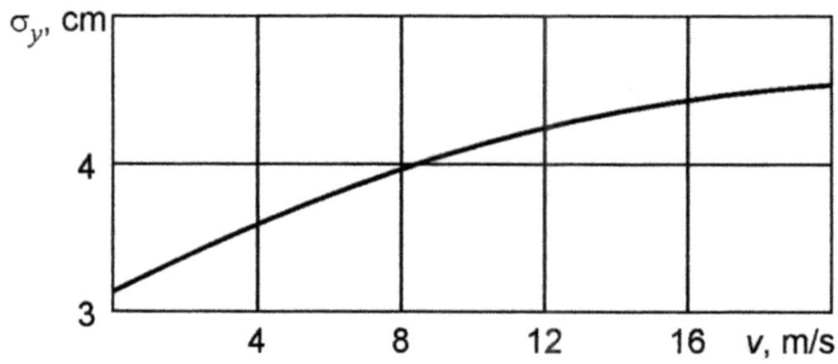

Fig. 5.20.

For example, at a travel speed of 30 km/h (8.3 m/sec) the probability that dynamic motion of a suspension will be exceeded equals $p = 0.02$. This probability is the probability of occurrence of the limit state (the suspensions "break-down"), which we cannot consider a mass event, i.e. such probability does not allow us to judge the quality of a structure. But if we compare two structures, the optimum one will be that whose "break-down" probability is lower.

The properly designed normal conditions ought not to result in "break-downs" or rather the "break-down" probability should be a very small quantity depending on specific operation conditions and structure reliability requirements. The random vibrations of the suspension cause random stresses in its elastic elements, governing the structures life. Random stresses varying in time lead to the continuous accumulation of fatigue damage. Therefore, to make an estimation of life we must know at least the average number of overshooting the given level of stresses, for example, the fatigue strength, during a known period of time t_k. With linear characteristic of elastic elements of

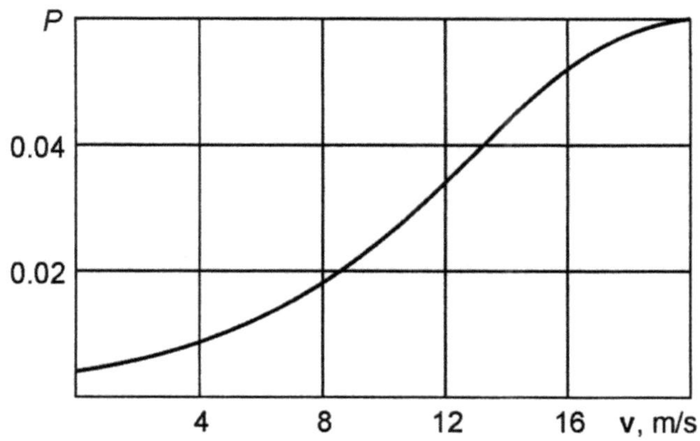

Fig. 5.21.

the suspension the stresses are proportional to its relative displacement, i.e. the overshooting of the given stress level is equivalent to the overshooting of the given relative suspension displacement, for example y_{n_1} in Fig. 5.22.

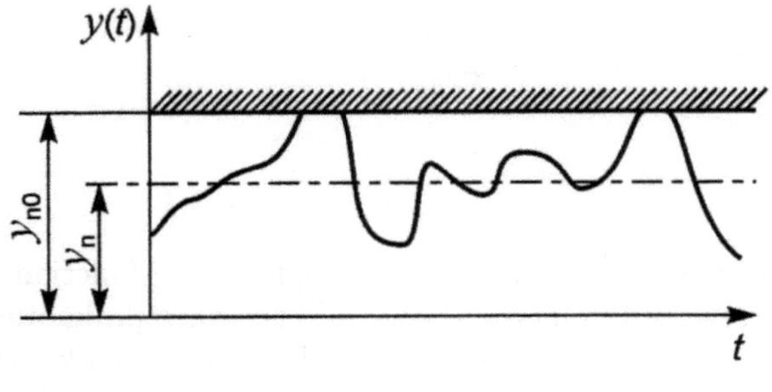

Fig. 5.22.

Let us consider a differentiable random process with the duration t_k (Fig. 5.23). Such realization has a finite number of maxima and minima with various values $x_{\max}(t)$ and $x_{\min}(t)$ on a finite interval of time. The plot of the variation of a random function $X(t)$ can more than once intersect the fixed level x_0 both from the bottom upwards (with a positive derivative), and from the top down (with a negative by derivative). The first intersection (the function $X(t)$ reaches the given level for the first time) occurs at an instant t_1. If the random function $X(t)$ intersects the level x_0 from the bottom upwards, it is usual to call this a positive overshoot and if the intersection has a top-down trends this is generally referred to as a negative overshoot.

Let us determine the average number of the intersections of a random process $X(t)$ with a given curve $a(t)$ (See Fig. 5.23). A special case of this problem is the overshooting problem, conditioned by

$$a(t) = x_0 = \text{const}.$$

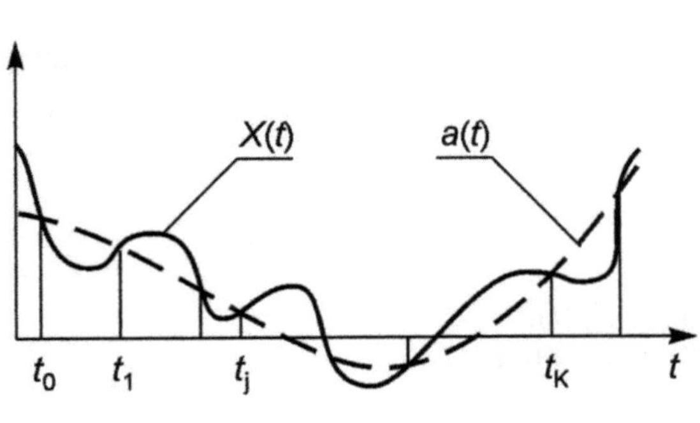

Fig. 5.23.

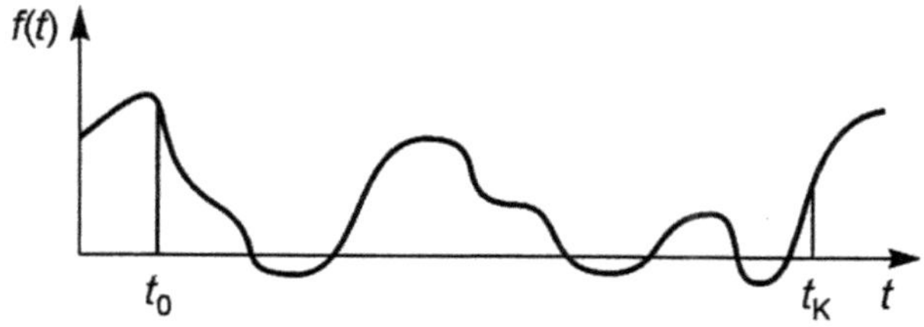

Fig. 5.24.

Henceforth we shall consider $a(t)$ as a continuous single-valued function of t; the random function $X(t)$ as a differentiable one; and the joint probability density $f(x, \dot{x})$ for the function $X(t)$ and its first derivative $\dot{X}(t)$ as a known probability. Let us show that the number of zeros for some function $f(t)$ on an interval $(t_0,\ t_k)$ (Fig. 5.24) is determined by the formula

$$n = \int_{t_0}^{t_k} |f(t)|\, \delta\,[f(t)]\,\mathrm{d}t. \tag{5.161}$$

Let us change the variable, setting $f(t) = z$.

Then we can transform expression (5.161) to the form (as $dz = f\mathrm{d}t$):

$$n = \begin{cases} \displaystyle\int_{z_1}^{z_2} \delta(z)\,\mathrm{d}z & 0 \le f(t) < \infty; \\ \displaystyle -\int_{z_1}^{z_2} \delta(z)\,\mathrm{d}z & -\infty < f(t) \le 0, \end{cases} \tag{5.162}$$

where

$$z_1 = f(t_0); \qquad z_2 = f(t_k). \tag{5.163}$$

Integrals (5.162), (5.163) are equal to zero at $z \neq 0$. At $z_i = 0$ integral (5.162) is equal to

$$\int_{z_i-\varepsilon}^{z_i+\varepsilon} \delta(z)\,\mathrm{d}z = 1,$$

where z_i is a zero of the function $f(t)$.

Similarly, integral (5.163) at $z_i = 0$ is

$$-\int_{z_i-\varepsilon}^{z_i+\varepsilon} \delta(z)\,\mathrm{d}z = \int_{z_i+\varepsilon}^{z_i-\varepsilon} \delta(z)\,\mathrm{d}z = 1.$$

Therefore the full number of zeros of the function $f(t)$ on the interval $(t_0,\ t_k)$ is equal to the sum of integrals, i.e.

$$n = \sum_{i=1}^{n} \int_{z_i-\varepsilon}^{z_i+\varepsilon} \delta(z)\,\mathrm{d}z - \sum_{i=1}^{n} \int_{z_i-\varepsilon}^{z_i+\varepsilon} \delta(z)\,\mathrm{d}z = n_+ + n_-,$$

where n_+ is the number of zeros on the interval $(t_0,\ t_k)$ with a positive derivative ($\dot{f}(t) > 0$) and n_- is the number of zeros on the interval $(t_0,\ t_k)$ with a negative derivative ($\dot{f}(t) < 0$). Let us take advantage of the obtained results in order to determine the average number of the intersections of a random function $x(t)$ with a nonrandom curve $a(t)$. For this purpose, let us enter the function

$$f(t) = \varepsilon(t) - a(t).$$

The intersections of the random function $X(t)$ and the given curve $a(t)$ on the interval $(t_1,\ t_2)$ coincide with the zeros of the random function $\varepsilon(t)$, therefore

$$n = \int_{t_0}^{t_k} |\dot{x} - \dot{a}|\,\delta(x-a)\mathrm{d}t.$$

The number of zeros n will be different for different realizations of the process, hence, n is a random variable randomly varying from one realization to another, therefore the average number of zeros is equal to the expectation of n, i.e.

$$N = M\,[n] = \int_{t_0}^{t_k} \left[\iint_{-\infty}^{\infty} |\dot{x} - \dot{a}|\,\delta(x-a) f(x,\dot{x})\,\mathrm{d}x\,\mathrm{d}\dot{x} \right] \mathrm{d}t. \tag{5.164}$$

The joint probability density $f(x,\dot{x})$ of the random function X and its first derivative $\dot{x}$ enters in expression (5.164). The methods of determining $f(x,\dot{x})$ are presented in [31]. In the general case, determination of a joint probability density (for non- stationary random processes) involves great difficulties because it requires a large body of information on the behavior of a random function. The problem of obtaining a joint probability density becomes easier to solve, if we know that the random process is normal.

For a normal random process, the joint probability density (at an instant t) is

$$f(x,\dot{x}) = \frac{1}{2\pi\sigma_x\sigma_{\dot{x}}\sqrt{1-k_{x\dot{x}}}} \times \exp\left\{ -\frac{1}{2(1-k_{x\dot{x}}^2)} \left[\frac{(x-m_x)^2}{\sigma_x^2} - \frac{2k_{x\dot{x}}(x-m_x)(\dot{x}-m_{\dot{x}})}{\sigma_x\sigma_{\dot{x}}} - \frac{(\dot{x}-m_{\dot{x}})^2}{\sigma_{\dot{x}}^2} \right] \right\}, \tag{5.165}$$

where $k_{x\dot{x}}$ is a normalized cross-correlation function.

For a stationary process, the cross-correlation function $k_{x\dot{x}}$ (if x and $\dot{x}$ are taken at the same instant) is equal to zero (see Sect. 3.3), therefore the joint probability density takes the form

$$f(x,\dot{x}) = \frac{1}{2\pi\sigma_x\sigma_{\dot{x}}}\exp\left\{-\frac{1}{2}\frac{(x-m_x)^2}{\sigma_x^2}\right\}\exp\left\{-\frac{(\dot{x}-m_{\dot{x}})^2}{2\sigma_{\dot{x}}^2}\right\} = F(x)F(\dot{x}). \tag{5.166}$$

We can simplify the expressions for N (5.164) by integrating over x (using the property of the delta function), i.e.

$$\int_{-\infty}^{\infty} f(x)\delta(x-a)\mathrm{d}x = f(a).$$

Some manipulations give us

$$N = \int_{t_0}^{t_k}\left[\int_{-\infty}^{\infty}|\dot{x}-\dot{a}|\, f(a,\,\dot{x})d\dot{x}\right]dt.$$

Let us assume that t_i are instants corresponding to an intersection of the random process by the curve $a(t)$ on an interval $(t_0,\, t_k)$, i.e. $x(t_i) = a(t_i)$. As has been shown above, the total number of intersections n is equal to

$$n = n_+ + n_-,$$

where

$$n_+ = \int_{t_0}^{t_k}(\dot{x}-\dot{a})\delta(x-a)\mathrm{d}t \quad (\dot{x} > 0);$$

$$n_- = \int_{t_0}^{t_k}(\dot{x}-\dot{a})\delta(x-a)\mathrm{d}t \quad (\dot{x} < 0).$$

The average number of intersections is: with a positive derivative

$$N_+ = M\left[n_+\right] = \int_{t_0}^{t_k}\left[\int_{0}^{\infty}\dot{\varepsilon}(t)f(a,\,\dot{a}+\dot{\varepsilon})\mathrm{d}\dot{\varepsilon}\right]dt;$$

with a negative derivative is

$$N_- = M\left[n_-\right] = -\int_{t_0}^{t_k}\left[\int_{-\infty}^{0}\dot{\varepsilon}(t)f(a,\,\dot{a}+\dot{\varepsilon})\mathrm{d}\dot{\varepsilon}\right]dt.$$

The total average number of intersections is

$$N = N_+ + N_-.$$

Let us determine the average number of intersections of the given level, if $a(t) = \text{const} = x_0$ (see Fig. 5.23). In this case

$$n_+ = \int_{t_0}^{t_k} \dot{\varepsilon}\delta(\varepsilon - x_0)\mathrm{d}t \quad (\dot{\varepsilon} > 0);$$

$$n_- = -\int_{t_0}^{t_k} \dot{\varepsilon}\delta(\varepsilon - x_0)\mathrm{d}t \quad (\dot{\varepsilon} < 0).$$

The average number of intersections is

$$\begin{aligned} N_+ &= \int_{t_0}^{t_k} \left[\int_0^{\infty} \dot{\varepsilon} f(x_0, \dot{\varepsilon})\mathrm{d}\dot{\varepsilon} \right] \mathrm{d}t; \\ N_- &= \int_{t_0}^{t_k} \left[\int_{-\infty}^{0} \dot{\varepsilon} f(x_0, \dot{\varepsilon})\mathrm{d}\dot{\varepsilon} \right] \mathrm{d}t. \end{aligned} \tag{5.167}$$

If $x_0 = 0$, then the average number of zeros of the process $x(t)$ on the interval of time (t_0, t_k) is

$$N = N_+ + N_- = \int_{t_0}^{t_k} \left[\int_0^{\infty} \dot{\varepsilon}\, f(x_0, \dot{\varepsilon})\mathrm{d}\dot{\varepsilon} \right] \mathrm{d}t - \int_{t_0}^{t_k} \left[\int_{-\infty}^{0} \dot{\varepsilon}\, f(x_0, \dot{\varepsilon})\mathrm{d}\dot{\varepsilon} \right] \mathrm{d}t. \tag{5.168}$$

In order to obtain a numerical result from formula (5.168) we must know the probability density distribution law of the random function $\varepsilon(t)$ derivative. For stationary random processes the expression for N_+ and N_- become simpler (because the internal integrals in (5.168) do not depend on time):

$$\begin{aligned} N_+ &= (t_k - t_0) \int_0^{\infty} \dot{\varepsilon} f(a_0, \dot{\varepsilon})\mathrm{d}\dot{\varepsilon}; \\ N_- &= -(t_k - t_0) \int_{-\infty}^{0} \dot{\varepsilon} f(a_0, \dot{\varepsilon})\mathrm{d}\dot{\varepsilon}. \end{aligned} \tag{5.169}$$

Expressions (5.169) allow us to determine the average number of intersections of the given level and the random stationary function $x(t)$ in a unit of time:

$$\tilde{N} = \frac{N_+ + N_-}{t_k - t_0} = \int_0^{\infty} \dot{\varepsilon} f(x_0, \dot{\varepsilon}) \mathrm{d}\dot{\varepsilon} - \int_{-\infty}^{0} \dot{\varepsilon}(x_0, \dot{\varepsilon}) \mathrm{d}\dot{\varepsilon}. \tag{5.170}$$

Let us determine the average number of intersections of a normal stationary process $x(t)$ with a horizontal straight line $x_0 = \text{const}$, using relations (5.169) and joint distribution law (5.165). In this case $k_{x\dot{x}} = 0$, therefore from (5.167) we obtain

$$\begin{aligned} N_+ &= \frac{(t_k - t_0)}{2\pi\sigma_x\sigma_{\dot{x}}} \exp\left\{-\frac{(x_0 - m_x)^2}{2\sigma_x^2}\right\} \int_0^{\infty} \dot{x} \exp\left\{-\frac{\dot{x}^2}{2\sigma_{\dot{x}}^2}\right\} \mathrm{d}\dot{x}; \\ N_- &= \frac{(t_k - t_0)}{2\pi\sigma_x\sigma_{\dot{x}}} \exp\left\{-\frac{(x_0 - m_x)^2}{2\sigma_x^2}\right\} \int_{-\infty}^{0} \dot{x} \exp\left\{-\frac{\dot{x}^2}{2\sigma_{\dot{x}}^2}\right\} \mathrm{d}\dot{x}. \end{aligned} \tag{5.171}$$

It follows from (5.171) that $N_+ = N_-$, hence the total average number of intersections of the given level and the normal stationary process is $N = 2N_+$.

By calculations we obtain

$$N = \frac{(t_k - t_0)\sigma_{\dot{x}}}{\pi\sigma_x} \exp\left\{-\frac{(x_0 - m_x)^2}{2\sigma_x^2}\right\}. \tag{5.172}$$

Let us determine the average duration of the stay of the random function $x(t)$ above the level x_0 [31]. The probability of this is

$$P(X > x_0) = \int_{a_0}^{\infty} f\left[x(t)\right] \mathrm{d}x.$$

Let us divide up an interval (t_k, t_0) into n equal small intervals Δt_t each of which is located near the instants t_t. Let us consider the intervals Δt_t so small that it is possible to neglect cases where the function $(x - a_0)$ reverses sign within them. Let us introduce a system of random variables Δ_t each of them being equal to the corresponding interval Δt_t or 0 depending on whether the random function in this interval is more or less than x_0. Then, the total time of the random function being above the given level x_0 is

$$t_{a_0} = \sum_{i=1}^{n} \Delta_i.$$

The average time of the random function being above the given level x_0 for a time $(t_k - t_0)$ is

$$T_a = M\left|t_{x_0}\right| = \sum_{i=1}^{n} M\left[\Delta_i\right]; \quad \left(M\left[\Delta_i\right] = \Delta t_i \int_{x_0}^{\infty} f(x) \mathrm{d}x\right). \tag{5.173}$$

In the limit at $n \to \infty$ we obtain from (5.173)

$$T_a = \int_{t_0}^{t_k} \int_{x_0}^{\infty} f(x) \mathrm{d}x \, \mathrm{d}t. \tag{5.174}$$

Let us determine the average duration of an overshoot, dividing the expression (5.174) by the average number of positive overshoots N_+, i.e.

$$\tau_a = \frac{T_a}{N_+} = \frac{\int_{t_0}^{t_k} \int_{x_0}^{\infty} f(x) \mathrm{d}x \, \mathrm{d}t}{\int_{t_0}^{t_k} \left[\int_0^{\infty} \dot{x} f(x_0, x) \mathrm{d}\dot{x} \right] \mathrm{d}t}. \tag{5.175}$$

For a stationary random process the expressions for T_a and τ_a take the form:

$$T_a = (t_k - t_0) \int_{x_0}^{\infty} f(x) \mathrm{d}x; \tag{5.176}$$

$$\tau_a = \frac{\int_{x_0}^{\infty} f(x) \mathrm{d}x}{\int_0^{\infty} \dot{x} f(x_0, \dot{x}) \mathrm{d}\dot{x}}. \tag{5.177}$$

For a normal stationary random process

$$T_a = \frac{(t_k - t_0)}{2\pi\sigma_x} \left[1 - \Phi\left(\frac{x_0 - m_x}{\sigma_x} \right) \right]; \tag{5.178}$$

$$\tau_a = \frac{\pi\sigma_x}{\sigma_{\dot{x}}} \exp\left\{ -\frac{(x_0 - m_x)^2}{2\sigma_x^2} \right\} \left[1 - \Phi\left(\frac{x_0 - m_x}{\sigma_x} \right) \right]. \tag{5.179}$$

Example 5.6. The correlation function of an angle φ (see Fig. 5.13) at the normal stationary vibrations of a trailer ($m_\varphi = 0$) is

$$K_\varphi(\tau) = \sigma_\varphi^2 \mathrm{e}^{-\alpha|\tau|} \left(\cos\beta\tau + \frac{\alpha}{\beta} \sin\beta\,|\tau| \right).$$

It is required to determine the average number of exceedances by the angle φ of the allowable value $a_0 = \varphi_0 = \pm 5°$ for 10 minutes and the average duration of an overshoot τ_a. The numerical values of parameters are known: $\sigma_\varphi = 2°$; $\alpha = 0.1$ 1/sec; $\beta = 0.6$ 1/sec.

As

$$\sigma_{\dot{\varphi}}^2 = -\frac{d^2 K_\varphi}{d\tau^2}\,|_{\tau=0} = \sigma_\varphi^2(\alpha^2 + \beta^2),$$

then using formula (5.172), we obtain

$$N = \frac{600\,\sqrt{\alpha^2+\beta^2}}{\pi} \exp\left\{-\frac{\varphi_0^2}{2\sigma_\varphi^2}\right\} = \frac{600}{\pi} 0.61 \cdot 0.043 \approx 5.$$

The average duration of the overshoot is

$$\tau_a = \frac{\pi}{\sqrt{\alpha^2+\beta^2}} \exp\left\{\frac{\varphi_0^2}{2\sigma_\varphi^2}\right\} \left[1 - \Phi\left(\frac{\varphi_0}{\sigma_\varphi}\right)\right].$$

By substituting numerical values, we determine

$$\tau_a = \frac{\pi}{0.615} 23.104\,(1 - 0.9948) \approx 0.7\,\text{sec}.$$

5.5 Nonlinear Random Vibrations

Nonlinear mechanical systems loaded with random forces are extensively practiced in mechanical engineering and, among other things, include numerous shock-absorption and damping devices for machines, instruments and structures; nonlinear control systems; and nonlinear problems of the dynamics of flying vehicles. Solving nonlinear problems, as a rule, involves great difficulties. It is an open secret that we cannot obtain a solution to a general-type nonlinear equation in an analytical form. This is true even for the simplest second-order equation, not to mention the system of nonlinear equations of the motion of mechanical systems loaded with deterministic random forces.

It was usual to believe not so long ago that any solution of nonlinear equations can only be deterministic at deterministic loads and random at random loads. Recent investigations of nonlinear dynamics carried out in years with the use of computers have made it possible to establish new physical phenomena that earlier seemed absolutely impossible in the field of traditional mechanics. It has been discovered that chaotic (unpredictable) motions are possible in a deterministic nonlinear system. This means that nonlinear systems can themselves, without external random actions, generate random processes. Moreover, approximated numerical methods of finding a solution, generally based on various simplifications and assumptions, are not only fraught with quantitative errors but can produce qualitatively different results as well. Therefore, the earlier developed approximate numerical methods of solving equations of nonlinear random vibrations (for example, the method of statistical linearization and that of moment functions) cannot always ensure the required accuracy of a solution. At small nonlinearities, however, approximate numerical methods of solution can be effective. Methods using the theory of Markov processes and the method of statistical tests allow us to obtain exact numerical solutions to nonlinear equations of statistical dynamics without a linearization of initial equations. This paragraph is devoted to three methods of solving nonlinear equations of the first and second order — the method of a statistical linearization, the method based on Markov processes and the method of statistical trials (Monte-Carlo method) that enjoy the greatest popularity in computational practice.

5.5.1 The Method of Statistical Linearization

Figure 5.25 *a* shows a one degree of freedom system, whose elastic characteristic of the spring is a nonlinear function of a displacement x (Fig. 5.25 *b*). The resistance force $F_2(\dot{x})$ (the friction force between the mass and the guide) has a nonlinear dependence on a motion velocity $\dot{x}$.

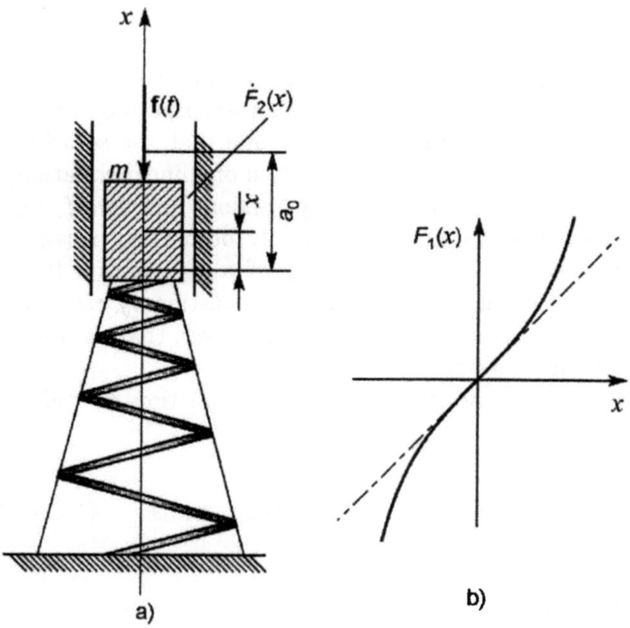

Fig. 5.25.

The equation of the forced oscillations of the mass m is

$$m\ddot{x} + F(x, \dot{x}) = f_\circ(t), \tag{5.180}$$

where $F(x, \dot{x}) = F_1(x) + F_2(\dot{x})$.

In each case there is an explicit specific dependence of functions F_1 and F_2 respectively on x and $\dot{x}$. For example, we may assume that for the conic spring, shown in Fig. 5.25,

$$F_1(x) = cx + c_1x^3 \tag{5.181}$$

up to a specific value of x (Fig. 5.25 *b*).

At the quadratic resistance law we may present the function F_2 as

$$F_2(\dot{x}) = \alpha\dot{x}^2 \text{sign}\dot{x}.$$

Let us consider the method of statistical linearization, which consists in the replacement of a nonlinear random function $F(x, \dot{x})$ by a linear one, i.e.

$$F(x, \dot{x}) \approx F_* = a_1 + a_2x_0 + a_3\dot{x}_0, \tag{5.182}$$

where x_0 and $\dot{x}_0$ are centered random functions; a_i are arbitrary coefficients, determined from the condition of the minimum variance of a random function

$$\Delta F = F(x, \dot{x}) - F_*. \tag{5.183}$$

The statistical linearization method is applicable to stationary nonlinear vibrations that are possible only in cases, where an external force $f_\circ(t)$ is a stationary random function. In addition, we assume that the solution $(x, \dot{x})$ has a normal distribution. Any stationary solution of a linear equation (for example, of equation (5.26)) with a stationary random function $f_\circ(t)$, having a normal distribution law, its would necessarily be normal because linear transformations do not change the normality of a distribution law. The solution of a linear equation at a normal law for a stationary random function $f_\circ(t)$ is not normal. It is possible to suppose, however, that at small nonlinearities it differs little from a normal one.

If vibrations are stationary, the variance of the random function ΔF takes the form

$$M\left[(\Delta F)^2\right] = \iint\limits_{-\infty}^{\infty} \left[F(x, \dot{x}) - a_1 - a_2x_0 - a_3\dot{x}_0\right]^2 f(x, \dot{x})\mathrm{d}x\,\mathrm{d}\dot{x}, \tag{5.184}$$

where $f(x, \dot{x})$ is the joint probability density of x and $\dot{x}$ that does not depend on time for stationary random functions.

Since $f(x, \dot{x})$ is an unknown function in formula (5.184), we have to assume that the function $f(x, \dot{x})$ is close to a two-dimensional normal distribution law of independent random functions, i.e. we may consider that

$$f(x, \dot{x}) = \frac{1}{2\pi\sigma_x\sigma_{\dot{x}}} \exp\left\{-\left[\frac{(x - m_x)^2}{2\sigma_x^2} + \frac{\dot{x}^2}{2\sigma_{\dot{x}}^2}\right]\right\}, \tag{5.185}$$

as for a stationary process $m_{\dot{x}} = 0$.

Let us determine the arbitrary parameters a_1, a_2 and a_3 from the conditions

$$\frac{\partial M\left[(\Delta F)^2\right]}{\partial a_i}=0 \qquad (i=1,\,2,\,3)\,. \tag{5.186}$$

By transforming we obtain the following values of coefficients from (5.186):

$$\begin{aligned} a_1 &= \int_{-\infty}^{\infty}\int_{-\infty}^{\infty} F(x,\dot{x})\, f(x,\dot{x})\,\mathrm{d}x\,\mathrm{d}\dot{x};\\ a_2 &= \frac{1}{\sigma_x^2}\int_{-\infty}^{\infty} F(x,\dot{x})\,(x-m_x)\, f(x,\,\dot{x})\,\mathrm{d}x\,\mathrm{d}\dot{x};\\ a_3 &= \frac{1}{\sigma_{\dot{x}}^2}\int_{-\infty}^{\infty}\int_{-\infty}^{\infty} F(x,\,\dot{x})\,\dot{x} f(x,\dot{x})\,\mathrm{d}x\,\mathrm{d}\dot{x}. \end{aligned} \tag{5.187}$$

From (5.187) we determine the relationships

$$a_i = a_i\left(m_x, \sigma_x, \sigma_{\dot{x}}\right). \tag{5.188}$$

As a result of the averaging operation of the function $F(x,\,\dot{x})$ we obtain the following linear equation in place of equation (5.180)

$$\ddot{x}_0+\frac{a_3}{m}\dot{x}_0+\frac{a_2}{m}x_0=\frac{1}{m}\,\mathring{f}_0(t)+\frac{m_f-a_1}{m}. \tag{5.189}$$

As x_0 and $\mathring{f}_0$ are centered random functions, by performing an operation of mathematical expectation on both parts of equation (5.189), we obtain

$$m_f-a_1=0, \qquad \left(a_1=a_1\left(m_x,\,\sigma_x,\,\sigma_{\dot{x}}\right)\right). \tag{5.190}$$

Finally we obtain the equation

$$\ddot{x}_0+\frac{a_3}{m}\dot{x}_0+\frac{a_2}{m}x_0=\frac{1}{m}\,\mathring{f}_0(t)\,. \tag{5.191}$$

The spectral densities of x and $\dot{x}$ will be:

$$S_x(\omega)=|W(i\omega)|^2 S_f(\omega)\,, \qquad S_{\dot{x}}(\omega)=|W(i\omega)|^2\omega^2 S_f(\omega)\,,$$

therefore the root-mean-square values of x and $\dot{x}$ are determined from the relations:

$$\sigma_x^2=\int_{-\infty}^{\infty}|W(i\omega)|^2 S_f\mathrm{d}\omega; \tag{5.192}$$

$$\sigma_{\dot{x}}^2 = \int_{-\infty}^{\infty} |W(i\omega)|^2 \omega^2 S_f d\omega. \tag{5.193}$$

As a result, we obtain three equations (5.190), (5.192), (5.193) for the determination of three unknowns m_x, σ_x and $\sigma_{\dot{x}}$ as a function of m_f and σ_f.

Example 5.7. It is required to determine the average number of exceedances N of a level a_0 by the center of gravity of a mass m for a given time t_k. The characteristic of the spring is nonlinear. Let us consider the stationary random vibrations of the mass presented in Fig. 5.25 *a*, assuming that the characteristic of the spring $F_1(x)$ can be presented as (5.181), and the resistance force linearly depends on $\dot{x}$.

The stationary random force of the normal white noise type acts on the mass m i.e.

$$m_{f_0} = \text{const}, \quad S_{f_0} = s_0.$$

The equation of the motion of the mass m at a nonlinear resistance force takes the form

$$\ddot{x} + 2n\dot{x} + p_0^2 x + \mu x^3 = \frac{1}{m}(\overset{\circ}{f_0} + m_{f_\circ}), \quad \left(\mu = \frac{c_1}{m}\right). \tag{5.194}$$

At $\mu = 0$ equation (5.194) becomes linear, and its solution is a normal stationary process. At $\mu \neq 0$ the solution of equation (5.194) is not a normal process, but at small values of μ we may assume that it differs little from a normal process, which enables us to use formulas (5.187) for the determination of coefficients a_1, a_2 and a_3. Having evaluated integrals (5.187) we obtain

$$a_1 = \mu m_x \left(m_x^2 + 3\sigma_x^2\right); \quad a_2 = 3\mu\left(m_x^2 + \sigma_x^2\right); \quad a_3 = 2n. \tag{5.195}$$

The linearized equation takes the form

$$\ddot{x}_0 + 2n\dot{x}_0 + p_0^2(x_0 + m_x) + a_2 x_0 + a_1 = \frac{\overset{\circ}{f_0} + m_{f_0}}{m}. \tag{5.196}$$

From (5.196) we obtain

$$p_0^2 m_x + a_1 = \frac{m_{f_0}}{m}.$$

Using expression (5.195) for a_1, we obtain the first equation relating m_f to m_x and σ_x :

$$\mu m_x \left(m_x^2 + 3\sigma_x^2\right) = \frac{m_{f_0}}{m} - p_0^2 m_x. \tag{5.197}$$

Having isolated the constant components from equation (5.196) we obtain the following equation with respect to the centered random functions

$$\ddot{x}_0 + 2n\dot{x}_0 + \left(p_0^2 + a_2\right) x_0 = \frac{\overset{\circ}{f}_0}{m}. \qquad (5.198)$$

In the considered case the spectral densities of x and $\dot{x}$ are equal to

$$S_x = \frac{s_0}{m^2 \left|(i\omega)^2 + 2ni\omega + p_0^2 + a_2\right|^2};$$

$$S_{\dot{x}} = \frac{s_0\omega^2}{m^2 \left|(i\omega)^2 + 2ni\omega + p_0^2 + a_2\right|^2},$$

therefore from (5.192) and (5.193) we obtain

$$\sigma_x^2 = \frac{s_0}{4nm^2\left[p_0^2 + 3\mu\left(m_x^2 + \sigma_x^2\right)\right]}; \quad \sigma_{\dot{x}}^2 = \frac{s_0}{4nm^2}. \qquad (5.199)$$

Let us determine m_x, σ_x and $\sigma_{\dot{x}}$ from equations (5.197) and (5.199). In a particular case when $m_f = 0$, we obtain

$$\sigma_x^2 = -\frac{p_0^2}{6\mu} + \sqrt{\frac{p_0^4}{36\mu} + \frac{s_0}{12nm^2\mu}} = \sigma_{x_0}^2\left(-1 + \sqrt{1 + 6\mu}\right)\frac{1}{3\mu_1} = \sigma_{x_0}^2 h_1, \qquad (5.200)$$

where $\sigma_{x_0}^2 = \dfrac{1}{2ap_0^2}$ is the variance of the solution at $\mu = 0$; $a = \dfrac{2nm^2}{s_0}$; $\mu_1 = \dfrac{\mu}{ap_0^4}$ is the non-dimensional small parameter.

Using formula (5.172), we determine the number of the level a_0 exceedances, (the number of these exceedances is half as large as the number of intersections of the level a_0):

$$N = \frac{t_k\sigma_{\dot{x}}}{2\pi\sigma_x}\exp\left\{-\frac{\left(a_0 - m_x\right)^2}{2\sigma_x^2}\right\}.$$

Example 5.8. Figure 5.26 shows the outlines of an accelerometer representing a mass m elastically attached to springs with a linear characteristic (with a total rigidity c). A stationary excitation $\overset{\circ}{f}(t)$ setting up the vibrations of the mass m is acting on the latter besides a slowly varying inertial force (that is to be measured by the accelerometer). In order to decrease the influence of a random disturbance on the indications of the accelerometer the mass m is placed in a cavity filled with a liquid. During the motion of the mass the resistance force is proportional to the square of a velocity $\dot{x}$. As the nominal force acting on the mass m, varies in time very slowly, the motion velocity of the mass is low in nominal conditions and we can ignore the resistance force, therefore the later primarily influences random vibrations.

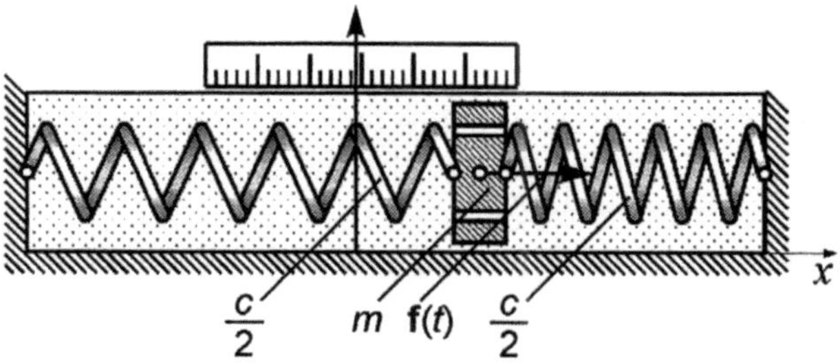

Fig. 5.26.

Denoting by Δx the displacement of the mass m from its nominal position let us write the following expression for the resistance force

$$F_c = \alpha_1 \Delta\dot{x}^2 \mathrm{sign} \Delta\dot{x}.$$

We consider the random disturbance probability characteristics to be known:

$$m_{f_0} = \mathrm{const}; \quad S_{f_0} = \frac{2\alpha\sigma_{f_0}^2}{\alpha^2 + \omega^2}, \quad \left(K_f = \sigma_{f_0}^2 \mathrm{e}^{-\alpha|\tau|}\right).$$

It is required to determine the expectation $m_{\Delta x}$, the standard deviation $\sigma_{\Delta x}$ of the random displacement of the mass m, considering that stationary random vibrations occur. The equation of the motion of the mass m takes the form

$$\Delta\ddot{x} + n_1 \Delta\dot{x}^2 \mathrm{sign} \Delta\dot{x} + p_0^2 \Delta x = \frac{f}{m}, \quad \left(n_1 = \frac{\alpha_1}{m}\right). \tag{5.201}$$

Let us replace the nonlinear resistance force $n_1 \Delta\dot{x}^2 \mathrm{sign} \Delta\dot{x}$ by a linear one:

$$n_1 \Delta\dot{x}^2 \mathrm{sign} \Delta\dot{x} \approx a_1 + a_3 \Delta\dot{x}_0.$$

Let us determine a_1 and a_3 from formulas (5.187). By calculations we obtain

$$a_1 = 0; \quad a_3 = \frac{4n_1}{\sqrt{2\pi}} \sigma_{\Delta x}; \quad m_{\Delta x} = \frac{m_{f_0}}{c}.$$

As a result, we obtain a linear equation in a centered random displacement

$$\Delta\ddot{x}_0 + a_3 \Delta\dot{x}_0 + p_0^2 \Delta x_0 = \frac{\overset{\circ}{f_0}}{m}. \tag{5.202}$$

The variances Δx_0 and $\Delta\dot{x}_0$ are respectively equal to:

$$D_{\Delta x} = \sigma_{\Delta x}^2 = \frac{\sqrt{\pi} \cdot 2\sigma_f^2 \left(\alpha + \frac{4n_1}{\sqrt{2\pi}} \sigma_{\Delta \dot{x}} \right)}{p_0^2 \sqrt{2} \sigma_{\Delta \dot{x}} \left(p_0^2 + \frac{4n_1}{\sqrt{2n}} \sigma_{\Delta x} \alpha + \alpha^2 \right)}; \tag{5.203}$$

$$D_{\Delta \dot{x}} = \sigma_{\Delta \dot{x}}^2 = \frac{\sqrt{\pi} \alpha \sigma_f^2}{\sqrt{2} \sigma_{\Delta \dot{x}} \left(p_0^2 + \frac{4n_1}{\sqrt{2n}} \sigma_{\Delta x} \alpha + \alpha^2 \right)}. \tag{5.204}$$

We determine $\sigma_{\Delta \dot{x}}$ from expression (5.204) and then $\sigma_{\Delta x}$ from formula (5.203).

The method of statistical linearization presented above produces an approximate solution to the simplest problems of the dynamics of nonlinear systems that is true at a number of limitations imposed on the input action and the mechanical system. These limitations include: the smallness of nonlinear terms in the left-hand side of equation (5.180) and the supposition that the distribution law of the solution is close to normal. They substantially reduce the information on a random process, allowing us to obtain only approximate values of the probability characteristics of a solution. The method of statistical linearization is not applied to cases where the nonlinearities cannot be considered small, as well as to analyses of non-stationary processes.

In order to answer the question about the true distribution law of the solution of (5.180) type equation with the stationary right-hand side, we must use the theory of Markov processes.

5.5.2 The Solution of the Nonlinear Equations with the use of Markov Processes

The method of statistical linearization, described in the previous item, allows us to reduce a nonlinear equation to a linear one. However, it is impossible to answer the question of how the accurate the obtained solution is. For this purpose, we must know the exact solution, which, for example, in a number of cases can be obtained, using the theory of Markov processes.

Let us consider equation (5.194), which we may present as a system of two equations of the first order, putting $\dot{x} = y_1, \quad x = y_2$ (Let us confine ourselves to the case where $m_f = 0$, and f_0 is an excitation of the white noise type with $K_f = s_0 \delta(\tau)$):

$$\dot{y}_1 + 2ny_1 + p_0^2 y_2 + \mu y_2^3 = \frac{f_0}{m}; \quad \dot{y}_2 - y_1 = 0. \tag{5.205}$$

Kolmogorov's second equation (4.73) for a stationary process is

$$y_1 \frac{\partial f}{\partial y_2} - 2nf - 2ny_1 \frac{\partial f}{\partial y_1} - (p_0^2 y_2 + \mu y_2^3) \frac{\partial f}{\partial y_1} - \frac{s_0}{2} \frac{\partial^2 f}{\partial y_1^2} = 0. \tag{5.206}$$

We may present equation (5.206) as

$$\frac{\partial}{\partial y_1}\left[\left(p_0^2 y_2 + \mu y_2^3\right) f + \frac{s_0}{4nm^2}\frac{\partial f}{\partial y_2}\right] + \left[-\frac{\partial}{\partial y_2} + 2n\frac{\partial}{\partial y_1}\right]\left(y_1 f + \frac{s_0}{4nm^2}\frac{\partial f}{\partial y_1}\right) = 0. \quad (5.207)$$

For the solution of equation (5.207) let us put

$$f\left(y_1, y_2\right) = f_1\left(y_1\right) f_2\left(y_2\right),$$

i.e. we assume that y_1 and y_2 are independent.

In this case, the two-dimensional distribution law can be presented as a product of one- dimensional distribution laws (1.52). As a result, we obtain the equation

$$\frac{\partial}{\partial y_1}\left\{f_1\left[\left(p_0^2 y_2 + \mu y_2^3\right) f_2 + \frac{s_0}{4nm^2}\frac{\partial f_2}{\partial y_2}\right]\right\} + \left(-\frac{\partial}{\partial y_2} + 2n\frac{\partial}{\partial y_1}\right)\left[f_2\left(y_1 f_1 + \frac{s_0}{4nm^2}\frac{\partial f_1}{\partial y_1}\right)\right] = 0, \quad (5.208)$$

which becomes an identity if the functions f_1 and f_2 meet the equations

$$\begin{aligned} &\frac{\mathrm{d}f_2}{\mathrm{d}y_2} + \frac{\left(p_0^2 + \mu y_2^3\right) 4nm^2}{s_0} f_2 = 0; \\ &\frac{\mathrm{d}f_1}{\mathrm{d}y_1} + \frac{4nm^2}{s_0} y_1 f_1 = 0. \end{aligned} \quad (5.209)$$

The solutions of equations (5.209) takes the form:

$$\begin{aligned} &f_1 = c_1 \exp\left\{-\frac{4nm^2}{2s_0} y_1^2\right\}; \\ &f_2 = c_2 \exp\left\{-\frac{4nm^2}{2s_0} p_0^2 y_2^2\right\} \exp\left\{-\mu\frac{4nm^2}{4s_0} y_2^4\right\}. \end{aligned} \quad (5.210)$$

The arbitrary constants c_1 and c_2 are determined from the conditions

$$1) \int_{-\infty}^{\infty} f_1\left(y_1\right) \mathrm{d}y_1 = 1; \quad 2) \int_{-\infty}^{\infty} f_2\left(y_2\right) \mathrm{d}y_2 = 1. \quad (5.211)$$

From the first condition (5.211) we determine

$$c_1 = \sqrt{\frac{2nm^2}{\pi s_0}}. \quad (5.212)$$

It follows from the obtained expressions for f_1 and f_2 that the distribution law of y_1 (first derivative of $\dot{y}$) is normal, but the distribution law of y_2 is not normal according to the earlier mentioned supposition of the statistical linearization method. Only at $\mu = 0$ f_2 transforms to a normal law.

Knowing the distribution laws of f_1 and f_2 we determine the variances

$$D_{y_1} = \sigma_{\dot{x}}^2 = c_1 \int_{-\infty}^{\infty} y_1^2 f_1(y_1)\,\mathrm{d}y_1; \tag{5.213}$$

$$D_{y_2} = \sigma_x^2 = c_2 \int_{-\infty}^{\infty} y_2^2 f_2(y_2)\,\mathrm{d}y_2. \tag{5.214}$$

From expression (5.213) we obtain an expression for the variance $\sigma_{\dot{x}}^2$ that exactly coincides with expression (5.199) obtained by the statistical linearization method. Expression (5.208) enables us to determine σ_x^2 numerically at any μ as opposed to expression (5.200) that is true only for small values of μ.

Let us consider the algorithm of solving the example 5.8 with the use of the theory of Markov processes. The key feature of this example consists in a fact, that the random disturbance is not a white noise because the spectral density S_f depends on ω.

We may always present any process with the correlation function

$$K_f = \sigma_f^2 \mathrm{e}^{-\alpha|\tau|}$$

as a result of the passage of the white noise through a linear first order system of the form (Sect. 4.3)

$$\dot{f}_0 + \alpha f_0 = \sqrt{2\alpha}\,\sigma_f \varepsilon(t), \tag{5.215}$$

where $\varepsilon(t)$ is a stationary random disturbance of the white noise type with $m_\varepsilon = 0$, $K_\varepsilon = s_0 \delta(\tau)$.

From equation (5.215) we obtain the spectral density of the random function f, which coincides with the example 5.8:

$$S_f = \frac{2\alpha\sigma_f^2}{\alpha^2 + \omega^2}.$$

Equation (5.215) should be considered together with equation (5.201) that can be presented as a system of two equations of the first order. As a result, we obtain a system of three equations:

$$\begin{aligned} &\dot{y}_1 + n_1 y_1^2 \operatorname{sign} y_1 + p_0^2 y_2 = \frac{1}{m} y_3; \\ &\dot{y}_2 - y_1 = 0; \\ &\dot{y}_3 + \alpha y_3 = \sqrt{2\alpha}\sigma_f \varepsilon(t), \end{aligned} \tag{5.216}$$

where $y_f = \Delta\dot{x}$, $y_2 = \Delta x$, $y_3 = f_0$.

Set of equations (5.216) describes a three-dimensional Markov process. Considering that the stationary vibrations of a mass m occur, Kolmogorov's second equation for this case takes the form (special case of equation (4.88)):

$$\frac{\partial}{\partial y_1}\left[\left(\frac{1}{m}y_3 - n_1 y_1^2 \text{sign} y_1 - p_0^2 y_2\right) f\right] \\ + \frac{\partial}{\partial y_2}(y_1 f) - \frac{\partial}{\partial y_3}(\alpha y_3 f) - \frac{1}{2} 2\alpha\sigma_f^2 \frac{\partial^2 f}{\partial y_3^2} = 0. \quad (5.217)$$

If the coefficients a_i linearly depend on y_i and provided that b_{ij} are constants, we can obtain a solution of equations of (5.217) type in an analytical form. A method of solving equations of multidimensional Markov processes for the case where

$$a_i = a_{0i} + \sum_{j=1}^{n} \alpha_{ij} y_j; \quad b_{ij} = \text{const},$$

is presented in [35]. It will be recalled that a_i and b_{ij} are coefficients of equation (5.217).

5.5.3 The Method of Statistical Trials (Monte-Carlo Method)

In Sects. 5.5.1 and 5.5.2 two methods of solving nonlinear equations of the first and second order were presented — the method of statistical linearization and the method using Markov processes. It has already been mentioned that the former is an approximate method and for this reason does not allow us to estimate the accuracy and certainty of its results.To secure ourselves against rough results we make a reservation to the effect that this method produces acceptable results at small nonlinearities, for example, at small μ entering equation (5.194). We cannot, however, establish the interval of μ variation, when it may be considered small, as well as the error of solution for this interval.

Another limitation of the statistical linearization method is the necessity of postulating the unknown distribution laws of a solution. As we have no information on the distribution laws of the "output" before obtaining a solution, we cannot but introduce probability hypotheses, for example, consider that the distribution laws of the "output" are normal. It is impossible to substantiate the reliability of such assumption. The method of statistical linearization enables us to determine a solutions accuracy only when we know the exact solution. But if we have it, any approximate solution becomes superfluous. Therefore, the sphere of using this method in applied problems (when one bears the responsiblility for calculation results) is very limited. It is expedient, however, to be aware of its main idea — the replacement of nonlinear terms in an equation by linear ones, because this principle is also applied when linearizing nonlinear equations of deterministic vibrations (generally those of the second order). The use of the statistical linearization method to solve systems of nonlinear equations, when we have to introduce multidimensional distribution laws into the algorithm of a solution, leads to practically insurmountable computation difficulties.

The method of Markov processes allows us (theoretically) to obtain at any instant exact distribution laws for the components of the state vector of a nonlinear dynamic system whatever its dimensionality may be and exact values of vector-of-the-state components probability characteristics. Unfortunately, this is far from being so in practice. In fact, we cannot produce the exact solution of a Kolmogorov's equation for a real nonlinear mechanical system with several degrees of freedom, especially when it comes to the necessity of taking into account real random disturbances (and not the white noise). Therefore, again we have at our disposal only approximate methods of solving Kolmogorov's equation that make us introduce simplifications and assumptions into the algorithm of a solution. And that brings about a discrepancy between approximate and exact solutions the way it happens when we use the statistical linearization method. This discrepancy cannot be estimated because we have no exact solution. These disadvantages are absent from the method of statistical trials (Monte-Carlo method) based on the numerical solution of initial nonlinear equations without simplifying them.

Let us dwell at length on the algorithm of solving nonlinear equations by this method and consider as an example an elementary system (see Fig. 5.27 *a*) with one input x and one output y. Having obtained n solutions for n realizations of a random function $x(t)$, we determine the expectation and the variance of a solution from the following formulas of mathematical statistics:

$$\tilde{m}_y(t) = \frac{1}{n}\sum_{i=1}^{n} y_i(t);$$

$$D_y(t) = \frac{1}{n-1}\sum_{i=1}^{n} (y_i - \tilde{m}_y)^2.$$

The accuracy of the obtained probability characteristics of a solution for mathematical expectation and variance depends on the number of solutions n and is estimated by standard deviations of estimations using the mathematical statistics formulas:

$$\sigma_{\tilde{m}_y}(t) = \sqrt{\left(\tilde{D}_y(t)\right)/n}; \tag{5.218}$$

$$\sigma_{\tilde{D}_y} = \sqrt{\frac{2}{n-1}\tilde{D}_y(t)}. \tag{5.219}$$

A more comprehensive estimate of the accuracy of the results produced by a solution can be obtained on the basis of on evaluating the confidence probabilities of the various deviations of $\tilde{m}_y$ and $\tilde{D}_y$ estimations from corresponding true probability characteristics. Given that the distribution laws of the estimations are close to normal, these probabilities can be approximately estimated according to the following formulas:

$$P_1 = P\left(|\tilde{m}_y - m_y| < \varepsilon_1\right) = 2\Phi\left(\frac{\varepsilon_1}{\sigma_{\tilde{m}_y}}\right); \tag{5.220}$$

$$P_2 = P\left(\left|\tilde{D}_y - D_y\right| < \varepsilon_2\right) = 2\Phi\left(\frac{\varepsilon_2}{\sigma_{\tilde{D}_y}}\right), \tag{5.221}$$

where $P_1,\ P_2$ are the confidence probabilities; $\varepsilon_1,\ \varepsilon_2$ are the given boundaries of the deviation of estimations; $\Phi\left(\varepsilon_1/\sigma_{\tilde{m}_y}\right)$ and $\Phi\left(\varepsilon_2\big/\sigma_{\tilde{D}_y}\right)$ are probability integrals.

Using the relations (5.218) and (5.219), we obtain

$$P_1 = 2\Phi\left(\frac{\varepsilon_1\sqrt{n}}{\sqrt{\tilde{D}_y}}\right);$$

$$P_2 = 2\Phi\left(\frac{\varepsilon_2\sqrt{(n-1/2)}}{\tilde{D}_y}\right).$$

The figures given in Tables 5.2 and 5.3 represent the values of the number of trials (solutions) n necessary for producing results with the given relative deviations $\nu_1 = \varepsilon_1\Big/\sqrt{\tilde{D}_y},\ \nu_2 = \varepsilon_2\big/\tilde{D}_y$ and accuracy P_1 and P_2.

The above tables show that when the requirements to the accuracy of the probability characteristics $\tilde{m}_y(t)$ and $\tilde{D}_y(t)$ become stricter, the necessary number of solutions n increases considerably. In the past, this circumstance impeded the widespread adoption of the Monte-Carlo method in design practice, because then there were no high-speed computers. Today, the performance of large number of computation experiments offers no serious difficulties.

Table 5.2. Number of trials for the determination of mathematical expectations at various values of ν_1.

P_1 \ ν_1	0.2	0.15	0.10	0.05	0.01
0.6	18	31	70	281	7000
0.7	27	47	108	431	10800
0.8	41	73	164	651	16400
0.9	68	121	272	1090	27200

Table 5.3. Number of trials for the determination of variances at various ν_2.

P_1 \ ν_2	0.2	0.15	0.10	0.05	0.01
0.6	37	63	141	563	14000
0.7	55	95	217	863	21600
0.8	83	147	239	1300	32800
0.9	137	243	545	2180	54400

If we need to obtain a distribution law of the output quantity y (Fig. 5.27 *a*) at an instant t_k, we break up the obtained numerical values $y_i(t_k)$ corresponding to random functions x_i (Fig. 5.27 *b*) into division groups (intervals of $y_\nu(t)$ values) and count the number k_j of values $y_i(t_k)$, falling within the ν-th interval. Then we divide this number by the total number of solutions and obtain the frequency $\tilde{P}_\nu$ of the occurrence of the solution numerical value $y_i(t_k)$ corresponding to the given interval $(y_\nu, y_{\nu+1})$ $\left(P_\nu = \frac{k_\nu}{n}\right)$. As a result, we get the histogram (Fig. 5.28). Increasing the number of solutions n we can reduce the size of the intervals $(y_\nu, y_{\nu+1})$, and increase their number. This will lead to the histogram approaching some curve that can be regarded as the probability density of a function $y(t_k)$ at an instant t_k. When considering a mechanical system with a finite number of degrees of freedom and one random function at the "input", we can obtain one-dimensional distribution laws for generalized coordinates and their first derivatives at any instant.

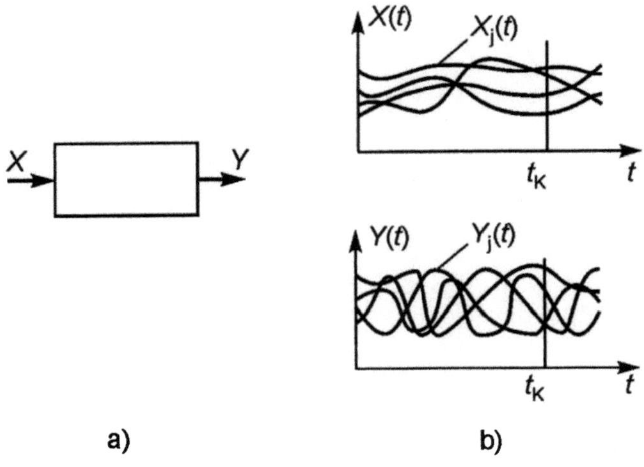

Fig. 5.27.

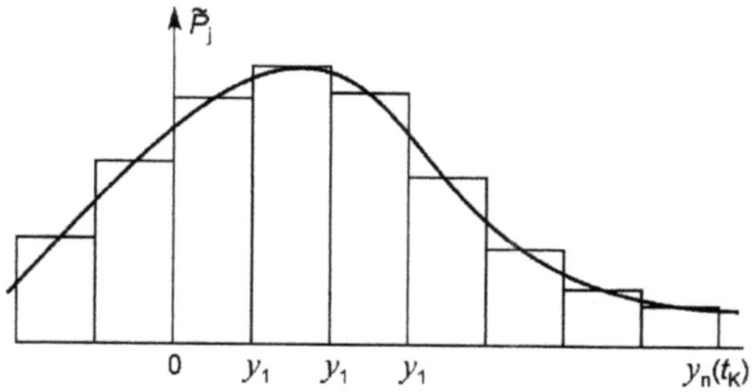

Fig. 5.28.

The information on the distribution laws of the generalized coordinates and their velocities obtained by the method of statistical trials is sufficient for the solution of complicated applied problems when we do not need to know the multidimensional distribution laws of the mechanical system state vector components.

The method of statistical trials can be applied both to nonlinear systems, where it is especially effective, and linear ones irrespective of the dimensionality of both. When using this method with respect to nonlinear systems, we should take into account all acting random disturbances in each mathematical realization, since the principle of superposition does not work for nonlinear systems.

6. Random Vibrations of Systems with Finite Number of Degrees of Freedom

6.1 Free Random Vibrations of Linear Systems

As a rule, the presentation of a real mechanical system as a system with one degree of freedom is approximate. For example, the system shown in Fig. 6.1 usually considered that with one degree of freedom is a system with two degrees of freedom (if we neglect the inertia of the rod and examine the motion of the mass only in the plane of the drawing), as the real lumped … tral axis that is

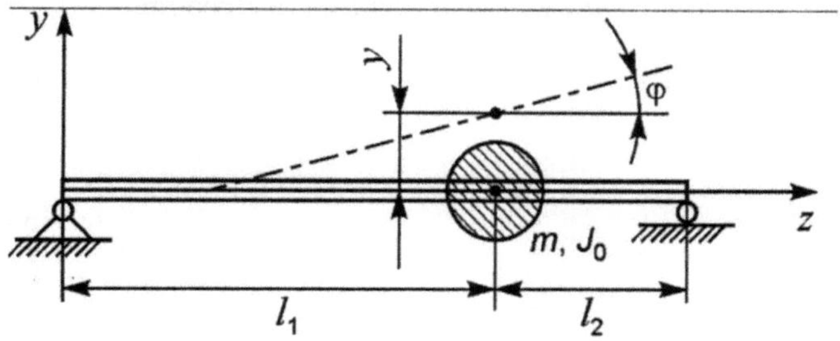

Fig. 6.1.

Using the force method we obtain the following two equations of motion (Fig. 6.2):

$$\begin{aligned} y &= \delta_{11}(-m\ddot{y}) + \delta_{12}(-J_o\ddot{\varphi}); \\ \varphi &= \delta_{21}(-m\ddot{y}) + \delta_{22}(-J_o\ddot{\varphi}), \end{aligned} \tag{6.1}$$

where

$$\delta_{11} = \frac{(l_1 l_2)^2}{3EJ_x(l_1+l_2)}; \quad \delta_{12} = \delta_{21} = \frac{l_1^2 l_2(l_1-l_2)}{3EJ_x(l_1+l_2)^2}; \quad \delta_{22} = \frac{l_1^3+l_2^3}{3EJ_x(l_1+l_2)^2}.$$

The system of equations (6.1) can be presented in a vector form as:

$$\Delta\mathbf{M}\ddot{\mathbf{y}} + \mathbf{y} = 0, \tag{6.2}$$

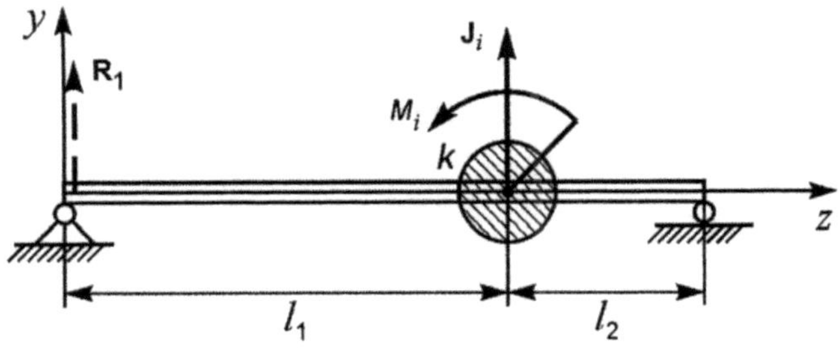

Fig. 6.2.

where

$$\Delta = \begin{bmatrix} \delta_{11} & \delta_{12} \\ \delta_{21} & \delta_{22} \end{bmatrix}; \quad \mathrm{M} = \begin{bmatrix} m & 0 \\ 0 & J_0 \end{bmatrix}; \quad \mathbf{y} = \begin{bmatrix} y \\ \varphi \end{bmatrix}.$$

The vector form of presentation substantially simplifies the study of oscillatory systems with many degrees of freedom because the equation (6.2) is also true for systems with any final number of degrees of freedom:

$$\mathrm{M}\ddot{\mathbf{y}} + \mathrm{C}\mathbf{y} = 0, \quad \left(\mathrm{C} = \Delta^{-1}\right) \tag{6.3}$$

or

$$\ddot{\mathbf{y}} + \mathrm{A}_1\mathbf{y} = 0, \tag{6.4}$$

where

$$\mathrm{A}_1 = \mathrm{M}^{-1}\mathrm{C}.$$

For a computer solution of equation (6.4) it is necessary to present it as a system of two equations of the first order which can be done by introducing new unknown vectors $\mathbf{z}_1 = \dot{\mathbf{y}}$, $\mathbf{z}_2 = \mathbf{y}$. As a result, we obtain a system

$$\begin{aligned} &\dot{\mathbf{z}}_1 + \mathrm{A}_1\mathbf{z}_2 = 0; \\ &\dot{\mathbf{z}}_2 - \mathbf{z}_1 = 0; \end{aligned} \tag{6.5}$$

or

$$\dot{\mathbf{Z}} + \mathrm{A}\mathbf{Z} = 0, \tag{6.6}$$

where

$$\mathrm{A} = \begin{bmatrix} 0 & \mathrm{A}_1 \\ -\mathrm{E} & 0 \end{bmatrix}; \quad \mathbf{Z} = \begin{bmatrix} \mathbf{z}_1 \\ \mathbf{z}_2 \end{bmatrix}, \quad \left(\mathbf{Z} = (\dot{y},\, \dot{\varphi},\, y,\, \varphi)^{\mathrm{T}}\right).$$

The vector $\mathbf{Z}$ has dimensionality 4 and is referred to as a system state vector. The first n $(1, 2, \ldots, n)$ components of the vector $\mathbf{Z}$ are derivatives of generalized coordinates with respect to t, and the subsequent n components $(n+1,\, n+2, \ldots, 2n)$ are generalized coordinates.

The system of equations (6.5) ignores the forces of resistance. In order to take them into account (if we assume that the resistance forces are proportional to the first derivatives of the generalized coordinates), it is necessary to introduce one more $\mathrm{B}\dot{\mathbf{Z}}$ term into the equation (6.3), where B is a matrix whose elements b_{ij} are coefficients of the friction forces.

With due account of resistance forces the equation (6.3) becomes

$$\mathrm{M}\ddot{\mathbf{y}} + \mathrm{B}\dot{\mathbf{y}} + \mathrm{C}\mathbf{y} = 0. \tag{6.7}$$

If we include the friction forces in the equation (6.6), only the matrix A changes, namely

$$\mathrm{A} = \begin{bmatrix} \mathrm{M}^{-1}\mathrm{B} & \mathrm{A}_1 \\ -\mathrm{E} & 0 \end{bmatrix}.$$

Further on, we denote the elements of the matrix $\mathrm{M}^{-1}\mathrm{B}$ by b_{ij}. During random free vibrations the initial data at $t = 0$ represent random quantities, whose probability characteristics we consider to be known, i.e. we know their mathematical expectations $m_{z_{i0}}$ and variances $D_{z_{i0}}$. In the more general case where the components of the vector $\mathbf{Z}$ at the initial instant of time correlate with one another, we know the correlation moments $K_{z_{i0}z_{j0}}$. The solution of the equation (6.6) takes the following vector form

$$\mathbf{Z} = \mathrm{K}(t)\mathbf{C}, \tag{6.8}$$

where $\mathrm{K}(t)$ is the fundamental matrix of solutions of the system of homogeneous equations and $\mathbf{C}$ is an arbitrary vector.

When numerically solving the equation (6.6), we may choose such algorithm of getting the columns of the matrix $\mathrm{K}(t)$ that at $t = 0$ makes the matrix $\mathrm{K}(0)$ equal to the identity matrix, namely, $\mathrm{K}(0) = \mathrm{E}$. In this case, the arbitrary vector $\mathbf{C}$ equals the initial values vector $\mathbf{Z}_0$, i.e.

$$\mathbf{Z} = \mathrm{K}(t)\mathbf{Z}_0. \tag{6.9}$$

All previous mathematics and representation of the solution of the equation (6.6) in the form (6.9) hold true for deterministic initial data. Let us determine the mathematical expectations as well as the correlation and cross-correlation functions of the components of the vector $\mathbf{Z}$.

The mathematical expectation of a vector $\mathbf{Z}$ is

$$M[\mathbf{Z}] = \mathbf{m}_z = \mathrm{K}(t)\mathbf{m}_{z0}, \tag{6.10}$$

where $\mathbf{m}_z$ is a vector whose components are the mathematical expectations of the components of the vector $\mathbf{Z}$ and $\mathbf{m}_{z0}$ is a vector whose components are the values of m_{z_j}.

The mathematical expectations of the components of the vector **Z** are

$$m_{z_i} = \sum_{j=1}^{2n} k_{ij} m_{z_{0j}}. \tag{6.11}$$

In order to determine the correlation and cross-correlation functions of the components of the vector of solutions let us present the solution (6.9) in a scalar form

$$z_i(t) = \sum_{j=1}^{2n} k_{ij}(t) z_{z_{j0}}. \tag{6.12}$$

Knowing the mathematical expectations m, we can obtain the expression for the centered random components of the solutions vector:

$$\begin{aligned} \dot{z}_i(t) &= z_i(t) - m_{z_i}(t) = \sum_{j=1}^{2n} k_{ij} \dot{z}_{j0}; \\ \mathrm{M}\left[\overset{\circ}{z}_l(t)\, \overset{\circ}{z}_l(t')\right] &= \mathrm{K}_{z_l\, z_k} = \mathrm{M}\left[\left(\sum_{j=1}^{2n} k_{ij}(t)\, \overset{\circ}{x}_{j0}\right)\left(\sum_{v=1}^{2n} k_{k\,v}(t')\, \overset{\circ}{z}_{\nu 0}\right)\right]. \end{aligned} \tag{6.13}$$

For the special case of $\mathrm{K}_{z_{i0} z_{k0}} = 0$, we have

$$\begin{aligned} \mathrm{K}_{z_i z_k} &= \sum_{v=1}^{2n} k_{iv}(t)\, k_{kv}(t')\, D_{z_{v0}}; \\ \mathrm{K}_{z_i z_i} &= \sum_{v=1}^{2n} k_{iv}(t)\, k_{kv}(t')\, D_{z_{v0}}; \\ \mathrm{D}_{z_i} &= \sum_{v=1}^{2n} k_{iv}^2(t) D_{z_{v0}}. \end{aligned} \tag{6.14}$$

As a result, we have obtained all probability characteristics of the components of the state vector of a linear system with a finite number of degrees of freedom at random free vibrations induced by random initial data.

In the design we are most interested in the random dynamic stresses that arise in structure elements, for example, the normal stress arising in the section k of the rod (Fig. 6.2). Let us determine its probability characteristics (Fig. 6.2)

$$\sigma_{\max} = \frac{R_1 l_1}{W_x}, \tag{6.15}$$

where R_1 is the reaction in the hinge.

Using d'Alembert's principle, we get

$$R_1 = \frac{l}{(l_1 + l_2)} \left(-J_i + M_i\right), \tag{6.16}$$

where J_i, M_i are respectively the inertia force and the moment,

$$J_i = -m\ddot{y}, \quad M_i = -J_0\ddot{\varphi}.$$

From the equations (6.1) we determine:

$$\begin{aligned} \ddot{y} &= \frac{\delta_{22}}{m\Delta} y + \frac{\delta_{12}}{m\Delta}\varphi \qquad (\Delta = \delta_{11}\delta_{22} - \delta_{12}\delta_{21}); \\ \ddot{\varphi} &= \frac{\delta_{21}}{J_0\Delta} y - \frac{\delta_{11}}{J_0\Delta}\varphi. \end{aligned} \tag{6.17}$$

Performing transformations with due account of (6.16) and (6.17), we obtain

$$\sigma_{\max} = b_1 y + b_2 \varphi, \tag{6.18}$$

where

$$b_1 = -\frac{(\delta_{21} + \delta_{22} l_2) l_1}{W_x (l_1 + l_2) \Delta}; \quad b_2 = -\frac{(\delta_{11} + \delta_{12} l_2) l_1}{W_x (l_1 + l_2) \Delta}.$$

In order to determine the greatest possible value of $\sigma_{\max}$ that depends on two random functions y and φ we must known the joint distribution law $f(y,\ \varphi,\ t)$. Considering the joint distribution law of y and φ a normal distribution law, we can determine the parameters of distribution law of the random function $\sigma_{\max}$, the latter being linearly dependent on the random functions y and φ (6.18). In this case, the $\sigma_{\max}$ distribution will also be a normal one with the following parameters:

$$m_{\sigma_{\max}}(t) = b_1 m_y(t) + b_2 m_\varphi(t); \tag{6.19}$$

$$\left(\sigma_{\sigma_{\max}}\right)^2 = b_1^2 \sigma_y^2(t) + b_2^2 \sigma_j^2(t) + 2 b_1 b_2 \mathrm{K}_{y\varphi}(t), \tag{6.20}$$

where

$$\mathrm{K}_{y\varphi} = \mathrm{K}_{z_3 z_4}(t,\ t). \tag{6.21}$$

The cross-correlation function (6.21) is a special case of (6.12) at $n = 2$ (if the initial data are independent).

Having determined normal distribution parameters for $\sigma_{\max}$, we find the greatest possible value of the normal stress $\sigma_{\max}$ at each instant

$$\max\left(\sigma_{\max}(t)\right) = m_{s_{\max}}(t) + 3\sigma_{\sigma_{\max}}(t). \tag{6.22}$$

6.2 Vibrations at Random Pulse Loading

When solving applied problems we often have to consider the motion of a system induced by the action of a random impulse of force (or a system of forces). A case when a vehicle drives into a single asperity markedly differing in height from other road asperities is shown in Fig. 6.3. At a high speed of the vehicle horizontal motion the action of this asperity is equivalent to a single impulse of force, which results in a sharp change of total velocity vector $v = v_0\mathbf{i}_1 + \dot{y}\mathbf{i}_2$ and the angular velocity $\omega = \dot{\varphi}$.

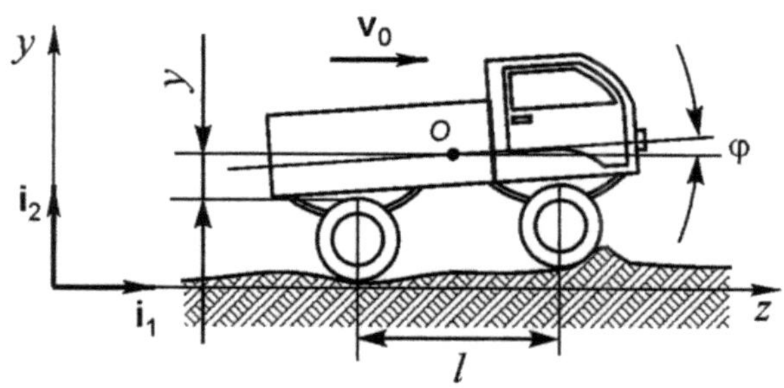

Fig. 6.3.

During the vehicles motion we consider that the displacement of the center of gravity (point 0) and the rotation of the vehicle as a rigid body occur in the plane of the drawing.

In the general case, a wheel may collide with the asperity, resulting in an onset of spatial vibrations (if there were some vibrations in the plane of the drawing before the collision). The ultimate goal of the design may vary depending on the specific conditions of a problem and the designation of the structure. For example, in the design of the structure shown in Fig. 6.4 the following two points may be of concern to its designers, provided that the structures pulse loading is $\mathbf{J}_j$: the greatest possible displacements of the masses from the vertical position and the greatest possible dynamic stresses arising in the structure. Let us consider each of them in succession. During the action of the pulse loads $\mathbf{J}_i$ on the masses m_i (Fig. 6.4) the latter obtain velocities equal to (at $\mathbf{J}_i = J_i\mathbf{i}_1$)

$$\dot{x}_{1i}(0) = \frac{J_i}{m_i} \quad (i = 1,\ 2,\ \ldots,\ k) \tag{6.23}$$

after the action of the impulse has terminated.

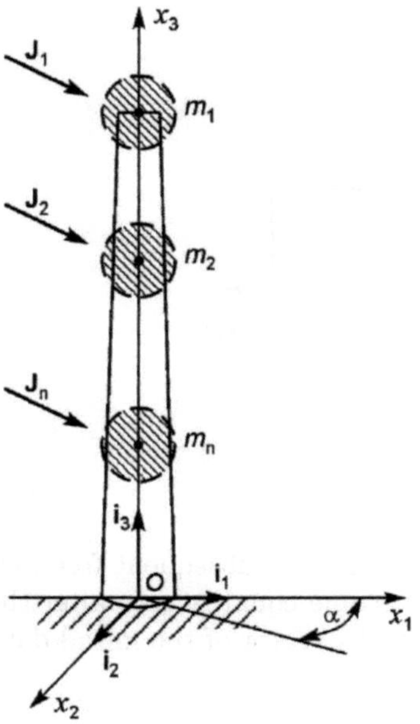

Fig. 6.4.

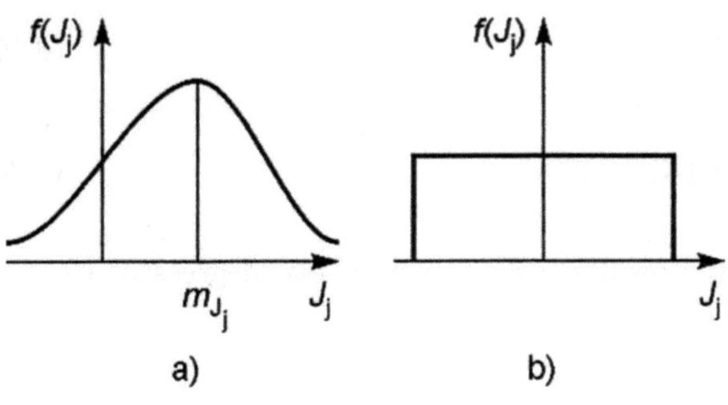

Fig. 6.5.

We assume that the distribution laws of $\mathbf{J}_i$ are known (Fig. 6.5 or 6.6).

Let us first consider the simplest case of $\mathbf{J}_i$ directions being known, for example, that of $\alpha = 0$. After the action termination of the impulses of forces has come to an end, masses obtain the following velocities (for example, at $\mathbf{J}_i = J_i\mathbf{i}_1$)

$$\dot{x}_{1i} = \frac{J_i}{m_i}. \tag{6.24}$$

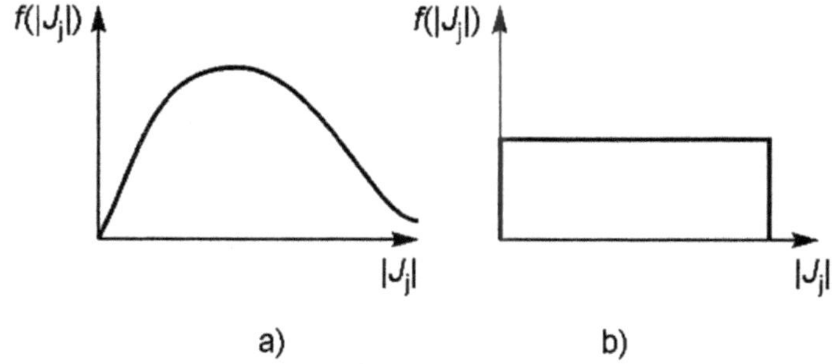

Fig. 6.6.

They can be considered the initial data of the systems subsequent free motion. (At pulse loading the displacements x_{ij} become equal to zero after the termination of the impulse action). In this case, the vector of the initial data takes the form

$$\mathbf{X}_0 = (\dot{x}_{11}, \dot{x}_{12}, \ldots, \dot{x}_{1n}, 0 \ldots 0)^{\mathrm{T}},$$

i.e. the given problem can be considered a special case of the previous problem. Therefore the probability characteristics of the components of the vector **Z** are equal (at $k = n$) to

$$m_{z_i} = \sum_{j=1}^{n} k_{ij} m_{z_{0j}} = \sum_{j=1}^{n} k_{ij} \frac{m_{1_i}}{m_j}; \tag{6.25}$$

$$K_{z_i z_j} = \sum_{k=1}^{n} k_{ij}(t) k_{iv}(t') K_{z_{k0} z_{v0}}. \tag{6.26}$$

Example 6.1. Let us consider a system with two degrees of freedom ignoring resistance forces (Fig. 6.7).

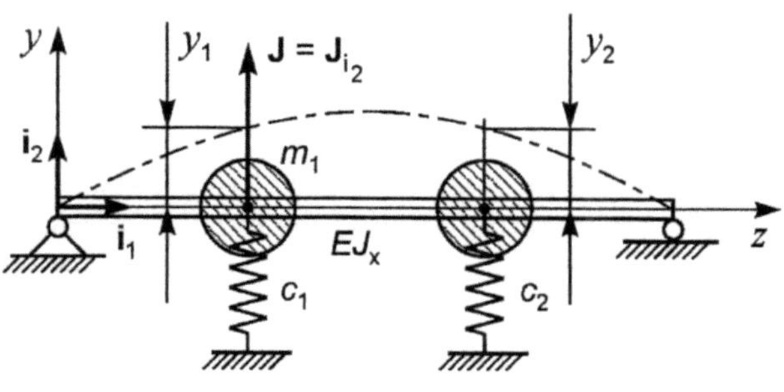

Fig. 6.7.

A random short-term impulse J (with known $m_j$and D_j) has suddenly acted on a mass m_1 at $t = 0$. After the termination of the impulse action the mass m_1 has obtained a random velocity y_{10}.

It is required to determine the mathematical expectation and variance of the displacement of the mass m_j at an arbitrary instant of time, i.e. to find $m_{y_1}(t)$ and $D_{y_1}(t)$. Making use of the force method, we obtain the following equations of motion:

$$\begin{aligned} y_1 &= \delta_{11}\left(-m_1\ddot{y}_1 - c_1 y_1\right) + \delta_{12}\left(-m_2\ddot{y}_2 - c_2 y_2\right); \\ y_2 &= \delta_{21}\left(-m_1\ddot{y}_1 - c_1 y_1\right) + \delta_{22}\left(-m_2\ddot{y}_2 - c_2 y_2\right); \end{aligned} \tag{6.27}$$

or

$$\begin{aligned} a_{11}\ddot{y}_1 + a_{12}\ddot{y}_2 + b_{11}y_1 + b_{12}y_2 &= 0; \\ a_{21}\ddot{y}_1 + a_{22}\ddot{y}_2 + b_{21}y_1 + b_{22}y_2 &= 0, \end{aligned} \tag{6.28}$$

where

$$a_{11} = \delta_{11}m_1; \quad a_{12} = \delta_{12}m_2; \quad b_{11} = 1 + \delta_{11}c_1;$$
$$b_{12} = \delta_{12}c_2; \quad a_{21} = \delta_{21}m_1; \quad a_{22} = \delta_{22}m_2;$$
$$b_{21} = \delta_{21}c_1; \quad b_{22} = 1 + \delta_{22}c_2.$$

The system of equations (6.28) can be reduced to a system of first-order equations, but in this particular case we can obtain the solution for the initial system right away. Assuming that

$$y_1 = A_1 \cos pt; \quad y_2 = A_2 \cos pt \tag{6.29}$$

or

$$y_1 = B_1 \sin pt; \quad y_2 = B_2 \sin pt \tag{6.30}$$

we obtain from the system (6.28)

$$\begin{aligned} \left(b_{11} - a_{11}p^2\right) A_1 + \left(b_{12} - a_{12}p^2\right) A_2 &= 0; \\ \left(b_{21} - a_{21}p^2\right) A_1 + \left(b_{22} - a_{22}p^2\right) A_2 &= 0. \end{aligned} \tag{6.31}$$

Having determined p_1 and p_2 from (6.31), we find the following coefficients of distribution

$$\begin{aligned} k_1 &= \frac{A_{21}}{A_{11}} = -\frac{b_{12} - a_{12}p_1^2}{b_{11} - a_{11}p_2^2}; \\ k_2 &= \frac{A_{22}}{A_{12}} = -\frac{b_{12} - a_{12}p_2^2}{b_{11} - a_{11}p_2^2}. \end{aligned} \tag{6.32}$$

As a result, we obtain the solution of the system (6.27)

$$\begin{aligned} y_1 &= A_1 \cos p_1 t + B_1 \sin p_1 t + A_2 \cos p_2 t + B_2 \sin p_2 t; \\ y_2 &= A_1 k_1 \cos p_1 t + B_1 k_1 \sin p_1 t + A_2 k_2 \cos p_2 t + B_2 k_2 \sin p_2 t. \end{aligned} \tag{6.33}$$

As at $t = 0$ $y_1 = y_2 = \dot{y}_2 = 0$, and $\dot{y}_1 = \dot{y}_{10}$ then, having determined the arbitrary constants A_i, B_i we obtain

$$y_1 = \frac{k_1 \dot{y}_{10} \sin p_1 t}{p_1 (k_2 - k_1)} - \frac{k_2 \dot{y}_{10} \sin p_2 t}{p_2 (k_2 - k_1)}. \tag{6.34}$$

The mathematical expectation and variance of y_1 are equal to

$$\begin{aligned} m_{y_1}(t) &= \left[\frac{k_1 \sin p_1 t}{p_1 (k_2 - k_1)} - \frac{k_2 \sin p_2 t}{p_2 (k_2 - k_1)} \right] m_{\dot{y}_{10}}; \\ D_{y_1}(t) &= \left[\frac{k_1 \sin p_1 t}{p_1 (k_2 - k_1)} - \frac{k_2 \sin p_2 t}{p_2 (k_2 - k_1)} \right]^2 D_{\dot{y}_{10}} \end{aligned} \tag{6.35}$$

respectively.

$m_{\dot{y}_{10}}$ and $D_{\dot{y}_{10}}$ entering the expressions (6.35) are related to the probability characteristics of the random impulse by the equation

$$m_{\dot{y}_{10}} = \frac{m_J}{m_1}; \quad D_{\dot{y}_{10}} = \frac{D_J}{m_1^2}.$$

Assuming that the displacements y_i and their first two derivatives have normal distribution, we obtain the greatest possible deviation of the mass m_1

$$\max(y_1) = \frac{1}{m_1} \left[\frac{k_1 \sin p_1 t}{p_1 (k_2 - k_1)} - \frac{k_2 \sin p_2 t}{p_2 (k_2 - k_1)} \right] (m_J + 3\sigma_J).$$

Let us consider a case of pulse loading of a structure, where the direction of impulses $\mathbf{J}_i$ is random (see Fig. 6.4), for example, with vectors $\mathbf{J}_i$ being parallel to the plane $x_1 x_3$ and having arbitrary direction ($\alpha \neq 0$). The impulse of force acting on a mass m_i imparts it a velocity (at $\mathbf{J}_i = J_{ix_1} \mathbf{i}_1 + J_{ix_2} \mathbf{i}_2$)

$$\dot{\mathbf{u}}_{i0} = \dot{x}_{10} \mathbf{i}_1 + \dot{x}_{20} \mathbf{i}_2 = \frac{\mathbf{J}_i}{m_i}, \qquad (i = 1,\ 2,\ \ldots,\ k). \tag{6.36}$$

In the case of the vectors $\mathbf{J}_i$ being arbitrary directed in the plane $x_1 x_2$ we consider that the distribution laws of the modulus $|J_i|$ are known (see Fig. 6.5, 6.6). The projections of $\mathbf{J}_i$ on the axes x_1 and x_2 $(\mathbf{J}_{ix_1}, \mathbf{J}_{ix_2})$, which are related by an additional condition

$$(\mathbf{J}_{ix_1})^2 + (\mathbf{J}_{ix_2})^2 = |\mathbf{J}_i|^2. \tag{6.37}$$

will enter in the equations of motion. In this case we must know the distribution law of the modulus of a normally distributed random quantity whose probability density obeys the law

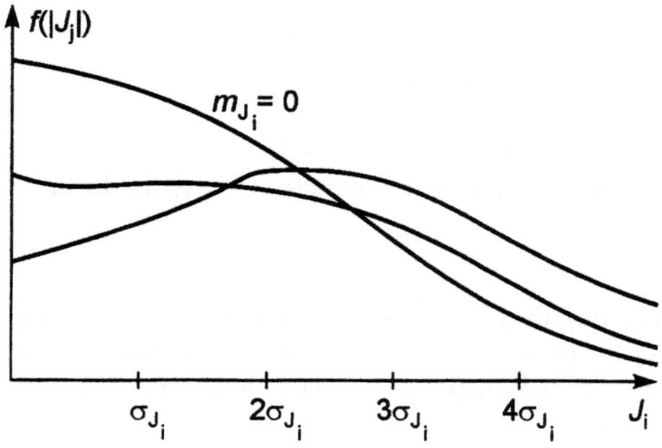

Fig. 6.8.

$$f\left(|\mathbf{J}_i|\right) = \frac{1}{\sigma_{j_i}\sqrt{2\pi}}\left(\exp\left\{-\frac{\left(J_i - m_{J_i}\right)^2}{2\sigma_{J_i}^2}\right\} + \exp\left\{-\frac{\left(J_i + m_{J_i}\right)^2}{2\sigma_{J_i}^2}\right\}\right), \quad (6.38)$$

where m_{J_i} and σ_{J_i} are the mathematical expectation and the standard deviation of the initial normal distribution law [30] for $\mathbf{J}_i$. The plots of $f\left(|\mathbf{J}_i|\right)$ for a number of m_{J_i} values are given in Fig. 6.8.

Considering that the components of the vector $\mathbf{Z}$ have normal distributions, we take advantage of the three sigma rule in order to determine their greatest possible values at any instant of time

$$\max\left(z_i\left(t\right)\right) = m_{J_1} + 3\sigma_{J_1} \quad (i = 1,\ 2,\ \ldots,\ 2n)\,. \quad (6.39)$$

The first n relations (6.39) are the greatest possible velocities of the masses m_j with subsequent $(n+1,\ n+2,\ \ldots,\ 2n)$ relations being the greatest possible deflections of the masses from the state of equilibrium. In order to obtain the greatest possible accelerations of the masses m_j (the overloads) from the equation (6.6) we determine (with due account of (6.9)) the derivative of the state vector

$$\dot{\mathbf{Z}} = -\mathrm{AKZ}_0. \quad (6.40)$$

The first n components of the vector $\dot{\mathbf{Z}}$ are the accelerations (in this example they represent the linear accelerations of the masses). The mathematical expectations and variances of the first n components of the vector $\dot{\mathbf{Z}}$ (at $\mathrm{K}_{z_{0j}z_{0k}} = 0$, if $i \neq k$) have the following values:

$$m_{\dot{z}_j} = \sum_{v=1}^{n} k_{jv}^{(1)} m_{z_{0v}} \qquad \left(m_{z_{0v}} = \frac{m_{J_v}}{m_v}\right);$$

$$D_{\dot{z}_j} = \sum_{v=1}^{n} k_{jv}^{(1)^2} D_v \qquad \left(D_v = \frac{D_{J_v}}{m_v^2} \right),$$

where $k_{jv}^{(1)}$ are the elements of the matrix $\mathrm{K}^{(1)} = -\mathrm{AK}$.

The greatest possible accelerations of the masses m_j take the form

$$\max(\dot{z}_j) = m_{\dot{z}_i} + 3\sigma_{\dot{z}_i}. \tag{6.41}$$

After the action of the impulses of forces has terminated, the motion of a system (see Fig. 6.4) with due account taken of viscous resistance forces is similar to that of the equation (6.7) or (if we go over to a system of first-order equations) the equation (6.6):

$$\dot{\mathbf{Z}} + \mathrm{A}\mathbf{Z} = 0, \tag{6.42}$$

where

$$\mathrm{A} = \begin{bmatrix} \mathrm{M}^{-1}\mathrm{B} & \mathrm{M}^{-1}\mathrm{C} \\ \mathrm{E} & 0 \end{bmatrix}, \qquad \mathbf{Z} = \begin{bmatrix} \mathbf{z}_1 \\ \mathbf{z}_2 \end{bmatrix} = \begin{bmatrix} \dot{\mathbf{x}} \\ \mathbf{x} \end{bmatrix}.$$

The solution of the equation (6.42) is of the form

$$\mathbf{Z} = \mathrm{K}(t)\,\mathbf{Z}_0, \quad (\mathrm{K}(0) = \mathrm{E}).$$

At a pulse loading we have the following initial data:

$$\mathbf{Z}_1(0) = \dot{\mathbf{X}}(0), \qquad \mathbf{Z}_2(0) = 0, \tag{6.43}$$

with the possibility of several components of the vector $\mathbf{Z}_1(0)$ being equal to zero (if $k < n$, where n is the number of degrees of freedom and k is the number of masses subjected to the action of impulses of forces).

The impulses $\mathbf{J}_i$ have random, coinciding directions parallel to the $x_1 0 x_2$ plane (Fig. 6.4). Let us assume that $\mathbf{J}_i$ are related by the equations

$$\mathbf{J}_i = \beta_i \mathbf{J}_1. \tag{6.44}$$

If, unlike the special case, considered above, each mass has two degrees of freedom (displacements along the axes x_1 and x_2), the system of equations (6.42) has the dimensionality of $4n$, because in this case the masses are allowed to move along the two axes x_1 and x_2. After the action of impulses has terminated each mass begins moving with a velocity (initial conditions) of

$$\dot{\mathbf{x}}_{0i} = \dot{x}_{1i0}\mathbf{i}_1 + \dot{x}_{2i0}\mathbf{i}_2 = \frac{\beta_i}{m_i} J_{1x_1}\mathbf{i}_1 + \frac{\beta_i}{m_i} J_{1x_2}\mathbf{i}_2, \tag{6.45}$$

where J_{1x_1}, J_{1x_2} are the projections of the vector $\mathbf{J}_1$ on the axes x_1 and x_2 (Fig. 6.9):

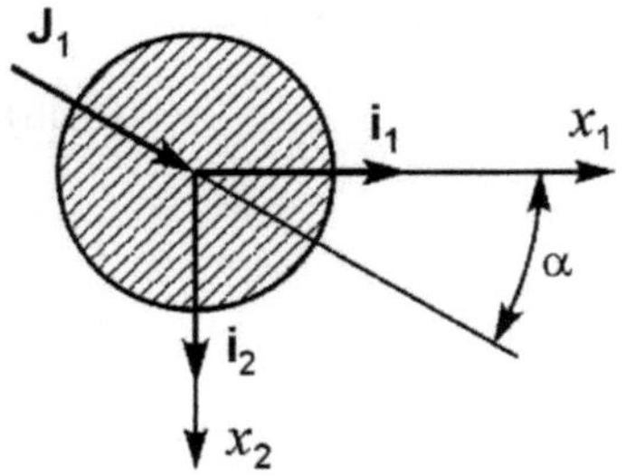

Fig. 6.9.

$$J_{1x_1} = |\mathbf{J}_1| \cos\alpha; \qquad J_{1x_2} = |\mathbf{J}_1| \sin\alpha, \tag{6.46}$$

where α is a random angle. The first $2n$ components of the initial conditions vector $\mathbf{X}_0$ are known $\left(\dot{X}_{10_j}, \dot{X}_{20_j}\right)$, $(j = 1, 2, \ldots, n)$ with the other $2n$ components of a vector $\mathbf{Z}$ being equal to zero.

We may present the components of the vector $\mathbf{Z}$ with due account taken of (6.45) and (6.46) as

$$z_1 = a_{i1} \cos\alpha \, |\mathbf{J}_1| + a_{i2} \sin\alpha \, |\mathbf{J}_1| \, . \tag{6.47}$$

(The explicit form of coefficients a_{i1}, a_{i2} is given below).

If we use the relations (6.47) to obtain the probability characteristics of the solution of equations of motion, we need information on the random angle α. To be more specific, we must know the distribution law of the angle α. For example, if we have found out that this angle follows the uniform distribution law on the interval $0 \leq \alpha \leq 2\pi$, we can determine all probability characteristics of z_i. This version of the solution, however, gives no answer to the question about the worst action of the impulses $\mathbf{J}_i$ on a system and that is precisely what designers want to know.

Let us consider a method of solving the equation (6.42) that involves the determination of the maximum values of the components of the state-of-the-system vector at any instant of time and their probability characteristics. The method presented below demands that we should know the distribution law of the random angle α, which makes it much easier for us to acquire statistical information on the input. In other words, our knowledge of the distribution law of the modulus of the impulse force vector (6.38) is quite sufficient here.

Let us consider the solution of the equation (6.42) implying a generalized solution to the considered case where the masses m_j have two degrees of freedom:

$$\mathbf{Z} = \mathrm{K}\,(\mathrm{t})\, \mathbf{Z}_0, \qquad \mathbf{Z}_0 = \left(\dot{x}_{10}^{(1)}, \dot{x}_{20}^{(1)}, \ldots, \dot{x}_{10}^{(n)}, \dot{x}_{20}^{(n)}, 0 \ldots 0\right)^{\mathrm{T}} .$$

With due account of (6.45) we may set the vector $\mathbf{Z}_0$ in writing as

$$\mathbf{Z}_0 = \beta \mathbf{J}, \tag{6.48}$$

where

$$\beta = \begin{bmatrix} \frac{\beta_1}{m_1} & 0 & \dots\dots & \dots & \dots & \dots 0 \\ 0 & \frac{\beta_1}{m_1} & 0 \dots & \dots & \dots & \dots 0 \\ 0 & \dots & \dots\dots & \dots & \dots & \dots 0 \\ 0 & \dots & \dots\dots & \dots & \dots & \dots 0 \\ 0 & \dots & \dots 0 & \frac{\beta_n}{m_n} & 0 & \dots 0 \\ 0 & \dots & \dots\dots & 0 & \frac{\beta_n}{m_n} & \dots 0 \\ 0 & \dots & \dots\dots & \dots & \dots & 0\ 0 \\ 0 & \dots & \dots\dots & \dots & \dots & \dots 0 \end{bmatrix};$$

$$\mathbf{J} = \left(J_{1x_1}, J_{1x_2}, J_{1x_1}, J_{1x_2}, J_{1x_1}, J_{1x_2}, 0 \dots 0\right)^{\mathrm{T}}.$$

Therefore

$$\mathbf{Z} = \mathrm{K}(t)\, \beta \mathbf{J}. \tag{6.49}$$

From (6.49) we obtain in a scalar form

$$z_i = a_{i1}(t)\, J_1 x_1 + a_{i2}(t)\, J_1 x_2, \tag{6.50}$$

where

$$a_{i1} = k_{i1}\frac{\beta_1}{m_1} + k_{i3}\frac{\beta_2}{m_2} + k_{i5}\frac{\beta_3}{m_3} + \ldots + k_{i(2n-1)}\frac{\beta_n}{m_n};$$
$$a_{i2} = k_{i2}\frac{\beta_1}{m_1} + k_{i4}\frac{\beta_2}{m_2} + k_{i6}\frac{\beta_3}{m_3} + \ldots + k_{i2n}\frac{\beta_n}{m_n}.$$

Let us present (6.50) as a scalar product

$$z_i = (\mathbf{a}_i \cdot \mathbf{J}_1), \tag{6.51}$$

where

$$a_i = a_{i1}\mathbf{i}_1 + a_{i2}\mathbf{i}_3, \qquad \mathbf{J}_1 = J_{1x_1}\mathbf{i}_1 + J_{1x_3}\mathbf{i}_3.$$

The projections of the pulse $\mathbf{J}_1$ meet the condition

$$\frac{J_{1x_1}^2}{|\mathbf{J}_1|^2} + \frac{J_{1x_2}^2}{|\mathbf{J}_1|^2} = 1, \tag{6.52}$$

which we may present as

$$(\mathbf{CJ}_1 \cdot \mathbf{J}_1) = 1, \quad \left(\mathrm{C} = \begin{bmatrix} \dfrac{1}{|\mathbf{J}_1|^2} & 0 \\ 0 & \dfrac{1}{|\mathbf{J}_1|^2} \end{bmatrix} \right). \tag{6.53}$$

Let us determine expressions for the maximum values of the components (6.51) with an additional constraint (6.53), using a Lagrangian multiplier

$$J = (\mathbf{a}_i \cdot \mathbf{J}_1) - \lambda \left[(\mathrm{C}\mathbf{J} \cdot \mathbf{J}) - 1 \right] = \max, \tag{6.54}$$

where λ is the Lagrangian multiplier.

The functional J maximum is determined from the condition

$$\frac{\partial J}{\partial \mathbf{J}_1} = 0 \tag{6.55}$$

or

$$\mathbf{a}_i = \lambda \mathrm{C}\mathbf{J}_1. \tag{6.56}$$

Let us multiply (6.56) by the matrix C^{-1} :

$$\mathrm{C}^{-1}\mathbf{a}_i = \lambda \mathbf{J}_1. \tag{6.57}$$

Scalarly multiplying the left- and right-hand sides of the equations (6.56) and (6.57), we obtain (with due account of the relationship (6.53))

$$\lambda^2 = \left(\mathrm{C}^{-1}\mathbf{a}_i \cdot \mathbf{a}_i \right), \tag{6.58}$$

wherefrom we determine the multiplier

$$\lambda = \sqrt{(\mathrm{C}^{-1}\mathbf{a}_i \cdot \mathbf{a}_i)}. \tag{6.59}$$

Knowing the Lagrangian multiplier, we determine the vector from the relationship (6.57)

$$\mathbf{J}_{1x} = \frac{\mathrm{C}^{-1}\mathbf{a}_i}{\lambda} = \frac{\mathrm{C}^{-1}\mathbf{a}_i}{\sqrt{(\mathrm{C}^{-1}\mathbf{a}_i \cdot \mathbf{a}_i)}}. \tag{6.60}$$

The maximum value of the solution vector component z_i is

$$\max(z_i) = (\mathbf{a}_i \cdot \mathbf{J}_1) = \sqrt{(\mathrm{C}^{-1}\mathbf{a}_i \cdot \mathbf{a}_i)} = \lambda \tag{6.61}$$

or

$$\max(z_i) = |\mathbf{J}_1| \sqrt{a_{i1}^2 + a_{i2}^2}. \tag{6.62}$$

The expression (6.62) for $\max z_i$ allows us to determine the probability characteristics of m_{zi} and σ_{zi} which are the maximum values of the components of the state-of-the-system vector:

$$\max(m_{zi}) = \sqrt{a_{i1}^2(t) + a_{i2}^2(t)}\,\tilde{m}_{J_i}; \tag{6.63}$$

$$\max(\sigma_{zi}) = \sqrt{a_{i1}^2(t) + a_{i2}^2(t)}\,\tilde{\sigma}_{J_i}; \tag{6.64}$$

where

$$\tilde{m}_{J_i} = 2\left[m_{J_i}\Phi_0\left(\frac{m_{J_i}}{\sigma_{J_i}}\right) + \sigma_{J_i}\varphi\left(\frac{m_{J_i}}{\sigma_{J_i}}\right)\right]; \tag{6.65}$$

$$\Phi_0\left(\frac{m_{J_i}}{\sigma_{J_i}}\right) = \frac{1}{\sqrt{2\pi}}\int_0^{m_{J_i}/\sigma_{J_i}} \exp\left\{-\frac{t^2}{2}\right\} dt;$$

$$\varphi\left(\frac{m_{J_i}}{\sigma_{J_i}}\right) = \frac{1}{\sqrt{2\pi}}\exp\left\{\frac{m_{J_i}^2}{2\sigma_{J_i}^2}\right\}; \tag{6.66}$$

$$\tilde{\sigma}_{J_1}^2 = \sigma_{J_1}^2 + m_{J_1}^2 - \tilde{m}_{J_1}^2.$$

(m_{J_1}, σ_{J_1} are the parameters of a normal distribution law).

Let us obtain the distribution laws of the maximum values of the components $\max(Z_i)$ that linearly depend on the random modulus of the impulse-of-force vector (the relationship (6.62)). According to the general theory of determining the distribution laws of the monotonously varied functions of random arguments, we have

$$f_{i1}(\max(z_i), t) = \frac{1}{|b_i(t)|\sigma_{J_i}\sqrt{2\pi}} \times \left[\exp\left\{-\frac{(z_{im} - b_i(t)\,m_{J_1})^2}{2b_i^2(t)\,\sigma_{J_i}^2}\right\} + \exp\left\{-\frac{(z_{im} + b_i(t)\,m_{J_1})^2}{2b_i^2(t)\,\sigma_{J_i}^2}\right\}\right], \tag{6.67}$$

where

$$b_i(t) = \sqrt{a_{i1}^2 + a_{i2}^2}, \qquad z_{im} = \max(z_i).$$

From a condition that is similar to the three sigma rule we determine the greatest possible value of the maximum component of the solution vector z_i (for example, the displacements of the mass m_j):

$$P(0 < z_{im} < z_{im}^*) = 0.99 \tag{6.68}$$

or

$$\int_0^{z_{i\,\max}^*} f_{i1}(z_{im}, t)\, dz_{im} = 0.99. \tag{6.69}$$

We determine z_{im}^* for any fixed instant of time from (6.69). The presented method enables us to obtain the greatest possible values of each component of a system state vector, for example, the maximum displacement of the k-th mass $\max(x_{jk})$ or its maximum velocities $\max(\dot{x}_{ik})$. With a knowledge of the distribution law of the maximum displacements of the masses of a system, we can determine the probability that the maximum displacement of the i–th mass meets the condition (6.62). We deal with similar problems at the design of the suspension system of vehicles (Fig. 6.3) in cases where a collision with a unit asperity occurs. This probability takes the form

$$P_1\left(0 < z_{im} < \Delta\right) = \int_0^{\Delta} f_{i1}\left(z_{im},\ t\right) \mathrm{d}z_{im}, \qquad \left(y_{im} = y_0\right). \tag{6.70}$$

For a motor vehicle (see Fig. 6.3) the expression (6.70) that permits us to determine y_0 (y_0 is the vertical displacement of a suspension), holds true only for the interval of time $t_k = l/v_0$ (up to the moment when the rear wheels drive into an asperity).

The expression obtained for the above probability depends on time t. We can get the instant t, when P_1 reaches the maximum value, from the condition

$$\frac{\mathrm{d}P_1}{\mathrm{d}t} = 0. \tag{6.71}$$

It is possible to obtain the relationship $P_1(t)$ for a number of discrete values, and then, using spline functions, get an analytical expression for $P_1(t)$.

A similar method also allows us to determine the maximum values of the accelerations (the first n components of the vector $\dot{\mathbf{Z}}$)

$$\max\left(\dot{z}_i\right) = \left|\mathbf{J}_1\right| b_{i1}\left(t\right). \tag{6.72}$$

By analogy with (6.67) we obtain the distribution law of the maximum values of the accelerations of the systems masses

$$f_{i2}\left(\dot{z}_{im},\ t\right) = \frac{1}{\left|b_i\left(t\right)\right| \sigma_{J_i}\sqrt{2\pi}} \times \left[\exp\left\{-\frac{\left(z_{im} - b_i m_{J_1}\right)^2}{b_i^2\left(t\right) 2\sigma_{J_i}^2}\right\} + \exp\left\{-\frac{\left(z_{im} + b_i m_{J_1}\right)^2}{b_i^2\left(t\right) 2\sigma_{J_i}^2}\right\}\right]. \tag{6.73}$$

The greatest possible value of the acceleration ($\max\left(\max \dot{z}_i\right) = \max \dot{z}_i^*$) is determined from the condition

$$\int_0^{\dot{z}^*_{i\,\max}} f_{i2}\left(\dot{z}_{im},\ t\right) \mathrm{d}\left(z_{im}\right) = 0.99. \tag{6.74}$$

Knowing the distribution law of the accelerations, we can determine the probability that the overloads acting on the lumped masses of a system, will not go beyond the given interval of values:

$$P_2\left(0 < \dot{z}_{im} < \dot{z}_{i\ddot{a}}\right) = \int\limits_0^{\dot{z}_{i\ddot{a}}} f_{i2}\left(\dot{z}_{im},\ t\right) \mathrm{d}\left(\dot{z}_{im}\right), \tag{6.75}$$

where $\dot{z}_{i\ddot{a}}$ is the upper value of allowable accelerations. We find the moment of time at which P_2 reaches the maximum value, from the following condition

$$\frac{\mathrm{d}P_2}{\mathrm{d}t} = 0.$$

As an illustration of the presented method consider a system with two degrees of freedom (Fig. 6.10) that is subjected to the action of a random, arbitrary directed impulse in the plane x_1x_2 (at the point k). Using the force method, we obtain the following equations of small vibrations of a system (ignoring resistance forces):

$$\begin{aligned} y_2 &= \delta_{11}\left(-M\ddot{x}_2\right) + \delta_{12}\left(-J_0\ddot{\varphi}\right); \\ \varphi &= \delta_{21}\left(-M\ddot{x}_2\right) + \delta_{22}\left(-J_0\ddot{\varphi}\right). \end{aligned} \tag{6.76}$$

In this example there is no need to pass to a system of first order equations as the solution can be obtained in an analytical form. We determine the initial velocities of a system after the termination of the impulse action from the following relationship

$$\begin{aligned} J_{x_2} &= -M\dot{x}_{20}; \\ J_{x_1}H &= -J_0\dot{\varphi}_0. \end{aligned}$$

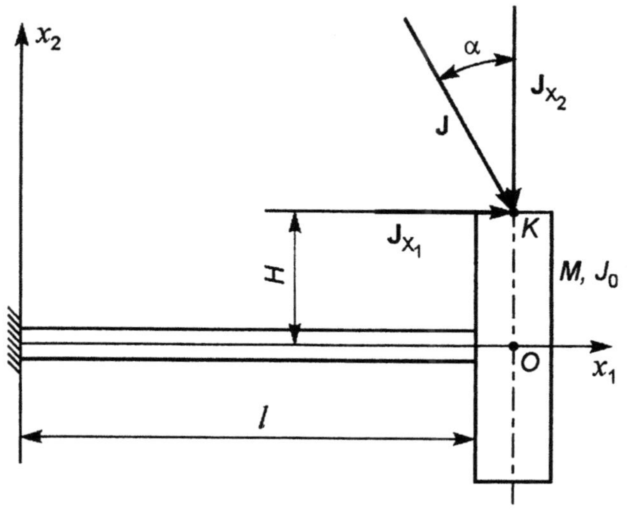

Fig. 6.10.

The solution of the system (6.76) is of the form

$$x_2 = c_1 \sin p_1 t + c_2 \cos p_1 t + c_3 \sin p_2 t + c_4 \cos p_2 t;$$
$$\varphi = c_1 \alpha_1 \sin p_1 t + c_2 \alpha_1 \cos p_1 t + c_3 \alpha_2 \sin p_2 t + c_4 \alpha_2 \cos p_2 t.$$

At $t = 0$ $x_2 = \varphi = 0$, therefore $c_2 = c_4 = 0$. The arbitrary constants are

$$c_1 = \frac{\left(\frac{Hp_2}{J_0}\right) J_{x_1} - p_2 \left(\frac{\alpha_2}{I} - \frac{l}{J_0}\right) J_{x_2}}{p_1 p_2 (\alpha_2 - \alpha_1)};$$
$$c_3 = \frac{\left(\frac{Hp_1}{J_0}\right) J_{x_1} + p_1 \left(\frac{l}{J_0} - \frac{\alpha_1}{I}\right) J_{x_2}}{p_1 p_2 (\alpha_2 - \alpha_1)}.$$

As a result we obtain the following expressions:

$$x_2 = a_1 J_{x_1} + a_2 J_{x_2};$$
$$\varphi = b_1 J_{x_1} + b_2 J_{x_2},$$

where

$$a_1 = \gamma_{11} \sin p_1 t - \gamma_{21} \sin p_2 t;$$
$$a_2 = \gamma_{22} \sin p_2 t - \gamma_{12} \sin p_1 t;$$
$$b_1 = \gamma_{11} \alpha_1 \sin p_1 t - \gamma_{21} \alpha_2 \sin p_2 t;$$
$$b_2 = \gamma_{22} \sin p_2 t - \gamma_{12} \alpha_1 \sin p_1 t;$$

$$\gamma_{11} = \frac{Hp_2}{J_0 p_1 p_2 (\alpha_2 - \alpha_1)}; \quad \gamma_{12} = \frac{\left(\frac{\alpha_2}{I} - \frac{l}{J_0}\right)}{p_1 (\alpha_2 - \alpha_1)};$$
$$\gamma_{21} = \frac{Hp_1}{J_0 (\alpha_2 - \alpha_1) p_1 p_2}; \quad \gamma_{22} = \frac{\left(\frac{l}{J_0} - \frac{\alpha_1}{I}\right)}{p_2 (\alpha_2 - \alpha_1)}.$$

Let us obtain expressions for the maximum value of the displacement x_2 and $\dot{x}_2$. In accordance with the previously given algorithm, we consider a functional

$$J = (\mathbf{a} \cdot \mathbf{J}) - \lambda \left[(\mathbf{C}\mathbf{J} \cdot \mathbf{J}) - 1 \right],$$

where

$$(\mathbf{a} \cdot \mathbf{J}) = \sum_{i=1}^{2} a_i J_{x_i}, \quad \mathrm{C} = \begin{bmatrix} \frac{1}{|\mathbf{J}|^2} & 0 \\ 0 & \frac{1}{|\mathbf{J}|^2} \end{bmatrix}.$$

From the condition (6.55) we obtain an equation

$$\mathbf{a} = \lambda \mathrm{C} \cdot \mathbf{J}.$$

Using the condition (6.53), we obtain

$$\lambda = \sqrt{|\mathbf{J}|^2 \left(a_1^2 + a_2^2\right)} = |\mathbf{J}| \sqrt{\left(a_1^2 + a_2^2\right)},$$

or

$$\lambda = |\mathbf{J}| \sqrt{c_{11} \sin^2 p_1 t + c_{22} \sin^2 p_2 t + c_{12} \sin p_1 t \cdot \sin p_2 t},$$

where

$$c_{11} = \gamma_{11}^2 + \gamma_{12}^2; \qquad c_{22} = \gamma_{21}^2 + \gamma_{22}^2;$$
$$c_{12} = -\left(\gamma_{11}\gamma_{21} + \gamma_{22}\gamma_{12}\right).$$

The maximum value of x_2 (6.55) is

$$\max\left(x_2\right) = \lambda = |\mathbf{J}| \sqrt{a_1^2 + a_2^2} =$$
$$= |\mathbf{J}| \sqrt{c_{11} \sin^2 p_1 t + c_{12} \sin p_1 t \sin p_2 t + c_{22} \sin^2 p_2 t}.$$

According to (6.67) we obtain the distribution law for $\max\left(x_2\right)$ and $\max\left(\dot{x}_2\right)$

$$f\left(x_{2m},\ t\right) = \frac{1}{\sigma_{J_i}\sqrt{2\pi}\, b}\left[\exp\left\{-\frac{\left(x_{2m} - b m_J\right)^2}{b^2\, 2\sigma_J^2}\right\} + \exp\left\{-\frac{\left(x_{2m} + b m_J\right)^2}{b^2 2\sigma_J^2}\right\}\right]. \tag{6.77}$$

where

$$b = \sqrt{a_1^2 + a_2^2}.$$

Carrying out similar calculations for $\dot{x}_2$

$$\dot{x}_2 = \dot{a}_1 J_{x_1} + \dot{a}_2 J_{x_2},$$

we determine $\max \dot{x}_2$ and the corresponding distribution law

$$\max \dot{x}_2 = |\mathbf{J}| \sqrt{a_1^2 + a_2^2},$$

$$f\left(\dot{x}_{2m},\ t\right) = \frac{1}{\sigma_0\sqrt{2\pi}\, b_1}\left[\exp\left\{-\frac{\left(\dot{x}_{2m} - b_1 m_J\right)^2}{2b_1^2\sigma_J^2}\right\} + \exp\left\{-\frac{\left(\dot{x}_{2m} + b_1 m_J\right)^2}{2b_1^2\sigma_J^2}\right\}\right], \tag{6.78}$$

where

$$b_1 = \sqrt{\dot{a}_1^2 + \dot{a}_2^2}; \quad x_{2m} = \max(x_2); \quad \dot{x}_{2m} = \max(\dot{x}_2).$$

Distribution laws continuously vary as time changes, and at discrete instants the distribution laws of x_{2m}, $\dot{x}_{2m}$ are similar to those of the random impulse modulus (see Fig. 6.8).

We find the greatest possible displacement $\max(x_{2m})$ and the velocity $\max(\dot{x}_{2m})$ from the conditions

$$p_{x_2} = \int_0^{x_{2m}^*} f_{x_2}(x_{2m}^*, t)\,\mathrm{d}x_{2m} = 0.99; \tag{6.79}$$

$$p_{\dot{x}_2} = \int_0^{x_{2m}^*} f_{x_2}(\dot{x}_{2m}^*, t)\,\mathrm{d}\dot{x}_{2m} = 0.99, \tag{6.80}$$

where

$$x_{2m}^* = \max(x_{2m}), \quad \dot{x}_{2m}^* = \max(\dot{x}_{2m}).$$

For a number of specific values of time t we can determine x_{2m}^* and $\dot{x}_{2m}^*$ from (6.79) and (6.80) only numerically.

The given method of analyzing mechanical systems under the action of random arbitrary directed impulses allows us to determine the greatest possible dynamic effects in a system and minimize them through a choice of its parameters.

6.3 Non-Stationary Random Vibrations of Linear Systems

As a rule, systems with one degree of freedom enable us to obtain the solution of equations of motion in an analytical form, which essentially simplifies the subsequent determination of the probability characteristics of the output at known probability characteristics of the input. In a number of cases we can obtain probability characteristics of the output for equations with constant coefficients in an analytical form convenient for making an analysis. For systems with a finite number of degrees of freedom, e.g. for linear systems with constant parameters, the solution can basically be obtained in an analytical form, but this gives us no substantial benefits in comparison with a numerical solution because of the cumbersome formulas. For this reason numerical methods of random vibrations analysis are usually preferred.

To introduce the reader to the section of non-stationary random vibrations of linear systems we shall examine a system with three degrees of freedom

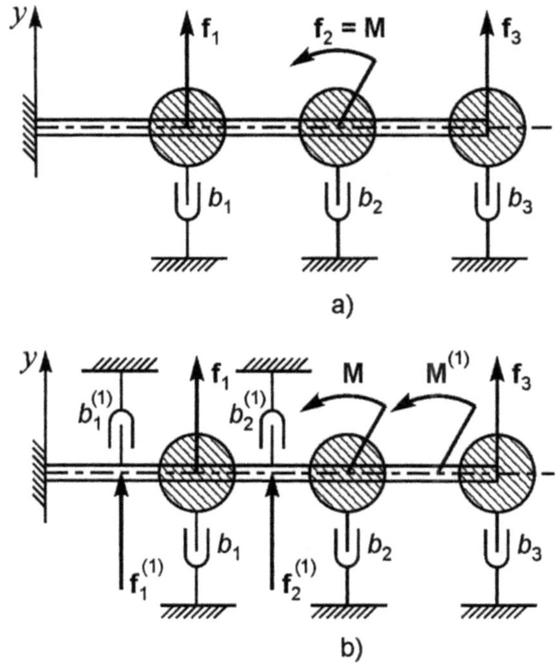

Fig. 6.11.

(Fig. 6.11 *a*), in which masses m_j are considered to be point. In Fig. 6.11 *a* the number of external forces is equal to the number of degrees of freedom, but there are cases, when the number of excitations is smaller than that of degrees of freedom or exceeds it, as illustrated by Fig. 6.11 *b*, where excitations are applied to massless points. Also possible are mechanical systems (damping systems) in which elements that realize the concentrated forces of viscous friction $(\varkappa_j \dot{y}_j)$ are related to massless points (Fig. 6.11 *b*).

Further on we shall assume that all necessary information on random forces (the probability characteristics of random excitations) is known. In this context, the principal difficulty associated with the obtainment of these probability characteristics, is not considered, which is acceptable when presenting mathematical methods of solving differential equations, but out of question as regards specific problems that engineers have to solve in their everyday work.

In most cases we do not have the probability characteristics of random forces (or other types of random excitations) that can act on a designed structure in real operational conditions. Besides, random excitations may depend on the structure itself, which is also not yet available. Therefore prior to solving equations enormous preparatory work is done to acquire information on random excitations by experimental methods (whenever possible) and through processing experimental data to obtain probability characteristics of the input, an especially laboursome effort when it comes to the necessity of getting information on probability characteristics of non-stationary random processes.

This laboriousness substantially limits the capabilities of the theory of non-stationary random vibrations during the solution of applied problems.

Preparatory work on determining probability characteristics of the input takes incommensurably more time, than the solution of equations of motion at the input known in probability sense. Eventually, the reliability of a mechanical systems operation wholly depends on how accurately the input given in the solution reflects the real physics of the process. In a number of cases the probability characteristics of the input do not depend on the structure itself and are well studied random functions. For example, the probability characteristics of road irregularities [28], aerodrome pavements irregularities, wind velocities for various areas, etc. have been obtained as a result of processing large quantities of experimental data. In this case we can use known probability characteristics of the input when analyzing the motion of a system.

The principal methods of solving the equations of motion at non-stationary random excitations are partially set forth in Sect. 2.5.

Using the force method, we obtain the system of equations of small vibrations (Fig. 6.11 *a*):

$$\begin{aligned} y_1 &= \delta_{11}\left(-m_1\ddot{y}_1 - b_1\dot{y}_1\right) + \delta_{12}\left(-m_2\ddot{y}_2 - b_2\dot{y}_2\right) \\ &\quad + \delta_{13}\left(-m_3\ddot{y}_3 - b_3\dot{y}_3\right) + \delta_{14}M + \delta_{11}f_1 + \delta_{13}f_3; \\ y_2 &= \delta_{21}\left(-m_1\ddot{y}_1 - b_1\dot{y}_1\right) + \delta_{22}\left(-m_2\ddot{y}_2 - b_2\dot{y}_2\right) \\ &\quad + \delta_{23}\left(-m_3\ddot{y}_3 - b_3\dot{y}_3\right) + \delta_{21}f_1 + \delta_{23}f_3 + \delta_{24}M; \\ y_3 &= \delta_{31}\left(-m_1\ddot{y}_1 - b_1\dot{y}_1\right) + \delta_{32}\left(-m_2\ddot{y}_2 - b_2\dot{y}_2\right) \\ &\quad + \delta_{33}\left(-m_3\ddot{y}_3 - b_3\dot{y}_3\right) + \delta_{31}f_1 + \delta_{33}f_3 + \delta_{34}M, \end{aligned} \tag{6.81}$$

or in a vector form (after transformations)

$$\mathrm{M}\ddot{\mathbf{y}} + \mathrm{B}\dot{\mathbf{y}} + \mathrm{C}\mathbf{y} = D_1 f, \qquad \left(\mathrm{D}_1 = \Delta^{-1}\Delta_1\right), \tag{6.82}$$

where

$$\Delta = \begin{bmatrix} \delta_{11} & \delta_{12} & \delta_{13} \\ \delta_{21} & \delta_{22} & \delta_{23} \\ \delta_{31} & \delta_{32} & \delta_{33} \end{bmatrix}, \qquad \mathrm{M} = \begin{bmatrix} m_1 & 0 & 0 \\ 0 & m_2 & 0 \\ 0 & 0 & m_3 \end{bmatrix},$$

$$\Delta_1 = \begin{bmatrix} \delta_{11} & \delta_{12} & \delta_{14} \\ \delta_{21} & \delta_{23} & \delta_{43} \\ \delta_{31} & \delta_{33} & \delta_{34} \end{bmatrix}, \quad \mathbf{f} = \begin{bmatrix} f_1 \\ f_2 \\ M \end{bmatrix}, \quad \mathrm{B} = \begin{bmatrix} b_1 & 0 & 0 \\ 0 & b_2 & 0 \\ 0 & 0 & b_3 \end{bmatrix}.$$

In a more general case the forces of viscous friction can also depend on relative velocities. Then the matrix is

$$\mathrm{B} = \begin{bmatrix} b_{11} & \dots & b_{1n} \\ \dots & \vdots & \dots \\ b_{n1} & \dots & b_{nn} \end{bmatrix}.$$

Let us transform the equation (6.82) to the form

$$\dot{\mathbf{Z}} + \mathbf{AZ} = \mathbf{DF}, \tag{6.83}$$

where

$$\mathrm{A} = \begin{bmatrix} \mathrm{M}^{-1}\mathrm{B} & \mathrm{M}^{-1}\mathrm{C} \\ -\mathrm{E} & 0 \end{bmatrix}, \quad \mathbf{Z} = \begin{bmatrix} \dot{\mathbf{y}} \\ \mathbf{y} \end{bmatrix}, \quad \mathrm{D} = \begin{bmatrix} \mathrm{D}_1 & 0 \\ 0 & 0 \end{bmatrix}, \quad \mathbf{F} = \begin{bmatrix} \mathbf{f} \\ 0 \end{bmatrix}.$$

The components f_j of the vector $\mathbf{F}$ are random functions with known probability characteristics (m_{f_j} and $K_{f_i f_j}$ are known).

Further on we assume that the number of components f_j of the vector $\mathbf{F}$ is less than or equal to the number of degrees of freedom of a system. The solution of the equation (6.83) at zero initial conditions takes the form

$$\mathbf{Z} = \int_0^t \mathrm{G}(t,\ \tau)\mathrm{D}(\tau)\,\mathbf{F}(\tau)\,\mathrm{d}\tau; \tag{6.84}$$

$$\mathrm{G}(t,\ \tau) = \mathrm{K}(t)\,\mathrm{K}^{-1}(\tau),$$

where $\mathrm{K}(t)$ is the fundamental matrix of the solutions of the homogeneous equation (6.83). The algorithm of determining the Green matrix $\mathrm{G}(t,\ \tau)$ was given in Sect. 2.19. For equations with constant coefficients the Green matrix is

$$\mathrm{G}(t,\ \tau) = \mathrm{G}(t-\tau) = \mathrm{K}(t-\tau).$$

The probability characteristics of the components z_i of the system state vector $\mathbf{Z}$ (6.84) (provided that the number of excitations f_i is equal to the number of degrees of freedom) are (at $\mathrm{D}(\tau) = \mathrm{E}$):

$$\begin{aligned} m_{z_i}(t) &= \sum_{j=0}^{n} \int_0^{t_i} g_{ij}(t,\tau) m_{f_j}(\tau)\,\mathrm{d}\tau; \\ K_{z_i z_k}(t,t') &= \sum_{j=2}^{n}\sum_{v=1}^{n} \int_0^{t}\int_0^{t'} g_{ij}(t,\tau) g_{kv}(t',\tau') K_{f_i f_k}\,\mathrm{d}\tau\,\mathrm{d}\tau'. \end{aligned} \tag{6.85}$$

where $g_{ij}(t,\ \tau)$ are the elements of the matrix $\mathrm{G}(t,\ \tau)$.

Let us consider two special cases: 1) f_i are independent random functions and 2) f_i are independent random quantities.

If f_i are independent random functions, then

$$K_{z_i z_k}(t,t') = \sum_{v=1}^{n} \int_0^t \int_0^{t'} g_{iv}(t,\tau) g_{kv}(t',\tau') K_{f_v f_v} \mathrm{d}\tau \, \mathrm{d}\tau'. \tag{6.86}$$

The autocorrelation functions and variances of the components are

$$K_{z_i z_i}(t,t') = \sum_{v=1}^{n} \int_0^t \int_0^{t'} g_{iv}(t,\tau) g_{iv}(t',\tau') K_{f_v f_v} \mathrm{d}\tau \, \mathrm{d}\tau'. \tag{6.87}$$

$$D_{z_i}(t) = \sum_{v=1}^{n} \int_0^t \int_0^t g_{iv}(t,\tau) g_{iv}(t,\tau') K_{f_v f_v} \mathrm{d}\tau \, \mathrm{d}\tau'. \tag{6.88}$$

In the particular case of the excitations being random quantities, i.e. at

$$m_{f_j} = \text{const}; \qquad K_{f_i f_v}(\tau,\ \tau') = K_{iv} = \text{const},$$

the expressions for the mathematical expectations and cross-correlation functions of the solution take the form

$$\begin{aligned} K_{z_i z_k}(t,t') &= \sum_{i=1}^{n} \sum_{v=1}^{n} K_{f_j f_v} \int_0^t \int_0^{t'} g_{ij}(t,\tau) g_{kv}(t',\tau') \, \mathrm{d}\tau \, \mathrm{d}\tau'; \\ m_{z_i}(t) &= \sum_{i=1}^{n} m_{f_j} \int_0^t g_{ij} \mathrm{d}\tau; \\ D_{z_i} &= \sum_{v=1}^{n} \sum_{i=1}^{n} K_{f_j f_v} \int_0^t \int_0^t g_{ij}(t,\tau) g_{iv}(t,\tau') \, \mathrm{d}\tau \mathrm{d}\tau'. \end{aligned} \tag{6.89}$$

Let us consider another particular case where the excitations $f_i(t)$ are random functions of the white noise type

$$K_{f_i f_j} = S_{ij} \delta(t - t_1). \tag{6.90}$$

Therefore

$$K_{z_j z_k}(t,\ t') = \sum_{i=1}^{n} \sum_{v=1}^{n} S_{iv} \int_0^t g_{ij}(t,\tau) \left[\int_0^{t'} g_{kv}(t',\tau') \delta(\tau' - \tau) \mathrm{d}\tau' \right] \mathrm{d}\tau$$

or

$$K_{z_j z_k}(t,\ t') = \sum_{i=1}^{n} \sum_{v=1}^{n} S_{iv} \int_0^t g_{ij}(t,\tau) g_{kv}(t',\tau) \mathrm{d}\tau. \tag{6.91}$$

We obtain the following expressions for the variances of the components z_i of the vector $\mathbf{Z}$

$$D_{z_i}(t) = \sum_{i=1}^{n}\sum_{v=1}^{n} S_{iv} \int_0^t g_{ij}(t,\tau) g_{kv}(t,\tau)\,\mathrm{d}\tau. \tag{6.92}$$

For independent f_i we have

$$K_{f_i f_j}(t,\, t') = \begin{cases} 0, & i \neq j \\ \sigma_i^2 \delta(t-t'), & j = i, \end{cases}$$

therefore

$$\begin{aligned} K_{z_i z_k}(t,\, t') &= \sum_{v=1}^{n} \sigma_v^2 \int_0^t g_{iv}(t,\tau) g_{kv}(t',\tau)\,\mathrm{d}\tau; \\ D_{z_i}(t) &= \sum_{v=1}^{n} \sigma_v^2 \int_0^t \left[g_{iv}(t,\tau)\right]^2 \mathrm{d}\tau. \end{aligned} \tag{6.93}$$

6.4 The Method of Principal Coordinates in Non-Stationary Vibrations Analysis

Let us consider a method of solving a system of nonhomogeneous linear equations with constant coefficients at non-stationary random vibrations that allows us to obtain solutions in analytical form.

If resistance forces can be neglected or if they satisfy certain conditions, we can use the method of principal coordinates to solve the system of equations (6.81).

Let us first consider the problem of determining eigenvalues p_j and eigenvectors for a system of homogeneous equations of small vibrations ignoring the resistance forces. The equation of free vibrations (at $b_{ij} = 0$) takes the form (a special case of the equation (6.29))

$$\mathbf{M}\ddot{\mathbf{y}} + \mathbf{C}\mathbf{y} = 0. \tag{6.94}$$

We seek a solution to the equation (6.94) in the form

$$\mathbf{y} = \mathbf{u} \cos pt.$$

As a result, we obtain

$$p^2 \mathbf{M}\mathbf{u} + \mathbf{C}\mathbf{u} = 0$$

or

$$\left[\mathrm{C} - p^2\mathrm{M}\right]\mathbf{u} = 0. \tag{6.95}$$

The frequencies p_i are determined from the equation

$$\det\left[\mathrm{C} - p^2\mathrm{M}\right] = 0. \tag{6.96}$$

Knowing the frequencies, we determine corresponding vectors $\mathbf{u}^{(i)}$ that satisfy algebraic homogeneous equations (we assume that all frequencies are distinct and differ from zero):

$$\left[\mathrm{C} - p_i^2\mathrm{M}\right]\mathbf{u}^{(i)} = 0. \tag{6.97}$$

or in scalar form

$$\sum_{v=1}^{n} c_{kv}^{(i)} u_v^{(i)} = 0 \qquad (k = 1, 2, \ldots, n), \tag{6.98}$$

where $c_{kv}^{(i)}$ are the elements of the matrix $\mathrm{C}^{(i)}$

$$\mathrm{C}^{(i)} = \left[\mathrm{C} - p_i^2\mathrm{M}\right].$$

It is known from the general theory of linear algebraic equations, that the system (6.98) is always compatible and has a non-trivial solution (provided that the rank of the matrix of a system was lower than the number of unknowns). For example, we can express the $n-1$ component of the vector $\mathbf{u}^{(i)}$ in terms of the component $u_n^{(i)}$ taking the first $n-1$ equations of the system (6.98):

$$u_j^{(i)} = d_j^{(i)} u_n^{(i)} \qquad (j = 1, \ldots, n-1). \tag{6.99}$$

As a result, for each of the frequencies p_i we obtain

$$\mathbf{y}^{(i)} = \mathbf{u}^{(i)} \cos p_i t, \tag{6.100}$$

where the $n-1$ components of the vector $\mathbf{u}^{(i)}$ are determined from the system of equations (6.99). The component $u_n^{(i)}$ of the vector $\mathbf{u}^{(i)}$ can be taken arbitrary and, in particular, that equal to unity. The vectors $\mathbf{u}^{(i)}$ satisfy the condition

$$\left(p_i^2 - p_j^2\right)\left(\mathrm{M}\mathbf{u}^{(i)} \cdot \mathbf{u}^{(j)}\right) = 0, \tag{6.101}$$

or (at $p_i^2 \neq p_j^2$)

$$\left(\mathrm{M}\mathbf{u}^{(i)} \cdot \mathbf{u}^{(j)}\right) = 0 \qquad (i \neq j), \tag{6.102}$$

which is referred to as the orthogonality condition of the vectors $\mathbf{u}^{(i)}$ and $\mathbf{u}^{(j)}$. Let us form a matrix U out of the components of the vectors $\mathbf{u}^{(i)}$:

$$\mathrm{U} = \begin{bmatrix} u_1^{(1)} & \ldots\ldots & u_1^{(n)} \\ u_2^{(1)} & \ldots\ldots & u_2^{(n)} \\ \vdots & \vdots \ddots & \vdots \\ u_n^{(1)} & \ldots\ldots & u_n^{(n)} \end{bmatrix}, \quad \left(u_n^{(i)} = 1; \quad i = 1,\ 2,\ \ldots,\ n\right)$$

and consider a transformation

$$\mathbf{y} = \mathrm{U}\mathbf{q}, \tag{6.103}$$

where $\mathbf{q}$ is the vector of new unknowns.

Substituting (6.103) in equation (6.82) (ignoring the resistance forces), we obtain

$$\mathrm{MU}\ddot{\mathbf{q}} + \mathrm{CU}\mathbf{q} = \mathrm{D}_1\mathbf{f}. \tag{6.104}$$

Let us multiply equation (6.104) by the transposed matrix

$$\mathrm{U^TMU}\ddot{\mathbf{q}} + \mathrm{U^TCU}\mathbf{q} = \mathrm{U^TD}_1\mathbf{f}. \tag{6.105}$$

Now we can show that matrixes $\mathrm{U^TMU}$ and $\mathrm{U^TCU}$ are diagonal if due account is taken of orthogonality condition (6.102):

$$\mathrm{U^TMU} = \mathrm{M}^{(1)}, \quad \mathrm{U^TCU} = \mathrm{A}^{(1)}.$$

The diagonal elements of the matrixes $\mathrm{M}^{(1)}$ and $\mathrm{A}^{(1)}$ are respectively equal to

$$m_{ii}^1 = \sum_{j=1}^{n} \left(u_j^{(i)}\right)^2 m_j, \qquad a_{ii}^1 = p_i^2 \sum_{j=1}^{n} \left(u_j^{(i)}\right)^2 m_j. \tag{6.106}$$

With due account of orthogonality condition (6.101) we obtain from (6.105) in scalar form

$$\ddot{q}_i + p_i^2 q_i = \frac{\sum\limits_{v=1}^{n} e_{iv} f_v(t)}{\sum\limits_{j=1}^{n} \left(u_j^{(i)}\right)^2 m_j}, \tag{6.107}$$

where $e_{iv} = \sum\limits_{k=1}^{n} \left(u_{ik} d_{kv}^{(1)}\right)$, $d_{kv}^{(1)}$ are elements of the matrix D_1.

If the system is loaded only with the forces f_i (the random moments are absent) applied to the masses m_j (Fig. 6.11 *a*), the matrix D_1 is equal to the identity matrix, i.e. $\mathrm{D}_1 = \mathrm{E}$ and $d_{kv}^{(1)} = 0$ at $v \neq k$.

The introduced new unknowns q_i are referred to as the principal coordinates. The solution of the equations (6.107) at zero initial conditions takes the form

$$q_i = \frac{1}{p_i} \int_0^t \sin p_i(t-\tau) \left[\sum_{v=1}^{n} \beta_{iv} f_v(\tau) \right] \mathrm{d}\tau, \tag{6.108}$$

or

$$q_i = \sum_{v=1}^{n} \int_0^t g_{iv}^0 f_v \mathrm{d}\tau,$$

where

$$\beta_{iv} = \frac{e_{iv}}{\sum\limits_{i=1}^{n} u_{ij} m}; \qquad g_{iv}^0 = \frac{1}{p_i} \sin p_i \left(t-\tau\right) \beta_{iv}.$$

Knowing q_i we determine the solution of the equation (6.82) ignoring the resistance forces

$$y_k = \sum_{i=1}^{n} u_i^{(k)} q_i = \sum_{v=1}^{n} \int_0^t g_{kv} f_v \mathrm{d}\tau, \tag{6.109}$$

where $g_{kv} = \sum\limits_{i=1}^{n} u_{ki} g_{iv}^0$. Given viscous resistance forces the equations of the motion of a system (6.82) after the substituting $\mathrm{U}\mathbf{q}$ for $\mathbf{y}$ and multiplication by the transposed matrix U^{T}take the form

$$\mathrm{U^T M U}\ddot{\mathbf{q}} + \mathrm{U^T B U}\dot{\mathbf{q}} + \mathrm{U^T C U}\mathbf{q} = \mathrm{U^T D^1}\mathbf{f}. \tag{6.110}$$

When resistance forces are taken into account, the matrix $U^T BU$ must also be diagonal, which it is at

$$\mathrm{B} = 2n\mathrm{M} + 2\lambda\mathrm{C}, \tag{6.111}$$

where $2n$ and 2λ are scalar multipliers.

Multiplying (6.111) from the left by the matrix $\mathrm{U^T}$ and from the right by the matrix U, we obtain

$$\mathrm{U^T B U} = 2n(\mathrm{U^T M U}) + 2\lambda(\mathrm{U^T C U}). \tag{6.112}$$

In scalar form, we obtain the following equations from (6.110) with due account taken of (6.112)

$$\ddot{q}_i + 2n_i\dot{q}_i + p_i^2 q_i = \frac{\sum\limits_{v=1}^{n} e_{iv} f_v(t)}{\sum\limits_{j=1}^{n} \left(u_j^{(i)}\right)^2 m_j}, \tag{6.113}$$

where

$$n_i = n + \lambda p_i^2. \tag{6.114}$$

As the distribution of friction forces is practically unknown and we only know (by experiment) the integral effect of their action, we may adopt any hypothesis about the distribution of friction forces whose actions effect is equivalent to the integral one. Therefore we may reasonably assume that the friction force is distributed so that the condition (6.114) is satisfied, i.e. principal coordinates for a conservative system also remain principal for a nonconservative system that takes into account the force of viscous friction. The solution of the equation (6.113) is

$$q_i = \mathrm{e}^{-n_i t}\left(c_1^{(i)} \cos p_i t + c_2^{(i)} \sin p_i t\right) + \frac{1}{p_i}\int_0^t \mathrm{e}^{-n_i(t-v)} \sin p_i(t-\tau)\left(\sum_{v=1}^{n} \beta_{iv} f_v\right) \mathrm{d}\tau, \tag{6.115}$$

or in a more compact form

$$q_i = \mathrm{e}^{-n_i t}\left(c_1^{(i)} \cos p_i t + c_2^{(i)} \sin p_i t\right) + \sum_{v=1}^{n}\int_0^t g_{iv}^{(2)}(t,\tau) f_v(\tau)\mathrm{d}\tau, \tag{6.116}$$

where

$$g_{iv}^{(2)} = \frac{1}{p_i}\mathrm{e}^{-n_i(t-\tau)} \sin p_i(t-\tau)\beta_{iv}, \qquad \beta_{iv} = \frac{e_{iv}}{\sum_{i=1}^{n}\left(u_j^{(i)}\right)^2 m_j}.$$

Having determined q_i we find the components of the vector $\mathbf{y}$ and of the first two derivatives:

$$\begin{aligned} y_k &= \sum_{j=1}^{n} u_{kj} q_j; \\ \dot{y}_k &= \sum_{j=1}^{n} u_{kj} \dot{q}_j; \\ \ddot{y}_k &= \sum_{j=1}^{n} u_{kj} \ddot{q}_j. \end{aligned} \tag{6.117}$$

At zero initial data we have

$$y_k = \sum_{v=1}^{n}\int_0^t g_{iv}^{(2)} f_v \mathrm{d}\tau. \tag{6.118}$$

For example, through transformations we obtain the following expression for y_k (at zero initial data), similar to (6.109):

$$y_k = \sum_{v=1}^{n} \int_0^t g_{kv}^{(3)} f_v \mathrm{d}\tau, \tag{6.119}$$

where $g_{kv}^{(3)} = \sum_{i=1}^{n} u_k^{(i)} g_{iv}^{(2)}$.

Then we find the probability characteristics of the components of the state vector $\mathbf{y}$:

$$m_{y_k} = \sum_{v=1}^{n} \int_0^t g_k^{(3)} m_{f_v} \mathrm{d}\tau; \tag{6.120}$$

$$K_{y_k y_j} = \sum_{v=1}^{n} \sum_{\rho=1}^{n} \int_0^t \int_0^{t'} g_{kv}^{(3)}(t) g_{j\rho}^{(3)}(t') K_{f_v f_\rho} \mathrm{d}\tau\, \mathrm{d}\tau'. \tag{6.121}$$

Let us consider non-stationary vibrations at suddenly applied random forces that are constant in time, using Fig. 6.12 as an example. A high-rise structure (Fig. 6.12) is presented as a system of lumped masses m_j, connected by inertialess elastic rods. Let us consider two cases of the vibrations of the structure: 1) random forces f_j lie in the plane of the drawing and are parallel to the axis x_1 and 2) the random forces are parallel to the plane $x_1 0 x_2$ and are parallel to one another, but have arbitrary directions the way it was in the considered case of pulse loading (Fig. 6.4).

In the first case, if the directions of the principal axes of the rod sections coincide with the directions of the axes x_1 and x_2 the vibrations of the masses m_i occur in the plane $x_1 0 x_2$. As f_i are parallel, it is possible to put

$$\mathbf{f}_i = \beta_i \mathbf{f}_1. \tag{6.122}$$

The equation of small vibrations is similar to the equation (6.82) ($\mathrm{D}_1 = \mathrm{E}$):

$$\mathrm{M}\ddot{\mathbf{x}}_1 + \mathrm{B}\dot{\mathbf{x}}_1 + \mathrm{C}\mathbf{x}_1 = \mathbf{f} \qquad \left(\mathbf{f} = (f_1,\ f_2,\ \ldots,\ f_n)^{\mathrm{T}}\right), \tag{6.123}$$

where B is the diagonal matrix. As the moduli of the forces $\mathbf{f}_i$ do not depend on time we obtain the probability characteristics of the components of the vector $\mathbf{x}_1$ from (6.89) with due account taken of (6.122):

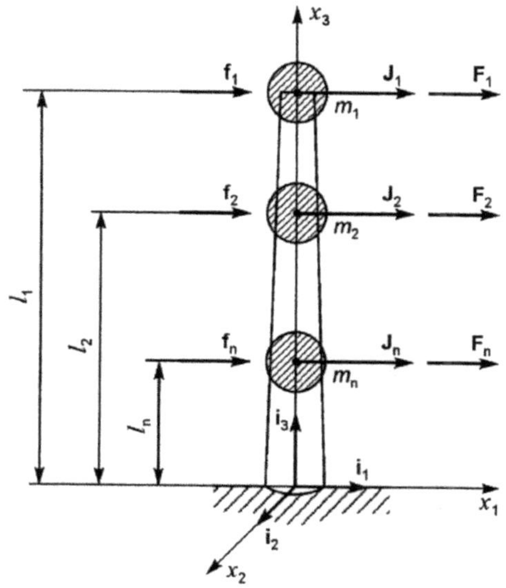

Fig. 6.12.

$$m_{x_k} = m_{f_1} \sum_{v=1}^{n} \beta_v \int_0^t g_{kv} \mathrm{d}\tau ;$$

$$K_{x_i x_k} = D_{f_1} \sum_{j=1}^{n} \sum_{v=1}^{n} \beta_j \beta_v \int_0^t \int_0^{t'} g_{ij}(t, \tau) g_{kv}(t', \tau') \, \mathrm{d}\tau \, \mathrm{d}\tau' ; \tag{6.124}$$

$$D_{x_k}(t) = \sum_{v=1}^{n} \sum_{j=1}^{n} K_{f_j f_v} \int_0^t \int_0^t g_k(t, \tau) g_{kv}(t, \tau') \, \mathrm{d}\tau \, \mathrm{d}\tau' .$$

We can also use the expressions (6.120) and (6.121) obtained with the help of the principal coordinates. Assuming that the components of the vector x_1 have normal distribution, we can determine their greatest possible values at each instant of time using the three sigma rule:

$$\max x_k = m_{x_k} + 3\sigma_{x_k} . \tag{6.125}$$

At the design, the stresses arising in a structure at its loading by random forces are of prime interest. For example, in the structure shown in Fig. 6.12 where the cross-section of the rod is constant and the structure is loaded with random unidirectional forces f_i, the dangerous will be the clamped section. The bending moment at the clamped section due to external forces acting on the masses m_i (including the viscous resistance forces $F_{i\,mp}$) is

$$M = \sum_{\rho=1}^{n} l_\rho (f_\rho + y_\rho + F_\rho) = \sum_{\rho=1}^{n} l_\rho (f_\rho - m_\rho \ddot{x}_{1\rho} - \alpha_\rho \dot{x}_{1\rho}) . \tag{6.126}$$

Using the equation of the motion of the mass m_ρ, we obtain

$$M = \sum_{\rho=1}^{n} \left(\sum_{j=1}^{n} c_{\rho j} x_{1j} \right) . \tag{6.127}$$

The expressions for the components of the system state vector are

$$x_{1j} = \sum_{v=1}^{n} \int_0^t g_{jv}^{(3)} f_v \mathrm{d}\tau . \tag{6.128}$$

Substituting (6.128) in (6.127), we obtain

$$M = \sum_{v=1}^{n} \int_0^t \left(\sum_{\rho=1}^{n} l_\rho q_{\rho v}^{(4)} \right) f_v \mathrm{d}\tau , \tag{6.129}$$

where

$$q_{\rho v}^{(4)} = \sum_{j=1}^{n} c_{\rho j} q_{jv}^{(3)} .$$

The maximum normal stress at the clamped end is

$$\tilde{\sigma} = \frac{M}{W_{x_2}} = \sum_{v=1}^{n} \int_0^t q_v f_v \mathrm{d}\tau ;$$

$$q_v = \frac{1}{W_{x_2}} \sum_{\rho=1}^{n} l_\rho q_{\rho v}^{(4)} .$$

At the loading by random forces that are constant in time, we obtain

$$\tilde{\sigma} = \sum_{v=1}^{n} \int_0^t q_v \mathrm{d}\tau \cdot f_v . \tag{6.130}$$

Considering, that $\tilde{\sigma}$ has a normal distribution, we obtain its greatest possible value (at $f_v = \beta_v f_1$)

$$\tilde{\sigma}^* = m_{\tilde{\sigma}} + 3\sigma_{\tilde{\sigma}} , \tag{6.131}$$

where

$$m_{\tilde{\sigma}} = \sum_{v=1}^{n} \int_0^t q_v \mathrm{d}\tau \, m_{f_v} \mathrm{d}\tau \qquad (m_{f_v} = \beta_v m_{j_1}) \, ;$$
$$\sigma_{\tilde{\sigma}}^2 = D_{f_1} \sum_{v=1}^{n} \sum_{\rho=1}^{n} \beta_v \beta_\rho \int_0^t \int_0^t q_v(\tau) q_\rho(\tau') \mathrm{d}\tau \, \mathrm{d}\tau' . \tag{6.132}$$

Example 6.2. A random force $P = P_0 H(t)$, where $H(t)$ is the Heaviside function, has suddenly acted on a mass m_2 $(m_2 = m_1)$ (Fig. 6.13). The probability characteristics of the random quantity P_0 are known: m_{p_0} D_{p_0}. It is required to determine the mathematical expectation and variance of the maximum normal stress at clamped end. The equations of the small vibrations of the masses are

$$y_1 = \delta_{11} (-m_1 \ddot{y}_1) + \delta_{12} (-m_2 \ddot{y}_2) + \delta_{12} P_0 H(t);$$
$$y_2 = \delta_{21} (-m_1 \ddot{y}_1) + \delta_{22} (-m_2 \ddot{y}_2) + \delta_{22} P_0 H(t).$$

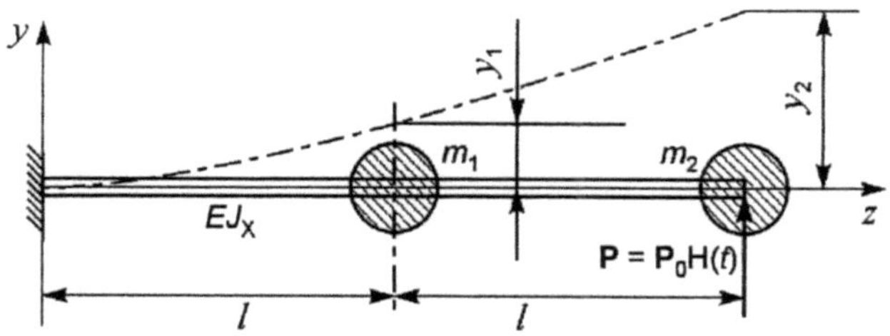

Fig. 6.13.

We determine the frequencies

$$p_1 = 0.585 \sqrt{\frac{EJ_x}{m_1 l^3}} ; \qquad p_2 = 3.881 \sqrt{\frac{EJ_x}{m_1 l^3}}$$

and the eigenvectors (putting $u_{11} = u_{12} = 1$)

$$\mathbf{u}^{(1)} = (1, \ u_{21}) \, , \qquad \mathbf{u}^{(2)} = (1, \ u_{22}) \, ,$$

where $u_{21} = 3.132$; $u_{22} = -0.319$.

Let us turn to the principal coordinates q_1 and q_2:

$$y_1 = u_{11} q_1 + u_{12} q_2 ;$$
$$y_2 = u_{21} q_1 + u_{22} q_2 . \tag{6.133}$$

As a result, we obtain the following equations of motion:

$$q_1 + p_1^2 q_1 = \frac{\sum\limits_{j=1}^{2} u_{j1} f_j}{\sum\limits_{j=1}^{2} u_{j1}^2 m_j} = \frac{4.132 P_0 H}{m_1 \left(1 + 3.132^2\right)};$$
$$q_2 + p_2^2 q_2 = \frac{\sum\limits_{j=1}^{2} u_{j2} f_j}{\sum\limits_{j=1}^{2} u_{j2}^2 m_j} = \frac{0.681 P_0 H}{m_2 \left(1 + 0.319^2\right)}. \tag{6.134}$$

The solutions of the equations (6.134) at zero initial data are

$$q_1 = \frac{0.382 P_0}{m_1 p_1^2} \left(1 - \cos p_1 t\right);$$
$$q_2 = 0.618 \frac{P_0}{m_1 p_2^2} \left(1 - \cos p_2 t\right). \tag{6.135}$$

Having determined q_j, we find y_1 and y_2 (6.133):

$$y_1 = \frac{0.382 P_0}{m_1 p_1^2} \left(1 - \cos p_1 t\right) + \frac{0.618 P_0}{m_1 p_2^2} \left(1 - \cos p_2 t\right);$$
$$y_2 = \frac{1.196 P_0}{m_1 p_1^2} \left(1 - \cos p_1 t\right) - 0.191 \frac{P_0}{m_1 p_2^2} \left(1 - \cos p_2 t\right).$$

The bending moment at the clamped end is

$$M = l\left(-m_1 \ddot{y}_1\right) + 2l\left(-m_1 \ddot{y}_1 + P_0\right)$$

or

$$M = (0.382 \cos p_1 t + 0.618 \cos p_2 t + 2.392 \cos p_1 t$$
$$- 0.382 \cos p_2 t + 2) P_0 l = F(t) P_0 l.$$

The maximum normal stress at the clamped end is

$$\sigma_{\max} = \frac{M}{W_x} = \frac{F(t)}{W_x} l P_0.$$

The mathematical expectation and the variance are respectively equal to

$$m_{\sigma \max} = \frac{F(t)}{W_x} l m_{p_0},$$
$$D_{\sigma \max} = \frac{F^2(t)}{W_2^2} l^2 D_{p_0}.$$

Considering that $\sigma_{\max}$ has a normal law of distribution, we determine the greatest possible value of the stress at the clamped end

$$\max(\sigma_{\max}) = \frac{F(t)}{W_x} l(m_{p_0} + 3\sigma_{p_0}).$$

Let us consider the case of suddenly applied time-constant forces $\mathbf{f}_j$ being parallel to one another, i.e. fulfilling the condition $\mathbf{f}_j = \beta_j \mathbf{f}_1$ and to the plane $x_1 0 x_2$, with their directions being arbitrary. The vectors $\mathbf{f}_j$ meet conditions similar to those of (6.53):

$$\left(\mathrm{C}^{(i)}\mathbf{f}_i \cdot \mathbf{f}_i\right) = 1, \qquad (\mathbf{f}_i = f_{ix_1}\mathbf{i}_1 + f_{ix_2}\mathbf{i}_2), \tag{6.136}$$

where n is the number of random forces taken equal to the number of lumped masses. Provided that the directions of the principal axes of the cross-sections of the structure coincide with the directions of the axes x_1 and x_2, the equations of masses small vibrations in planes $x_1 0 x_3$ and $x_1 0 x_2$ are independent, therefore we have respectively two vector equations of the type

$$\mathrm{M}\ddot{\mathbf{x}}_1 + \mathrm{B}^{(1)}\dot{\mathbf{x}}_1 + \mathrm{C}^{(1)}\mathbf{x}_1 = \mathbf{f}_{x_1}; \tag{6.137}$$

$$\mathrm{M}\ddot{\mathbf{x}}_2 + \mathrm{B}^{(2)}\dot{\mathbf{x}}_2 + \mathrm{C}^{(2)}\mathbf{x}_2 = \mathbf{f}_{x_2}, \tag{6.138}$$

$$\mathbf{x}_1 = \begin{bmatrix} x_{11} \\ x_{12} \\ \vdots \\ x_{1n} \end{bmatrix}; \quad \mathbf{x}_2 = \begin{bmatrix} x_{21} \\ x_{22} \\ \vdots \\ x_{2n} \end{bmatrix}; \quad \mathbf{f}_{x_1} = \begin{bmatrix} f_{1x_1} \\ f_{2x_1} \\ \vdots \\ f_{nx_1} \end{bmatrix}; \quad \mathbf{f}_{x_2} = \begin{bmatrix} f_{1x_2} \\ f_{2x_2} \\ \vdots \\ f_{nx_2} \end{bmatrix}.$$

Let us consider the algorithm of the numerical solution of the initial equations (6.137) and (6.138) without using the principal coordinates which are of little use during this solution. The solutions of the equations at zero initial data are of the form

$$\mathbf{x}_1 = \int_0^t G^{(1)} \mathrm{d}\tau \mathbf{f}_{x_1}; \tag{6.139}$$

$$\mathbf{x}_2 = \int_0^t G^{(2)} d\tau \mathbf{f}_{x_2}. \tag{6.140}$$

In this case f_{ix_1} and f_{ix_2} are dependent since they are the projections of the vectors $\mathbf{f}_i$. These projections of the vectors $\mathbf{f}_i$ meet the conditions (6.136). As the vectors $\mathbf{f}_i$ are parallel, we have

$$\beta_i^2 \left(C^{(1)}\mathbf{f}_i \cdot \mathbf{f}_i\right) = 1. \tag{6.141}$$

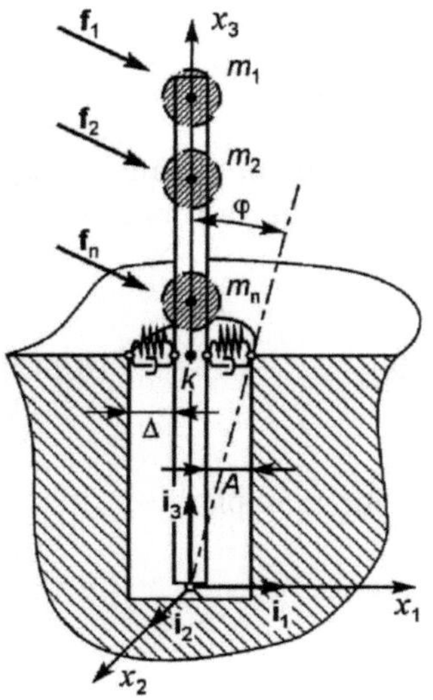

Fig. 6.14.

Let us consider problems arising during an analysis of the vibrations of mechanical systems at the action of suddenly applied random time-constant forces having an arbitrary direction. For example, the problem of designing a damping system with a limited deflection is classified among such problems. The damping system of an object that allows a limited angular displacement dependent on Δ before the impact of the object against the wall of the structure is shown in Fig. 6.14. During the design of the damping system it is required to determine the greatest possible displacement of a point k (Fig. 6.14) at the worst action of $\mathbf{f}_i$ in the given direction which should be less than Δ. In the general statement of the problem the task is to determine the greatest possible displacement (or velocity) of the mass m_i in the given direction determined by a single vector $\mathbf{e}$ (Fig. 6.15), i.e. to find the maximum of the functional

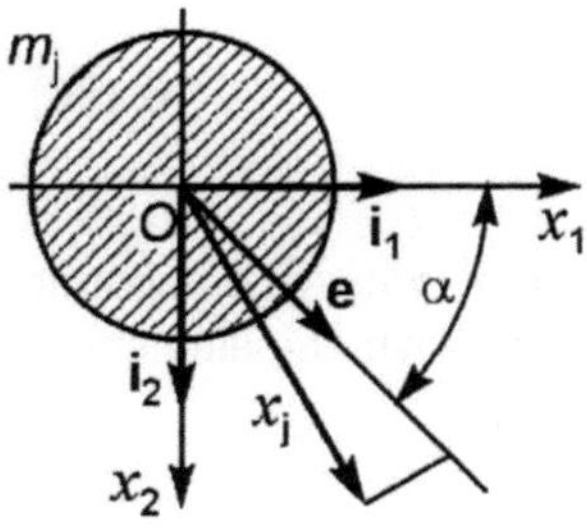

Fig. 6.15.

$$J_j = (\mathbf{x}_j \cdot \mathbf{e}), \qquad (\mathbf{x}_j = x_1\mathbf{i}_1 + x_2\mathbf{i}_2, \ \mathbf{e} = e_1\mathbf{i}_1 + e_2\mathbf{i}_2),$$

where $\mathbf{x}_j$ is the vector of the displacement of the mass m_j in a plane $x_1 0 x_2$. From the general solution (6.137)–(6.138) we can obtain the vectors x_j of the form

$$\mathbf{x}_j = \sum_{i=1}^{n} \int_0^t G_i^{(j)}(t,\tau)\mathbf{f}_i \mathrm{d}\tau, \quad (\mathbf{f}_i = f_{ix_1}\mathbf{i}_1 + f_{ix_2}\mathbf{i}_2), \tag{6.142}$$

where $G_i^{(j)}(t,\tau)$ are the diagonal matrixes (2×2). For example,

$$\mathbf{G}_1^{(j)} = \begin{bmatrix} \left(g_{11}^{(j)}\right)_1 & 0 \\ 0 & \left(g_{22}^{(j)}\right)_1 \end{bmatrix},$$

where $g_{j1}^{(1)}$, $g_{j1}^{(2)}$ are respectively the elements of the matrixes $G^{(1)}$ and $G^{(2)}$. With due account of $\mathbf{f}_i = \beta_i \mathbf{f}_1$ we have

$$\mathbf{x}_j = \sum_{i=1}^{n} \int_0^t G_i^{(j)}(t,\tau)\mathrm{d}\tau \, \beta_i \mathbf{f}_1 \tag{6.143}$$

and the condition, which the vector $\mathbf{f}_1$ satisfies

$$(C^{(1)}\mathbf{f}_1 \cdot \mathbf{f}_1) = 1. \tag{6.144}$$

Let us consider the special case of the vector $\mathbf{e}$ coinciding with one of the vectors of the basis $\{\mathbf{i}_j\}$, e.g., with $\mathbf{i}_1$. In this case

$$J_j = x_{j_1} \tag{6.145}$$

or, if we use the expression (6.143),

$$J_j = x_{j_1} = \int_0^t G^{(j)\mathrm{T}}\mathbf{i}_1 \mathrm{d}\tau \cdot \mathbf{f}_1, \quad \left(G^{(j)} = \sum_{i=1}^{n} G_i^{(j)}\beta_i\right). \tag{6.146}$$

It is required to determine the maximum value of (6.146) with due account of the condition (6.144). The matrixes $\mathbf{G}^{(j)}$ are diagonal with the elements

$$g_{11}^{(j)} = \sum_{i=1}^{n} g_{ji}^{(1)}\beta_i, \qquad g_{22}^{(j)} = \sum_{i=1}^{n} g_{ji}^{(2)}\beta_i,$$

therefore $G^{(j)T} = G^{(j)}$.

Using the Lagrangian multiplier, we obtain a functional

$$J_{j1} = J_j - \frac{1}{2}\lambda_1 \left[(C^{(1)}\mathbf{f}_1 \cdot \mathbf{f}_1) - 1 \right]. \tag{6.147}$$

From (6.147) we obtain an equation for the determination of the vector $\mathbf{f}_1^*$, at which the functional J_{j1} reaches the maximum value at an instant t:

$$\frac{\partial J_{j1}}{\partial \mathbf{f}_1} = 0$$

or

$$\int_0^t G^{(j)}\mathbf{i}_1 \mathrm{d}\tau - \lambda_1 C^{(1)}\mathbf{f}_1 = 0. \tag{6.148}$$

The relationships (6.148) and (6.144) enable us to determine the Lagrangian multipliers

$$\lambda_1^2 = (C^{(1)})^{-1} \int_0^t G^{(j)}\mathbf{i}_1 \mathrm{d}\tau \cdot \int_0^t G^{(j)}\mathbf{i}_1 \mathrm{d}\tau. \tag{6.149}$$

As $G^{(1)-1} = |f_1|^2 E$, then

$$\lambda_1 = |f_1| \left| \int_0^t G^{(j)}\mathbf{i}_1 \mathrm{d}\tau \right|. \tag{6.150}$$

Having determined λ_1 from the relationship (6.148), we find

$$\mathbf{f}_1^* = \frac{1}{\lambda_1} \left(C^{(1)} \right)^{-1} \int_0^t G^{(j)}\mathbf{i}_1 \mathrm{d}\tau. \tag{6.151}$$

With due account of (6.151) we obtain

$$\mathbf{f}_1^* = |\mathbf{f}_1| \frac{\int_0^t G^{(j)}\mathbf{i}_1 \mathrm{d}\tau}{\left| \int_0^t G^{(j)}\mathbf{i}_1 \mathrm{d}\tau \right|} = |\mathbf{f}_1| \, \mathbf{i}_1, \tag{6.152}$$

as one would expect, because the equations of motion (6.137) and (6.138) are independent.

Having substituted the expression for $\mathbf{f}_1^*$ in (6.146), we obtain the maximum value of the displacement of the mass m_j in the direction of the axis x_1

$$\max x_{j1} = |\mathbf{f}_1| \int_0^t G^{(j)} \mathbf{i}_1 \mathrm{d}\tau \, \mathbf{i}_1 = |\mathbf{f}_1| \int_0^t G^{(j)} \mathrm{d}\tau. \tag{6.153}$$

Let us determine the probability characteristics of the greatest possible displacement of the mass x_{j1}

$$m_{x_{j1}} = a_j m_{f_1}, \quad D_{x_{j1}} = a_j^2 D_{f_1}, \quad \left(a_j = \int_0^t G^{(j)} \mathbf{i}_1 \mathrm{d}\tau \, \mathbf{i}_1 \right). \tag{6.154}$$

The law of distribution of the random quantity $\max x_{j1} = x_{j1}^*$ takes the form (as is the case with (6.67))

$$f_1\left(x_{j1}^*, t\right) = \frac{1}{|a_j(t)| \, \sigma_{f_1} \sqrt{2\pi}} \times$$
$$\times \left(\exp\left\{ -\frac{\left(x_{j1}^* - a_j m_{j_1}\right)^2}{a_j^2 2\sigma_{f_1}^2} \right\} + \exp\left\{ -\frac{\left(x_{j1}^* - a_j m_{j_n}\right)^2}{a_j^2 2\sigma_{f_1}^2} \right\} \right). \tag{6.155}$$

Knowing the distribution law of the $\max x_{j1}$ and taking advantage of condition (6.69), we can determine the greatest possible value x_1 at an arbitrary fixed instant of time

$$\int_0^{x_{j1\,\max}^*} f_j\left(x_{j1}^*, t\right) \mathrm{d}x_{j1}^* = 0.99. \tag{6.156}$$

Having determined $x_{j1\,\max}^*$ from (6.156) for a number of discrete instants of time t, we obtain a plot of variation of $x_{j1\,\max}^*$ with time that allows us to determine an instant of time t^* at which $x_{j1\,\max}^*$ attains the maximum value, i.e. $\left(\max x_{j1\,\max}^*\right)$.

Let us consider the basic case, when it is necessary to determine the greatest possible value of the projection of the vector displacement $\mathbf{x}_j$ of a j-th mass onto the given direction determined by a vector $\mathbf{e}$ (Fig. 6.15), i.e. the task is to determine the maximum J_j at $\mathbf{f}_i = \beta_i \mathbf{f}_1$:

$$J_j = (\mathbf{x}_j \cdot \mathbf{e}) = x_{j1} \cos\alpha + x_{j2} \sin\alpha,$$

or with due account of (6.142)

$$J_j = \int_0^t \left(G^{(j)} \mathbf{e} \cdot \mathbf{f}_1 \right) \mathrm{d}\tau, \quad \left(G^{(j)} = \sum_{i=1}^n G_i^{(j)} \beta_i \right). \tag{6.157}$$

With due account of condition (6.144) we obtain a functional

$$J_j = \int_0^t \left(G^{(j)} \mathbf{e} \cdot \mathbf{f}_1 \right) \mathrm{d}\tau - \frac{1}{2} \lambda_1 \left[\left(G^{(j)} \mathbf{f}_1 \cdot \mathbf{f}_1 \right) - 1 \right]. \tag{6.158}$$

According to the earlier outlined method, we find $\mathbf{f}_1^*$

$$\mathbf{f}_1^* = \frac{\left(C^{(1)} \right)^{-1} \int_0^t G^{(j)T} \mathbf{e} \, \mathrm{d}\tau}{\sqrt{\left(C^{(1)} \right)^{-1} \int_0^t G^{(j)T} \mathbf{e} \, \mathrm{d}\tau \cdot \int_0^t G^{(j)T} \mathbf{e} \, \mathrm{d}\tau}}, \tag{6.159}$$

or on rearranging, the equation takes the form

$$\mathbf{f}_1^* = |\mathbf{f}_1| \, \mathbf{e}. \tag{6.160}$$

The maximum value of the functional (6.157) is

$$\begin{aligned} \max J_j &= \max \left(\mathbf{x}_j \cdot \mathbf{e} \right) = \lambda_1 \\ &= |\mathbf{f}_1| \sqrt{\left(\int_0^t g_{11}^{(j)} \cos\alpha \, \mathrm{d}\tau \right)^2 + \left(\int_0^t g_{22}^{(j)} \sin\alpha \, \mathrm{d}\tau \right)^2}. \end{aligned} \tag{6.161}$$

Having determined the maximum value of the projection of the displacement of the j–th mass on the given direction (6.161), we find the parameters of distribution law (6.155) for $\max(\mathbf{x}_j \cdot \mathbf{e}) = \max x_{je}$

$$\begin{aligned} f_j \left(x_{je}^*, t \right) &= \frac{1}{|c_j| \, \sigma_{f_1} \sqrt{2\pi}} \\ &\times \left(\exp\left\{ -\frac{\left(x_{je}^* - c_j m_j \right)^2}{c_j^2 2\sigma_{f_1}^2} \right\} + \exp\left\{ -\frac{\left(x_{je}^* + c_j m_{f_1} \right)^2}{2c_j^2 \sigma_{f_1}^2} \right\} \right), \end{aligned} \tag{6.162}$$

where

$$x_{je}^* = \max x_{je}, \quad c_j = \sqrt{\left(\int_0^t g_{11}^{(j)} \cos\alpha \, \mathrm{d}\tau \right)^2 + \left(\int_0^t g_{22}^{(j)} \sin\alpha \, \mathrm{d}\tau \right)^2}$$

We determine the greatest possible value x_{ji}^* from the condition

$$\int_0^{\max x_{je}^*} f_j \left(x_{je}^*, t \right) dx_{je}^* = 0.99. \tag{6.163}$$

The outlined methods of non-stationary vibrations of linear systems analysis under the action of random, suddenly applied time-constant forces, enables us to solve a number of practically important problems, including that of the probability of location of the masses (with reference to the object shown in Fig. 6.4) at the vibrations in the area of allowable displacements within the limits of a given interval of time at the worst external random actions on a system.

Let us consider as an example the launch of a flying vehicle (Fig. 6.16 *a*), assuming that the motion of the guide-rocket system occurs in and relative to the plane of the drawing. The problems connected with the launch of flying vehicles when only limited information is available on the random forces involved, are considered in Ref. [25]. Before the loss of contact with the guide the system has two degrees of freedom (angle φ in plane $x_1'0x_2'$ and angle ϑ in plane $x_1'0x_3'$. The start of the engine produces the thrust $\mathbf{R}$ that varies in time, as shown in Fig. 6.16 *b*. Ignoring the interval of time $(0, t_1)$ in the limit, we may assume for practical purposes that the thrust practically instantly reaches its nominal value R_0 (the worst case of the action on the system), i.e. the function $\mathbf{R}$ can be considered a suddenly applied time-constant force. In real systems, however, due to technological inaccuracies the engines thrust vector $\mathbf{R}$ does not coincide in its direction with that of the centerline of the flying vehicle (deviates from this line by a random angle ε) and, besides, shifts

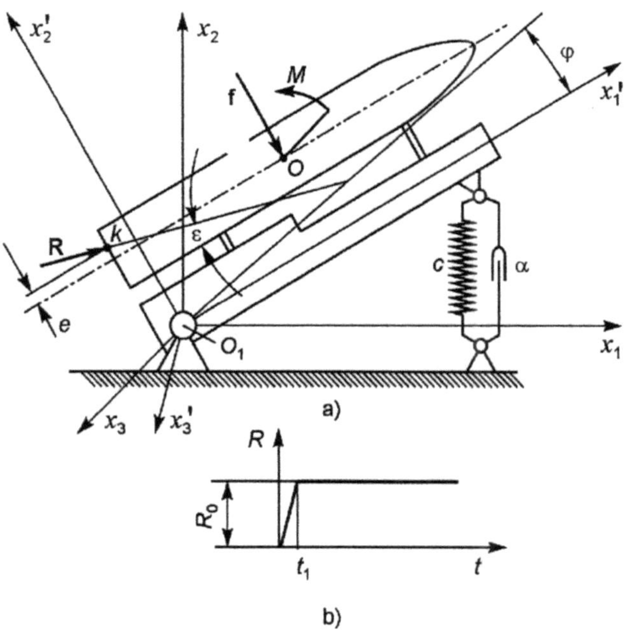

Fig. 6.16.

for some random distance e, which results in the occurrence of a random force $\mathbf{f}$ and a random moment $\mathbf{M}$ (see Fig. 6.16 *a*).

Generally, the point k, where the force $\mathbf{R}$ is applied, does not lie in the plane of the drawing (see Fig. 6.16 *a*). The arising random vectors $\mathbf{f}$ and $\mathbf{M}$, therefore, have arbitrary directions, i.e. the projections of these vectors onto the axes x_j are non-zero, which leads to vibrations in the system at the launch both in and relative to the plane of the drawing. Considering the motion of the system during the launch, we can obtain two linear equations (assuming the arising vibrations to be small) in the angles φ and ϑ in the coordinate system $x_j^{(1)}$ connected with the guide (see Fig. 6.16 *a*):

$$\ddot{\varphi} + a_{11}\dot{\varphi} + a_{12}\dot{\vartheta} + b_{11}\varphi + b_{12}\vartheta = c_{11}f_{x'_2} + c_{12}f_{x'_3} + d_{11}M_{x'_2} + d_{12}M_{x'_3};$$
$$\ddot{\vartheta} + a_{21}\dot{\varphi} + a_{22}\dot{\vartheta} + b_{21}\varphi + b_{22}\vartheta = c_{21}f_{x'_2} + c_{22}f_{x'_3} + d_{22}M_{x'_2} + d_{22}M_{x'_2},$$

or in vector form

$$\ddot{\boldsymbol{\vartheta}} + B\dot{\boldsymbol{\vartheta}} + C\boldsymbol{\vartheta} = D_1\mathbf{f} + D_2\mathbf{M} \quad \left(\boldsymbol{\vartheta} = (\varphi, \vartheta)^{\mathrm{T}}\right). \tag{6.164}$$

The section of a rocket and random excitations (force $\mathbf{f}$ and moment $\mathbf{M}$), reduced to its center of masses are shown in Fig. 6.17 in a system of coordinates connected with the guide. It follows from Fig. 6.17 that:

$$\mathbf{f} = -R_0\varepsilon\cos\alpha\mathbf{i}'_3 - R_0\varepsilon\sin\alpha\mathbf{i}'_2 = f_{x'_3}\mathbf{i}'_3 + f_{x'_2}\mathbf{i}'_2;$$
$$\mathbf{M} = -R_0 e\cos\alpha\mathbf{i}'_3 + R_0 e\sin\alpha\mathbf{i}'_2 = M_{x'_3}\mathbf{i}'_3 + M_{x'_2}\mathbf{i}'_2,$$

where α is a random angle.

The projections of the random force $\mathbf{f}$ and the random moment $\mathbf{M}$ satisfy the conditions similar to the condition (6.144):

$$\left(\frac{f^2_{x'_2}}{|R_0\varepsilon|^2} + \frac{f^2_{x'_3}}{|R_0\varepsilon|^2}\right) = 1, \quad \left(\frac{M^2_{x'_2}}{|R_0 e|^2} + \frac{M^2_{x'_3}}{|R_0 e|^2}\right) = 1.$$

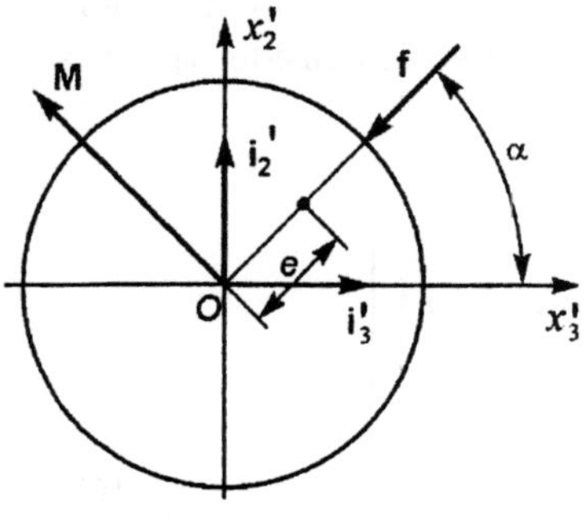

Fig. 6.17.

During the launch (at the moment t_k when the flying vehicle loses contact with the guide) of interest, among other things, is the greatest possible value of the functional

$$J = a_1\dot{\varphi}(t_k) + a_2\varphi(t_k),$$

which enables us to estimate the rockets range scatter induced by initial excitations ($\dot{\varphi}(t_k)$) and ($\varphi(t_k)$). Considering that the laws of distribution of the moduli $|e|$ and $|\varepsilon|$ are known, the greatest possible value of the functional J is determined in accordance with the preceding algorithm.

6.5 Forced Stationary Random Vibrations of Linear Systems

In the case of the time of the process being much greater than that of the transient process, the vibrations of a system may be considered as steady-state, or, stationary if they are induced by stationary random forces.

The vector equation of the forced vibrations of a system with n degrees of freedom is of the form (6.82)

$$\mathrm{M}\ddot{\mathbf{y}} + \mathrm{B}\dot{\mathbf{y}} + \mathrm{C}\mathbf{y} = \mathrm{D}_1\mathbf{f}. \tag{6.165}$$

We consider that the probability characteristics of the components of the vector $\mathbf{f}(t)$, including spectral densities $S_{f_i}(\omega)$ and $S_{f_i f_j}$, are known.

First, let us find out under what additional conditions a centered stationary random function $\overset{\circ}{f}_k(t)$ can be presented in the form of the Fourier integral

$$\overset{\circ}{f}_k(t) = \int\limits_{-\infty}^{\infty} \varPhi_k(\omega)\mathrm{e}^{i\omega t}\mathrm{d}\omega$$

or in vector form (when there is a system of random functions)

$$\overset{\circ}{\mathbf{f}}(t) = \int\limits_{-\infty}^{\infty} \mathbf{\Phi}(\omega)\mathrm{e}^{i\omega t}\mathrm{d}\omega. \tag{6.166}$$

The correlation function of a stationary random function should depend on the difference of instants of time (see Sect. 3.1). Let us consider the correlation function

$$\begin{aligned} K_{f_k}(t,t') &= M\left[\overset{\circ}{f}_k(t), \overset{\circ}{f}{}^*_k(t')\right] \\ &= M\left[\left(\int\limits_{-\infty}^{\infty} \varPhi_k \mathrm{e}^{i\omega t}\mathrm{d}\omega\right)\left(\int\limits_{-\infty}^{\infty} \varPhi^*_k \mathrm{e}^{-i\omega' t'}\mathrm{d}\omega'\right)\right] \\ &= \iint\limits_{-\infty}^{\infty} \mathrm{e}^{i(\omega t-\omega' t')} K\left[\varPhi_k\varPhi^*_k\right]\mathrm{d}\omega\,\mathrm{d}\omega', \end{aligned} \tag{6.167}$$

whose integrand will depend on the difference of instants of time, if the correlation function $K[\Phi_k, \Phi_k^*]$ satisfies the condition

$$K[\Phi_k(\omega), \Phi_k^*(\omega')] = S_{f_k}(\omega')\delta(\omega' - \omega). \tag{6.168}$$

In this case, having integrated with respect to ω, we obtain from (6.168)

$$K_{f_k}(t, t') = \int_{-\infty}^{\infty} S_{f_k}(\omega) e^{i\omega\tau} d\omega, \tag{6.169}$$

where $\tau = t - t'$; $S_{f_k}(\omega)$ is the spectral density of the components of the vector $\mathbf{f}$.

Similarly we obtain the following expression for cross-correlation functions

$$K_{f_k f_v}(t, t') = \int_{-\infty}^{\infty}\int_{-\infty}^{\infty} e^{i(\omega t - \omega' t')} K[\Phi_k(\omega')\Phi_v^*(\omega)] \, d\omega \, d\omega'. \tag{6.170}$$

It follows from relationships (6.170), that in the general case stationary random functions can be related non-stationarily as their cross-correlation function depends on two moments of time t and t' and not on their difference. However, if

$$K[\Phi_k(\omega)\Phi_v^*(\omega')] = S_{f_k f_v}(\omega')\delta(\omega' - \omega), \tag{6.171}$$

where $S_{f_k f_v}(\omega)$ is the cross-spectral density, the correlation function $K_{f_k f_v}$ depends on $t - t'$. Indeed, having substituting (6.171) in (6.170) and integrating with respect to ω, we obtain

$$K_{f_k f_v}(t, t') = \int_{-\infty}^{\infty} e^{i\omega\tau} S_{f_k f_v}(\omega) \, d\omega. \tag{6.172}$$

We shall seek the solution of equation (6.165) in the form (stationary solution)

$$\mathbf{y} = \int_{-\infty}^{\infty} \mathbf{y}_0 e^{i\omega\tau} d\omega. \tag{6.173}$$

Having substituted (6.173) and (6.166) in the equation (6.165), we obtain (Sect. 3.6)

$$\left[-\omega^2 \mathrm{M} + i\omega \mathrm{B} + \mathrm{C}\right]\mathbf{y}_0 = \mathrm{D}_1 \mathbf{\Phi}, \tag{6.174}$$

whence

$$\mathbf{y}_0 = \mathrm{W}(i\omega)\,\mathbf{\Phi}, \tag{6.175}$$

where

$$\mathrm{W}(i\omega) = \left[-\omega^2\mathrm{M} + i\omega\mathrm{B} + \mathrm{C}\right]^{-1}\mathrm{D}_1.$$

In scalar form

$$y_{k0}(\omega) = \sum_{j=1}^{n} w_{kj}(\omega)\Phi_j(\omega). \tag{6.176}$$

where $w_{kj}(\omega)$ are the elements of the matrix W.

From (6.173) we find the solution in scalar form

$$y_k = \int_{-\infty}^{\infty} y_{k0}(\omega)\mathrm{e}^{i\omega t}\mathrm{d}\omega. \tag{6.177}$$

Having substituted into (6.177) expressions for y_{k0} we obtain

$$y_k = \int_{-\infty}^{\infty} \sum_{j=1}^{n} w_{kj}(\omega)\Phi_j(\omega)\mathrm{e}^{i\omega t}\mathrm{d}\omega. \tag{6.178}$$

The cross-correlation functions of the components of the vector $\mathbf{Y}$ are equal to

$$K_{y_k y_v}(t,\, t') = M\left[Y_k(t)\, Y_v^*(t')\right] = \int_{-\infty}^{\infty}\int_{-\infty}^{\infty} M\left[y_{k_0}(\omega)\, y_{v_0}^*(\omega')\right]\mathrm{e}^{i(\omega t - \omega' t')}\mathrm{d}\omega\,\mathrm{d}\omega'.$$

The cross-correlation function $K_{y_k y_v}$ depends on the difference of instants of time $t - t'$ only in cases, where the condition similar to that of (6.171) is satisfied

$$M\left[Y_{k_0}(\omega)Y_{v0}^*(\omega')\right] = S_{y_k y_v}(\omega)\,\delta(\omega' - \omega). \tag{6.179}$$

Therefore from (6.179) we obtain

$$K_{y_k y_v}(\tau) = \int_{-\infty}^{\infty} S_{y_k y_v}(\omega)\mathrm{e}^{i\omega\tau}\mathrm{d}\omega. \tag{6.180}$$

Let us obtain an expression for the autocorrelation function $K_{y_k y_v}(\tau)$, dependent on the input spectral densities

$$K_{y_k y_v}(t,t') = M\left[Y_k(t)\,Y_v^*(t')\right]$$
$$= M\left[\left(\int\limits_{-\infty}^{\infty}\sum_{j=1}^{0} w_{kj}(\omega)\,\Phi_j(\omega)\,\mathrm{e}^{i\omega t}\mathrm{d}\omega\right)\right.$$
$$\left.\times\left(\int\limits_{-\infty}^{\infty}\sum_{j=1}^{n} w_{v\rho}^*(\omega')\,\Phi_\rho^*(\omega')\,\mathrm{e}^{-i\omega' t'}\mathrm{d}\omega'\right)\right] \qquad (6.181)$$
$$= \int\limits_{-\infty}^{\infty}\int\limits_{-\infty}^{\infty}\sum_{j=1}^{n}\sum_{\rho=1}^{n} w_{kj}(\omega)\,w_{v\rho}^*(\omega')\,M\left[\Phi_j(\omega)\,\Phi_\rho^*(\omega')\right]\mathrm{e}^{i\omega t - i\omega t'}\mathrm{d}\omega'\,\mathrm{d}\omega.$$

The expression (6.181) depends on the difference $t-t'$ if condition (6.171) is satisfied

$$M\left[\Phi_j(\omega)\Phi_\rho^*(\omega')\right] = S_{f_j f_\rho}(\omega)\,\delta(\omega'-\omega)\,.$$

Therefore from (6.181) we have

$$K_{y_k y_v}(\tau) = \int\limits_{-\infty}^{\infty}\sum_{j=1}^{n}\sum_{\rho=1}^{n} w_{k_j}(\omega')w_{v\rho}^* S_{f_\rho f_j}(\omega)\mathrm{e}^{i\omega\tau}\mathrm{d}\omega. \qquad (6.182)$$

Having equated the obtained expressions for cross-correlation functions (6.180) and (6.182) we obtain

$$\int\limits_{-\infty}^{\infty}\left[S_{y_k y_v}(\omega) - \sum_{j=1}^{n}\sum_{\rho=1}^{n} w_{k_j}(\omega)w_{v\rho}^*(\omega)S_{f_j f_\rho}(\omega)\right]\mathrm{e}^{i\omega\tau}\mathrm{d}\omega \equiv 0. \qquad (6.183)$$

The identical equality of the left-hand side of relationship (6.183) to zero will be fulfilled if (as $\mathrm{e}^{i\omega\tau} \neq 0$)

$$S_{y_k y_v}(\omega) = \sum\sum w_{k_j}(\omega)w_{v\rho}^*(\omega)S_{f_j f_\rho}(\omega). \qquad (6.184)$$

From obtained relationship (6.184) we determine the spectral densities of each of the component ($k = v$)

$$S_{y_k}(\omega) = \sum_{j=1}^{n}\sum_{\rho=1}^{n} w_{k_j}(\omega)w_{k\rho}^*(\omega)S_{f_j f_\rho}(\omega). \qquad (6.185)$$

In the specific case of f_j being independent

$$S_{f_j f_\rho}(\omega) = \begin{cases} 0 & j \neq \rho; \\ S_{f_j} & j = \rho, \end{cases}$$

relationships (6.184) and (6.185) take the form

$$S_{y_k y_v}(\omega) = \sum_{j=1}^{n} w_{kj} w_{vj} S_{f_j}(\omega);$$
$$S_{y_k}(\omega) = \sum_{j=1}^{n} w_{kj} w_{kj}^{*} S_{f_j}(\omega) = \sum_{j=1}^{n} |w_{kj}|^2 S_{f_j}(\omega). \tag{6.186}$$

The variances of the components $y_k(t)$ of the solution vector $\mathbf{y}$ are

$$D_{y_k} = \frac{1}{2\pi} \int_{-\infty}^{\infty} S_{y_k}(\omega) \mathrm{d}\omega = \frac{1}{2\pi} \int_{-\infty}^{\infty} \sum_{j=1}^{n} \sum_{v=1}^{n} w_{kj} w_{k\rho}^{*} S_{f_j f_\rho}(\omega) \mathrm{d}\omega \tag{6.187}$$

or in the specific case of

$$S_{f_j f_\rho}(\omega) = \begin{cases} 0 & j \neq v; \\ S_{f_j} & j = v, \end{cases}$$

$$D_{y_k} = \frac{1}{2\pi} \int_{-\infty}^{\infty} \left[\sum_{j=1}^{n} |w_{kj}|^2 S_{f_j}(\omega) \mathrm{d}\omega \right]. \tag{6.188}$$

Knowing spectral densities, we can determine the correlation functions of the components of the vector $\mathbf{y}$:

$$K_{y_k y_k} = \int_{-\infty}^{\infty} S_{y_k}(\omega) \mathrm{e}^{i\omega\tau} \mathrm{d}\tau;$$
$$K_{y_k y_v} = \int_{-\infty}^{\infty} S_{y_k y_v}(\omega) \mathrm{e}^{i\omega\tau} \mathrm{d}\tau. \tag{6.189}$$

Relationships (6.189) allow us to determine the parameters of the joint multidimensional normal distribution law of the components y_k

$$\sigma_{y_k}^2 = K_{y_k y_k}(0) = \int_{-\infty}^{\infty} S_{y_k}(\omega) \mathrm{d}\omega;$$

$$r_{y_k y_v} = \frac{K_{y_k y_v}(0)}{\sigma_{y_k} \sigma_{y_v}} = \frac{1}{\sigma_{y_k} \sigma_{y_v}} \int_{-\infty}^{\infty} S_{y_k y_v}(\omega) \mathrm{d}\omega.$$

For the solution of equation (6.165) we can use the method of principal coordinates (the way we did it in the earlier considered case of non-stationary random vibrations). With due account taken of the forces of viscous friction we obtain for q_j the equations of the (6.113) type:

$$\ddot{q}_j + 2n_j\dot{q}_j + p_j^2 q_j = \frac{\sum_{v=1}^{n} d_{jv} f_v(t)}{\sum_{k=1}^{n} u_{kj}^2 m_k} = \sum_{v=1}^{n} d'_{jv} f_v(t). \tag{6.190}$$

Let us take advantage of the Laplace transform, then from (6.190)

$$Q_j(p) = \sum_{v=1}^{n} w_{jv}(p)\Phi_v(p), \quad \left(w_{jv} = d'_{jv} \Big/ \left(p^2 + 2n_j p + p_j^2\right)\right). \tag{6.191}$$

For the k–th component of the vector of solutions (passing to the Laplace transform) we have

$$y_k(p) = \sum_{j=1}^{n} u_{kj} Q_j(p). \tag{6.192}$$

Having substituted (6.191) in (6.192), we obtain

$$y_k(p) = \sum_{v=1}^{n} w_{kv}(p)\Phi_v(p), \quad \left(w_{kv} = \sum_{j=1}^{n} u_{kj} w_{vj}\right). \tag{6.193}$$

If we replace operator p by $i\omega$ in (6.193), we obtain the following expression

$$y_k(\omega) = \sum_{v=1}^{n} w_{kv}(\omega)\Phi_v(\omega). \tag{6.194}$$

Let us consider the special case of a system with two degrees of freedom (Fig. 6.18). The equations of motion take the form

$$\begin{aligned} y_1 &= \delta_{11}\left(-m_1\ddot{y}_1 - b_{11}\dot{y}_1\right) + \delta_{12}\left(-m_2\ddot{y}_2 - b_{22}\dot{y}_2\right) + \delta_{11} f_1 + \delta_{12} f_2; \\ y_2 &= \delta_{21}\left(-m_1\ddot{y}_1 - b_{11}\dot{y}_1\right) + \delta_{22}\left(-m_2\ddot{y}_2 - b_{22}\dot{y}_2\right) + \delta_{21} f_1 + \delta_{22} f_2. \end{aligned} \tag{6.195}$$

The spectral densities S_{f_1}, S_{f_2} and $S_{f_1 f_2}$ of the random stationary forces acting on a system, are considered to be unknown. In order to determine the spectral densities of the output S_{y_1}, S_{y_2} and the cross- spectral density $S_{y_1 y_2}$ we must obtain the expression for $w_{ik}(\omega)$.

According to the presented method of solution for the general case of the system with n degrees of freedom we have

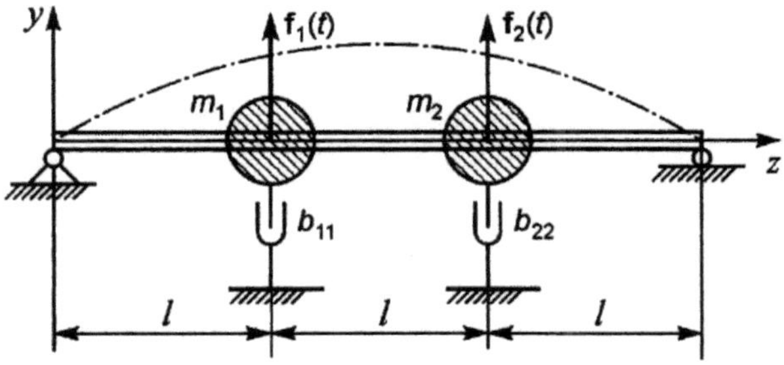

Fig. 6.18.

$$y_1 = \int_{-\infty}^{\infty} y_{10} e^{i\omega t} d\omega; \qquad y_2 = \int_{-\infty}^{\infty} y_{20} e^{i\omega t} d\omega;$$
$$f_1 = \int_{-\infty}^{\infty} \Phi_1 e^{i\omega t} d\omega; \qquad f_2 = \int_{-\infty}^{\infty} \Phi_2 e^{i\omega t} d\omega. \tag{6.196}$$

In order to determine y_{10} and y_{20} we obtain the system of equations

$$\begin{gathered}\left(1 + b_{11} i\omega\delta_{11} - \omega^2 m\delta_{11}\right) y_{10} + \left(\delta_{12} i\omega b_{22} - \delta_{12} m_2\omega^2\right) y_{20} \\ = \delta_{11}\Phi_1 + \delta_{12}\Phi_2; \\ \left(\delta_{21} b_{11} i\omega - \delta_{21} m_1\omega^2\right) y_{10} + \left(1 + \delta_{22} i\omega b_{22} - \delta_{22} m_2\omega^2\right) y_{20} \\ = \delta_{21}\Phi_1 + \delta_{22}\Phi_2.\end{gathered} \tag{6.197}$$

From system (6.197) we determine

$$y_{10} = w_{11}(\omega)\Phi_1 + w_{12}(\omega)\Phi_2;$$
$$y_{20} = w_{21}(\omega)\Phi_1 + w_{22}(\omega)\Phi_2,$$

where $w_{ij} = {}^{\Delta ij}/_{\Delta}$

$$\Delta_{11} = \begin{vmatrix} \delta_{11} & \left(i\omega b_{22} - m_2\omega^2\right)\delta_{12} \\ \delta_{21} & \left(1 + \delta_{22} i\omega b_{22} - \delta_{22}\omega^2 m_2\right) \end{vmatrix};$$

$$\Delta_{12} = \begin{vmatrix} \delta_{12} & \left(i\omega b_{22} - m_2\omega^2\right)\delta_{12} \\ \delta_{22} & 1 + \delta_{22} i\omega b_{22} - \delta_{22} m_2\omega^2 \end{vmatrix};$$

$$\Delta_{21} = \begin{vmatrix} \left(1 + \delta_{11} i\omega b_{11} - \delta_{11} m_1\omega^2\right) & \delta_{11} \\ \left(i\omega b_{11} - m_1\omega^2\right)\delta_{21} & \delta_{21} \end{vmatrix};$$

$$\Delta_{22} = \begin{vmatrix} \left(1 + \delta_{11} i\omega b_{11} - \delta_{11} m_1 \omega^2\right) & \delta_{12} \\ \left(i\omega b_{11} - m_1\omega^2\right)\delta_{21} & \delta_{22} \end{vmatrix};$$

$$\Delta = \begin{vmatrix} \left(1 + \delta_{11} i\omega b_{11} - \delta_{11} m_1 \omega^2\right) & \left(i\omega b_{22} - m_2\omega^2\right)\delta_{12} \\ \left(i\omega b_{11} - m_1\omega^2\right)\delta_{21} & \left(1 + \delta_{22} i\omega b_{22} - \delta_{22} m_2\omega^2\right) \end{vmatrix}.$$

According to (6.183), we obtain the following expressions for the spectral densities of the solutions

$$\begin{aligned} S_{y_1}(\omega) &= w_{11}w_{11}^* S_{f_1} + w_{11}w_{12}^* S_{f_1 f_2} + w_{12}w_{11}^* S_{f_2 f_1} + w_{12}w_{12}^* S_{f_2}; \\ S_{y_2}(\omega) &= w_{21}w_{21}^* S_{f_1} + w_{21}w_{22}^* S_{f_1 f_2} + w_{22}w_{21}^* S_{f_2 f_1} + w_{22}w_{22}^* S_{f_2}; \\ S_{y_1 y_2}(\omega) &= w_{11}w_{21}^* S_{f_1} + w_{11}w_{22}^* S_{f_1 f_2} + w_{12}w_{21}^* S_{f_2 f_1} + w_{22}w_{22}^* S_{f_2}. \end{aligned} \tag{6.198}$$

Example 6.6. A stationary random force $f_1(t)$ with the known spectral density of

$$S_{f_1} = \frac{2D_{f_1}\alpha}{\alpha^2 + \omega^2} \qquad \left(K_{f_1}(\tau) = D_{f_1} \mathrm{e}^{-\alpha|\tau|}\right)$$

acts on a mass m_1 (Fig. 6.19).

It is required to determine the variance of a deflection y_1 as a function of the masses ratio m_2/m_1 The equations of motion of the masses m_1 and m_2 (a special case of the equations (6.195)) are

$$\begin{aligned} &m_1\delta_{11}\ddot{y}_1 + m_2\delta_{12}\ddot{y}_2 + \delta_{11}b_{11}\dot{y}_1 + y_1 = \delta_{11} f_1; \\ &m_1\delta_{21}\ddot{y}_1 + m_2\delta_{22}\ddot{y}_2 + \delta_{21}b_{11}\dot{y}_1 + y_2 = \delta_{21} f_1, \\ &\delta_{11} = \delta_{22} = \frac{4l^3}{9EJ_x}; \qquad \delta_{12} = \delta_{21} = \frac{7l^3}{18EJ_x}. \end{aligned} \tag{6.199}$$

Going over to the non-dimensional time $t_1 = p_0 t$, where p_0 is the frequency of the vibrations of the mass m_1 (at $m_2 = 0$), we obtain the following system of equations:

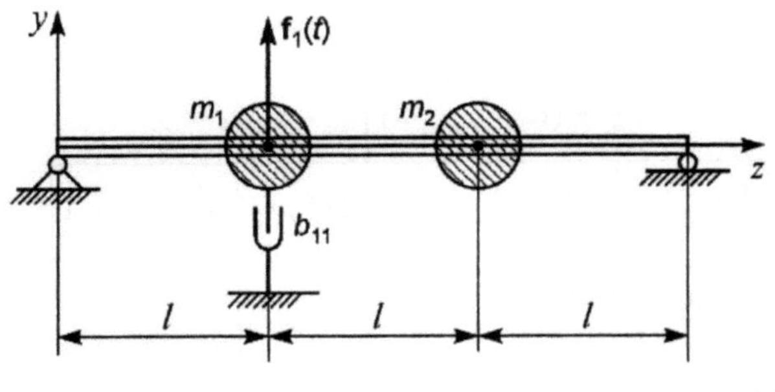

Fig. 6.19.

$$\ddot{y}_1 + \frac{7}{8} n_1 \ddot{y}_2 + \frac{2}{3} b_{11} \sqrt{\frac{l^3}{m_1 E J_x}}\, \dot{y}_1 + y_1 = \delta_{11} f_1;$$
$$\frac{7}{8} \ddot{y}_1 + n_1 \ddot{y}_2 + \frac{7}{12} b_{11} \sqrt{\frac{l^3}{m_1 E J_x}}\, \dot{y}_1 + y_2 = \delta_{21} f_1, \tag{6.200}$$
$$\left(n_1 = \frac{m_2}{m_1}\right).$$

It is possible to present the coefficient b_{11} as $b_{11} = m_1 p_0 n_2$, where n_2 is a non-dimensional factor. Let us go over to the non-dimensional time, then the correlation function (τ is the non-dimensional time) takes the form

$$K_{f_1}(\tau) = D_{f_1} \mathrm{e}^{-\alpha' |\tau|} \qquad (\alpha' = \alpha / p_0).$$

Let us obtain the spectral density expressed in terms of the non-dimensional frequency $\omega' = \dfrac{\omega}{p_0}$,

$$S_{f_1} = \frac{2 D_{f_1} \alpha'}{(\alpha'^2 + \omega'^2)\, p_0}.$$

Assuming that $\alpha' = n \rho_0$ where n is a non-dimensional parameter, we obtain

$$S_{f_1} = \frac{2 D_{f_1} n}{p_0 \left(n^2 + \omega'^2\right)} = \frac{2 D_{f_1} n}{p_0 \left| n + i\omega' \right|^2}.$$

Finally we obtain the system of equations

$$\ddot{y}_1 + \frac{7}{8} n_1 \ddot{y}_2 + n_2 \dot{y}_1 + y_1 = \delta_{11} f_1;$$
$$\frac{7}{8} \ddot{y}_1 + n_1 \ddot{y}_2 + \frac{7}{8} n_2 \dot{y}_1 + y_2 = \delta_{21} f_1. \tag{6.201}$$

The spectral density (a special case of (6.198)) is

$$S_{y_1}(\omega) = \left| w_{11} \right|^2 S_{f_1}, \tag{6.202}$$

where

$$w_{11} = \frac{l^3 \left[\dfrac{4}{9} + (i\omega')^2 \dfrac{15}{144} n_1 \right]}{E J_x \left[\dfrac{15}{64} n_1 (i\omega')^4 + \dfrac{15}{64} n_1 n_2 (i\omega')^3 + (n_1 + 1)(i\omega')^2 + n_2 i\omega + 1 \right]}.$$

The variance of the deflection amplitude is

$$D_{y_1} = \frac{1}{p_0} \int\limits_{-\infty}^{\infty} \left| w_{11} \right|^2 \frac{2 D_{f_1} n}{\left| n + i\omega' \right|^2} \mathrm{d}\omega'. \tag{6.203}$$

Through transformations we obtain the following expression for D_{y_1}, which allows us to use the tabulated integrals (see Appendix 2):

$$D_{y_1} = \frac{D_{f_1}}{\pi p_0} \int\limits_{-\infty}^{\infty} \frac{G(i\omega')}{|A(i\omega)|^2} \mathrm{d}\omega = \frac{2D_{f_1} n J_5}{p_0}, \tag{6.204}$$

where

$$\begin{aligned} G(i\omega') &= \frac{4}{9} + \frac{15}{144} n_1 (i\omega')^2; \\ A(i\omega) &= \frac{15}{64} n_1 (i\omega')^5 + \frac{15}{64} (n_1 n + n_1 n_2) (i\omega')^4 \\ &+ \left(1 + n_1 + \frac{15}{64} n_1 n_2 n\right) (i\omega')^3 + (n_1 n + n + n_2)(i\omega')^2 \\ &+ (1 + n n_2) i\omega' + n. \end{aligned}$$

Having used the values of the integral given in the Appendix 2, we obtain

$$J_5 = \frac{I_5}{2a_0 \Delta_5}, \tag{6.205}$$

where

$$\begin{aligned} &I_5 = a_0 b_3 (-a_0 a_3 + a_1 a_2) + \frac{a_0 b_4}{a_5} \left(-a_0 a_1 a_5 + a_0 a_3^2 + a_1^2 a_4 - a_1 a_2 a_3\right); \\ &\Delta_5 = a_0^2 a_5^2 - 2a_0 a_1 a_4 a_5 - a_0 a_2 a_3 a_5 + a_0 a_3^2 a_4 + a_1^2 a_4^2 + a_1 a_2^2 - a_1 a_2 a_3 a_4; \\ &a_1 = \frac{15}{64} n_1 (n + n_2); \qquad a_2 = 1 + n_1 + \frac{15}{64} n_1 n_2 n; \\ &a_3 = n_1 n + n + n_2 \qquad a_4 = 1 + n n_2; \qquad a_5 = n; \\ &b_0 = b_1 = b_2 = 0; \qquad b_3 = \frac{15}{144} n_1; \qquad b_4 = \frac{4}{9}; \qquad a_0 = \frac{15}{64} n_1. \end{aligned}$$

The plot of J_5 as a function of $n_1 = m_1/m_2$ for $n_2 = 0.2$ and $n_2 = 0.6$ at $n_3 = 0.05$ is shown in Fig. 6.20.

Let us consider the random vibrations of a mechanical system induced by stationary forces with a delay, for example

$$f_k(t) = f_1 (t - t_k) \qquad (k = 1, 2, \ldots, n), \tag{6.206}$$

where t_k is time of delay ($t_1 = 0$). Similar excitations are acting, for example, on a vehicle (Fig. 6.21), which moves along a road with irregularities. The excitations acting on the wheels, depend on the irregularities of the road. If at the instant of time t the irregularity was under the front wheels, then at the instant of time $t + t_2$, where $t_2 = L/v$, it will be under the rear wheels. In this case it is possible to present the random excitations as

$$f_k(t) = f_1(t - t_k) = \int_{-\infty}^{\infty} \Phi_1(\omega) e^{i\omega(t-t_k)} d\omega. \tag{6.207}$$

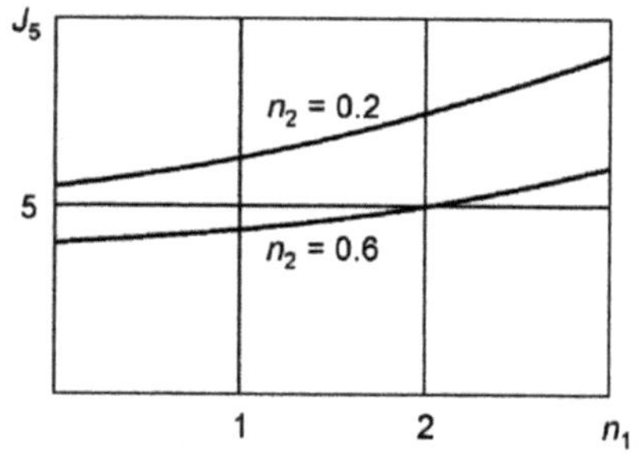

Fig. 6.20.

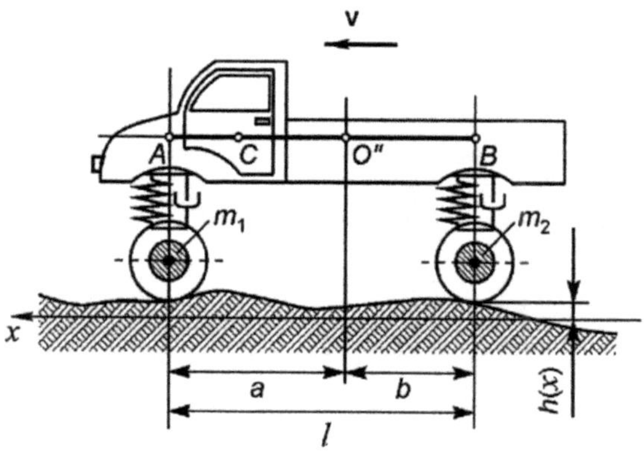

Fig. 6.21.

We determine the solution of equation (6.165) in the form of (6.177)

$$y_k(t) = \int_{-\infty}^{\infty} y_{k_0}(\omega) e^{i\omega t} d\omega. \tag{6.208}$$

Having substituted (6.206) and (6.207) in equation (6.165) (going on to vector form), we obtain

$$\left[-\omega^2 \mathrm{M} + i\omega \mathrm{B} + \mathrm{C}\right] \mathbf{y}_0 = \mathrm{D}_1 \mathrm{H} \mathbf{\Phi}, \tag{6.209}$$

where

$$H = \begin{bmatrix} 1 & 0 & 0 & \dots & 0 \\ 0 & e^{-i\omega t_2} & 0 & \dots & 0 \\ 0 & 0 & e^{-i\omega t_3} & 0 & 0 \\ \vdots & \vdots & \vdots & \ddots & \vdots \\ 0 & \dots & \dots & 0 & e^{-i\omega t_n} \end{bmatrix}; \qquad \mathbf{\Phi} = \begin{bmatrix} \Phi_1 \\ \Phi_1 \\ \vdots \\ \vdots \\ \Phi_1 \end{bmatrix}.$$

It follows from (6.209):

$$\mathbf{y}_0 = \mathrm{W}_1(\omega)\mathbf{\Phi}, \quad \left(\mathrm{W}_1(\omega) = \left|-\omega^2\mathrm{M} + \mathrm{i}\omega\mathrm{B} + \mathrm{C}\right|^{-1} D_1 H\right),$$

Or in scalar form

$$y_{k_0} = \sum_{j=1}^{n} w_{1k_j}\Phi_j = \left(\sum_{j=1}^{n} w_{1k_j}\right)\Phi_1 = w_k\Phi_1. \tag{6.210}$$

For the k–th component of the vector $\mathbf{y}\,(t)$ we obtain the expression

$$y_k(t) = \int_{-\infty}^{\infty} w_k\,(\omega)\,\Phi_1 e^{i\omega t} d\omega. \tag{6.211}$$

By transforming the way we did during the derivation of relationship (6.185), we obtain the following expressions for the spectral densities and cross-spectral densities:

$$\begin{aligned} S_{y_k} &= w_k w_k^* S_{f_1} = |w_k|^2 S_{f_1}; \\ S_{y_k y_\rho} &= w_k w_\rho^* S_{f_1}, \end{aligned} \tag{6.212}$$

as $S_{f_j f_v} = S_{f_1}$.

Let us consider the stationary vibrations of a vehicle moving on a road with random irregularities (Fig. 6.21).

Let us restrict our consideration to the case of the vibrations of the vehicle in the plane of the drawing (Fig. 6.21). The requirements specified for motor transport (carrying capacity, speed, cross-country capability etc.), substantially depend on the elastic characteristic of the suspension system. Suspension analysis is complicated by the fact that forces acting on it from the road are usually of random nature. Numerous experimental investigations of the effect of various types of roads on vehicles carried out in the past few years have made it possible to obtain the necessary information on random excitations that act on vehicles (in particular, the spectral densities of forces arising due to random irregularities of a road).

A design scheme (with due account of the masses of the suspension system) is given in Fig. 6.22.

When deriving equations of motion, let us take advantage of Lagrange equations of the second kind. To be able to do this let us obtain expressions for kinetic and potential energies, as well as for Rayleighs dissipative function:

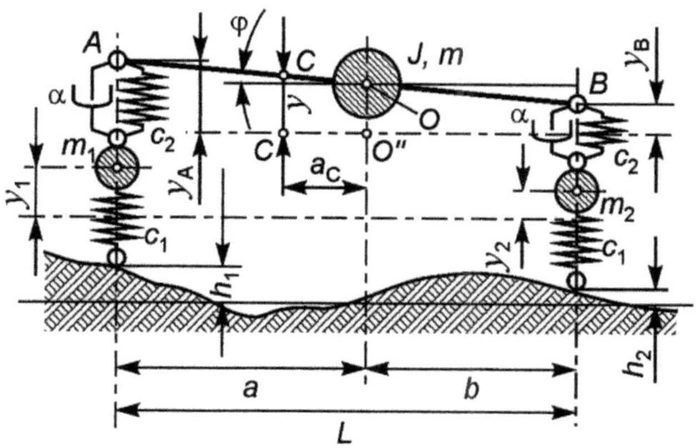

Fig. 6.22.

$$T = \frac{m\dot{y}^2}{2} + \frac{J\dot{\varphi}^2}{2} + \frac{m_1\dot{y}_1^2}{2} + \frac{m_2\dot{y}_2^2}{2};$$
$$\varPi = \frac{1}{2}c_2\left(y_A - y_1\right)^2 + \frac{1}{2}c_1\left(y_1 - h_1\right)^2 + \frac{1}{2}c_2\left(y_B - y_2\right)^2 + \frac{1}{2}c_1\left(y_2 - h_2\right)^2;$$
$$R = \frac{\alpha}{2}\left(\dot{y}_A - \dot{y}_1\right)^2 + \frac{\alpha}{2}\left(\dot{y}_B - \dot{y}_2\right)^2.$$

As

$$y_A = y + a\varphi, \qquad y_B = y - b\varphi,$$

it is possible to present the expression for kinetic energy by way of eliminating φ and y as

$$T = \frac{m}{2L^2}\left(b\dot{y}_A + a\dot{y}_B\right)^2 + \frac{J}{2L^2}\left(\dot{y}_A + \dot{y}_B\right)^2 + \frac{m_1\dot{y}_1^2}{2} + \frac{m_2\dot{y}_2^2}{2}.$$

Having substituted the expressions for T, $\varPi$ and R into the Lagrange equation of the second kind

$$\frac{\mathrm{d}}{\mathrm{d}t}\left(\frac{\partial T}{\partial \dot{q}_i}\right) - \frac{\partial T}{\partial q_i} + \frac{\partial \varPi}{\partial q_i} = -\frac{\partial R}{\partial \dot{q}_i}. \tag{6.213}$$

Through transformations we obtain the system of equations that take the form

$$\begin{aligned}
&\frac{mb^2 + J}{L^2}\ddot{y}_A + \frac{mab - J}{L^2}\ddot{y}_B + c_2\left(y_A - y_1\right) + \alpha\left(\dot{y}_A - \dot{y}_1\right) = 0;\\
&\frac{ma^2 + J}{L^2}\ddot{y}_B + \frac{mab - J}{L^2}\ddot{y}_A + c_2\left(y_B - y_2\right) + \alpha\left(\dot{y}_B - \dot{y}_2\right) = 0;\\
&m_1\ddot{y}_1 - c_2\left(y_A - y_1\right) + c_1\left(y_1 - h_1\right) - \alpha\left(\dot{y}_A - \dot{y}_1\right) = 0;\\
&m_2\ddot{y}_2 - c_2\left(y_B - y_2\right) + c_1\left(y_2 - h_2\right) - \alpha\left(\dot{y}_B - \dot{y}_2\right) = 0.
\end{aligned} \tag{6.214}$$

If the condition $J = mab$ is satisfied, the vibrations of the front and rear suspensions become independent. In this case, from the system (6.214) we obtain two independent systems of equations:

$$\begin{aligned} &m_A\ddot{y}_A + c_2(y_A - y_1) + \alpha(\dot{y}_A - \dot{y}_1) = 0; \\ &m_1\ddot{y}_1 + \alpha\dot{y}_1 + (c_1 + c_2)y_1 - \alpha\dot{y}_A - c_2y_A = c_2h_1(t), \end{aligned} \tag{6.215}$$

$$\begin{aligned} &m_B\ddot{y}_B + c_2(y_B - y_2) + \alpha(\dot{y}_B - \dot{y}_2) = 0, \\ &m_2\ddot{y}_2 + \alpha\dot{y}_2 + (c_1 + c_2)y_2 - \alpha\dot{y}_B - c_2y_B = c_1h_2(t - t_2), \end{aligned} \tag{6.216}$$

where

$$m_A = \frac{J + mb^2}{L^2}, \qquad m_B = \frac{J + ma^2}{L^2}.$$

Let us consider an example (Fig. 6.22) with the following specific numerical data: $a = 240$ cm; $b = 150$ cm; $m = 50\,\mathrm{kg}$; $m_1 = m_2 = 8\,\mathrm{kg}$; $c_2 = 4000\,\mathrm{H \cdot cm^{-1}}$; $c_1 = 10^4\,\mathrm{H \cdot cm^{-1}}$; $\alpha = 150\,\mathrm{H \cdot s \cdot cm^{-1}}$. The spectral density of the irregularities of the road h (for the given type of road) has the form

$$S_h = \frac{3v\left(\omega^2 + 0.6v^2\right)}{\left(\omega^2 - 0.6v^2\right)^2 + 3v^2\omega^2} + \frac{25.5v}{\omega^2 + 15v^2}, \tag{6.217}$$

where v is the speed of the vehicle, km/h. The numerical factors entering in the left-hand side of expression (6.217), are dimensional.

The laws of variation of the spectral density $S_h(\omega)$ as a function of ω for a number of motion velocity values v, km/h are shown in Fig. 6.23: I — $v = 20$; II — $v = 30$; III — $v = 40$; IV — $v = 50$; V — $v = 60$. Putting $\alpha = 0$ in equations (6.213), (6.214), let us determine four frequencies: for system

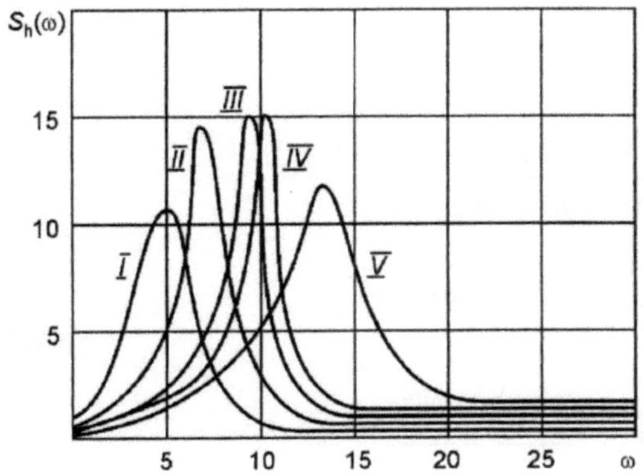

Fig. 6.23.

(6.215) we obtain $p_{11} = 12.2\ \mathrm{s}^{-1}$, $p_{12} = 42.6\ \mathrm{s}^{-1}$ and for system (6.216) we get $p_{21} = 9.6\ \mathrm{s}^{-1}$, $p_{22} = 40.2\ \mathrm{s}^{-1}$.

Let us obtain the mapping of equations (6.215) and (6.216) in the frequency domain

$$\begin{aligned}
&\left[m_A(i\omega)^2 + \alpha(i\omega) + c_2\right] Y_A(i\omega) - (c_2 + \alpha i\omega)\, Y_1(i\omega) = 0;\\
&-(c_2 + \alpha i\omega)\, Y_A(i\omega) + \left[m_1(i\omega)^2 + \alpha i\omega + c_1 + c_2\right] Y_1(i\omega) = c_2 H_1(i\omega);\\
&\left[m_B(i\omega)^2 + \alpha(i\omega) + c_2\right] Y_B(i\omega) - (c_2 + \alpha i\omega)\, Y_2(i\omega) = 0;\\
&-(c_2 + \alpha i\omega)\, Y_B(i\omega) + \left[m_2(i\omega)^2 + \alpha(i\omega) + c_1 + c_2\right] Y_2(i\omega) = c_1 H_2(i\omega),
\end{aligned}$$

where

$$H_2(i\omega) = H_1(i\omega)\mathrm{e}^{-i\omega t_2}.$$

From the obtained system of algebraic equations we determine $Y_A(i\omega)$, $Y_B(i\omega)$, $Y_B(i\omega)$, $Y_1(i\omega)$ and $Y_2\,(i\omega)$.

$$\begin{aligned}
&Y_A(i\omega) = W_A\,(i\omega)\,H_1\,(i\omega)\,, \qquad Y_B(i\omega) = W_B(i\omega)\mathrm{e}^{-i\omega t_2} H_1(i\omega),\\
&Y_1(i\omega) = W_1\,(i\omega)\,H_1\,(i\omega)\,, \qquad Y_2(i\omega) = W_2(i\omega)\mathrm{e}^{-i\omega t_2} H_1(i\omega),
\end{aligned} \tag{6.218}$$

where

$$\begin{aligned}
W_A &= \frac{\Delta_A}{\Delta_1}\\
&= \frac{c_1c_2 + i\alpha\omega c_1}{c_1c_2 + m_A m_1(i\omega)^4 + (m_A + m_1)\,\alpha(i\omega)^3 +} \Rightarrow\\
&\Rightarrow \overline{+(m_Ac_2 + m_1c_1 + m_Ac_2)(i\omega)^2 + \alpha c_1(i\omega)^2};\\
W_1 &= -\frac{c_1\left(m_A\,(i\omega)^2 + \alpha i\omega + c_2\right)}{\Delta_1};
\end{aligned} \tag{6.219}$$

$$\begin{aligned}
W_B &= \frac{\Delta_B}{\Delta_2}\\
&= \frac{(i\alpha\omega c_1 + c_1c_2)\,\mathrm{e}^{-i\omega t_2}}{m_Bm_2(i\omega)^4 + \alpha\,(m_B + m_2)\,(i\omega)^3 +} \Rightarrow\\
&\Rightarrow \overline{(m_2c_2 + m_Bc_2 + m_Bc_1)(i\omega)^2 + \alpha c_1(i\omega)};\\
W_2 &= -\frac{c_1\left(m_B\,(i\omega)^2 + \alpha i\omega + c_2\right)\mathrm{e}^{-i\omega t_2}}{\Delta_2},
\end{aligned} \tag{6.220}$$

$(t_2 = L/v)$.

According to Sect. 3.6 we obtain the spectral densities of the displacements y_A, y_B, y_1, y_2 and of their first derivatives. Let us restrict our consideration to the determination of the spectral densities of the accelerations of the points A and B:

$$\begin{aligned} S_{\ddot{y}_A} &= |W_A|^2 \omega^4 S_h(\omega), \\ S_{\ddot{y}_B} &= \left|W_B^{(1)}\right|^2 \omega^4 S_h(\omega), \\ &\left(W_B^{(1)} = W_B \mathrm{e}^{-i\omega t_2}\right). \end{aligned} \tag{6.221}$$

The variation of the spectral densities of the accelerations of the points A and B as a function of ω for different velocities v is shown in Fig. 6.24–6.26. Let us determine the variance of the vertical acceleration of the point C (the driver seat), which characterizes the degree of riding comfort. According to Fig. 6.22, we have

$$y_c = y + \varphi a_c.$$

Or, going onto y_A and y_B,

$$y_c = K_A y_A + K_B y_B, \tag{6.222}$$

where

$$\begin{aligned} K_A &= \frac{b}{L} + \frac{a_c}{L}; \\ K_B &= \frac{a}{L} + \frac{a_c}{L}. \end{aligned}$$

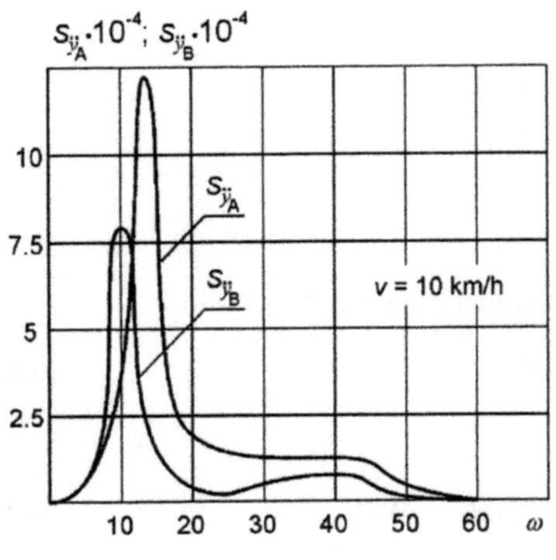

Fig. 6.24.

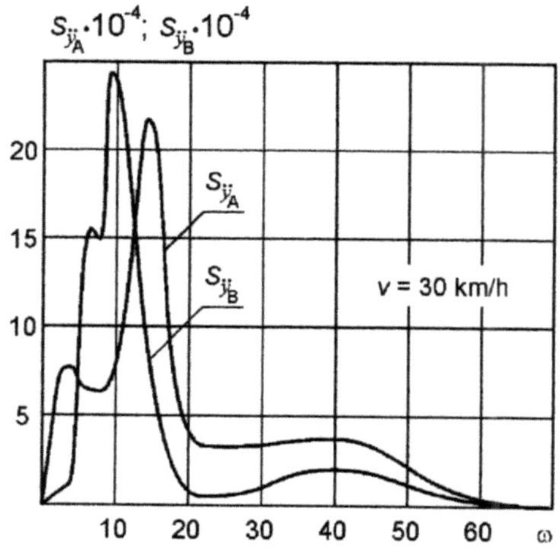

Fig. 6.25.

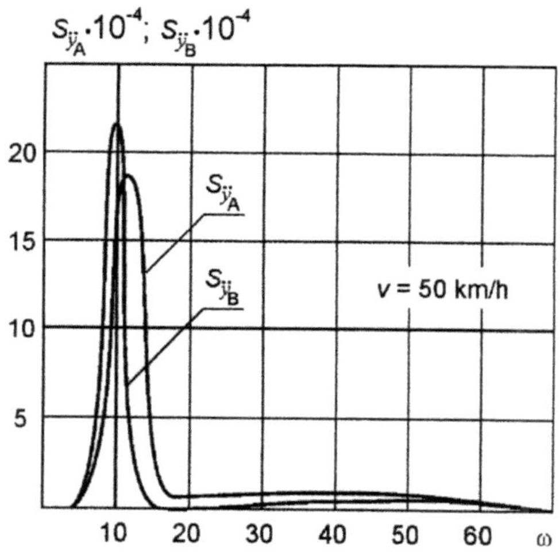

Fig. 6.26.

Going in (6.222) to the Fourier transformation we obtain

$$y_c(i\omega) = K_A y_A(i\omega) + K_B y_B(i\omega) \tag{6.223}$$

or (using relationships (6.218))

$$y_c(i\omega) = (K_A W_A + K_B W_B e^{-i\omega t_2})H_1 = W_c H_1. \tag{6.224}$$

The spectral density of the displacement of the point C with due account of relationship (6.224) is

$$S_{y_c} = |W_c|^2 \, S_h(v, \omega), \tag{6.225}$$

where $W_c = (K_A W_A + K_B W_B e^{-i\omega t_2})$.

The spectral density of the acceleration of the point C and the root-mean-square acceleration of the point C are respectively

$$S_{\ddot{y}_c} = |W_c|^2 \, \omega^4 S_h, \tag{6.226}$$

$$\sigma^2_{\ddot{y}_c} = \int_{-\infty}^{\infty} |W_c|^2 \, \omega^4 S_h(v, \omega) \, d\omega. \tag{6.227}$$

Variation of the root-mean-square value of the acceleration $\sigma_{\ddot{y}_c}$ of the point C as a function of velocity v for the example considered is shown in Fig. 6.27. It follows from the plot that during the motion along the given road (characterized by spectral density (6.218)) the most unfavorable velocity v is $v \approx 50$ km/h at which the root-mean-square values of accelerations acting on the driver reach the maximum.

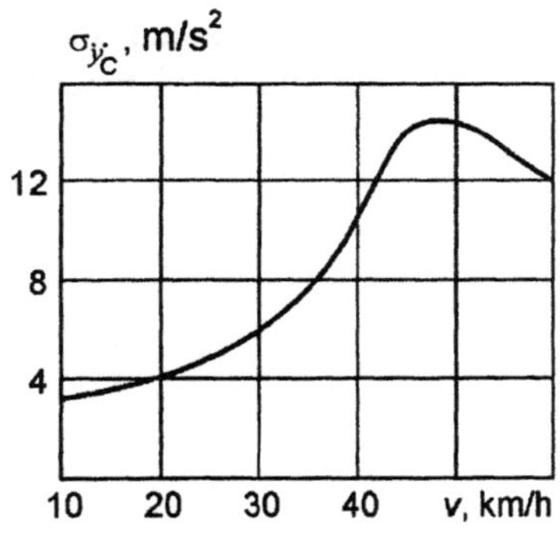

Fig. 6.27.

Let us bring the expression (W_c) to the form convenient for integration. It is well known, that any function $f\,(i\omega)$ depending on imaginary argument can be presented as

$$f(i\omega) = f_1(\omega) + i f_2(\omega).$$

Therefore

$$W_A = W_{A_1} + i W_{A_2}, \quad W_B = W_{B_1} + i W_{B_2}$$

and the expression for W_c is transformed to the form

$$W_c(i\omega) = W_1(\omega) + i W_2(\omega),$$

where

$$W_1 = K_A W_{A_1} + K_B W_{B_1} \cos \omega t_2 + K_B W_{B_2} \sin \omega t_2,$$
$$W_2 = K_A W_{A_2} + K_B W_{B_2} \cos \omega t_2 + K_B W_{B_1} \sin \omega t_2.$$

The square of the modulus $W_c(i\omega)$ is equal to

$$|W_c|^2 = W_1^2(\omega) + W_2^2(\omega).$$

As a result we obtain

$$\sigma_{\ddot{y}_c}^2(v) = \int_{-\infty}^{\infty} W_1^2 \omega^4 S_h(v,\omega) \mathrm{d}\omega + \int_{-\infty}^{\infty} W_2^2 \omega^4 S_h(v,\omega) \mathrm{d}\omega.$$

Considering, that $\ddot{y}_c$ has normal distribution, we obtain the greatest possible value of the random acceleration (at $m_h = 0$) acting on the driver

$$\max \ddot{y}_c(v) = 3\sigma_{\ddot{y}_c}(v).$$

As $S_h(v,\omega)$ (6.217) depends on the vehicles motion velocity v we can determine (numerically) such value of this velocity v_*, at which $\max \ddot{y}_c(v)$ reaches its maximum value

$$\max_v (\max \ddot{y}_c(v)) = \max \ddot{y}_c(v_*).$$

7. Random Vibrations of Strings; Longitudinal and Torsional Vibrations of Straight Rods

7.1 Introduction

It was considered in the preceding chapters devoted to random vibrations of mechanical systems with a finite number of degrees of freedom that elastic elements (for example, rod elements in Fig. 5.8, 5.9, 5.24, 6.7, 6.10) are inertialess, which, of course, is not quite so. This is true only in cases, where concentrated masses are considerably greater than the masses of elastic elements. Unfortunately, the term considerably greater does not relate to a specific numerical estimation and for this reason it is uncertain and sometimes unconvincing. Everything depends on the degree of accuracy imposed on the final numerical results of an analysis. For example, Figure 5.24 shows a concentrated mass m, connected with a spring that was considered massless (inertialess). The real spring, however, has a mass, which at vibrations leads to the occurence of inertia forces that can substantially change any calculation results obtained without regard to them.

A mast with an antenna is shown in Fig. 5.8. During the analysis of random vibrations of this structure, the inertia forces of the rod (mast) were not taken into account, which makes it impossible for us to estimate the accuracy of the obtained results. The only thing we may assert is that the greater the ratio $\frac{M}{m_1 l}$ (where m_1 is the mass of the unit length of the mast), the more accurate are the results. However, in order to obtain specific numerical results demonstrating the influence of this ratio on the accuracy of a solution, we must take the inertia forces of the mast into account in our analysis, which is only possible if we consider the mast to be a system with distributed parameters. Similar problems arise when dealing with systems that have several degrees of freedom (for example, see Figs. 6.7, 6.10). It is possible to estimate the accuracy of the results of an analysis of mechanical systems containing lumped masses connected by elastic elements in cases where the latter are considered massless only with respect to more accurate mathematical models that take into account the inertial properties of elastic elements. This means that it is necessary to consider these elements as systems with an infinite number of degrees of freedom (systems with distributed parameters). In this chapter we consider the random vibrations of simplest mechanical systems with distributed parameters. These systems include real objects, the design

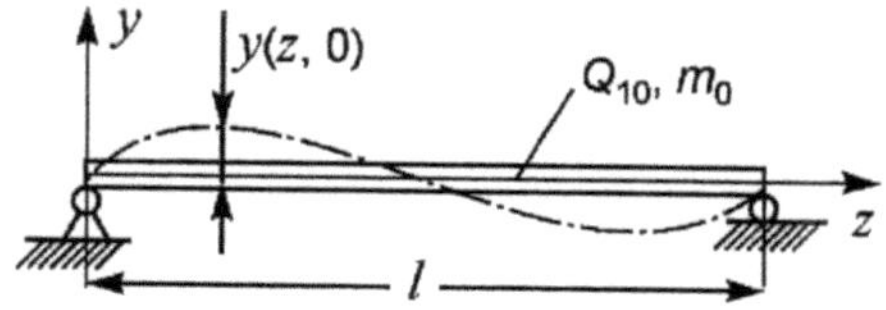

Fig. 7.1.

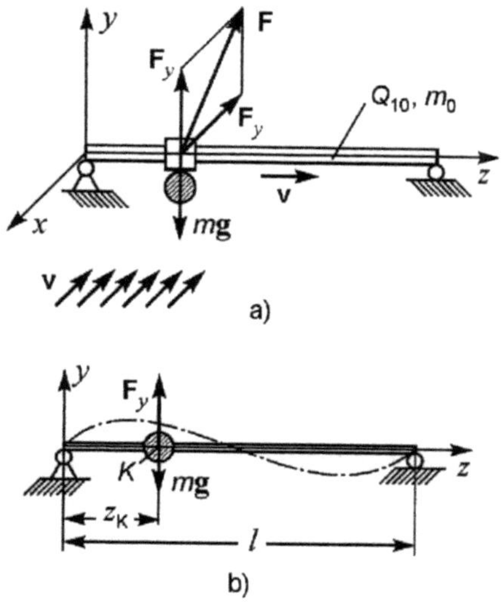

Fig. 7.2.

schemes of which can be presented as a string (a string is a rod whose bending and torsional rigidity can be ignored) (Fig. 7.1–7.5), as a rod when we consider longitudinal vibrations (Fig. 7.6 *a*) or as a shaft if we have to do with torsional vibrations (Fig. 7.6 *b*). The random vibrations of spatial-curved rods are considered in the following chapter.

A stretched string that is often used as a frequency gauge or as a low-frequency electromechanical filter (low-frequency in comparison with the frequency spectrum of electrical filters) is shown in Fig. 7.1.

A simplified mathematical model of a cable road is given in Fig. 7.2. A lumped mass subjected to the action of a random aerodynamic force F moves on a tensioned cable (string) with a velocity $\mathbf{v}$. As a result some spatial random vibrations of the mass m will take place. If, as an example, we confine ourselves to vibrations that occur only in the vertical plane ($Y0Z$), we can present the mathematical model of the system as it is shown in Fig. 7.2 *b*, where F_y is the vertical component of the aerodynamic force F. Random accelerations occurring at vibrations can be quite considerable. Having determined the greatest possible values of the accelerations acting on the mass m when it moves along the cable, we can reduce the force by changing the

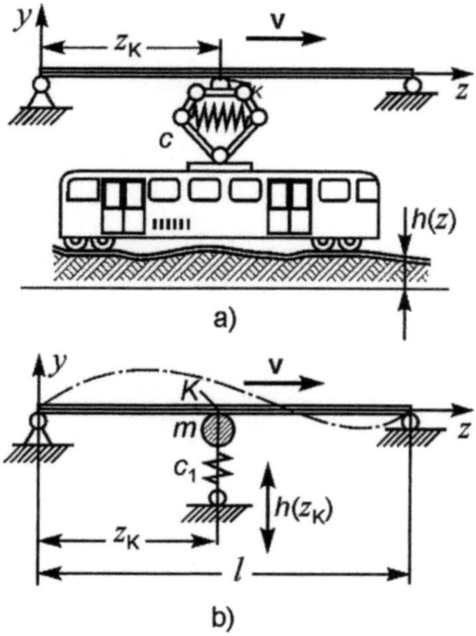

Fig. 7.3.

tension Q_{10} and the velocity of the motion of the mass. This can be done, however, if we have an analytical or numerical solution to the stated problem.

A section of a wire (string) that contacts, for example, with a moving trolleybus is shown in Fig. 7.3 *a*. The contact device (current collector) can be presented as a lumped mass m and a rigidity c_1 (Fig. 7.3 *b*). Because of the roads random irregularities (h) the point k obtains random vertical displacements which results in a kinematic excitation of the system. When vibrations take place, there is the possibility of cases where, depending on the probability characteristics of the road, motion velocity V and other parameters of the system (m, C_1, Q_{10}), the contact force between the wire and the mass m at discrete instants of time vanishes. This may happen because the constraint between the wire and the mass is unilateral. In real conditions a small sagging of the wire (dot-and-dash line in Fig. 7.3 *b*) is always there and this considerably increases the probability of loss of contact.

A tape drive mechanism of an information recording and representation system is shown in Fig. 7.4 *a*. The system rests on a foundation that has a random displacement $Y_0\,(t)$ (Fig. 7.4 *b*), as a result this causes random vibrations of the moving tape, which leads to the distortion of the information.

A hose section, inside which a liquid moves (a hose is a particular case of a pipeline whose bending and torsional rigidities are equal to zero, i.e. it can be considered as an absolutely flexible rod) is shown in Fig. 7.5. The hose has a local bilateral constraint in the section k (Fig. 7.5 *a*) with a random vertical

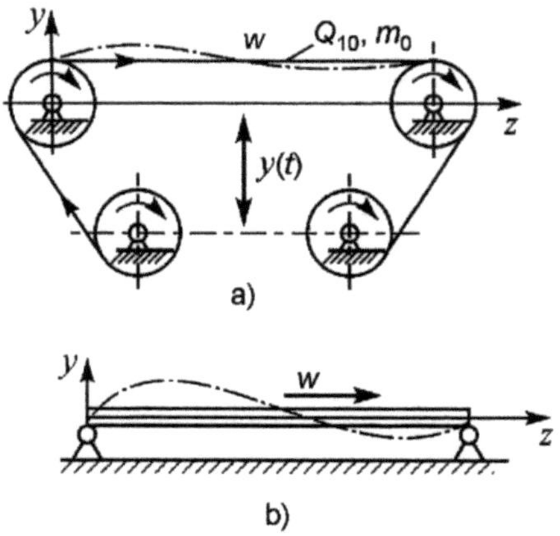

Fig. 7.4.

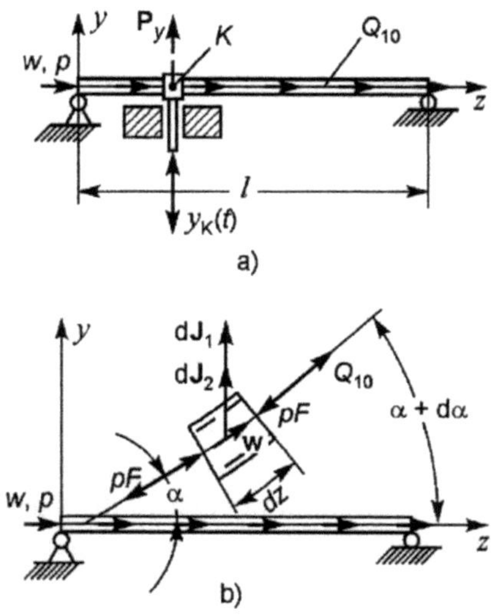

Fig. 7.5.

displacement that results in the kinematically-induced vibrations of the hose. The modes of random vibrations depend on the motion velocity of the liquid w and a pressure p. A rod with a variable cross-section and a lumped mass m, shown in Fig. 7.6 a, is subjected to the action of a random distributed force $q_z(t)$ and a concentrated force $P_z(t)$. The random longitudinal vibrations of the rod cause random stresses that must be incorporated in the analysis. For example, if the normal operation of the rod necessitates the fulfillment of a condition $(\sigma_y - \sigma_{\max}) > 0$, where σ_y is the yield limit of the material of the rod and $\sigma_{\max}$ is the maximum stress in the rod, then at random $\sigma_{\max}$ and with due account of the possible scatters σ_y it is required to determine the

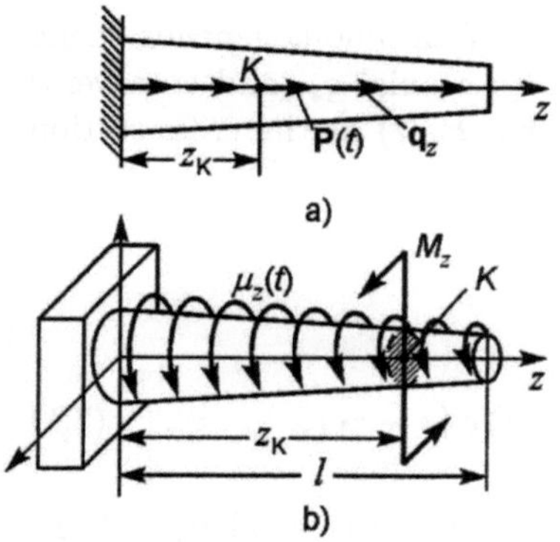

Fig. 7.6.

probability that the inequality will be satisfied, i.e. $P[(\sigma_y - \sigma_{\max}) > 0]$. We can get numerical value of only when we have the probability characteristics of $\sigma_{\max}$ that can be obtained solely from the solution of the equation of the random longitudinal vibrations of the rod.

A variable cross-section shaft with a lumped mass whose moment of inertia with respect to an axis z is equal to J_z is shown in Fig. 7.6 *b*. The shaft is loaded with a random distributed moment $\mu_z(t)$ and a random concentrated moment $M_z(t)$. At the random torsional vibrations of the shaft random shear stresses arise. To estimate the normal operation of the shaft, it is necessary to determine the probability of failure-free operation $P[(\tau_y - \tau_{\max}) > 0]$ the way we did it with respect to the previous example.

7.2 Equations of Small Vibrations

The equations of the small vibrations of a string, of the longitudinal vibrations of a rod and of the torsional vibrations of a shaft belong to the same class of the partial differential equations and are classified among equations of the hyperbolic type. Monographs and textbooks on equations of mathematical physics contain equations that do not take into account lumped masses and concentrated forces. These scientific papers and educational supplies are mainly devoted to wave equations or equations pertaining to cases where forces acting on a string, rod or shaft are distributed through to its whole length. The cited examples (Fig. 7.1–7.6) show that real problems can be much more intricate than the classical ones presented in mathematical literature, to say nothing of the problems that are usually considered in monographs on the theory of vibrations. These equations presented without a derivation for the most general cases, have the following form (with due account of viscous friction forces):

1. The equation of the transverse vibrations of an inhomogeneous string that has a lumped mass m and is loaded with an axial q_z, and transverse $q_y(z)$ distributed loads and a concentrated force $P_y(t)$ at the cross-section of the string z_1 takes the following form

$$m_1(z)\frac{\partial^2 y}{\partial t^2}+\alpha\frac{\partial y}{\partial t}+m\frac{\partial^2 y}{\partial t^2}\delta(z-z_k) = \frac{\partial}{\partial z}\left[Q_{10}(z)\frac{\partial y}{\partial z}\right]+q_y(z,t)+P_y\delta(z-z_1), \tag{7.1}$$

where y is the vertical displacement of the points of the axial line of the string, $\delta(z-z_1)$, $\delta(z-z_k)$ are the Dirac delta functions, α is the coefficient of the viscous friction force. This force approximately takes into account all energy losses at vibrations dependent on the resistance of the external medium, hysteresis etc.

The distributions of the loads q_z and q_y are not shown in Fig. 7.1 *a*.

The axial force $Q_{10}(z)$ and $m_1(z)$ mass of string unit length are considered known. Let us obtain the equation of small vibrations for a moving string (tape) (Fig. 7.4) and for a hose filled with a flow of fluid (Fig. 7.5).

An element of a liquid-filled hose is shown in Fig. 7.5 *b*. At vibrations the inertia forces dJ_1 and dJ_2 equal to

$$dJ_1=-m_1\frac{\mathrm{d}^2 y}{\mathrm{d}t^2}\mathrm{d}z, \qquad dJ_2=-m_2\frac{\mathrm{d}^2 y}{\mathrm{d}t^2}\mathrm{d}z \tag{7.2}$$

act on the element, where m_1 is the mass of the hoses unit length, m_2 is the mass of the liquid in the hoses unit length. By projecting all forces applied to the element of the hose on the axis Y, we obtain

$$-m_1\frac{\mathrm{d}^2 y}{\mathrm{d}t^2}\mathrm{d}z-m_2\frac{\mathrm{d}^2 y}{\mathrm{d}t^2}\mathrm{d}z+Q_{10}\mathrm{d}\alpha-pF\,\mathrm{d}\alpha=0, \tag{7.3}$$

where $\mathrm{d}\alpha=\dfrac{\partial^2 y}{\partial z^2}\mathrm{d}z$.

In order to obtain the partial differential equation we must pass to Lagrangian or Euler variables used in the continuum mechanics. When solving problems that involve relative motion of mediums, Euler variables are most effective. Passing in (7.2) to Euler variables, we obtain

$$m_1\frac{\mathrm{d}^2 y}{\mathrm{d}t^2}=m_1\frac{\partial^2 y}{\partial t^2};$$

$$m_2\frac{\mathrm{d}^2 y}{\mathrm{d}t^2}=m_2\left(\frac{\partial^2 y}{\partial t^2}+2w\frac{\partial^2 y}{\partial z\partial t}+w^2\frac{\partial^2 y}{\partial t^2}\right).$$

Therefore by transformations we obtain from (7.3) the following equation (with due account of the viscous friction force)

$$(m_1 + m_2)\frac{\partial^2 y}{\partial t^2} + \alpha\frac{\partial y}{\partial t} + 2wm_2\frac{\partial^2 y}{\partial z \partial t}$$
$$= \left[Q_{10} - \left(pF + m_2 w^2\right)\right]\frac{\partial^2 y}{\partial z^2}. \tag{7.4}$$

The term containing the mixed derivative is the Coriolis force. From equation (7.4) we obtain as a special case at $m_1 = 0$, $p = 0$ the equation of tape vibrations (Fig. 7.4 *b*)

$$m_2\frac{\partial^2 y}{\partial t^2} + \alpha\frac{\partial y}{\partial t} + 2wm_2\frac{\partial^2 y}{\partial z \partial t} = \left(Q_{10} - m_2 w^2\right)\frac{\partial^2 y}{\partial z^2}, \tag{7.5}$$

where m_2 is the mass of the tape unit length.

If the hose has a lumped mass m in the section with the coordinate z_k, and the force P_y is applied at the section z_p, the equation of the forced vibrations of the hose is

$$(m_1 + m_2)\frac{\partial^2 y}{\partial t^2} + \alpha\frac{\partial y}{\partial t} + 2wm_2\frac{\partial^2 y}{\partial z \partial t} + m\frac{\partial^2 y}{\partial z^2}\delta\left(z - z_k\right)$$
$$= Q_{10}^{(1)}\frac{\partial^2 y}{\partial z^2} + P_y\delta(z - z_p), \tag{7.6}$$

where $Q_{10}^{(1)} = Q_{10} - \left(pF + m_2 w^2\right)$.

2. The equation of the longitudinal vibrations of a variable cross-section rod with a lumped mass m and loaded with distributed and concentrated loads (Fig. 7.7 *a*) is

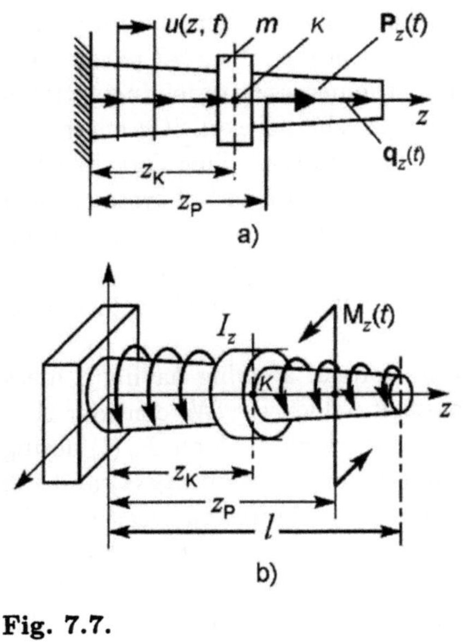

Fig. 7.7.

$$m_0(z)\frac{\partial^2 u}{\partial t^2}+\alpha\frac{\partial u}{\partial t}+m\frac{\partial^2 u}{\partial z^2}\delta(z-z_k)$$
$$=\frac{\partial}{\partial z}\left[F(z)E\frac{\partial u}{\partial z}\right]+q_z(z,t)+P_z\delta(z-z_p), \tag{7.7}$$

where u is the longitudinal displacement of the rods section (displacement of the points of the axial line of the rod at vibrations), F is the rod cross-sectional area, E is the elastic modulus of the first kind.

3. The equation of the torsional vibrations of a variable cross-section shaft with a lumped mass and distributed and concentrated moments (Fig. 7.7 *b*):

$$\rho J_z(z)\frac{\partial^2 \varphi}{\partial t^2}+\alpha\frac{\partial \varphi}{\partial t}+I_z\frac{\partial^2 \varphi}{\partial z^2}\delta(z-z_J)$$
$$=\frac{\partial}{\partial z}\left(J_z(z)G\frac{\partial \varphi}{\partial z}\right)+\mu_z(z,t)+M_z\delta(z-z_M), \tag{7.8}$$

where φ is the angle of rotation of cross-section of the shaft, $J_z(z)$ is the polar moment of inertia of the circular cross-section shaft (we have a similar geometric characteristic of the section J_k for a non-circular cross-section shaft), G is the elastic modulus of the second kind, I_z is the physical moment of inertia of the lumped mass with respect to the central axis coinciding with the axis Z, ρ is the density of the shafts material. If the external load is absent, equations (7.1) - (7.8) describe free vibrations caused, for example, by the displacement of the points of the axial line of the string at the initial time instant. It is very difficult to obtain the solutions of equations (7.1)–(7.8) in analytical form (except for special cases), particularly when the forces applied at arbitrary sections vary in time. The method of initial parameters generally used in solving problems related to the necessity of joining sections lest the lumped masses and concentrated forces should explicitly enter into the equations, results in very cumbersome final results. Therefore in what follows we use, during the analysis of the random vibrations of systems with distributed parameters, both exact methods allowing us to obtain a solution in analytical form and approximate methods, the latter being preferable because they make it possible to obtain the numerical values of the solutions of complex problems.

Let us consider the non-stationary vibrations of a string having a mass m (Fig. 7.2 *a*) that moves on it with a constant velocity v. We consider that the probability characteristics of the random aerodynamic force $F_y(t)$ acting on the mass m are known, i.e. the mathematical expectation $m_F(t)$ and the autocorrelation function $K_F(t, t')$ are known. It is required to determine the greatest possible accelerations of the mass m, assuming that the distribution law of the acceleration is normal. We obtain the equation of the small vibrations of the system from (7.1) at $m_1 = \text{const}$, $Q_{10} = \text{const}$, $q_y = 0$. The following three forces act on the mass m: 1) the force of gravity mg; 2) the aerodynamic force $F_y(t)$ and 3) the force of inertia $-m\left.\dfrac{d^2y}{dt^2}\right|_{z=z_k}$, therefore

$$m_0\frac{\partial^2 y}{\partial t^2}+\alpha\frac{\partial y}{\partial t}=Q_{10}\frac{\partial^2 y}{\partial z^2}+F_y\delta(z-z_k)+\left(-m\frac{d^2y}{dt^2}\right)\delta(z-z_k). \quad (7.9)$$

Passing to the Euler variables, we obtain ($\delta_k=\delta(z-z_k),\ z_k=vt$)

$$m\frac{d^2y}{dt^2}\delta_k=m\left(\frac{\partial^2 y}{\partial z^2}+2v\frac{\partial^2 y}{\partial z\,\partial t}+v^2\frac{\partial^2 y}{\partial z^2}\right)\delta_k.$$

Therefore from (7.9) we obtain

$$\begin{aligned}L(y)=(m_0+m\delta_k)\frac{\partial^2 y}{\partial t^2}+\alpha\frac{\partial y}{\partial t}+2mv\frac{\partial^2 y}{\partial z\,\partial t}\delta_k\\ -(Q_{10}-mv^2\delta_k)\frac{\partial^2 y}{\partial z^2}-F_y\delta_k=0.\end{aligned} \quad (7.10)$$

7.3 Solving Equations of Small Vibrations

Let us find an approximate solution of equation (7.10), assuming that

$$y=\sum_{j=1}^{n}f_j(t)\sin\frac{\pi jz}{l} \quad (7.11)$$

Let us take the possible displacements as

$$\delta_y=\sum_{j=1}^{n}\delta a_j\sin\frac{\pi jz}{l}. \quad (7.12)$$

Having used the virtual work principle, we obtain the system of equations

$$\int_0^l L(y)\sin\frac{\pi\nu z}{l}\,dz, \qquad (\nu=1,2,\ldots,n). \quad (7.13)$$

By rearrangements with due account of the properties of the delta-functions we obtain the system of equations in unknown functions $f_j(t)$

$$\sum_{j=1}^{n}\left[a_{\nu j}(t)\ddot{f}_j+b_{\nu j}(t)\dot{f}_j+c_{\nu j}(t)f_j\right]=b_\nu F_y, \quad (\nu=1,2,\ldots,n). \quad (7.14)$$

For example, confining ourselves to a two-term approximation we obtain the following expressions for coefficients $a_{\nu j}$, $b_{\nu j}$, $c_{\nu j}$ and b_ν.

$$
\begin{aligned}
a_{11} &= m_0\frac{l}{2} + m\sin\frac{2\pi vt}{l}; \quad a_{12} = m\sin\frac{2\pi vt}{l}\sin\frac{\pi vt}{l};\\
a_{21} &= m\sin\frac{\pi vt}{l}\sin\frac{2\pi vt}{l}; \quad a_{22} = m_0\frac{l}{2} + m\sin^2\frac{2\pi vt}{l};\\
b_{11} &= \alpha\frac{l}{2} + 2mv\left(\frac{\pi}{l}\right)\cos\frac{2\pi vt}{l}\sin\frac{\pi vt}{l};\\
b_{12} &= 2mv\left(\frac{2\pi}{l}\right)\cos\frac{2\pi vt}{l}\sin\frac{\pi vt}{l};\\
b_{21} &= 2mv\left(\frac{\pi}{l}\right)\cos\frac{\pi vt}{l}\sin\frac{\pi vt}{l};\\
b_{22} &= \alpha\frac{l}{2} + 2mv\left(\frac{2\pi}{l}\right)\cos\frac{2\pi vt}{l}\sin\frac{2\pi vt}{l};\\
c_{11} &= Q_{10}\left(\frac{\pi}{l}\right)^2 - mv^2\left(\frac{\pi}{l}\right)^2\sin^2\frac{\pi vt}{l};\\
c_{12} &= -mv^2\left(\frac{\pi}{l}\right)^2\sin\frac{2\pi vt}{l}\sin\frac{\pi vt}{l};\\
c_{21} &= -mv^2\left(\frac{v}{l}\right)^2\sin\frac{\pi vt}{l}\sin\frac{2\pi vt}{l};\\
c_{22} &= Q_{10}\frac{l}{2}\left(\frac{2\pi}{l}\right)^2 - mv^2\left(\frac{2\pi}{l}\right)^2\sin^2\frac{2\pi vt}{l};\\
b_1 &= F_y\sin\frac{\pi vt}{l}; \quad b_2 = F_y\sin\frac{2\pi vt}{l}.
\end{aligned}
$$

In vector form the system of equations (7.14) is

$$A(t)\,\ddot{\mathbf{f}} + B(t)\,\dot{\mathbf{f}} + C(t)\,\mathbf{f} = \mathbf{b}_0 \tag{7.15}$$

or

$$\dot{\boldsymbol{\Phi}} + D(t)\,\boldsymbol{\Phi} = \mathbf{b}, \tag{7.16}$$

where

$$
\begin{aligned}
&\boldsymbol{\Phi} = (\boldsymbol{\Phi}_1,\, \boldsymbol{\Phi}_2)^{\mathrm{T}}, \quad \boldsymbol{\Phi}_1 = (\dot{f}_1,\, \dot{f}_2)^{\mathrm{T}}, \quad \boldsymbol{\Phi}_2 = (f_1,\, f_2)^{\mathrm{T}},\\
&D(t) = \begin{vmatrix} A^{-1}B & A^{-1}C \\ -E & 0 \end{vmatrix}, \quad \mathbf{b} = (A^{-1}\mathbf{b}_0,\, 0)^{\mathrm{T}}.
\end{aligned}
$$

Equation (7.16) is a linear one with time-dependent coefficients. Therefore we can solve this equation only numerically. In the considered problem the time of process is limited ($0 \le t \le \frac{l}{v}$). That is why at any random stationary or non-stationary aerodynamic force the vibrations of the mass m are non-stationary.

The solution of equation (7.15) at zero initial data is

$$\mathbf{\Phi} = \int_0^t G(t,\, t_1)\, \mathbf{b}(t_1)\, \mathrm{d}t_1; \quad (G(t,\, t_1)) = K(t)\, K^{-1}(t_1), \tag{7.17}$$

where $K(t)$ is the fundamental matrix of the solutions of homogeneous equation (7.16). The matrixes $K(t)$ and $K^{-1}(t)$ are determined by numerical methods.

From equation (7.16) we determine the derivative of the vector $\mathbf{\Phi}$

$$\dot{\mathbf{\Phi}} = -\int_0^t D(t)\, G(t,\, t_1)\, \mathbf{b}(t_1)\, \mathrm{d}t_1 + \mathbf{b}(t),$$

Then we determine the second derivatives $\ddot{f}_1$ and $\ddot{f}_2$

$$\left(\dot{\mathbf{\Phi}} = \left(\ddot{f}_1,\, \ddot{f}_2,\, \dot{f}_1,\, \dot{f}_2\right)^{\mathrm{T}}\right),$$

$$\ddot{f}_1 = -\int_0^t \left[q_{11}(t, t_1)\, b_{01}(t) + q_{12}(t, t_1)\, b_{02}(t)\right] F_y(t_1)\, \mathrm{d}t_1 + b_{01} F_y(t);$$

$$\ddot{f}_2 = -\int_0^t \left[q_{21}(t, t_1)\, b_{01}(t) + q_{22}(t, t_1)\, b_{02}(t)\right] F_y(t_1)\, \mathrm{d}t_1 + b_{02} F_y(t),$$

where q_{ij} are the elements of the matrix $D(t)\, K(t)\, K^{-1}(t_1)$,

$$b_{01} = a_{11}^{(1)}(t_1) \sin \frac{\pi v t_1}{l} + a_{12}^{(1)}(t_1) \sin \frac{2\pi v t_1}{l};$$

$$b_{02} = a_{21}^{(1)}(t_1) \sin \frac{\pi v t_1}{l} + a_{22}^{(1)}(t_1) \sin \frac{2\pi v t_1}{l},$$

($a_{ij}^{(1)}(t_1)$ are the elements of the matrix $(A^{-1}(t_1))$.

The acceleration of the mass m is equal to

$$\ddot{y}(z,\, t) = \ddot{f}_1 \sin \frac{\pi z}{l} + \ddot{f}_2 \sin \frac{2\pi z}{l} \tag{7.18}$$

or

$$\ddot{y}(z,\, t) = -\int_0^t q(t,\, t', z)\, f_y(t_1)\, \mathrm{d}t_1 + b(t,\, z)\, F(t), \tag{7.19}$$

where

$$q(t,t_1,z) = \left(q_{11}\sin\frac{\pi z}{l} + q_{21}\sin\frac{2\pi z}{l}\right)b_{01} + \left(q_{21}\sin\frac{\pi z}{l} + q_{22}\sin\frac{2\pi z}{l}\right)b_{02},$$

$$b(t,z) = b_{01}\sin\frac{\pi z}{l} + b_{02}\sin\frac{2\pi z}{l}.$$

We determine the mathematical expectation of the acceleration $m_{\ddot{y}}(t,z)$, and the correlation function $K_{\ddot{y}}(t,t')$

$$m_{\ddot{y}}(t,z) = \int_0^t q(t,t_1,z)\, m_y(t_1)\, \mathrm{d}t_1 + b(t,z)\, m_y(t)\,, \tag{7.20}$$

$$\begin{aligned} K_{\ddot{y}}(t,\, t',\, z) = &\int_0^t\int_0^{t'} q(t,t_1,z)\, q(t',t_2)\, K_{\ddot{y}}(t_1,t_2)\, \mathrm{d}t_1\, \mathrm{d}t_2 \\ &- b(t',z)\int_0^t q(t,t_1,z)\, K_{\ddot{y}}(t',t_1)\, \mathrm{d}t_1 \\ &- b(t,z)\int_0^{t'} q(t',t_2,z)\, K_{\ddot{y}}(t,t_2)\, \mathrm{d}t_2 \\ &+ b(t_1,z)\, b(t',z)\, K_{\ddot{y}}(t,t')\,. \end{aligned} \tag{7.21}$$

Assuming that $t' = t$, we obtain the variance of the acceleration

$$D_{\ddot{y}}(t,z) = K_{\ddot{y}}(t,t',z)\,|_{t'=t}. \tag{7.22}$$

Assuming discrete values t_j $(0 \le t \le t_k)$, we numerically determine $m_{\ddot{y}}(t_j,z)$ and $\sigma_{\ddot{y}}(t_j,z)$.

Having used the three sigma rule, we obtain the maximum value of the acceleration for the given velocity

$$\max\ddot{y}(t_j,z) = m_{\ddot{y}}(t_j,z) + 3\sigma_{\ddot{y}}(t_j,z)\,. \tag{7.23}$$

The qualitative behaviour of the variation of the maximum acceleration with the coordinate Z (at fixed t_j) is shown in Fig. 7.8. In its turn, the maximum values of the acceleration $\max\ddot{y}(t_j,z)$ for each instant t_j reach their maximum also on the coordinate Z (Fig. 7.8). Therefore, during the motion of the mass m on the section $0 \le z \le l$ the greatest possible acceleration is $\ddot{y}_*$ that is equal (at a given velocity v) to

$$\ddot{y}_* = \max\left(\max \ddot{y}\left(t_j, t\right)\right);$$
$$0 \leq t \leq t;$$
$$0 \leq z \leq l.$$

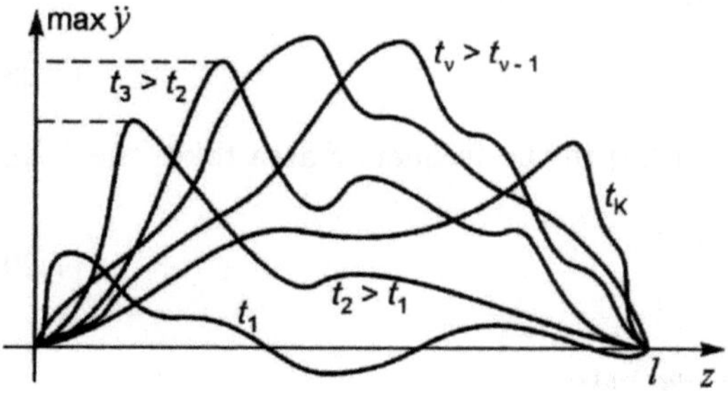

Fig. 7.8.

Let us consider the stationary vibrations of a hose (Fig. 7.5) caused by a stationary kinematic excitation with known probability characteristics: $m_{y_k} = 0$, $S_{y_k}(\omega)$. Let us determine the spectral density of the vertical displacements of the points of the axial line of the hose ($S_y(\omega, z)$) and the maximum value of a concentrated random force that arises in a section K, considering that this force has normal distribution. Let us introduce an unknown concentrated force $P_y(t)$ in the section K where the forced displacement occurs. Let us take advantage of equation (7.6), assuming that $m = 0$

$$(m_1 + m_2)\frac{\partial^2 y}{\partial t^2} + \alpha\frac{\partial y}{\partial t} + 2wm_2\frac{\partial^2 y}{\partial z\,\partial t} = Q_{10}^{(1)}\frac{\partial^2 y}{\partial z^2} + P_y(t)\,\delta(z - z_k), \quad (7.24)$$
$$\left(Q_{10}^{(1)} = Q_{10} - \left(pF + m_2 W^2\right)\right).$$

For the approximate solution of equation (7.24) we assume that

$$y = \sum_{j=1}^{n} f_j(t)\sin\frac{\pi j z}{l}. \quad (7.25)$$

Substituting (7.25) into equation (7.24) and using the virtual work principle, we obtain the following system of differential equations

$$\sum_{j=1}^{n}\left(a_{ij}\ddot{f}_j + b_{ij}\dot{f}_j + c_{ij}f_j\right) = b_i P_y(t), \quad (i = 1, 2, \ldots, n). \quad (7.26)$$

Taking advantage of the Fourier transform, we obtain a system of algebraic equations

$$\sum_{j=1}^{n} \left(a_{ij} (i\omega)^2 b_{ij} (i\omega) + c_{ij} (i\omega) \right) \Phi_j (i\omega) = b_i P_{0y} (i\omega) , \tag{7.27}$$

where b_{ij} and c_{ij} are constant coefficients dependent on the velocity of the liquid w and on the liquid pressure p.

Let us determine $\Phi_j (i\omega)$ from system of equations (7.27)

$$\Phi_j (i\omega) = W_j (i\omega) P_{0y} (i\omega) . \tag{7.28}$$

The approximate expression for (7.25) in the frequency area takes the form

$$Y (z, i\omega) = \sum_{j=1}^{n} \Phi_j (i\omega) \sin \frac{\pi j z}{l}. \tag{7.29}$$

Eliminating $\Phi_j (i\omega)$ from (7.29), we have

$$Y (z, i\omega) = \sum_{j=1}^{n} W_j (i\omega) P_{y_0} (i\omega) \sin \frac{\pi j z}{l}. \tag{7.30}$$

We determine the introduced unknown force (image of the force P_y) from the following condition

$$Y (z, i\omega)|_{z=z_k} = Y_k (i\omega) , \tag{7.31}$$

where $Y_k (i\omega)$ is the image of the displacement of the point k. We obtain the following equation for the determination of $P_{0y} (i\omega)$

$$Y_k (i\omega) = \left(\sum_{j=1}^{n} W_j (i\omega) \sin \frac{\pi j z_k}{l} \right) P_{0y} (i\omega) , \tag{7.32}$$

Therefore,

$$P_{0y} (i\omega) = \frac{Y_k (i\omega)}{\left(\sum_{j=1}^{n} W_j (i\omega) \sin \frac{\pi j z_k}{l} \right)}. \tag{7.33}$$

As a result we obtain $Y (z, i\omega)$

$$Y (z, i\omega) = \frac{\left(\sum_{j=1}^{n} W_j (i\omega) \sin \frac{\pi j z}{l} \right)}{\left(\sum_{j=1}^{n} W_j (i\omega) \sin \frac{\pi j z_k}{l} \right)} Y_k (i\omega) \tag{7.34}$$

or

$$Y(z, i\omega) = W(z, i\omega)\, Y_k(i\omega). \tag{7.35}$$

The spectral density of the vertical displacements of the points of the hose axial line and the variance of displacements are equal to

$$S_y(\omega, z) = |W(z, i\omega)|^2\, S_{y_k}(\omega), \tag{7.36}$$

$$D_y(z) = \int_{-\infty}^{\infty} |W(z, i\omega)|^2\, S_{y_k}(\omega) d\omega. \tag{7.37}$$

Let us determine the variance of the concentrated force P_y originating in the section k. We find the spectral density $S_{P_y}(\omega)$ from (7.33)

$$S_{P_y}(\omega) = |W_1|^2\, S_{y_k}(\omega), \tag{7.38}$$

where

$$W_1 = \frac{1}{\sum\limits_{j=1}^{n} W_j(i\omega) \sin \frac{\pi j z_k}{l}}.$$

The variance of the force P_y is equal to

$$D_{P_y} = \int_{-\infty}^{\infty} |W_1|^2\, S_{y_k}(\omega)\, d\omega. \tag{7.39}$$

The greatest possible value of P_y (at $m_{P_y} = 0$) is equal to

$$\max P_y = 3\sqrt{D_{P_y}}.$$

3. A missile (Fig. 7.9) moves with an increasing velocity in a bore of a barrel. The force of friction between the missile and the barrel is equal to $P = P_0 + \Delta P(t)$. Owing to the possible momentary sticking of the surface of the missile to the surface of the bore or to the reduction of the contact pressure the force P has a random component $\Delta P(t)$. One of the realizations of the stochastic function $\Delta P(t)$ is shown in Fig. 7.10.

It is required to determine the variance of the random component of the stress $\Delta\sigma$ in the bore at the moment of the missile exit. To simplify calculations we shall consider that the velocity of the missile linearly depends on time, i.e. $v = at$, where a is the constant acceleration.

At $t = 0$ the body is in the origin. Let us determine the correlation function $K_{\Delta P}(t)$ considering that the random function $\Delta P(t)$ is stationary. The product of two values of the function $\Delta P(t)$ and $\Delta P(t+\tau)$ depends on

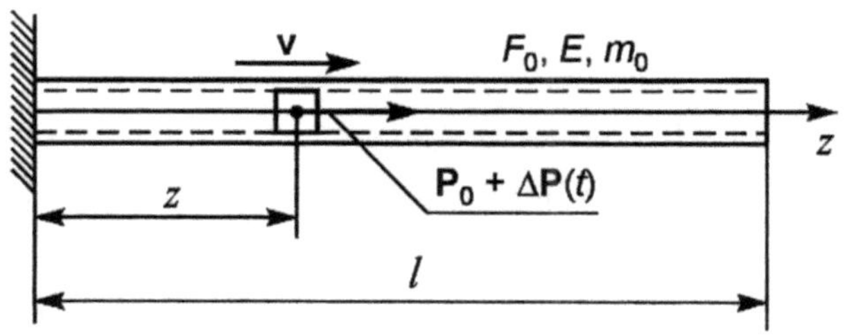

Fig. 7.9.

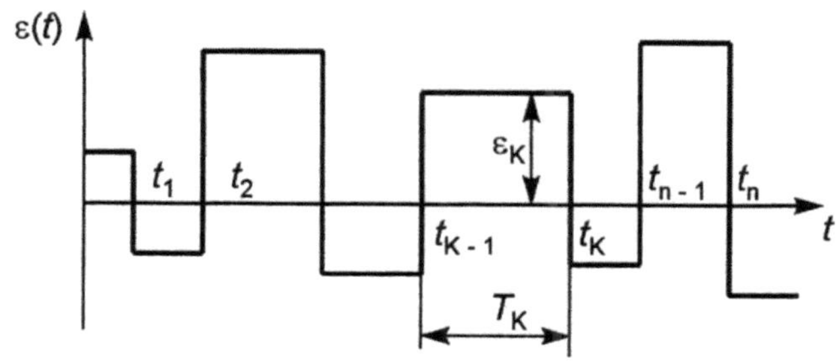

Fig. 7.10.

whether the time instants t and $t + \tau$ are in the same interval T_k or not. If they are,

$$\Delta P(t)\, \Delta P(t + \tau) = \Delta P_k^2, \tag{7.40}$$

and, if they are not,

$$\Delta P(t)\, \Delta P(t + \tau) = \Delta P_k \Delta P_i. \tag{7.41}$$

In order to obtain the correlation function we must add together products (7.40) and (7.41) averaged over a set of values and multiplied by appropriate probabilities, for which purpose it is necessary to find the probability that a section of the length τ completely falls within the interval T_k. This probability is equal to the probability that on a randomly taken span τ of the time axis there is not a single point of the discontinuity of the function $\Delta P(t)$. If we know the distribution function of the zeros (discontinuities) of the function $\Delta P(t)$ on the interval τ as a function of the length of this interval, it is possible to find the probability $P(n, \tau)$ of the number of zeros of the function $\Delta P(t)$, where n is the number of zeros. The probability that there will be not a single discontinuity of the function $\Delta P(t)$ on the interval τ is equal to $P(0, \tau)$. If we assume that the number of the discontinuities of the function $\Delta P(t)$ on any time span τ follows the Poisson distribution law

$$P(n, \tau) = \frac{(\mu\tau)^n}{n!} e^{-\mu\tau},$$

where μ is the average frequency of discontinuities which is considered known, the probability is $P(0,\tau) = \mathrm{e}^{-\mu\tau}$.

The probability that the interval τ exceeds the interval T_k (i.e. the time instants t and $t+\tau$ are in different intervals T_k) is equal to $1 - P(0,\tau)$.

Hence, the correlation function takes the form

$$K_{\Delta P}(\tau) = M\left[\Delta P_k^2\right] P(0,\tau) - M\left[\Delta P_k \Delta P_i\right]\left[1 - P(0,\tau)\right].$$

Considering that ΔP_k and ΔP_i are independent and have the same distribution, we obtain

$$K_{\Delta P}(\tau) = M\left[\Delta P_k^2\right] P(0,\tau) = D_{\Delta P}\mathrm{e}^{-\mu\tau}.$$

The spectral density ΔP is

$$S_{\Delta P}(\omega) = \int\limits_{-\infty}^{\infty} K_{\Delta P}(\tau)\cos\omega\tau\,\mathrm{d}\tau = \frac{2}{\pi}\frac{D_{\Delta P}\mu}{(\mu^2+\omega^2)}.$$

The equation of the longitudinal vibrations of a constant cross-section rod with due account of the moving concentrated force (we ignore the resistance force) takes the form

$$m_1\frac{\partial^2 u}{\partial t^2} = EF_0\frac{\partial^2 u}{\partial z^2} + \Delta P(t)\,\delta\left(z - \frac{at^2}{2}\right). \tag{7.42}$$

Assuming that in (7.42)

$$u = \sum_{j=1,3\ldots} f_j(t)\sin\frac{\pi j z}{2l},$$

$$\delta\left(z - \frac{at^2}{2}\right) = \sum_{j=1,3,\ldots} c_j\sin\frac{\pi j z}{2l}, \quad \left(c_j = \frac{2}{l}\sin\frac{\pi j a t^2}{4l}\right),$$

we obtain after transformations equations for the determination of the functions $f_j(t)$:

$$f_j + p_j^2 f_j = \frac{2}{m_1 l}\Delta P(t)\sin\frac{\pi j a t^2}{4l}, \quad \left(p_j^2 = \left(\frac{\pi j}{2l}\right)^2\frac{EF_0}{m_1}\right). \tag{7.43}$$

The solution of equation (7.43) at zero initial conditions is

$$f_j = \frac{2}{p_j l m_1}\int\limits_0^t \sin p_j(t-\tau)\sin\frac{\pi j a \tau^2}{4l}\Delta P(\tau)\,\mathrm{d}\tau. \tag{7.44}$$

As a result, we obtain the solution of equation (7.42)

$$u = \sum_{j=1,3\ldots}^{\infty} \frac{2}{p_j l m_1} \sin\frac{\pi j z}{2l} \int_0^t \sin p_j (t-\tau) \sin\frac{\pi j a \tau^2}{4l} \Delta P(\tau)\, \mathrm{d}\tau. \qquad (7.45)$$

The stress in the sections of the rod caused only by the random force $\Delta P(t)$ is

$$\Delta\sigma = E\frac{\partial u}{\partial z} = \sum_{j=1,3\ldots} \frac{E\pi j}{p_j l^2 m_1} \cos\frac{\pi j z}{2l} \int_0^t \sin p_j (t-\tau) \sin\frac{\pi j a \tau^2}{4l} \Delta P(\tau)\, \mathrm{d}\tau.$$

The variance of the stress is

$$D_{\Delta\sigma} = \sum_{j=1,3\ldots}^{\infty} \sum_{k=1,3\ldots}^{\infty} \frac{E^2 (\pi j)^2}{p_j p_k l^4 m_1^2} \cos\frac{\pi j z}{2l} \cos\frac{\pi k z}{2l} \int_0^t \int_0^t \sin p_j (t-\tau)$$
$$\times \sin\frac{\pi j a \tau^2}{4l} \sin p_k (t-\tau_1) \sin\frac{\pi k a \tau_1^2}{2l} K_{\Delta P} \mathrm{d}\tau\, \mathrm{d}\tau_1. \qquad (7.46)$$

In order to determine the variances of the stress at the instant of the missile exit out of the bore of the barrel we must take the upper limit of integration in expression (7.46) equal to

$$t = t_k = \sqrt{\frac{2l}{a}}.$$

4. In order to average the misalignments of the thrust R of the engine of a rocket the latter is set into rotation (Fig. 7.11) when leaving the guides. In order to set the rocket into rotation there are rotation engines *1* that are ignited at the moment of leaving (this moment can be taken for the initial one). The rotation engines produce a moment with respect to the longitudinal axis. This moment has a constant component M_0 and a random component $\Delta M(t)$ $(M = M_0 + \Delta M(t))$. The random component can be considered as

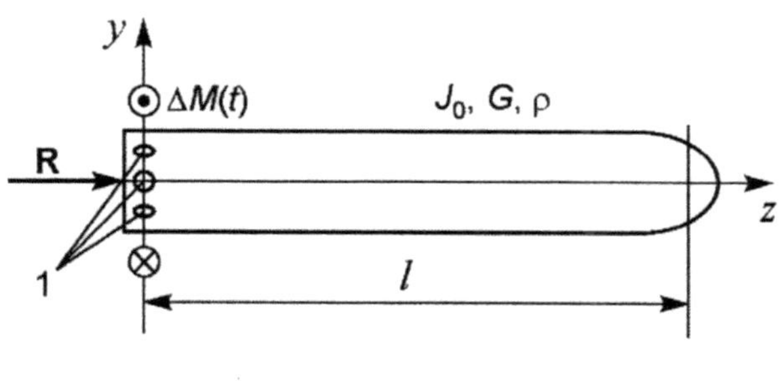

Fig. 7.11.

a random stationary function with a known spectral density $S(\omega)$. As a result of the action of a random twisting moment $\Delta M(t)$ some torsional vibrations of the rocket take place introducing errors in the readings of the instruments of the control system. In order to estimate the possible errors of the instruments due to the random torsional vibrations we must know the probability characteristics of the torsional vibrations and, in particular, the variance of the angular acceleration in those cross-sections of the rocket where appropriate instruments are located.

Assuming that the random torsional vibrations of the rocket are stationary, it is required to determine the variance of the angular acceleration of the rocket. When solving the problem, we shall confine ourselves to the simplest case where $\rho J_0 = \text{const}$ and $GJ_0 = \text{const}$. The equation of the random torsional vibrations of the rocket with due account of the viscous friction takes the form (a particular case of equation (4.57))

$$J_0\rho\frac{\partial^2\varphi}{\partial t^2} + \alpha\frac{\partial\varphi}{\partial t} = J_0 G\frac{\partial^2\varphi}{\partial z^2} + \Delta M(t)\,\delta(z). \tag{7.47}$$

Here the eigenfunctions of the free torsional vibrations of the rod are equal to $\cos\dfrac{\pi n z}{l}$ (as the boundary conditions take the form: 1) $z = 0$; $\dfrac{\partial\varphi}{\partial z} = 0$; 2) $z = l$; $\dfrac{\partial\varphi}{\partial z} = 0$. Therefore, assuming that in (7.47)

$$\varphi = \sum_{j=1}^{\infty} f_j(t)\cos\frac{\pi j z}{l}; \quad \delta(z) = \sum_{j=1}^{\infty} c_j\cos\frac{\pi j z}{l}; \quad c_j = (-1)^j\,\frac{2}{l}. \tag{7.48}$$

Substituting the expression (7.48) for φ into equation (7.47) and using the virtual work principle we obtain by rearrangements the following system of equations in $f_j(t)$

$$\ddot{f}_j + \frac{\alpha}{J_0\rho}\dot{f}_j + p_j^2 f_j = (-1)^i\,\frac{2}{J_0\rho l}\Delta M(t), \tag{7.49}$$

where

$$p_j^2 = \frac{G}{\rho}\left(\frac{\pi j}{l}\right)^2.$$

Obtained equations (7.48) are similar to those of (4.40), therefore the spectral density of solution (7.48) is

$$S_\varphi(\omega) = |W|^2\,S_{\Delta M}(\omega),$$

where

$$|W|^2 = \left|\sum_{j=1}^{\infty}\frac{2(-1)^j\cos\dfrac{\pi j z}{l}}{J_0\rho l\left(-\omega^2 + \dfrac{\alpha}{\rho}i\omega + p_j^2\right)}\right|^2.$$

The variance of the angular acceleration is

$$D_{\ddot{\varphi}} = \int_{-\infty}^{\infty} |W|^2 \, \omega^4 S_{\Delta M}(\omega) \, d\omega. \tag{7.50}$$

8. Random Vibrations of Rods

In the previous chapter the most simple problems related to systems with distributed parameters have been considered.

Rod elements of machines, devices and structures belong to more complex systems with distributed parameters as we usually have to consider systems of partial differential equations when solving problems of the statics and dynamics of rod systems. Elastic rod and structural elements considered as rods in design practice are given as examples in Figs. 8.1–8.4. Two elastic rod elements of devices: a cylindrical (Fig. 8.1 *a*) spring and a spiral (Fig. 8.1 *b*) spring are shown in Fig. 8.1 *a, b*. If the devices using these elements are placed in an object moving with acceleration, a random acceleration $\mathbf{a}_c(t)$ acts on these elements due to the scatter of the engines thrust. As real elastic elements have a mass, occuring random vibrations can lead to substantial errors in the operation of these devices. The shock absorption system of a mass m with the use of a conic spring is shown in Fig. 8.1 *c*. At a random kinematic excitation $(y_k(t))$ random vibrations take place. The order of the mass m can be the same as that of the mass of the spring. Therefore, in order to determine the probability characteristics of the displacement of the mass m and its first

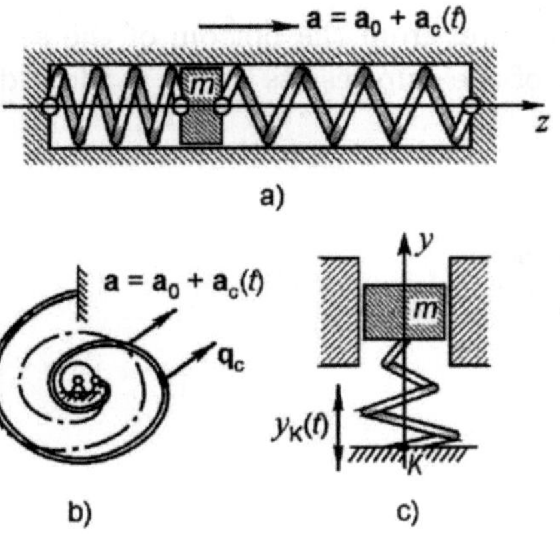

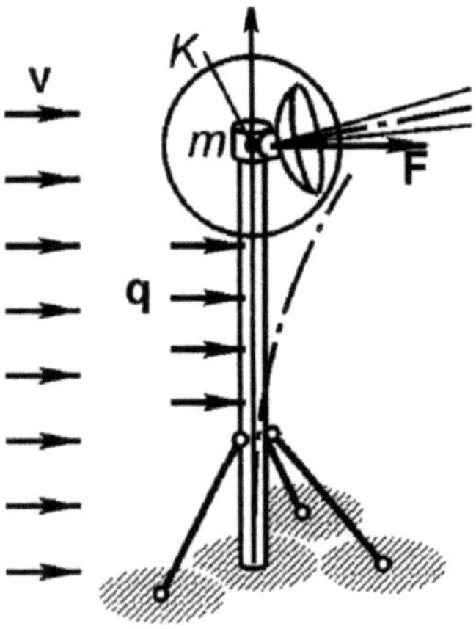

Fig. 8.2.

derivatives $y(\dot{y}, \ddot{y})$ it is necessary to consider the spring as a system with distributed parameters.

A mast with an aerial located in an air flow whose velocity has a random component is shown in Fig. 8.2. Random aerodynamic forces act on the mast and the aerial (a concentrated force **F** acts on the aerial and a distributed force **q** acts on the mast) which causes random vibrations of the system. The random characteristics of the rotation angle of the aerials beam are of interest in the process of designing because they influence the accuracy of the reception or transfer of signals.

A curved section of a pipeline filled with a moving liquid, for example, a section of an aircraft engines onboard feed system that involves pipeline-section mountings connected with different foundations, is shown in Fig. 8.3. The pipeline has a local hinged constraint (section k) with a random displacement ($u_k(t)$), which causes some random kinematic vibrations of the pipeline.

A pipeline intended for lifting structures from the bottom of the sea is shown in Fig. 8.4. The random rolling of the ship results in the random dis-

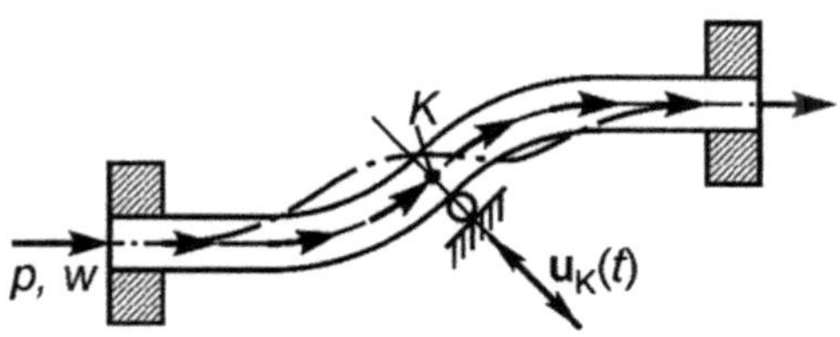

Fig. 8.3.

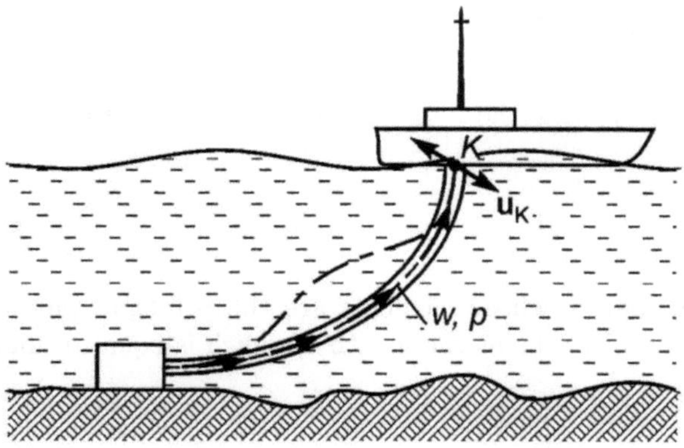

Fig. 8.4.

placement of a point $k\,(\mathbf{u}_k\,(t))$, which leads to the random kinematically induced vibrations of the pipeline.

8.1 Nonlinear Equations of Motion of Three-Dimensional Curvilinear Rods

A rod at an arbitrary time instant is shown in Fig. 8.5 *a* and an element of the rod with forces and moments applied to it is given in Fig. 8.5 *b*. In addition to these forces, an inertia force $\mathrm{d}\mathbf{J}_u$

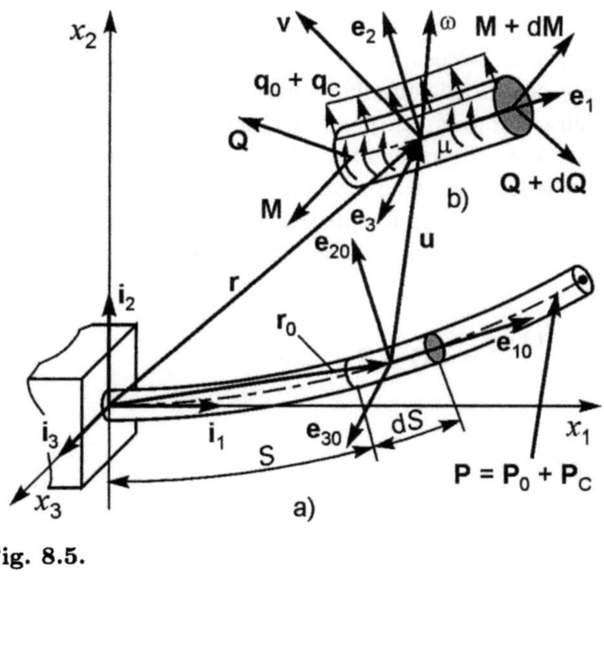

Fig. 8.5.

$$\mathrm{d}\mathbf{J}_u = -m_1 \frac{\mathrm{d}\mathbf{v}}{\mathrm{d}t}\mathrm{d}s, \qquad \left(\mathbf{v} = \frac{\mathrm{d}\mathbf{u}}{\mathrm{d}t}\right), \tag{8.1}$$

and a moment of inertia (if we take the rotary inertia of the rod element into account)

$$\mathrm{d}\boldsymbol{\mu}_a = -\frac{\mathrm{d}}{\mathrm{d}t}(J\boldsymbol{\omega})\,\mathrm{d}s, \tag{8.2}$$

where J is the diagonal matrix whose elements are the principal physical moments of inertia of the rods element the length of which is equal to unity

$$J = \begin{bmatrix} J_{11} & 0 & 0 \\ 0 & J_{22} & 0 \\ 0 & 0 & J_{33} \end{bmatrix} \tag{8.3}$$

act on the element of the rod at its motion.

For a variable cross-section rod J_{ii} depends on an arc coordinate s. Using the Lagrangian variables, we obtain (8.1) and (8.2) in partial derivatives

$$\mathrm{d}\mathbf{J}_u = -\mathrm{d}m\frac{\partial \mathbf{v}}{\partial t}; \quad \mathbf{v} = -\frac{\partial \mathbf{u}}{\partial t}; \quad \mathrm{d}\mu_u = -\frac{\partial}{\partial t}(J\boldsymbol{\omega})\,\mathrm{d}s. \tag{8.4}$$

Taking advantage of d'Alembert's principle we obtain the following vector equations of the translatory motion and rotation of the rod element (confining ourselves to one concentrated force **P** and one concentrated moment **T**):

$$m_1\frac{\partial \mathbf{v}}{\partial t} = \frac{\partial \mathbf{Q}}{\partial s} + \mathbf{q} + \mathbf{P}\delta(s - s_i); \tag{8.5}$$

$$\frac{\partial}{\partial t}(J\boldsymbol{\omega}) = \frac{\partial \mathbf{M}}{\partial s} + \mathbf{e}_1 \times \mathbf{Q} + \boldsymbol{\mu} + \mathbf{T}\delta(s - s_v), \tag{8.6}$$

where **Q** is the vector of internal forces , **M** is the vector of internal moments, $\boldsymbol{\omega}$, **v** are respectively the vector of the angular velocity and the vector of the linear velocity; δ is the Dirac function, $\mathbf{q}$, $\boldsymbol{\mu}$ are vectors of distributed forces and moments respectively; **P** is the concentrated force applied in a section with coordinates s_i; **T** is the concentrated moment.

The matrix J has elements that do not depend on time only in the attached coordinate system, therefore in equations (8.5) and (8.6) we pass to local derivatives

$$m_0\left(\frac{\tilde{\partial}\mathbf{v}}{\partial t} + \boldsymbol{\omega} \times \mathbf{v}\right) = \frac{\tilde{\partial}\mathbf{Q}}{\partial s} + æ \times \mathbf{Q} + \mathbf{P}; \tag{8.7}$$

$$J\frac{\tilde{\partial}\boldsymbol{\omega}}{\partial t} + \boldsymbol{\omega} \times J\boldsymbol{\omega} = \frac{\tilde{\partial}\mathbf{M}}{\partial s} + æ \times \mathbf{M} + \mathbf{e}_1 \times \mathbf{Q} + \mathbf{T}, \tag{8.8}$$

where

$$\mathbf{P} = \mathbf{q} + \mathbf{P}\delta(s - s_i); \qquad \mathbf{T} = \boldsymbol{\mu} + \mathbf{T}\delta(s - s_v).$$

In the general case the distributed and concentrated forces and moments can be presented as

$$\mathbf{q} = \mathbf{q}_0 + \mathbf{q}_d + \mathbf{q}_c; \qquad \mathbf{P} = \mathbf{P}_0 + \mathbf{P}_d + \mathbf{P}_c;$$
$$\boldsymbol{\mu} = \boldsymbol{\mu}_0 + \boldsymbol{\mu}_d + \boldsymbol{\mu}_c; \qquad \mathbf{T} = \mathbf{T}_0 + \mathbf{T}_d + \mathbf{T}_c,$$

where $\mathbf{q}_0$, $\mathbf{P}_0$, $\boldsymbol{\mu}_0$ and $\mathbf{T}_0$ are the static loads; $\mathbf{q}_d$, $\mathbf{P}_d$, $\boldsymbol{\mu}_d$ and $\mathbf{T}_d$ are the dynamic deterministic loads; $\mathbf{q}_c$, $\mathbf{P}_c$, $\boldsymbol{\mu}_c$ and $\mathbf{T}_c$ are the dynamic random loads.

In what follows the tilde sign in the notation of the local derivative is dropped. The vector æ entering into equations (8.7), (8.8) is

$$æ = \sum_{i=1}^{3} æ_i \mathbf{e}_i,$$

where $æ_1$ is the twist of the axial line of the rod; $æ_2$, $æ_3$ are the curvatures of the projection of the rods axial line onto planes $(\mathbf{e}_1, \mathbf{e}_2)$ and $(\mathbf{e}_1, \mathbf{e}_3)$ that go through the principal axes of the cross-section of the rod.

Very often in applied problems the rotary inertia of the rod is ignored $(J\omega \approx 0)$, therefore

$$\frac{\partial \mathbf{M}}{\partial s} + æ \times \mathbf{M} + \mathbf{e}_1 \times \mathbf{Q} + \mathbf{T} = 0. \tag{8.9}$$

The vector $\mathbf{M}$ is connected with the vector æ by the equation

$$\mathbf{M} = A\left(æ - æ_0^{(1)}\right), \tag{8.10}$$

where

$$A = \begin{bmatrix} A_{11} & 0 & 0 \\ 0 & A_{22} & 0 \\ 0 & 0 & A_{33} \end{bmatrix}; \tag{8.11}$$

A_{11} is the torsional rigidity of the rod and A_{22}, A_{33} are its flexural rigidities.

The components of the vector $æ_0^{(1)}$ are the twist $æ_{10}$ and curvatures $æ_{20}$, $æ_{30}$ of the rod axial line in the natural state in the attached axes with them simultaneously being the principal axes of the section. In the natural axes the vector æ is the Darboux vector equal to

$$\bar{\boldsymbol{\Omega}} = \Omega_1 \mathbf{e}_1 + \Omega_3 \mathbf{e}_3, \tag{8.12}$$

where Ω_1 is the twist of the rods axial line; Ω_3 is its curvature. Therefore for a circular cross section rod (for which all axes are principal) we have

$$æ_1 = \Omega_1; \quad æ_2 = \Omega_2; \quad æ_3 = \Omega_3 = \frac{1}{\rho}, \tag{8.13}$$

where ρ is the radius of the curvature of the rods axial line.

For the displacements vector ($\mathbf{u} = \mathbf{r} - \mathbf{r}_0$) (Fig. 8.5) we have the following equation (in the attached axes)

$$\frac{\partial \mathbf{u}}{\partial s} + \boldsymbol{æ} \times \mathbf{u} = \mathbf{e} - \mathbf{e}_{10}. \tag{8.14}$$

Considering the derivatives of the unit vectors $\mathbf{e}_i$ and $\mathbf{e}_{i0}$ with respect to the axial coordinate, we can obtain an equation relating components of the vectors $\boldsymbol{æ}$ and $\boldsymbol{æ}_0$ with the angles ϑ_j:

$$\boldsymbol{æ} = L_1 \frac{\partial \boldsymbol{\vartheta}}{\partial s} + L\boldsymbol{æ}_0^{(1)} \qquad \left(\boldsymbol{\vartheta} = (\vartheta_1, \vartheta_2, \vartheta_3)^{\mathrm{T}}\right), \tag{8.15}$$

where

$$L = \begin{bmatrix} \cos\vartheta_2\cos\vartheta_3 & \cos\vartheta_2\sin\vartheta_3\cos\vartheta_1 + \\ & +\sin\vartheta_2\sin\vartheta_1 & \cos\vartheta_2\sin\vartheta_3\sin\vartheta_1 - \\ & & -\sin\vartheta_2\cos\vartheta_1 \\ -\sin\vartheta_3 & \cos\vartheta_1\cos\vartheta_3 & \cos\vartheta_3\sin\vartheta_1 \\ \sin\vartheta_2\cos\vartheta_3 & \sin\vartheta_2\sin\vartheta_3\cos\vartheta_1 - & -\sin\vartheta_2\sin\vartheta_3\sin\vartheta_1 + \\ & -\cos\vartheta_2\sin\vartheta_1 & +\cos\vartheta_2\cos\vartheta_1 \end{bmatrix}; \tag{8.16}$$

$$L_1 = \begin{bmatrix} \cos\vartheta_2\cos\vartheta_3 & 0 & -\sin\vartheta_2 \\ -\sin\vartheta_3 & 1 & 0 \\ \sin\vartheta_2\cos\vartheta_3 & 0 & \cos\vartheta_2 \end{bmatrix}. \tag{8.17}$$

Considering the derivatives of unit attached basis vectors with respect to time, we can obtain the following equation

$$\boldsymbol{\omega} = L_1 \frac{\partial \boldsymbol{\vartheta}}{\partial t}. \tag{8.18}$$

For an element of a moving rod we can obtain the following vector equation relating the vectors $\mathbf{v}$ and $\boldsymbol{\omega}$:

$$\frac{\mathrm{d}\mathbf{v}}{\partial s} = \boldsymbol{\omega} \times \mathbf{e}_1. \tag{8.19}$$

Let us reduce equations (8.7)–(8.10), (8.15), (8.18), (8.19) to non-dimensional form, assuming that

$$s = l\eta;\quad \tau = p_0 t;\quad \boldsymbol{\omega} = \tilde{\boldsymbol{\omega}} p_0;\quad \mathbf{v} = \tilde{\mathbf{v}} l p_0;$$
$$\mathbf{M} = \tilde{\mathbf{M}} \frac{A_{33}(0)}{l};\quad \boldsymbol{\mu} = \tilde{\boldsymbol{\mu}} \frac{A_{33}(0)}{l^2};$$
$$\mathbf{Q} = \tilde{\mathbf{Q}} \frac{A_{33}(0)}{l^2};\quad \mathbf{q} = \tilde{\mathbf{q}} \frac{A_{33}(0)}{l^3};\quad p_0 = \left[\frac{A_{33}(0)}{m_1(0)\, l^4}\right]^{\frac{1}{2}};$$
$$\tilde{J}_{ii}(\eta) = \frac{J_i(\eta)}{F_0 l^2};\quad \tilde{A}_{ii}(\eta) = \frac{A_{ii}(\eta)}{A_{33}(0)};\quad \tilde{\text{æ}} = \text{æ} l,$$

where $m_1(0)$ is the mass of the unit length of the rod in the origin; $A_{33}(0)$ is the rigidity of the rod in the origin; $\tilde{\boldsymbol{\omega}}$, $\tilde{\mathbf{v}}$, etc. are the non-dimensional quantities. We may present the mass of the unit length of the rod in its arbitrary section in terms of $m_1(0)$ as

$$m_1(\eta) = m_1(0)\, n_1(\eta) = \rho F_0 n_1(\eta),$$

where $n_1(\eta)$ is the non-dimensional function; F_0 is the area of the rod's section in the origin.

We obtain the following system of the differential nonlinear equations of motion of the rod in non-dimensional form (dropping the tilde sign in the notation of local derivatives and non-dimensional quantities):

$$n_1(\eta)\left(\frac{\partial \mathbf{v}}{\partial \tau} + \boldsymbol{\omega} \times \mathbf{v}\right) - \frac{\partial \mathbf{Q}}{\partial \eta} - \text{æ} \times \mathbf{Q} - \mathbf{P} = 0; \tag{8.20}$$
$$J(\eta)\frac{\partial \boldsymbol{\omega}}{\partial \tau} + \boldsymbol{\omega} \times J(\eta)\,\boldsymbol{\omega} - \frac{\partial \mathbf{M}}{\partial \eta} - \text{æ} \times \mathbf{M} - \mathbf{e}_1 \times \mathbf{Q} - \mathbf{T} = 0 \tag{8.21}$$
$$\mathbf{M} = A\left(\text{æ} - \text{æ}_0^{(1)}\right); \tag{8.22}$$
$$L_1 \frac{\partial \boldsymbol{\vartheta}}{\partial \eta} + L\text{æ}_0^{(1)} - \text{æ} = 0; \tag{8.23}$$
$$\frac{\partial \boldsymbol{\vartheta}}{\partial \eta} + \text{æ} \times \mathbf{v} - \boldsymbol{\omega} \times \mathbf{e}_1 = 0; \tag{8.24}$$
$$\frac{\partial \mathbf{u}}{\partial \eta} + \text{æ} \times \mathbf{u} - (1 - l_{11})\,\mathbf{e}_1 - l_{21}\mathbf{e}_2 - l_{31}\mathbf{e}_3 = 0 \tag{8.25}$$
$$L_1 \frac{\partial \boldsymbol{\vartheta}}{\partial \tau} - \boldsymbol{\omega} = 0, \tag{8.26}$$

where ℓ_{ij} are the elements of the matrix L.

From equations (8.20)-(8.26) we can obtain as a special case the following nonlinear equations of the rods equilibrium:

$$\frac{d\mathbf{Q}}{d\eta} + æ \times \mathbf{Q} + \mathbf{P} = 0;$$
$$\frac{d\mathbf{M}}{d\eta} + æ \times \mathbf{M} + \mathbf{e} \times \mathbf{Q} + \mathbf{T} = 0;$$
$$\mathbf{M} = A\left(æ - æ_0^{(1)}\right); \tag{8.27}$$
$$L_1 \frac{d\boldsymbol{\vartheta}}{d\eta} + Læ_0^{(1)} - æ = 0;$$
$$\frac{d\mathbf{u}}{d\eta} + æ \times \mathbf{u} - \mathbf{e}_1 + \mathbf{e}_{10} = 0.$$

Let us consider expression (8.22) for the moment $\mathbf{M}$ in more detail. In the presented form $æ_0^{(1)}$ is a vector whose components are curvatures describing the natural state of the rod. If the vector $æ_0^{(1)}$ characterizes the static state of the rod, then

$$\mathbf{M} = A\left(æ - æ_0^{(1)}\right) + \mathbf{M}_0, \tag{8.28}$$

where $\mathbf{M}_0$ is the static moment.

8.2 Equations of the Motion of a Rod in the Attached Coordinate System

8.2.1 Equation of Space Motion of a Rod

In tensor form system of equations (8.20)–(8.26) takes the form

$$n_1(\eta)\left(\frac{\partial v_k}{\partial \tau} + \varepsilon_{kij}\omega_i v_i\right) - \frac{\partial Q_k}{\partial \eta} - \varepsilon_{kij}æ_i Q_j - P_k = 0;$$
$$J_{k\gamma}^0 \frac{\partial \omega_\gamma}{\partial \tau} + \varepsilon_{kij}\omega_i J_{j\gamma}^0 \omega_\gamma - \frac{\partial M_k}{\partial \eta} - \varepsilon_{kij}æ_i M_j + \varepsilon_{kij} Q_i - T_k = 0;$$
$$l_{(1)kj}\frac{\partial \vartheta_j}{\partial \eta} - æ_k + l_{kj}æ_{0j} = 0;$$
$$\frac{\partial u_k}{\partial \eta} + \varepsilon_{kij}æ_i u_j - \delta_{1k} + l_{k1} = 0; \tag{8.29}$$
$$\frac{\partial v_k}{\partial \eta} + \varepsilon_{kij}æ_i u_j - \varepsilon_{ki1}\omega_i = 0;$$
$$l_{(1)kj}\frac{\partial \vartheta_j}{\partial \tau} - \omega_k = 0.$$

Written in an expanded form in projections ontj the attached axes the system is:

$$n_1(\eta)\left(\frac{\partial v_1}{\partial t}+v_3\omega_2-v_2\omega_3\right)-\frac{\partial Q_1}{\partial \eta}-Q_3æ_2+Q_2æ_3-P_1=0;$$
$$n_1(\eta)\left(\frac{\partial v_2}{\partial t}+v_1\omega_3-v_3\omega_1\right)-\frac{\partial Q_2}{\partial \eta}-Q_1æ_3+Q_3æ_1-P_2=0; \quad (8.30)$$
$$n_1(\eta)\left(\frac{\partial v_3}{\partial t}+v_2\omega_1-v_1\omega_2\right)-\frac{\partial Q_3}{\partial \eta}-Q_2æ_1+Q_1æ_2-P_3=0,$$

$$J_{11}\frac{\partial \omega_1}{\partial \tau}+(J_{33}-J_{22})\,\omega_2\omega_3-\frac{\partial M_1}{\partial \eta}+æ_3M_2-æ_2M_3-T_1=0;$$
$$J_{22}\frac{\partial \omega_2}{\partial \tau}+(J_{11}-J_{33})\,\omega_1\omega_3-\frac{\partial M_2}{\partial \eta}+æ_1M_3-æ_3M_1+Q_3-T_2=0; \quad (8.31)$$
$$J_{33}\frac{\partial \omega_3}{\partial \tau}+(J_{22}-J_{11})\,\omega_1\omega_2-\frac{\partial M_3}{\partial \eta}+æ_2M_1-æ_1M_2-Q_2-T_3=0,$$

$$M_1=A_{11}(æ_1-æ_{10})\,;\quad M_2=A_{22}(æ_2-æ_{20})\,;\quad M_3=A_{31}(æ_3-æ_{30})\,, \quad (8.32)$$

$$l_{(1)11}\frac{\partial \vartheta_1}{\partial \eta}+l_{(1)13}\frac{\partial \vartheta_3}{\partial \eta}-æ_1-l_{11}æ_{10}+l_{12}æ_{20}+l_{13}æ_{30}=0;$$
$$l_{(1)21}\frac{\partial \vartheta_1}{\partial \eta}+\frac{\partial \vartheta_3}{\partial \eta}-æ_2+l_{21}æ_{10}+l_{22}æ_{20}-l_{23}æ_{30}=0; \quad (8.33)$$
$$l_{(1)31}\frac{\partial \vartheta_1}{\partial \eta}+l_{(1)33}\frac{\partial \vartheta_3}{\partial \eta}-æ_3+l_{31}æ_{10}+l_{32}æ_{20}+l_{33}æ_{30}=0,$$

$$\frac{\partial v_1}{\partial \eta}+v_3æ_2-v_2æ_3=0;$$
$$\frac{\partial v_2}{\partial \eta}+v_1æ_3-v_3æ_1=\omega_3; \quad (8.34)$$
$$\frac{\partial v_3}{\partial \eta}+v_2æ_1-v_1æ_2=-\omega_2.$$

$$\frac{\partial u_1}{\partial \eta}+æ_2u_3-æ_3u_2-1+l_{11}=0;$$
$$\frac{\partial u_2}{\partial \eta}+æ_3u_1-æ_1u_3+l_{21}=0; \quad (8.35)$$
$$\frac{\partial u_3}{\partial \eta}+æ_1u_2-æ_2u_1+l_{31}=0.$$

$$
\begin{aligned}
&l_{(1)11}\frac{\partial \vartheta_1}{\partial \tau}+l_{(1)13}\frac{\partial \vartheta_3}{\partial \tau}-\omega_1=0;\\
&l_{(1)21}\frac{\partial \vartheta_1}{\partial \tau}+\frac{\partial \vartheta_2}{\partial \tau}-\omega_2=0;\\
&l_{(1)31}\frac{\partial \vartheta_1}{\partial \tau}+j_{(1)33}\frac{\partial \vartheta_3}{\partial \tau}-\omega_3=0.
\end{aligned}
\qquad (8.36)
$$

8.2.2 Equation of Plane Motion of a Rod

Let us consider the special case of a rod in its natural state having the axial line that lies in a plane and one of the principal axes of its section being perpendicular to this plane. During the plane motion of the rod a number of the components of the vectors entering into equations (4.39)-(4.44), vanish:

$$
\begin{aligned}
&u_3=v_3=\omega_1=\omega_2=0; \qquad \vartheta_1=\vartheta_2=0;\\
&æ_1=æ_2=æ_{10}=æ_{20}=0;\\
&Q_3=M_1=M_3=0; q_3=\mu_1=\mu_3=0.
\end{aligned}
$$

From systems (4.30)-(4.36) we obtain the following equations:

$$
\begin{aligned}
&n_1(\eta)\left(\frac{\partial v_1}{\partial \tau}-v_2\omega_3\right)-\frac{\partial Q_1}{\partial \eta}+æ_3Q_2-P_1=0;\\
&n_1(\eta)\left(\frac{\partial v_2}{\partial \tau}+v_1\omega_3\right)-\frac{\partial Q_2}{\partial \eta}-æ_3Q_1-P_2=0.
\end{aligned}
\qquad (8.37)
$$

$$
J_{33}\frac{\partial \omega_3}{\partial \tau}-\frac{\partial M_3}{\partial \eta}-Q_2-T_3=0. \qquad (8.38)
$$

$$
M_3=A_{33}\left(æ_3-æ_{30}\right). \qquad (8.39)
$$

$$
\begin{aligned}
&\frac{\partial \vartheta_3}{\partial \eta}-æ_3=0;\\
&\frac{\partial v_1}{\partial \eta}-v_2æ_3=0;\\
&\frac{\partial v_2}{\partial \eta}+v_1æ_3=\omega_3;\\
&\left(\omega_3=\frac{\partial \vartheta_3}{\partial \tau}\right).
\end{aligned}
\qquad (8.40)
$$

From system (8.35) we obtain two equations for the determination of displacements in the attached frame

$$
\begin{aligned}
&\frac{\partial u_1}{\partial \eta}-æ_3u_2-1+\cos\vartheta_3=0;\\
&\frac{\partial u_2}{\partial \eta}+æ_3u_1+\sin\vartheta_3=0.
\end{aligned}
\qquad (8.41)
$$

8.2.3 Rods Having Lumped Masses

A rod having lumped masses, a point mass m_1 and inertial mass m_2 is shown in Fig. 8.6. during the vibrations of the rod the lumped masses are subjected to the action of forces of inertia $\mathbf{J}_u^{(i)}$ and a moment of inertia $\mathbf{M}_u^{(2)}$ that we can enter into equations of motion using delta functions the way we did it with concentrated forces. Changed into non-dimensional form the force of inertia $\mathbf{J}_u^{(i)}$ and the moment $\mathbf{M}_u^{(2)}$ are

$$\mathbf{J}_u^{(1)} = -m_1^0 \frac{\partial \mathbf{v}}{\partial \tau} \delta(\eta - \eta_1), \qquad \mathbf{J}_u^{(2)} = -m_2^0 \frac{\partial \mathbf{v}}{\partial \tau} \delta(\eta - \eta_2); \tag{8.42}$$

$$\mathbf{M}_u^{(2)} = -\frac{\partial}{\partial \tau}\left(J^{(2)}\boldsymbol{\omega}\right) \delta(\eta - \eta_2). \tag{8.43}$$

In the attached frame we have (the sign of local derivative is dropped)

$$\mathbf{J}_u^{(1)} = -m_1^0 \left(\frac{\partial \mathbf{v}}{\partial \tau} + \boldsymbol{\omega} \times \mathbf{v}\right) \delta(\eta - \eta_1); \tag{8.44}$$

$$\mathbf{J}_u^{(2)} = -m_2^0 \left(\frac{\partial \mathbf{v}}{\partial \tau} + \boldsymbol{\omega} \times \mathbf{v}\right) \delta(\eta - \eta_2); \tag{8.45}$$

$$\mathbf{M}_u^{(2)} = -\left(J^{(2)}\frac{\partial \boldsymbol{\omega}}{\partial \tau} + \boldsymbol{\omega} \times J^{(2)}\boldsymbol{\omega}\right) \delta(\eta - \eta_2). \tag{8.46}$$

In given expressions (8.42)-(8.46) m_i^0 are non-dimensional masses equal to

$$m_i^0 = \frac{m_i}{m_0 l},$$

where m_0 is the mass of the rods unit length; l is the length of the rod; $J^{(2)}$ is the matrix, whose elements are the non-dimensional moments of inertia of the mass m_2. If the principal axes of the sections of the rod coincide with

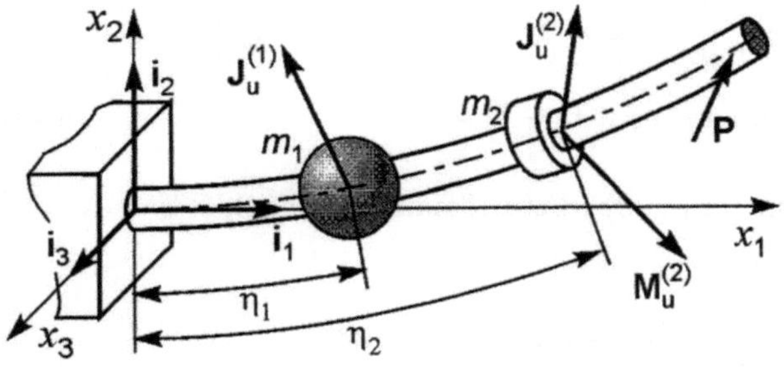

Fig. 8.6.

those of the lumped mass m_2, the matrix $J^{(2)}$ is a diagonal one (similar to that of (8.3)). If the principal axes of the sections of the rod do not coincide with those of the lumped mass (the movable axes are connected with the principal axes of the rod's sections) the matrix $J^{(2)}$ takes the form

$$J^{(2)} = \begin{bmatrix} J_{11}^{(2)} & J_{12}^{(2)} & J_{13}^{(2)} \\ J_{21}^{(2)} & J_{22}^{(2)} & J_{23}^{(2)} \\ J_{31}^{(2)} & J_{32}^{(2)} & J_{33}^{(2)} \end{bmatrix}.$$

The non-dimensional elements of the matrix $J^{(2)}$ are connected with its dimensional ones by the following relationships

$$J_{ij}^{(2)} = \frac{\left(J_{ij}^{(2)}\right)_1}{m_0 l^3},$$

where $J_{ij}^{(2)}$, $\left(J_{ij}^{(2)}\right)_1$ are respectively non-dimensional and dimensional quantities. The presented expressions for $\mathbf{J}_u^{(2)}$ and $\mathbf{M}_u^{(2)}$ are true provided that we can ignore the size of the mass m_2 along the coordinate η in comparison with the total length of the rod. On this assumption we can put a number of the elements of the matrix $J^{(2)}$ equal to zero, namely

$$J_{12}^{(2)} = J_{13}^{(2)} = J_{21}^{(2)} = J_{31}^{(2)} = 0.$$

Having included the concentrated forces of inertia $\mathbf{J}_u^{(2)}$ and $\mathbf{M}_u^{(2)}$ in equations (8.20) and (8.21), we obtain equations of motion of the rod that take the lumped masses into account:

$$\left[n_1(\eta) + m_1^0 \delta(\eta - \eta_1) + m_2^0 \delta(\eta - \eta_2)\right] \left(\frac{\partial \mathbf{v}}{\partial \tau} + \boldsymbol{\omega} \times \mathbf{v}\right) - \frac{\partial \mathbf{Q}}{\partial \eta} - \text{æ} \times \mathbf{Q} - \mathbf{P} = 0; \tag{8.47}$$

$$\left[J(\eta) + J^{(2)} \delta(\eta - \eta_2)\right] \frac{\partial \boldsymbol{\omega}}{\partial \tau} + \boldsymbol{\omega} \times \left[J(\eta) + J^{(2)} \delta(\eta - \eta_2)\right] \boldsymbol{\omega} - \frac{\partial \mathbf{M}}{\partial \eta} - \text{æ} \times \mathbf{M} - \mathbf{e}_1 \times \mathbf{Q} - \mathbf{T} = 0. \tag{8.48}$$

Other equations of system (8.20)-(8.26) remain unaltered.

8.3 Equation of Small Vibrations of Rods

Let us obtain equations of the small vibrations of a rod about an equilibrium state, assuming that additional internal forces, displacements and angles of rotation that occur at vibrations are small, which is possible at small external dynamic loads.

Let us assume that:

$$\begin{aligned}&\mathbf{Q} = \mathbf{Q}_0 + \Delta\mathbf{Q}; \quad \mathbf{M} = \mathbf{M}_0 + \Delta\mathbf{M}; \quad \text{æ} = \text{æ}_0 + \Delta\text{æ}; \quad \boldsymbol{\vartheta} = \Delta\boldsymbol{\vartheta};\\ &\mathbf{P} = \mathbf{P}_0 + \Delta\mathbf{P}; \quad \mathbf{T} = \mathbf{T}_0 + \Delta\mathbf{T}; \quad \mathbf{q} = \mathbf{q}_0 + \Delta\mathbf{q};\\ &\boldsymbol{\mu} = \boldsymbol{\mu}_0 + \Delta\boldsymbol{\mu}; \quad \mathbf{v} = \Delta\mathbf{v}; \quad \mathbf{u} = \Delta\mathbf{u}; \quad \boldsymbol{\omega} = \Delta\boldsymbol{\omega},\end{aligned} \tag{8.49}$$

where $\mathbf{Q}_0$, $\mathbf{M}_0$, æ_0, $\mathbf{q}_0$, $\boldsymbol{\mu}_0$ are static components and $\Delta\mathbf{Q}$, $\Delta\mathbf{M}$, $\Delta\text{æ}$, $\Delta\boldsymbol{q}$, $\Delta\boldsymbol{\mu}$, $\Delta\mathbf{v}$, $\Delta\mathbf{u}$, $\Delta\boldsymbol{\omega}$ are vectors whose components are considered to be small quantities, therefore we can ignore their products (vector and scalar ones). We consider small vibrations about an equilibrium state, therefore

$$\mathbf{M} = \mathbf{M}_0 + \Delta\mathbf{M}, \tag{8.50}$$

where $\Delta\mathbf{M} = A \cdot \Delta\text{æ}$; $\Delta\text{æ} = \text{æ} - \text{æ}_0^{(1)}$.

Since

$$\mathbf{v} = \frac{\partial \mathbf{u}}{\partial \tau} + \boldsymbol{\omega} \times \mathbf{u},$$

then at small vibrations we can put $\boldsymbol{\omega} \times \mathbf{u} \approx 0$. Therefore

$$\mathbf{v} = \frac{\partial \mathbf{u}}{\partial \tau} = \frac{\tilde{\partial} \mathbf{u}}{\partial \tau}.$$

At the small angles of rotation ϑ_i of the attached axes with respect to their position in statics the vector is

$$\boldsymbol{\omega} = \frac{\tilde{\partial} \boldsymbol{\vartheta}}{\partial \tau}. \tag{8.51}$$

Let us obtain equations of small vibrations of a rod in the attached axes using equations (8.20)-(8.26). Substituting expression (8.49) into these equations and retaining only the terms linearly dependent on small quantities, we obtain the following vector equations in the attached frame

$$\begin{aligned}&n_1(\eta)\frac{\partial^2 \mathbf{u}}{\partial \tau^2} - \frac{\partial \Delta\mathbf{Q}}{\partial \eta} - \Delta\text{æ} \times \mathbf{Q}_0 - \text{æ}_0 \times \Delta\mathbf{Q} = \mathbf{P};\\ &J\frac{\partial^2 \boldsymbol{\vartheta}}{\partial \tau^2} - \frac{\partial \Delta\mathbf{M}}{\partial \eta} - \Delta\text{æ} \times \mathbf{M}_0 - \text{æ}_0 \times \Delta\mathbf{M} - \mathbf{e}_1 \times \Delta\mathbf{Q} = \mathbf{T};\\ &\Delta\mathbf{M} = A\Delta\text{æ};\\ &\frac{\partial \boldsymbol{\vartheta}}{\partial \eta} + \text{æ}_0 \times \boldsymbol{\vartheta} - \Delta\text{æ} = 0;\\ &\frac{\partial \mathbf{u}}{\partial \eta} + \text{æ}_0 \times \mathbf{u} - \vartheta_3\mathbf{e}_2 + \vartheta_2\mathbf{e}_3 = 0,\end{aligned} \tag{8.52}$$

where $\mathbf{P} = \mathbf{q} + \mathbf{P}^{(i)}\delta(\eta - \eta_i)$; $\mathbf{T} = \boldsymbol{\mu} + \mathbf{T}^{(v)}\delta(\eta - \eta_v)$.

Let us consider vector products $\varDelta æ \times \mathbf{Q}_0$, $\varDelta æ \times \mathbf{M}_0$, that can be presented as

$$\varDelta æ \times \mathbf{Q}_0 = A_Q \varDelta æ; \qquad \varDelta æ \times \mathbf{M}_0 = A_M \varDelta æ, \tag{8.53}$$

where

$$A_Q = \begin{bmatrix} 0 & Q_{30} & -Q_{20} \\ -Q_{30} & 0 & Q_{10} \\ Q_{20} & -Q_{10} & 0 \end{bmatrix}; \qquad A_M = \begin{bmatrix} 0 & M_{30} & -M_{20} \\ -M_{30} & 0 & M_{10} \\ M_{20} & -M_{10} & 0 \end{bmatrix}. \tag{8.54}$$

Similarly we obtain expressions for the other vector products entering into system (8.52). By transformations (having eliminated $\varDelta æ$) we obtain the following system of equations of small vibrations of a three-dimensional curvilinear rod

$$\begin{aligned}
&n_1(\eta)\frac{\partial^2 \mathbf{u}}{\partial \tau^2} - \frac{\partial \Delta \mathbf{Q}}{\partial \eta} - A_Q A^{-1} \Delta \mathbf{M} + A_æ \Delta \mathbf{Q} = \mathbf{P};\\
&J\frac{\partial^2 \boldsymbol{\vartheta}}{\partial \tau^2} - \frac{\partial \Delta \mathbf{Q}}{\partial \eta} - A_M A^{-1} \Delta \mathbf{M} - A_æ \Delta \mathbf{M} - A_1 \Delta \mathbf{Q} = \mathbf{T};\\
&\frac{\partial \boldsymbol{\vartheta}}{\partial \eta} + A_æ \boldsymbol{\vartheta} - A^{-1}\Delta \mathbf{M} = 0;\\
&\frac{\partial \mathbf{u}}{\partial \eta} + A_æ \mathbf{u} + A_1 \boldsymbol{\vartheta} = 0,\\
&(\varDelta \mathbf{M} = A \varDelta æ)\,.
\end{aligned} \tag{8.55}$$

where

$$A_æ = \begin{bmatrix} 0 & -æ_{30} & æ_{20} \\ æ_{30} & 0 & -æ_{10} \\ -æ_{20} & æ_{10} & 0 \end{bmatrix}; \qquad A_1 = \begin{bmatrix} 0 & 0 & 0 \\ 0 & 0 & -1 \\ 0 & 1 & 0 \end{bmatrix}.$$

We can write system of equations (4.127) in the form of one vector equation (having eliminated $\varDelta æ$)

$$A^{(1)}\frac{\partial^2 \mathbf{Z}}{\partial \tau^2} + \frac{\partial \mathbf{Z}}{\partial \eta} + A^{(2)}\mathbf{Z} = \boldsymbol{\Phi}, \tag{8.56}$$

where

$$\mathbf{Z} = \begin{bmatrix} \Delta \mathbf{Q} \\ \Delta \mathbf{M} \\ \boldsymbol{\vartheta} \\ \mathbf{u} \end{bmatrix}; \quad \boldsymbol{\Phi} = \begin{bmatrix} -\mathbf{P} \\ -\mathbf{T} \\ 0 \\ 0 \end{bmatrix}; \quad A^{(1)} = \begin{bmatrix} 0 & 0 & 0 & -n_1 E \\ 0 & 0 & -J & 0 \\ 0 & 0 & 0 & 0 \\ 0 & 0 & 0 & 0 \end{bmatrix};$$

$$A^{(2)} = \begin{bmatrix} A_{æ} & A_Q A^{-1} & 0 & 0 \\ A_1 & A_M A^{-1} + A_{æ} & 0 & 0 \\ 0 & -A^{-1} & A_{æ} & 0 \\ 0 & 0 & A_1 & A_{æ} \end{bmatrix}.$$

If we put $\dfrac{\partial}{\partial \tau} = 0$, then from (8.56) we obtain the following linear equation of a rods equilibrium

$$\frac{\partial \mathbf{Z}}{\partial \eta} + A^{(2)}\mathbf{Z} = \mathbf{\Phi}_0, \tag{8.57}$$

where $\mathbf{\Phi}_0$ depends on the static load.

8.3.1 Equations of Small Vibrations in the Attached Coordinate Frame

From equations (8.55) we obtain equations of small forced random vibrations in the attached coordinate frame:

$$\begin{aligned} n_1\frac{\partial^2 u_1}{\partial \tau^2} - \frac{\partial \Delta Q_1}{\partial \eta} + Q_{20}\Delta æ_3 - Q_{30}\Delta æ_2 + æ_{30}\Delta Q_2 - æ_{20}\Delta Q_2 &= P_1; \\ n_1\frac{\partial^2 u_2}{\partial \tau^2} - \frac{\partial \Delta Q_2}{\partial \eta} + Q_{30}\Delta æ_1 - Q_{10}\Delta æ_3 + æ_{10}\Delta Q_3 - æ_{30}\Delta Q_1 &= P_2; \\ n_1\frac{\partial^2 u_3}{\partial \tau^2} - \frac{\partial \Delta Q_3}{\partial \eta} + Q_{10}\Delta æ_2 - Q_{20}\Delta æ_1 + æ_{20}\Delta Q_1 - æ_{10}\Delta Q_2 &= P_3, \end{aligned} \tag{8.58}$$

$$\begin{aligned} J_{11}\frac{\partial^2 \vartheta_1}{\partial \tau^2} - \frac{\partial \Delta M_1}{\partial \eta} - æ_{20}\Delta M_3 + æ_{30}\Delta M_2 - M_{30}\Delta æ_2 + M_{20}\Delta æ_3 &= T_1; \\ J_{22}\frac{\partial^2 \vartheta_2}{\partial \tau^2} - \frac{\partial \Delta M_2}{\partial \eta} - æ_{30}\Delta M_1 + æ_{10}\Delta M_3 - M_{10}\Delta æ_3 + M_{30}\Delta æ_1 + \Delta Q_3 &= T_2; \\ J_{33}\frac{\partial^2 \vartheta_3}{\partial \tau^2} - \frac{\partial \Delta M_3}{\partial \eta} - æ_{10}\Delta M_2 + æ_{20}\Delta M_1 - M_{20}\Delta æ_1 + M_{10}\Delta æ_2 - \Delta Q_2 &= T_3, \end{aligned} \tag{8.59}$$

$$\begin{aligned} \frac{\partial \vartheta_1}{\partial \eta} + æ_{20}\vartheta_3 - æ_{30}\vartheta_2 - \Delta æ_1 &= 0; \\ \frac{\partial \vartheta_2}{\partial \eta} + æ_{30}\vartheta_1 - æ_{10}\vartheta_3 - \Delta æ_2 &= 0; \\ \frac{\partial \vartheta_3}{\partial \eta} + æ_{10}\vartheta_2 - æ_{20}\vartheta_1 - \Delta æ_3 &= 0, \end{aligned} \tag{8.60}$$

$$\frac{\partial u_1}{\partial \eta} + æ_{20} u_3 - æ_{30} u_2 = 0;$$
$$\frac{\partial u_2}{\partial \eta} + æ_{30} u_1 - æ_{10} u_3 - \vartheta_3 = 0; \tag{8.61}$$
$$\frac{\partial u_3}{\partial \eta} + æ_{10} u_2 - æ_{20} u_1 + \vartheta_2 = 0,$$
$$\Delta M_1 = A_{11} \Delta æ_1, \quad \Delta M_2 = A_{22} \Delta æ_2, \quad \Delta M_3 = A_{33} \Delta æ_3. \tag{8.62}$$

8.3.2 Equations of Small Vibrations About a Natural State

Let us present the equations of small vibrations of rods about a natural state (non-loaded). In this special case we must put $A_Q = A_M = 0$ in equation (8.55). As a result we obtain:

1) Vector equations

$$n_1(\eta) \frac{\partial^2 \mathbf{u}}{\partial \tau^2} - \frac{\partial \Delta \mathbf{Q}}{\partial \eta} - A_æ \Delta \mathbf{Q} = \mathbf{P};$$
$$J \frac{\partial^2 \boldsymbol{\vartheta}}{\partial \tau^2} - \frac{\partial \Delta \mathbf{M}}{\partial \eta} - A_æ \Delta \mathbf{M} - A_1 \Delta \mathbf{Q} = \mathbf{T};$$
$$\frac{\partial \boldsymbol{\vartheta}}{\partial \eta} + A_æ \boldsymbol{\vartheta} - \Delta \boldsymbol{æ} = 0; \tag{8.63}$$
$$\frac{\partial \mathbf{u}}{\partial \eta} + A_æ \mathbf{u} + A_1 \boldsymbol{\vartheta} = 0;$$
$$\Delta \mathbf{M} = A \Delta \boldsymbol{æ}.$$

Let us obtain equations of the small random vibrations of the rods the axial line of which is a plane curve. A spiral spring whose axial line both in a natural state ($\mathbf{q} = 0$) and in a loaded state ($\mathbf{q} \neq 0$) is a plane curve is shown as an example in Fig. 8.1 *b*. If we deflect the spring from its equilibrium state, it will begin to vibrate. If it is deflected in the plane of the drawing, small vibrations will occur in this plane; if the deflection of the spring is relative to the plane, small spatial vibrations will take place. The corresponding equations can be obtained from system (8.58)-(8.62), if we put

$$Q_{30} = 0, \quad M_{10} = M_{20} = 0, \quad æ_{10} = æ_{20} = 0.$$

As a result, from the system of equations (8.58)-(8.62) we obtain the following equations of random forced vibrations in the attached frame:

$$n_1 \frac{\partial^2 u_1}{\partial \tau^2} - \frac{\partial \Delta Q_1}{\partial \eta} + Q_{20} \Delta æ_3 + æ_{30} \Delta Q_2 = P_1;$$
$$n_1 \frac{\partial^2 u_2}{\partial \tau^2} - \frac{\partial \Delta Q_2}{\partial \eta} - Q_{10} \Delta æ_3 - æ_{30} \Delta Q_1 = P_2; \tag{8.64}$$
$$n_1 \frac{\partial^2 u_3}{\partial \tau^2} - \frac{\partial \Delta Q_3}{\partial \eta} + Q_{10} \Delta æ_2 - Q_{20} \Delta æ_1 = P_3;$$

$$J_{11}\frac{\partial^2\vartheta_1}{\partial\tau^2} - \frac{\partial \Delta M_1}{\partial\eta} + æ_{30}\Delta M_2 - M_{30}\Delta æ_2 = T_1;$$
$$J_{22}\frac{\partial^2\vartheta_2}{\partial\tau^2} - \frac{\partial \Delta M_2}{\partial\eta} - æ_{30}\Delta M_1 + M_{30}\Delta æ_1 + \Delta Q_3 = T_2; \tag{8.65}$$
$$J_{33}\frac{\partial^2\vartheta_3}{\partial\tau^2} - \frac{\partial \Delta M_3}{\partial\eta} - \Delta Q_2 = T_3,$$

$$\frac{\partial\vartheta_1}{\partial\eta} - æ_{30}\vartheta_2 - \Delta æ_1 = 0;$$
$$\frac{\partial\vartheta_2}{\partial\eta} + æ_{30}\vartheta_1 - \Delta æ_2 = 0; \tag{8.66}$$
$$\frac{\partial\vartheta_3}{\partial\eta} - \Delta æ_3 = 0;$$

$$\frac{\partial u_1}{\partial\eta} - æ_{30}u_2 = 0;$$
$$\frac{\partial u_2}{\partial\eta} + æ_{30}u_1 - \vartheta_3 = 0; \tag{8.67}$$
$$\frac{\partial u_3}{\partial\eta} + \vartheta_2 = 0,$$
$$\Delta M_i = A_{ii}\Delta æ_i.$$

If a random load [for example, q_c (See Fig. 8.1 *b*)] acts in the plane of the drawing, i.e.

$$\mathbf{q}_c = q_{1c}\mathbf{e}_1 + q_{2c}\mathbf{e}_2 = q_{x_1}\mathbf{i}_1 + q_{x_2}\mathbf{i}_2,$$

system of equations (8.64)–(8.67) falls apart into two independent systems:

1) In the plane of the drawing:

$$n_1\frac{\partial^2 u_1}{\partial\tau^2} - \frac{\partial \Delta Q_1}{\partial\eta} + Q_{20}\Delta æ_3 + æ_{30}\Delta Q_2 = P_1;$$
$$n_1\frac{\partial^2 u_2}{\partial\tau^2} - \frac{\partial \Delta Q_2}{\partial\eta} - Q_{10}\Delta æ_3 - æ_{30}\Delta Q_1 = P_2;$$
$$J_{33}\frac{\partial^2\vartheta_3}{\partial\tau^2} - \frac{\partial \Delta M_3}{\partial\eta} - \Delta Q_2 = T_3;$$
$$\frac{\partial\vartheta_3}{\partial\eta} - \Delta æ_3 = 0; \tag{8.68}$$
$$\frac{\partial u_1}{\partial\eta} - æ_{30}u_2 = 0;$$
$$\frac{\partial u_2}{\partial\eta} + æ_{30}u_1 - \vartheta_3 = 0;$$
$$\Delta M_3 = A_{33}\Delta æ_3.$$

2) Relative to the plane of the drawing:

$$
\begin{aligned}
& n_1 \frac{\partial^2 u_3}{\partial \tau^2} - \frac{\partial \Delta Q_3}{\partial \eta} + Q_{10} \Delta æ_2 - Q_{20} \Delta æ_1 = 0; \\
& J_{11} \frac{\partial^2 \vartheta_1}{\partial \tau^2} - \frac{\partial \Delta M_1}{\partial \eta} + æ_{30} \Delta M_2 - M_{30} \Delta æ_2 = 0; \\
& J_{22} \frac{\partial^2 \vartheta_2}{\partial \tau^2} - \frac{\partial \Delta M_2}{\partial \eta} - æ_{30} \Delta M_1 + M_{30} \Delta æ_1 + \Delta Q_3 = 0; \\
& \frac{\partial \vartheta_1}{\partial \eta} - æ_{30} \vartheta_2 - \Delta æ_1 = 0; \\
& \frac{\partial \vartheta_2}{\partial \eta} + æ_{30} \vartheta_1 - \Delta æ_2 = 0; \\
& \frac{\partial u_3}{\partial \eta} + \vartheta_2 = 0; \\
& \Delta M_1 = A_{11} \Delta æ_1, \quad \Delta M_2 = A_{22} \Delta æ_2,
\end{aligned}
\tag{8.69}
$$

In the case of free random vibrations of a non-loaded rod (at $Q_{10} = Q_{20} = M_{30} = 0$) from equations (8.68) and (8.69) we obtain the following equations:

a) In the plane of the drawing:

$$
\begin{aligned}
& n_1 \frac{\partial^2 u_1}{\partial \tau^2} - \frac{\partial \Delta Q_1}{\partial \eta} + æ_{30} \Delta Q_2 = P_1; \\
& n_1 \frac{\partial^2 u_2}{\partial \tau^2} - \frac{\partial \Delta Q_2}{\partial \eta} - æ_{30} \Delta Q_1 = P_2; \\
& J_{33} \frac{\partial^2 \vartheta_3}{\partial \tau^2} - \frac{\partial \Delta M_3}{\partial \eta} - \Delta Q_2 = 0; \\
& \frac{\partial \vartheta_3}{\partial \eta} - \Delta æ_3 = 0; \\
& \frac{\partial u_1}{\partial \eta} - æ_{30} u_2 = 0; \quad \frac{\partial u_2}{\partial \eta} + æ_{30} u_1 - \vartheta_3 = 0; \\
& \Delta M_3 = A_{33} \Delta æ_3.
\end{aligned}
\tag{8.70}
$$

b) Relative to the plane of the drawing:

$$
\begin{aligned}
&n_1\frac{\partial^2 u_3}{\partial\tau^2}-\frac{\partial\Delta Q_3}{\partial\eta}=0;\\
&J_{11}\frac{\partial^2\vartheta_1}{\partial\tau^2}-\frac{\partial\Delta M_1}{\partial\eta}-æ_{30}\Delta M_2=0;\\
&J_{22}\frac{\partial^2\vartheta_2}{\partial\tau^2}-\frac{\partial\Delta M_2}{\partial\eta}-æ_{30}\Delta M_1+\Delta Q_3=0;\\
&\frac{\partial\vartheta_1}{\partial\eta}-æ_{30}\vartheta_2-\Deltaæ_1=0;\\
&\frac{\partial\vartheta_2}{\partial\eta}+æ_{30}\vartheta_1-\Deltaæ_2=0;\\
&\frac{\partial u_3}{\partial\eta}+\vartheta_2=0;\\
&\Delta M_1=A_{11}\Deltaæ_1,\quad \Delta M_2=A_{22}\Deltaæ_2.
\end{aligned}
\tag{8.71}
$$

8.4 Determination of Eigenvalues and Eigenvectors

For the approximate numerical solution of equations of the free and forced random vibrations of rods we must know eigenvectors characterizing the small free vibrations of rods under specific boundary conditions.

Let us consider the free vibrations of a rod (without considering resistance forces) relative to a static state of stress and strain, using equation (8.56) (having put $P=T=0$):

$$A^{(1)}\frac{\partial^2\mathbf{Z}}{\partial\tau^2}+\frac{\partial\mathbf{Z}}{\partial\eta}+A^{(2)}\mathbf{Z}=0. \tag{8.72}$$

We find the solution of equation (8.72) as

$$\mathbf{Z}=\mathbf{Z}_0(\eta)\,\mathrm{e}^{i\lambda\tau}. \tag{8.73}$$

Substituting (8.73) into equation (8.72), we obtain

$$\frac{\mathrm{d}\mathbf{Z}_0}{\mathrm{d}\eta}+B(\eta,\lambda)\,\mathbf{Z}_0=0, \tag{8.74}$$

where

$$B(\eta,\lambda)=\begin{bmatrix} A_æ & A_QA^{-1} & 0 & \lambda^2 n_1E\\ A_1 & A_MA^{-1}+A_æ & J\lambda^2 & 0\\ 0 & -A^{-1} & A_æ & 0\\ 0 & 0 & A_1 & A_æ \end{bmatrix}$$

In that specific case of considering vibrations relative to a rods natural state ($\mathbf{Q}_0=\mathbf{M}_0=0$), the matrix $B(\eta,\lambda)$ takes the form

$$B(\eta, \lambda) = \begin{bmatrix} A_{æ} & 0 & 0 & \lambda^2 n_1 E \\ A_1 & A_{æ} & \lambda^2 J & 0 \\ 0 & -A^{-1} & A_{æ} & 0 \\ 0 & 0 & A_1 & A_{æ} \end{bmatrix}.$$

Equation (8.74) is solved numerically, for example, by the method of initial parameters with a subsequent refinement of the fundamental matrix of solutions. As a result, we obtain

$$\mathbf{Z}_0 = K(\eta, \lambda)\,\mathbf{C}, \quad (K(0, \lambda) = E), \tag{8.75}$$

where $K(\eta, \lambda)$ is the fundamental matrix of the solutions of homogeneous equations (8.74).

Solution (8.75) should satisfy the boundary conditions. For a space-curved rod we have twelve boundary conditions (in groups of six conditions at $\eta = 0$ and $\eta = 1$). For example, for a rod fixing, shown in Fig. 8.6, we have the following boundary conditions: 1) $\eta = 0$; $\mathbf{u} = \boldsymbol{\vartheta}_0 = 0$; 2) $\eta = 1$; $\Delta\mathbf{Q}_0 = \Delta\mathbf{M}_0 = 0$. In order to fulfill the boundary conditions at $\eta = 0$, we must put $c_7 = c_8 = \cdots = c_{12} = 0$, where c_i are the components of the vector $\mathbf{C}$. From the boundary conditions at $\eta = 1$ we obtain the following system of homogeneous equations

$$\begin{aligned} &k_{11}c_1 + k_{12}c_2 + \cdots + k_{16}c_6 = 0; \\ &\cdots\cdots\cdots\cdots\cdots\cdots\cdots\cdots \\ &\cdots\cdots\cdots\cdots\cdots\cdots\cdots\cdots \\ &k_{61}c_1 + k_{62}c_2 + \cdots + k_{66}c_6 = 0. \end{aligned} \tag{8.76}$$

The values of λ_i, at which the determinant of system (8.76) is equal to zero, are non-dimensional frequencies.

Having determined λ_j, we find $c_1^{(j)}$, $c_2^{(j)}$, $c_3^{(j)}$, $c_4^{(j)}$ and $c_5^{(j)}$ from system (8.76) as a function of $c_6^{(j)}$ (we may put $c_6^{(j)}$ equal to zero):

$$c_k^{(j)} = \alpha_k^{(j)} c_6^{(j)} \quad (k = 1, 2, 3, 4, 5). \tag{8.77}$$

We solve equation (8.74) for each λ_j

$$\frac{\mathrm{d}\mathbf{Z}_0^{(j)}}{\mathrm{d}\eta} + B(\eta, \lambda_j)\,\mathbf{Z}_0^{(j)} = 0 \tag{8.78}$$

and find

$$\mathbf{Z}_0^{(j)} = K(\eta, \lambda_j)\,\mathbf{C}^{(j)}, \tag{8.79}$$

where $\mathbf{C}^{(j)} = \left(\alpha_1^{(j)}, \alpha_2^{(j)}, \alpha_3^{(j)}, \alpha_4^{(j)}, \alpha_5^{(j)}, 1, 0, 0 \ldots 0\right)^{\mathrm{T}}$.

When solving equations of the random vibrations of rods by approximate methods it is convenient to present the vectors $\mathbf{Z}_0^{(j)}$ as

$$\mathbf{Z}_0^{(j)} = \left(\boldsymbol{\psi}^{(j)}, \boldsymbol{\varphi}^{(j)}\right)^{\mathrm{T}}, \tag{8.80}$$

where $\boldsymbol{\psi}^{(j)} = \left(\Delta\mathbf{Q}_0^{(j)}, \Delta\mathbf{M}_0^{(j)}\right)$; $\boldsymbol{\varphi}^{(j)} = \left(\boldsymbol{\vartheta}_0^{(j)}, \mathbf{u}_0^{(j)}\right)^{\mathrm{T}}$.

The vectors $\boldsymbol{\psi}^{(j)}$ and $\boldsymbol{\varphi}^{(j)}$ characterize respectively the state of stress and the state of strain of a rod at vibrations with the frequency λ_j.

8.5 Non-Stationary Random Vibrations of Rods

The vectors

$$\mathbf{P} = \mathbf{q}_c + \mathbf{P}_c\delta\left(\eta - \eta_P\right);$$
$$\mathbf{T} = \boldsymbol{\mu}_c + \mathbf{T}_c\delta\left(\eta - \eta_M\right).$$

entering in the right-hand sides of the first two equations of system (8.55) are random non-stationary vectors with the known probability characteristics of the components. Let us enter the force of viscous friction $\alpha\frac{\partial\mathbf{u}}{\partial\tau}$ into the first equation of system (8.55) :

$$n_1\left(\eta\right)\frac{\partial^2\mathbf{u}}{\partial\tau^2} + \alpha\frac{\partial\mathbf{u}}{\partial\tau} - \frac{\partial\Delta\mathbf{Q}}{\partial\eta} - A_Q\Delta\boldsymbol{æ} - A_{æ}\Delta\mathbf{Q} = \mathbf{P}. \tag{8.81}$$

For the numerical solution of system of equations (8.55) it is more convenient to present it with due account of the forces of viscous friction in the form of one equation similar to equation (8.56)

$$\mathbf{L} = A^{(1)}\frac{\partial^2\mathbf{Z}}{\partial\tau^2} + A^{(3)}\frac{\partial\mathbf{Z}}{\partial\tau} + \frac{\partial\mathbf{Z}}{\partial\eta} + A^{(2)}\mathbf{Z} - \boldsymbol{\Phi} = 0, \tag{8.82}$$

where

$$A^{(3)} = \begin{bmatrix} 0 & 0 & 0 & \alpha E \\ 0 & 0 & 0 & 0 \\ 0 & 0 & 0 & 0 \\ 0 & 0 & 0 & 0 \end{bmatrix}.$$

For the solution of equation (8.82) let us take advantage of generalized virtual work principle [26], restricting ourselves to the following two-term approximation

$$\mathbf{Z} = \mathbf{Z}_0^{(1)}\left(\eta\right)f_1\left(\tau\right) + \mathbf{Z}_0^{(2)}\left(\eta\right)f_2\left(\tau\right). \tag{8.83}$$

We can take vectors proportional to the eigenvectors $\mathbf{Z}_0^{(j)}$ as generalized displacements i.e.

$$\delta \mathbf{Z}_0 = \delta A_1\, E_0 \mathbf{Z}_0^{(1)} + \delta A_2\, E_0 \mathbf{Z}_0^{(2)}, \tag{8.84}$$

where δA_j are independent arbitrary small quantities;

$$E_0 = \begin{bmatrix} 0 & 0 & 0 & E \\ 0 & 0 & E & 0 \\ 0 & E & 0 & 0 \\ E & 0 & 0 & 0 \end{bmatrix}.$$

By generalized virtual "displacements" we mean not only the variations of linear $\delta \mathbf{u}$ and angular $\delta \boldsymbol{\vartheta}$ displacements, but also the variations of internal forces $\delta \Delta \mathbf{Q}$ and moments $\delta \Delta \mathbf{M}$, i.e. the variations of all components of vectors $\mathbf{Z}_0^{(j)}$.

The matrix E_0 is introduced in order that all scalar products $\left(\mathbf{Z} E_0 \mathbf{Z}_0^{(j)}\right)$ have the dimensionality of work (if we consider dimensional equations (8.56)). Since

$$E_0 \mathbf{Z}_0^{(j)} = \left(\mathbf{u}_0^{(j)}, \boldsymbol{\vartheta}_0^{(j)}, \Delta \mathbf{M}_0^{(j)}, \Delta \mathbf{Q}_0^{(j)}\right)^{\mathrm{T}},$$

the scalar product $\left(\mathbf{Z}^{(i)} \cdot E_0 \mathbf{Z}_0^{(j)}\right)$ in a more comprehensive form is

$$\begin{aligned}\left(\mathbf{Z}^{(i)} \cdot E_0 \mathbf{Z}_0^{(j)}\right) &= \left(\Delta \mathbf{Q}_0^{(j)} \cdot \mathbf{u}_0^{(j)}\right) + \left(\Delta \mathbf{M}_0^{(j)} \cdot \boldsymbol{\vartheta}_0^{(j)}\right) \\ &\quad + \left(\boldsymbol{\vartheta}_0^{(j)} \cdot \Delta \mathbf{M}_0^{(j)}\right) + \left(\mathbf{u}_0^{(j)} \cdot \boldsymbol{\vartheta}_0^{(j)}\right).\end{aligned}$$

Having substituted (8.83) into equation (8.82), we obtain

$$\mathbf{L}\left(\mathbf{Z}_0^{(1)}, \mathbf{Z}_0^{(2)}, f_1, f_2\right) = \tilde{\boldsymbol{\delta}} \quad \left(\tilde{\boldsymbol{\delta}} \neq 0\right). \tag{8.85}$$

Let us require that the integral of the scalar product of vectors $\tilde{\boldsymbol{\delta}}$ and $\delta \mathbf{Z}_0$ be equal to zero:

$$\int_0^1 \left(\tilde{\boldsymbol{\delta}} \cdot \delta \mathbf{Z}_0\right) \mathrm{d}\eta = 0. \tag{8.86}$$

In view of independence of δA_j we obtain two equations from (8.86)

$$\int_0^1 \left(\tilde{\boldsymbol{\delta}} \cdot E_0 \mathbf{Z}_0^{(1)}\right) \mathrm{d}\eta = 0, \quad \int_0^1 \left(\tilde{\boldsymbol{\delta}} \cdot E_0 \mathbf{Z}_0^{(2)}\right) \mathrm{d}\eta = 0. \tag{8.87}$$

By transformations we obtain from (8.87) the equations for the determination of $f_j(\tau)$:

$$
\begin{aligned}
&a_{11}\ddot{f}_1 + a_{12}\ddot{f}_{21} + b_{11}\dot{f}_1 + b_{12}\dot{f}_2 + c_{11}f_1 + c_{12}f_2 = b_1;\\
&a_{21}\ddot{f}_1 + a_{22}\ddot{f}_{21} + b_{21}\dot{f}_1 + b_{22}\dot{f}_2 + c_{21}f_1 + c_{22}f_2 = b_2,
\end{aligned} \tag{8.88}
$$

where

$$
\begin{aligned}
a_{ij} &= \int_0^1 \left(A^{(1)}\mathbf{Z}_0^{(j)} \cdot E_0\mathbf{Z}_0^{(i)} \right) \mathrm{d}\eta;\\
b_{ij} &= \int_0^1 \left(A^{(3)}\mathbf{Z}_0^{(j)} \cdot E_0\mathbf{Z}_0^{(i)} \right) \mathrm{d}\eta;\\
c_{ij} &= \int_0^1 \left(\mathbf{Z}_0^{(j)} + A^{(2)}\mathbf{Z}_0^{(j)} \right) \cdot E_0\mathbf{Z}_0^{(i)}\, \mathrm{d}\eta,\\
b_1 &= \int_0^1 \left(\mathbf{\Phi} \cdot E_0\mathbf{Z}_0^{(1)} \right) \mathrm{d}\eta, \quad b_2 = \int_0^1 \left(\mathbf{\Phi} \cdot E_0\mathbf{Z}_0^{(2)} \right) \mathrm{d}\eta.
\end{aligned}
$$

In a more comprehensive form

$$
\begin{aligned}
b_1 &= -\int_0^1 \left(\mathbf{P} \cdot \mathbf{u}_0^{(1)} \right) \mathrm{d}\eta - \int_0^1 \left(\mathbf{T} \cdot \boldsymbol{\vartheta}_0^{(1)} \right) \mathrm{d}\eta;\\
b_2 &= -\int_0^1 \left(\mathbf{P} \cdot \mathbf{u}_0^{(2)} \right) \mathrm{d}\eta - \int_0^1 \left(\mathbf{T} \cdot \boldsymbol{\vartheta}_0^{(2)} \right) \mathrm{d}\eta.
\end{aligned}
$$

With due account of expressions for **P** and **T** we obtain

$$
\begin{aligned}
b_1 &= -\int_0^1 \left(\mathbf{q}_c \cdot \mathbf{u}_0^{(1)} \right) \mathrm{d}\eta - \int_0^1 \left(\boldsymbol{\mu}_c \cdot \boldsymbol{\vartheta}_0^{(1)} \right) \mathrm{d}\eta - \mathbf{P}_c\mathbf{u}_0^{(1)}\left(\eta_P\right) - \mathbf{T}_c \cdot \boldsymbol{\vartheta}_0^{(1)}\left(\eta_M\right);\\
b_2 &= -\int_0^1 \left(\mathbf{q}_c \cdot \mathbf{u}_0^{(2)} \right) \mathrm{d}\eta - \int_0^1 \left(\boldsymbol{\mu}_c \cdot \boldsymbol{\vartheta}_0^{(2)} \right) \mathrm{d}\eta - \mathbf{P}_c \cdot \mathbf{u}_0^{(2)}\left(\eta_P\right) - \mathbf{T}_c \cdot \boldsymbol{\vartheta}_0^{(2)}\left(\eta_M\right).
\end{aligned} \tag{8.89}
$$

Let us consider random loads of the form

$$
\begin{aligned}
&\mathbf{q}_c = \mathbf{q}_{0c}\left(\eta\right) f_q\left(\tau\right); \quad \mathbf{P}_c = \mathbf{P}_{0c}\left(\eta\right) f_P\left(\tau\right);\\
&\boldsymbol{\mu}_c = \boldsymbol{\mu}_{0c}\left(\eta\right) f_\mu\left(\tau\right); \quad \mathbf{T}_c = \mathbf{T}_{0c}\left(\eta\right) f_M\left(\tau\right).
\end{aligned} \tag{8.90}
$$

Let us restrict ourselves to the case of $\mathbf{q}_{0c}$, $\mathbf{P}_{0c}$, $\boldsymbol{\mu}_{0c}$ and $\mathbf{T}_{0c}$ being known, and f_q, f_P, f_μ and f_M being independent random functions of time with known probability characteristics.

From (8.89) after transformations with account of (8.90) we obtain

$$\begin{aligned} b_1 &= -d_{11}f_q - d_{12}f_\mu - d_{13}f_P - d_{14}f_M; \\ b_2 &= -d_{21}f_q - d_{22}f_\mu - d_{23}f_P - d_{24}f_M; \end{aligned} \tag{8.91}$$

where

$$d_{i1} = \int_0^1 \left(\mathbf{q}_{0c}\mathbf{u}_0^{(i)}\right) \mathrm{d}\eta; \quad d_{i2} = \int_0^1 \left(\boldsymbol{\mu}_{0c}\boldsymbol{\vartheta}_0^{(i)}\right) \mathrm{d}\eta;$$

$$d_{i3} = \left(\mathbf{P}_{0c}\mathbf{u}_0^{(i)}(\eta_P)\right); \quad d_{i4} = \left(\mathbf{T}_{0c}\boldsymbol{\vartheta}_0^{(i)}(\eta_M)\right).$$

Passing to the vector form, from (8.88) we obtain an equation

$$A\ddot{\mathbf{f}} + B\dot{\mathbf{f}} + C\mathbf{f} = D\boldsymbol{\Phi}, \tag{8.92}$$

where

$$D = \begin{bmatrix} d_{11} & d_{12} & d_{13} & d_{14} \\ d_{21} & d_{22} & d_{23} & d_{24} \end{bmatrix}; \quad \boldsymbol{\Phi} = \left(f_q, f_P, f_\mu, f_M\right)^{\mathrm{T}}.$$

The solution of equation (8.92) with the subsequent determination of the probability characteristics of the components of the vectors $\mathbf{f}$ is given in Chapt. 3. The solution of equation (8.92) at zero initial conditions is

$$\mathbf{f} = \int_0^t G(t,\tau)\, A^{-1} D\boldsymbol{\Phi}\, \mathrm{d}\tau \tag{8.93}$$

or in scalar form

$$f_j = \int_0^t \left(k_{j1}(t,\tau) f_q + k_{j2}(t,\tau) f_P + k_{j3}(t,\tau) f_\mu + k_{j4}(t,\tau) f_M\right) \mathrm{d}\tau$$

$$(j = 1, 2),$$

where k_{ji} are the elements of the matrix $K = GA^{-1}D$.

It is possible to obtain the matrix $G(t,\tau)$ by solving equation (8.92) by the arbitrary constants variation method.

As a result, we obtain an approximate solution to equation (8.82):

$$\begin{aligned} \mathbf{Z} = &\int_0^t \left(\mathbf{Z}_0^{(1)} k_{11} + \mathbf{Z}_0^{(2)} k_{21}\right) f_q \mathrm{d}\tau + \int_0^t \left(\mathbf{Z}_0^{(1)} k_{12} + \mathbf{Z}_0^{(2)} k_{22}\right) f_P \mathrm{d}\tau \\ &+ \int_0^t \left(\mathbf{Z}_0^{(1)} k_{13} + \mathbf{Z}_0^{(2)} k_{23}\right) f_\mu \mathrm{d}\tau + \int_0^t \left(\mathbf{Z}_0^{(1)} k_{14} + \mathbf{Z}_0^{(2)} k_{24}\right) f_M \mathrm{d}\tau. \end{aligned} \tag{8.94}$$

The components of the vector $\mathbf{Z}$ are:

$$z_j = \int_0^t \left(z_{0j}^{(1)} k_{11} + z_{0j}^{(2)} k_{21}\right) f_q \mathrm{d}\tau + \int_0^t \left(z_{0j}^{(1)} k_{12} + z_{0j}^{(2)} k_{22}\right) f_P \mathrm{d}\tau$$
$$+ \int_0^t \left(z_{0j}^{(1)} k_{13} + z_{0j}^{(2)} k_{23}\right) f_\mu \mathrm{d}\tau + \int_0^t \left(z_{0j}^{(1)} k_{14} + z_{0j}^{(2)} k_{24}\right) f_M \mathrm{d}\tau. \qquad (8.95)$$

The mathematical expectation of the components z_j of the state vector $\mathbf{Z}$ and their variances for independent excitations are

$$m_{z_j} = \int_0^t \left(z_{0j}^{(1)} k_{11} + z_{0j}^{(2)} k_{21}\right) m_{f_q} \mathrm{d}\tau$$
$$+ \cdots + \int_0^t \left(z_{0j}^{(1)} k_{14} + z_{0j}^{(2)} k_{24}\right) m_{f_M} \mathrm{d}\tau;$$
$$D_{z_j} = \int_0^t \int_0^t \left[z_{0j}^{(1)} k_{11}(t,\tau) + z_{0j}^{(2)} k_{21}(t,\tau)\right] \cdot \left[z_{0j}^{(1)} k_{11}(t,\tau')\right.$$
$$\left. + z_{0j}^{(2)} k_{21}(t,\tau')\right] K_{f_q}(\tau,\tau')\, \mathrm{d}\tau\, \mathrm{d}\tau'$$
$$+ \cdots + \int_0^t \int_0^t \left[z_{0j}^{(1)} k_{14}(t,\tau) + z_{0j}^{(2)} k_{24}(t,\tau)\right] \cdot \left[z_{0j}^{(1)} k_{14}(t,\tau')\right.$$
$$\left. + z_{0j}^{(2)} k_{24}(t,\tau')\right] K_{f_M}(\tau,\tau')\, \mathrm{d}\tau\, \mathrm{d}\tau'. \qquad (8.96)$$

Assuming that the components z_j have normal distributions, we can determine their maximum values for an arbitrary time instant τ and a coordinate η

$$\max z_j = m_{z_j}(\eta,\tau) + 3\sigma_{z_j}(\eta,\tau). \qquad (8.97)$$

8.6 Stationary Random Vibrations of Rods

Let us consider equation (8.82) of small forced vibrations with due account of viscous resistance forces, confining ourselves to the case of a concentrated stationary force $\mathbf{P}_c$ and a concentrated moment $\mathbf{T}_c$ ($\mathbf{q}_c = \boldsymbol{\mu}_c = 0$) acting on the rod. We find the solution of equation (8.82) in the form of two-term approximation (8.83). Having used the generalized virtual work principle, by manipulations we obtain an equation similar to that of (8.92)

$$A\ddot{\mathbf{f}} + B\dot{\mathbf{f}} + C\mathbf{f} = D^{(1)}\mathbf{P}_c + D^{(2)}\mathbf{T}_c, \tag{8.98}$$

where

$$D^{(1)}_{\eta_p} = \begin{bmatrix} u^{(1)}_{01} & u^{(1)}_{02} & u^{(1)}_{03} \\ \\ u^{(2)}_{01} & u^{(2)}_{02} & u^{(2)}_{03} \end{bmatrix}; \quad D^{(2)}_{\eta_m} = \begin{bmatrix} \vartheta^{(1)}_{01} & \vartheta^{(1)}_{02} & \vartheta^{(1)}_{03} \\ \\ \vartheta^{(2)}_{01} & \vartheta^{(2)}_{02} & \vartheta^{(2)}_{03} \end{bmatrix}.$$

Having used the Fourier transform, we obtain from (8.98) the image of the vector $\mathbf{f}(\omega)$ in frequency area.

$$\mathbf{f}(\omega) = W(i\omega)\, D^{(1)}\mathbf{P}_c(\omega) + W(i\omega)\, D^{(2)}\mathbf{T}_c(\omega),$$

where

$$W(i\omega) = \left[\omega^2 A + i\omega B + C\right]^{-1}. \tag{8.99}$$

In scalar form we have

$$f^{(1)}(\omega) = \sum_{k=0}^{3} u^{(1)}_{0k}(w_{11}P_k) + \sum_{k=0}^{3} u^{(2)}_{0k}(w_{12}P_k) + \sum_{k=0}^{3} \vartheta^{(1)}_{0k}(w_{11}T_k) + \sum_{k=0}^{3} \vartheta^{(2)}_{0k}(w_{12}T_k); \tag{8.100}$$

$$f^{(2)}(\omega) = \sum_{k=0}^{3} u^{(1)}_{0k}(w_{21}P_k) + \sum_{k=0}^{3} u^{(2)}_{0k}(w_{22}P_k) + \sum_{k=0}^{3} \vartheta^{(1)}_{0k}(w_{21}T_k) + \sum_{k=0}^{3} \vartheta^{(2)}_{0k}(w_{22}T_k). \tag{8.101}$$

We consider that the spectral densities S_{P_k}, S_{T_k} of the components of the vectors $\mathbf{P}_c$ and $\mathbf{T}_c$ are known. Let us confine ourselves to a case where it is possible to consider the components of the vectors to be independent random functions. Then the cross-spectral densities are equal to zero. For the case of the vectors $\mathbf{P}_c$ and $\mathbf{T}_c$ being directionally invariable, but random in absolute value, we have

$$P_j = |\mathbf{P}_c|(\mathbf{e}_P \cdot \mathbf{e}_j), \quad T_j = |T_c|(\mathbf{e}_T \cdot \mathbf{e}_j), \tag{8.102}$$

where $\mathbf{e}_P$, $\mathbf{e}_T$ are unit vectors directionally coinciding with those of $\mathbf{P}_c$ and $\mathbf{T}_c$. If the moduli of the forces are random stationary functions, it is sufficient to know only their spectral densities $S_P(\omega)$, $S_T(\omega)$. Let us obtain the spectral densities $f^{(1)}$ and $f^{(2)}$ for the general case where the spectral densities of the components of the vectors P_j and T_j are different.

The algorithm of determining the spectral densities is presented in Sect. 6.3. Having used this algorithm, we obtain from (8.100) and (8.101) (provided that $\mathbf{P}_c$ and $\mathbf{T}_c$ are independent) the spectral densities $S_{f^{(1)}}$; and $S_{f^{(2)}}$

$$S_{f^{(1)}}(\omega) = \sum_{k=1}^{3} \beta_k^{(1)}(i\omega) S_{P_k}(\omega) + \sum_{k=1}^{3} \gamma_k^{(1)}(i\omega) S_{T_k}(\omega);$$
$$S_{f^{(2)}}(\omega) = \sum_{k=1}^{3} \beta_k^{(2)}(i\omega) S_{P_k}(\omega) + \sum_{k=1}^{3} \gamma_k^{(2)}(i\omega) S_{T_k}(\omega), \tag{8.103}$$

where

$$\beta_k^{(1)}(i\omega) = \left(u_{0k}^{(1)}\right)^2 w_{11}w_{11}^* + u_{0k}^{(1)}u_{0k}^{(2)}w_{11}w_{12}^* + u_{0k}^{(2)}u_{0k}^{(1)}w_{12}w_{11}^* + \left(u_{0k}^{(2)}\right)^2 w_{12}w_{12}^*;$$
$$\gamma_k^{(1)}(i\omega) = \left(\vartheta_{0k}^{(1)}\right)^2 w_{11}w_{11}^* + \vartheta_{0k}^{(1)}\vartheta_{0k}^{(2)}w_{11}w_{12}^* + \vartheta_{0k}^{(2)}\vartheta_{0k}^{(1)}w_{12}w_{11}^* + \left(\vartheta_{0k}^{(2)}\right)^2 w_{12}w_{12}^*;$$
$$\beta_k^{(2)}(i\omega) = \left(u_{0k}^{(1)}\right)^2 w_{21}w_{21}^* + u_{0k}^{(1)}u_{0k}^{(2)}w_{21}w_{22}^* + u_{0k}^{(2)}u_{0k}^{(1)}w_{22}w_{21}^* + \left(u_{0k}^{(2)}\right)^2 w_{22}w_{22}^*;$$
$$\gamma_k^{(2)}(i\omega) = \left(\vartheta_{0k}^{(1)}\right)^2 w_{21}w_{21}^* + \vartheta_{0k}^{(1)}\vartheta_{0k}^{(2)}w_{21}w_{22}^* + \vartheta_{0k}^{(2)}\vartheta_{0k}^{(1)}w_{22}w_{21}^* + \left(\vartheta_{0k}^{(2)}\right)^2 w_{22}w_{22}^*. \tag{8.104}$$

The cross-spectral densities are

$$S_{f^{(1)}f^{(2)}} = \sum_{k=1}^{3} \beta_k^{(3)} S_{P_k} + \sum_{k=1}^{3} \gamma_k^{(3)} S_{T_k};$$
$$S_{f^{(2)}f^{(1)}} = \sum_{k=1}^{3} \beta_k^{(4)} S_{P_k} + \sum_{k=1}^{3} \gamma_k^{(4)} S_{T_k}, \tag{8.105}$$

where

$$\begin{aligned}\beta_k^{(3)} &= \left(u_{0k}^{(1)}\right)^2 w_{21}^* w_{11} + u_{0k}^{(1)} u_{0k}^{(2)} w_{21}^* w_{12} \\ &+ u_{0k}^{(2)} u_{0k}^{(1)} w_{22}^* w_{11} + \left(u_{0k}^{(2)}\right)^2 w_{22}^* w_{12}; \\ \gamma_k^{(3)} &= \left(\vartheta_{0k}^{(1)}\right)^2 w_{21}^* w_{11} + \vartheta_{0k}^{(1)} \vartheta_{0k}^{(2)} w_{21}^* w_{12} \\ &+ \vartheta_{0k}^{(2)} \vartheta_{0k}^{(1)} w_{22}^* w_{11} + \left(\vartheta_{0k}^{(2)}\right)^2 w_{22}^* w_{12};\end{aligned} \tag{8.106}$$

$$\begin{aligned}\beta_k^{(4)} &= \left(u_{0k}^{(1)}\right)^2 w_{11}^* w_{21} + u_{0k}^{(1)} u_{0k}^{(2)} w_{12}^* w_{21} \\ &+ u_{0k}^{(2)} u_{0k}^{(1)} w_{11}^* w_{22} + \left(u_{0k}^{(2)}\right)^2 w_{12}^* w_{22}; \\ \gamma_k^{(4)} &= \left(\vartheta_{0k}^{(1)}\right)^2 w_{11}^* w_{21} + \vartheta_{0k}^{(1)} \vartheta_{0k}^{(2)} w_{12}^* w_{21} \\ &+ \vartheta_{0k}^{(2)} \vartheta_{0k}^{(1)} w_{11}^* w_{22} + \left(\vartheta_{0k}^{(2)}\right)^2 w_{12}^* w_{22}.\end{aligned} \tag{8.107}$$

The expressions for spectral densities $S_{f^{(i)}}(\omega)$ (8.103) and cross-spectral densities $S_{f^{(i)}f^{(k)}}(\omega)$ (8.105) can also be obtained, using the Wiener–Khintchin relationships that connect correlation and cross-correlation functions with spectral densities, as we did it in Chap. 6.

The products of the functions w_{ij} and w_{ik}^* dependent on $i\omega$ enter in the right-hand side of expression (8.104) for the coefficient $\beta_k^{(i)}$. The products of the functions with the same indexes are equal to the square of the modulus of an appropriate function, i.e.

$$w_{11} w_{11}^* = |w_{11}|^2 . \tag{8.108}$$

The square of the modulus of a complex function is a real function. The products of complex functions with different indexes are complex functions, for example,

$$\begin{aligned}(w_{11} w_{12}^*) &= \left(w_{11}^{(1)} + i w_{11}^{(2)}\right)\left(w_{12}^{(1)} - i w_{12}^{(2)}\right) = w_{11}^{(1)} w_{12}^{(1)} \\ &+ w_{11}^{(2)} w_{12}^{(2)} + i\left(w_{11}^{(2)} w_{12}^{(1)} - w_{11}^{(1)} w_{12}^{(2)}\right),\end{aligned} \tag{8.109}$$

where $w_{11}^{(1)}$, $w_{12}^{(1)}$ are real parts; $w_{11}^{(2)}$, $w_{12}^{(2)}$ are imaginary parts. The real parts of the complex functions are even functions of ω, while the imaginary ones are odd functions, i.e.

$$w_{11}^{(1)}(\omega) = w_{11}^{(1)}(-\omega); \quad w_{11}^{(2)}(\omega) = -w_{11}^{(2)}(-\omega) . \tag{8.110}$$

The product of odd functions is an even function. Therefore the real part of expression (8.109) is an even function, while the imaginary one is an odd function. These results will be used in determining the variance of the components of the vector $\mathbf{Z}$ (8.83).

The components of the vector $\mathbf{Z}$ at a two-term approximation are

$$z_j(\eta,\tau) = z_{0j}^{(1)}(\eta)\, f^{(1)}(\tau) + z_{0j}^{(2)}(\eta)\, f^{(2)}(\tau)\,. \tag{8.111}$$

Passing to the Fourier image, we obtain

$$z_j(\eta,\omega) = z_{0j}^{(1)}(\eta)\, f^{(1)}(\omega) + z_{0j}^{(2)}(\eta)\, f^{(2)}(\omega)\,. \tag{8.112}$$

We can obtain the spectral densities of the components z_j (and the cross-spectral densities of z_j, z_k) by using the algorithm of determining S_{y_k} and $S_{y_k y_v}$ that was presented in Sect. 6.5.

$$S_{z_j}(\eta,\omega) = \left(z_{0j}^{(1)}\right)^2 S_{f^{(1)}} + z_{0j}^{(1)} z_{0j}^{(2)} \left(S_{f^{(1)}f^{(2)}} + S_{f^{(2)}f^{(1)}}\right) + \left(z_{0j}^{(2)}\right)^2 S_{f^{(2)}}, \tag{8.113}$$

where $S_{f^{(1)}}$, $S_{f^{(2)}}$ and $S_{f^{(1)}f^{(2)}}$ are determined from relationships (8.103) and (8.105). We can present the coefficients $\beta_k^{(j)}$, $\gamma_k^{(j)}$ ($k = 1, 2, 3;\ j = 1, 2, 3, 4$) that enter into (8.105) and in (8.107) as

$$\beta_k^{(j)} = \beta_k^{(j)(1)} + i\beta_k^{(j)(2)},\ \gamma_k^{(j)} = \gamma_k^{(j)(1)} + i\gamma_k^{(j)(2)}, \tag{8.114}$$

where $\beta_k^{(j)(1)}$, $\gamma_k^{(j)(1)}$ are the even functions of ω; $\beta_k^{(j)(2)}$ and $\gamma_k^{(j)(2)}$ are the odd functions of ω. Therefore we obtain:

$$\begin{aligned}
S_{f^{(1)}}(i\omega) &= \sum_{k=1}^{3}\left(\beta_k^{(1)(1)} S_{P_k} + \gamma_k^{(1)(1)} S_{M_k}\right) \\
&\quad + i\left[\sum_{k=1}^{3}\left(\beta_k^{(1)(2)} S_{P_k} + \gamma_k^{(1)(2)} S_{M_k}\right)\right]; \\
S_{f^{(2)}}(i\omega) &= \sum_{k=1}^{3}\left(\beta_k^{(2)(1)} S_{P_k} + \gamma_k^{(2)(1)} S_{M_k}\right) \\
&\quad + i\left[\sum_{k=1}^{3}\left(\beta_k^{(2)(2)} S_{P_k} + \gamma_k^{(2)(2)} S_{M_k}\right)\right];
\end{aligned} \tag{8.115}$$

$$
\begin{aligned}
S_{f^{(1)}f^{(2)}}(i\omega) &= \sum_{k=1}^{3}\left(\beta_k^{(3)(1)}S_{P_k}+\gamma_k^{(3)(1)}S_{M_k}\right)\\
&\quad + i\left[\sum_{k=1}^{3}\left(\beta_k^{(3)(2)}S_{P_k}+\gamma_k^{(3)(2)}S_{M_k}\right)\right];\\
S_{f^{(2)}f^{(1)}}(i\omega) &= \sum_{k=1}^{3}\left(\beta_k^{(4)(1)}S_{P_k}+\gamma_k^{(4)(1)}S_{M_k}\right)\\
&\quad + i\left[\sum_{k=1}^{3}\left(\beta_k^{(4)(2)}S_{P_k}+\gamma_k^{(4)(2)}S_{M_k}\right)\right];
\end{aligned}
\tag{8.116}
$$

We can present relationships (8.115) as

$$
S_{f^{(1)}} = S^{(1)}_{f^{(1)}} + iS^{(2)}_{f^{(1)}}, \quad S_{f^{(2)}} = S^{(1)}_{f^{(2)}} + iS^{(2)}_{f^{(2)}}; \tag{8.117}
$$

$$
S_{f^{(1)}f^{(2)}} = S^{(1)}_{f^{(1)}f^{(2)}} + iS^{(2)}_{f^{(1)}f^{(2)}}, \; S_{f^{(2)}f^{(1)}} = S^{(1)}_{f^{(2)}f^{(1)}} + iS^{(2)}_{f^{(2)}f^{(1)}}; \tag{8.118}
$$

with

$$
\begin{aligned}
&S^{(1)}_{f^{(1)}} = S^{(1)}_{f^{(2)}}; \quad S^{(2)}_{f^{(1)}} = -S^{(2)}_{f^{(2)}},\\
&S^{(1)}_{f^{(1)}f^{(2)}} = S^{(1)}_{f^{(2)}f^{(1)}}; \quad S^{(2)}_{f^{(1)}f^{(2)}} = -S^{(2)}_{f^{(2)}f^{(1)}}.
\end{aligned}
\tag{8.119}
$$

The sum of cross-spectral densities enters into relationships (8.113) and with due account of (8.118) and (8.119) is equal to

$$
S_{f^{(1)}f^{(2)}} + S_{f^{(2)}f^{(1)}} = 2S^{(1)}_{f^{(1)}f^{(2)}}, \tag{8.120}
$$

i.e. is an even function of ω. The cross-spectral densities of the components of the vector **Z** are

$$
S_{z_j z_\nu} = z_{0j}^{(1)} z_{0\nu}^{(1)} S_{f^{(1)}} + z_{0j}^{(1)} z_{0\nu}^{(2)} \left(S_{f^{(1)}f^{(2)}} + S_{f^{(2)}f^{(1)}}\right) + z_{0j}^{(2)} z_{0\nu}^{(2)} S_{f^{(2)}}.
$$

The variances of the components z_j can be expressed as

$$
D_{z_j} = \int_{-\infty}^{\infty} S_{z_j}(\eta,\omega)\,\mathrm{d}\omega,
$$

or

$$
\begin{aligned}
D_{z_j} &= \left(z_{0j}^{(1)}(\eta)\right)^2 \int_{-\infty}^{\infty} S_{f^{(1)}}\mathrm{d}\omega\\
&\quad + 2z_{0j}^{(1)}(\eta)\,z_{0j}^{(2)}(\eta)\int_{-\infty}^{\infty} S^{(1)}_{f^{(1)}f^{(2)}}\mathrm{d}\omega + \left(z_{0j}^{(2)}(\eta)\right)^2 \int_{-\infty}^{\infty} S_{f^{(2)}}\mathrm{d}\omega.
\end{aligned}
\tag{8.121}
$$

As the integrals of the odd functions are equal to zero, we obtain from (8.121)

$$D_{z_j}(\eta) = \left(z_{0j}^{(1)}(\eta)\right)^2 \left[\sum_{k=1}^{3} \left(\beta_k^{(1)(1)} S_{P_k} + \gamma_k^{(1)(1)} S_{M_k}\right) d\omega\right]$$
$$+ z_{0j}^{(1)}(\eta)\, z_{0j}^{(2)}(\eta) \left[\sum_{k=1}^{3} \left(\beta_k^{(4)(1)} S_{P_k} + \gamma_k^{(4)(1)} S_{M_k}\right) d\omega\right]$$
$$+ \left(z_{0j}^{(2)}(\eta)\right)^2 \left[\sum_{k=1}^{3} \left(\beta_k^{(2)(1)} S_{P_k} + \gamma_k^{(2)(1)} S_{M_k}\right) d\omega\right]. \qquad (8.122)$$

The root-mean-square values of the components z_j take the form

$$\sigma_{z_j}(\eta) = \sqrt{D_{z_j}(\eta)}. \qquad (8.123)$$

Let us consider special cases.

1. The one-term approximation at $\mathbf{P}_c \neq 0$, $\mathbf{T}_c = 0$ may be written as

$$\mathbf{Z} = \mathbf{Z}_0^{(1)} f^{(1)}.$$

The equation for determining $f^{(1)}(\tau)$ has the form

$$\eta_{11}\ddot{f}^{(1)} + b_{11}\dot{f}^{(1)} + c_{11} f^{(1)} = \sum_{j=1}^{3} u_{0j}^{(1)}(\eta_1)\, P_1(\tau).$$

The frequency function is defined by the formula

$$w(i\omega) = w_{11}(i\omega) = \frac{1}{-\omega^2 a_{11} + i\omega b_{11} + c_{11}} = w_{11}^{(1)}(\omega) + i w_{11}^{(2)}(\omega).$$

The Fourier image of the function $f^{(1)}$ is

$$f^{(1)}(\omega) = \sum_{k=1}^{3} u_{0k}^{(1)}(\eta_1)\, w_{11}(\omega)\, P_k(\omega).$$

From (8.105) we obtain (at $S_{M_k} = 0$, $\beta_k^{(1)} = u_{0_k}^{(1)^2}$)

$$S_{f^{(1)}}(\omega) = \sum_{k=1}^{3} u_{0k}^{(1)}(\eta_1)\, |w_{11}|^2\, S_{P_k}(\omega).$$

The variances of $z_j(\eta, \omega)$ are equal to

$$D_{z_j}(\eta) = \left(z_{0j}^{(1)}(\eta)\right)^2 \left[\sum_{k=1}^{3} \int_{-\infty}^{\infty} \left(u_{0k}^{(1)}(\eta_1)\right)^2 |w_{11}|^2\, S_{P_k}(\omega)\, d\omega\right]. \qquad (8.124)$$

2. For the case of $\mathbf{P}_c = 0$, $\mathbf{T}_c \neq 0$ we have

$$D_{z_j}(\eta) = \left(z_{0j}^{(1)}(\eta)\right)^2 \times \left[\sum_{k=1}^{3} \int_{-\infty}^{\infty} \left(v_{0k}^{(1)}(\eta_2)\right)^2 |w_{11}|^2 S_{M_k}(\omega)\, d\omega\right]. \quad (8.125)$$

In some cases the integrals entering into D_{zj} can be integrated (Appendix 2). In those cases when this is impossible the integrals can be obtained numerically at finite limits that are specified from the requirement of final result given accuracy achievement. The expressions obtained for the variances and standard deviations of the components z_j (8.124), (8.125) allow to obtain their values at any cross-section of a bar including cross-sections where they attain maximum value. Using the three sigma rule we obtain the greatest possible value of z_j at $m_{z_j} = 0$ as

$$z_j^*(\eta) = 3\sigma_{z_j}(\eta). \quad (8.126)$$

Normal stresses $\tilde{\sigma}$ at an arbitrary cross-section with due account of static stresses are equal to

$$\tilde{\sigma}(\eta) = \frac{(z_{10} + z_1)}{F} + \frac{(z_{50} + z_5)x}{I_y} + \frac{(z_{60} + z_6)y}{I_x}, \quad (8.127)$$

where $z_{10} = Q_{10}$, $z_{50} = M_{20}$, $z_{06} = M_{30}$; I_x, I_y are cross-sections moments of inertia; x, y are principal central axes of cross-section.

By way of transformations we obtain

$$\tilde{\sigma}(\eta) = \tilde{\sigma}_0(\eta) + 3\left(\frac{\sigma_{\Delta Q_1}(\eta)}{F} + \frac{\sigma_{\Delta M_2}(\eta)x}{I_y} + \frac{\sigma_{\Delta M_3}(\eta)y}{I_x}\right), \quad (8.128)$$

where $\tilde{\sigma}_0 = \dfrac{Q_{10}}{F} + \dfrac{M_{20}x}{I_y} + \dfrac{M_{30}y}{I_x}$.

The maximum stresses at the dangerous cross-section of the bar are

$$\tilde{\sigma}_{\max}(\eta) = \max_{\eta} \max_{F} \left[\tilde{\sigma}_0 + 3\left(\frac{\sigma_{\Delta Q_1}}{F} + \frac{\sigma_{\Delta M_2}x}{I_y} + \frac{\sigma_{\Delta M_3}y}{I_x}\right)\right]$$

or

$$\tilde{\sigma}_{\max}(\eta) = \max_{\eta} \left[\tilde{\sigma}_0^{(1)} + 3\left(\frac{\sigma_{\Delta Q_1}}{F} + \frac{\sigma_{\Delta M_2}x^*}{I_y} + \frac{\sigma_{\Delta M_3}y^*}{I_x}\right)\right], \quad (8.129)$$

$$\left(\tilde{\sigma}_0^{(1)} = \frac{Q_{10}}{F} + \frac{M_{20}x^*}{I_y} + \frac{M_{30}y^*}{I_x}\right),$$

where x^*, y^* are coordinates of the dangerous point in the cross-section.

Normal stresses $\tilde{\sigma}_{\max}(\eta)$ depend on dimensionless coordinate η therefore there is a section $\eta = \eta^*$, where $\tilde{\sigma}_{\max}(\eta)$ attains maximum value, i.e.

$$\tilde{\sigma}_{\max}(\eta^*) = \tilde{\sigma}_0^{(1)}(\eta^*) + 3\left(\frac{\sigma_{\Delta Q_1}(\eta^*)}{F} + \frac{\sigma_{\Delta M_2}(\eta^*)x^*}{I_y} + \frac{\sigma_{\Delta M_3}(\eta^*)\,y^*}{I_x}\right). \tag{8.130}$$

Minimum stress in this section (at the point having coordinates (x^*, y^*)) is

$$\tilde{\sigma}_{\min}(\eta^*) = \tilde{\sigma}_0^{(1)}(\eta^*) - 3\left(\frac{\sigma_{\Delta Q_1}(\eta^*)}{F} + \frac{\sigma_{\Delta M_2}(\eta^*)x^*}{I_y} + \frac{\sigma_{M_3}(\eta^*)\,y^*}{I_x}\right). \tag{8.131}$$

Maximum (8.130) and minimum (8.131) normal stresses allow to estimate fatigue strength of the bar.

9. Fundamentals of Reliability Theory

9.1 Introduction

The previous chapters were devoted to the theoretical fundamentals of statistical mechanics and the theory of random vibrations. The readers attention was focused on mathematical methods of solving problems of the dynamics of mechanical systems loaded with random forces and the determination of the probability characteristics of the vector of the system state, or, what is the same, to the determination of the probability characteristics of the "output" given that the probability characteristics of the "input" are known. Mechanical systems with a finite number of degrees of freedom and systems with distributed parameters (structures or elements of structures reduced to a mathematical model of a rod) were considered. Methods making it possible to determine the probability characteristics of the stress-strain state of the structural members at non-stationary and stationary random forces were presented. It has been shown that methods of statistical dynamics allow the solution of many applied problems when the random components of loads cannot be ignored. However, questions of the "strength" of a structure at random loads were, in fact, not considered.

Before going on to methods of estimating the "strength" of structures in a probability formulation let us recall how traditional analysis in a determinate formulation is carried out. These methods include a limit state design method (based on the occurrence of plastic deformations, on fracture or on buckling) and an allowable stress design method.

The limit state design method compares the maximum acting load (more precisely, the load- induced stress-strain state) with the load corresponding to the limit state which determines the load-carrying capacity of a structure. This capacity is the onset of a limiting stress-strain state that corresponds to a structures loss of serviceability or failure. These limiting states embrace plastic deformations, loss of static stability, and failure. Let us consider some examples of limit state design.

A straight rod tensioned by a force P is shown in Fig. 9.1 *a*; in so doing the arising stresses should not exceed the elastic limit. In this case, the limiting state and the maximum force P, corresponding to it, will be the force P at which the stresses in the rod reach the yield point σ_y. Therefore to ensure the normal work of this element the condition

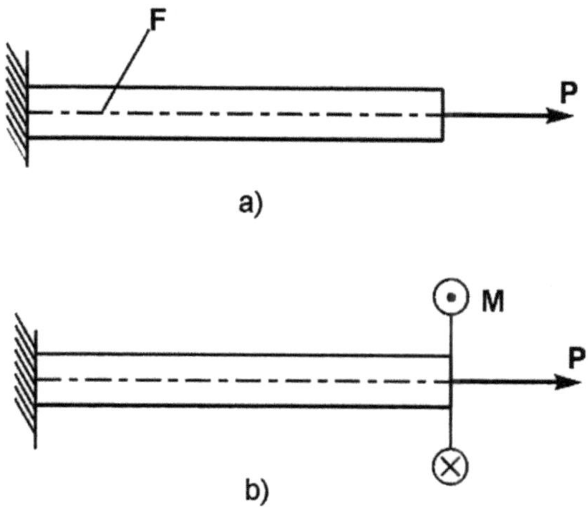

Fig. 9.1.

$$\sigma_y - \sigma > 0 \quad \left(\sigma = \frac{P}{F}\right). \tag{9.1}$$

must be satisfied.

A straight rod loaded with a tensile force P and a twisting moment M is shown in Fig. 9.1 *b*. A two-dimensional stress state arises in the rod. Therefore to avoid plastic deformations in the rod it is necessary to fulfill the condition

$$\sigma_y - \max \sigma_e (P, M) > 0, \tag{9.2}$$

where $\max \sigma_e = \max \sqrt{\sigma^2 + 4\tau^2}$ or $\max \sigma_e = \max \sqrt{\sigma^2 + 3\tau^2}$, $\max \tau = \dfrac{M}{W_k}$ depending on a strength criteria adopted at calculations. (W_k is the geometrical characteristic of the rod section; for example, for a round rod it is equal to $\pi D^3/16$).

A rocket in a launch position is shown in Fig. 9.2. A system of elastic constraints (a damping system) enables it to deflect at a shock wave passage by an angle ϑ (in order to reduce dynamic overloads), but the angle of rotation of the axial line of the rocket is bounded by the greatest possible angle ϑ_n, which depends on the structure of the dampers and their allowable linear displacement.

A drilling rig is shown schematically in Fig. 9.3. During the operation of the drill vibrations arise resulting in a situation where the rod, at large deflections from the axis of symmetry, can strike against the surface of the well, which is extremely undesirable. Therefore to ensure the normal operation of the system that would exclude contacts with the surface of the wall the displacements of the points of the rods axial line $u(z,t)$ at vibrations must satisfy the condition

$$\Delta - \max |u(z,t)| > 0. \tag{9.3}$$

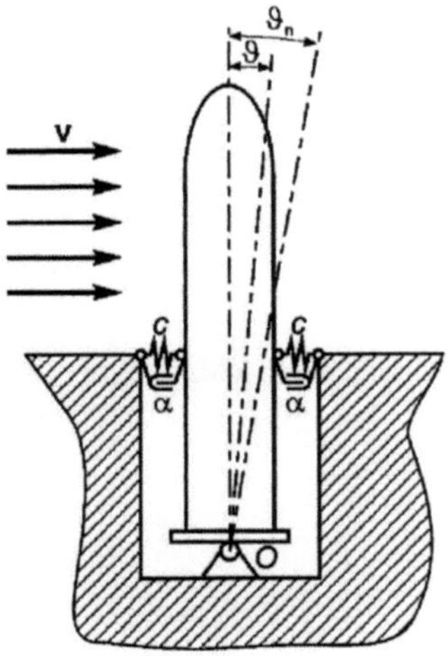

Fig. 9.2.

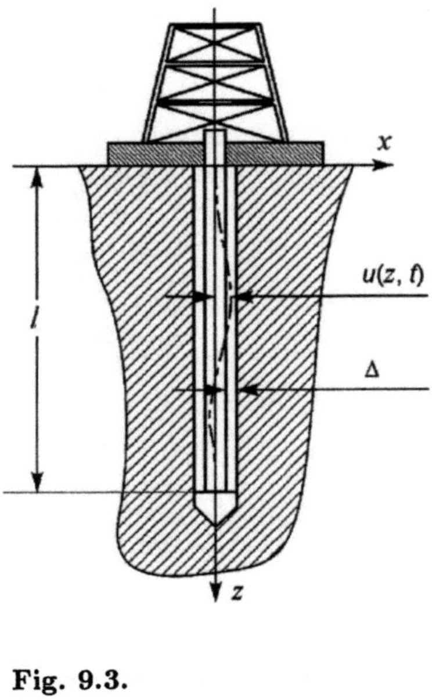

Fig. 9.3.

A straight rod loaded with a compressive force P is shown in Fig. 9.4. A limiting state in this case is the loss of static stability, therefore the critical force P_k should exceed the compressive force

$$P_k - P > 0. \tag{9.4}$$

Figure 9.5 schematically shows a turbojet engine that would not operate normally unless the elongations of the blades Δl appearing during its oper-

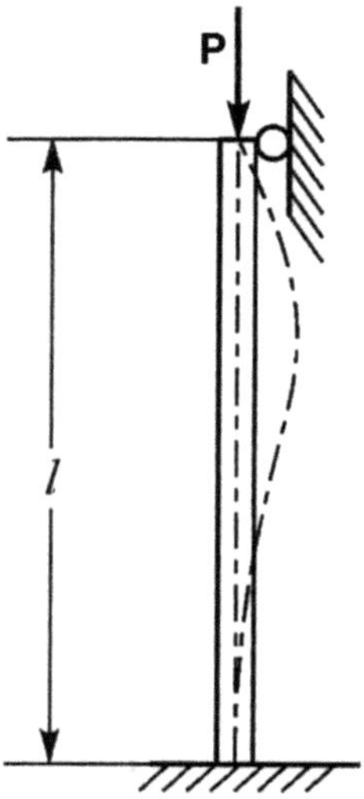

Fig. 9.4.

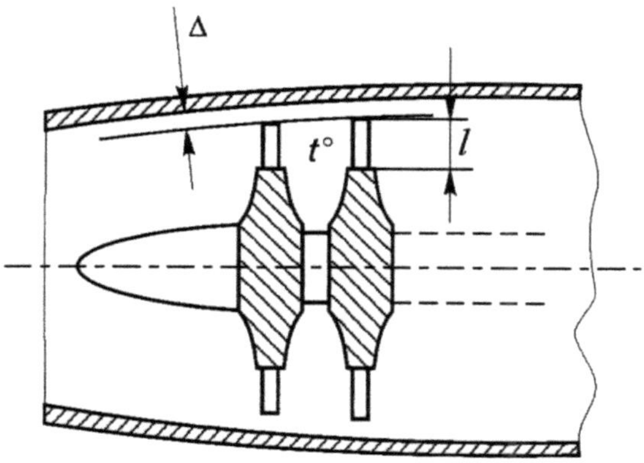

Fig. 9.5.

ation (at $\omega \neq 0$) are smaller than the gap Δ between the internal surface of the casing and the blades at $\omega = 0$

$$\Delta - \Delta l\,(\omega) > 0. \tag{9.5}$$

A variation of normal stress σ, arising at the work of a structures element, as a function of time is shown in Fig. 9.6. The normal operation of the element at a fully reversed loading cycle requires that the stress is at least smaller than the limiting stress corresponding to the fatigue strength of the given material

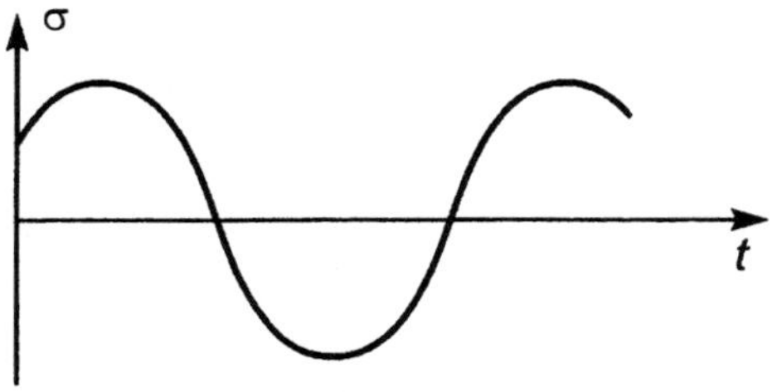

Fig. 9.6.

$$\sigma_{-1} - \sigma > 0, \tag{9.6}$$

where σ_{-1} is the fatigue strength at the symmetrical loading cycle.

Methods of limit state design allow us to take full advantage of the load carrying capacity of a structure.

Generally we have

$$S - F > 0, \tag{9.7}$$

where S is the function describing the limiting capabilities of the structure, F is the function describing the real state of the structure.

For example, S depends on the mechanical characteristics of the material of the structure

$$S = S\left(\sigma_y, \ \sigma_u, \ \mu, \ E, \ G\right),$$

where σ_u is the ultimate strength, μ is the Poisson ratio, E, G are elastic moduli of respectively the first and second kind. The function F depends on the stresses arising in the elements of the structure at loading

$$F = F\left(\sigma_x, \ \sigma_y, \ \sigma_z, \ \tau_{xy}, \ \tau_{xz}, \ \tau_{yz}\right).$$

Limiting states can be related not only to the strength properties of a structure. For example, Fig. 9.7 shows a rocket whose trajectory of motion should not go beyond an allowable "tube" of trajectories (for each instant t it represents some design closed limiting area D_d). Therefore the control system of the rocket should ensure the fulfillment of the condition

$$D_d(t) - D(t) > 0, \tag{9.8}$$

where $D(t)$ is the real area, inside which the rocket is during its motion.

In the design method that uses the allowable stresses the concept of a safety factor is introduced and the allowable stress σ_a is taken equal to

$$\sigma_a = [\sigma] = \frac{\sigma_n}{n}, \tag{9.9}$$

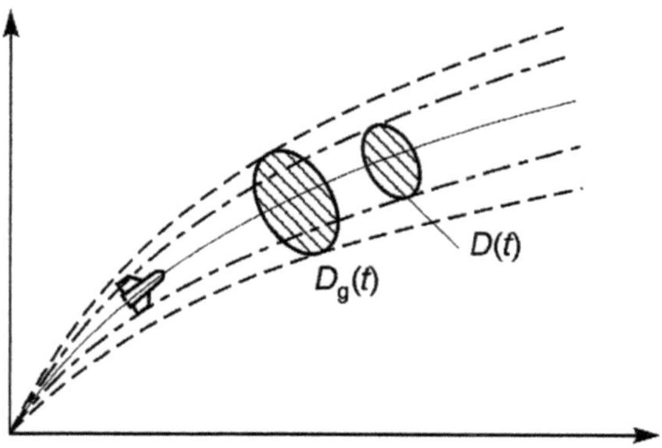

Fig. 9.7.

where σ_n is the limiting stress (yield point for plastic materials, ultimate strength for brittle materials or critical stress corresponding to loss of static stability); n is the factor of safety.

Therefore the conditions of "strength", for example for (9.1), (9.2), (9.4) and (9.5) at a design based on allowable states are

$$\begin{aligned} &\frac{\sigma_y}{n_1} - \sigma \geq 0, \\ &\frac{\sigma_y}{n_1} - \max \sigma_e (P, M) \geq 0, \\ &\frac{P_k}{n_2} - P \geq 0, \\ &\frac{\Delta}{n_3} - \Delta l(\omega) \geq 0 \qquad (n > 1). \end{aligned} \tag{9.10}$$

It is generally agreed that the fulfillment of conditions (9.10) at the determinate values of the quantities entered in them ensures the "strength" of these structures.

It is possible to present relationships (9.10) in a more general form as

$$S_d - S \geq 0, \qquad \left(S_d = \frac{R_c}{n} \right), \tag{9.11}$$

where R_c is the load carrying capacity of the structure or of its elements, S_d is the allowable state, for example, the allowable load, allowable stresses etc., S is the real state, n is the factor of safety.

The values R_c, S and n entered in the relationships (9.11) are considered determinate quantities. In a determinate statement the ultimate goal of a design is to check the fulfillment of inequalities (9.11). We may consider that the fulfillment of inequalities (9.11) assures the no-failure operation of the structure while it is in use. (In the general case any failure represents a malfunction of the structure, and not just the occurrence of a limiting state).

The factors n_j are chosen on the basis of the gained experience. Each field of technology has its own requirements for the designed equipment and traditional methods of analysis, which allow us to recommend the numerical values of the safety factors. They are often referred to as "normative factors of safety on strength" or "normative safety margins". For example, in aircraft technology it is recommended to take safety factors equal to $n = 1.5$, while in rocket and space technology they advice to take these factors equal to $n = 1.2$.

The chosen numerical values of the factors n depend on the importance of the designed equipment, the level of manufacturing technology, the properties of materials and the accuracy of specifying loads. Safety factors have been revised with due account of generalized many-year experience of designing in all industries, therefore each of them has its own "strength design codes" that are used in practical designing. These codes define the structure and volume of basic works performed at all stages of producing structures and necessary for the assurance of the required strength. Designing, where all strength requirements that should be realized in a developmental prototype are taken into account, including every "physical" feature of the structure and its real operation conditions, plays the leading role in making this structure.

When using design methods based on strength design codes during the determination of the load carrying capacity of a structure the function of safety factors is in the main reduced to the compensation of: 1) the discrepancy between the determinate form of presenting the results of a strength design and the possible scatters of the structures parameters and loads; 2) the deviation of the design scheme (mathematical model) from the real structure and real loading conditions, which results in a large factor of safety scatter even when designing structures of the same type. For example, in the process of designing guided projectiles in Britain the factor of safety n, equal to 1.33 is adopted, while in the USA during the designing of guided projectiles with a close **system** performance the factor n equal to 1.25 is used [3]. There is no rational explanation as to why the factors n differ, therefore specifying them involves some uncertainty , i.e. they are random in character.

In many cases the introduction of safety factors allows the obtainment of satisfactory structures. When designing new equipment, however, when the lack of experience and operational data makes it very difficult to choose a reasonable safety factor. The arbitrarily assigned safety factor can produce wrong solutions resulting either in the overestimated weight of structures or in emergencies. The principal difficulty in determining allowable stresses (or deformations) and a structures load carrying capacity is to coordinate design and actual data. The problem of choosing the specific value of safety factor in order to determine, for example, the allowable stress is complicated by the fact that the mechanical characteristics of the material, influencing the structure limit states, as well as real forces and geometrical sizes of structure elements, influencing the structure current states, have random scatters. Traditional design methods both at limit state design and at the design using the allowable stresses, do not explicitly take into account possible random scatters, i.e. do not consider the probability character of the limit states of a structure or that of its real state.

Therefore it is more logical to estimate the structure serviceability not on the basis of determinate inequalities (9.1)–(9.3), but in terms of the probability of their fulfillment, i.e.

$$P\left[(S-F)>0\right], \tag{9.12}$$

where P is the probability of no-failure operation.

The probability of failure is equal to

$$R=1-P, \qquad \left(R\left[(S-F)<0\right]\right). \tag{9.13}$$

Estimation of the "strength" with due account of the probability character of parameters and loads of the structure requires different methods based on the theory of probabilities and statistical mechanics. Therefore we introduce the concept of the systems reliability H which is estimated by the probability of fulfilling inequality (9.7)

$$H=P\left[(S-F)>0\right]. \tag{9.14}$$

By reliability we shall mean the ability of machines, devices and structures to operate trouble-free during a certain time interval. The no-failure operation of technical objects is considered to mean the fulfillment by them of all their functions in the given operating conditions. There is another, more detailed definition of reliability as: "the property of objects to retain in time, within the predetermined limits, the values of all the parameters characterizing the ability to fulfill the required functions in the preset operating conditions and conditions of use, maintenance, repair, storage and transportation".

Passing to the probability methods of "strength" estimation, when it is necessary to take into account the random character of loads and the parameters of the structure, we find the probability of the fulfillment of determinate inequalities (9.1)–(9.8) rather than the latter themselves. For general case (9.7) the estimation of trouble-free operation is related to the determination of the probability

$$P\left[(S-F)>0\right]. \tag{9.15}$$

The question arises about the extent of probability estimations of "strength" being better than traditional ones that use the factor of safety which is as unspecific as the probability itself. The point is that probability estimations of "strength" take into account objectively existing random scatters of loads, mechanical characteristics of materials, etc., and for this reason represent real conditions more completely. In what follows it will be shown that at factor $n = 1.5$ the probability of trouble-free operation can be lower than at $n = 1.2$, which seems strange as it is usual to assume that the greater the factor of safety, the greater the "strength" of a structure.

Probabilities of the no-failure operation of a system in themselves are of little use (for example, if $P = 0.9$ it is hard to say whether it is good or bad), but if we carry out an analysis for two versions of a material with due account of the probability properties of their mechanical characteristics and it turns out that the probability of no-failure operation is equal to 0.9 and 0.95 respectively, we can definitely state that the structure with the 0.95 probability of no-failure operation will be better. The account of random scatters produces qualitatively different estimations of "strength", which makes it possible to design more rational structures having greater reliability, durability and lifetime.

Assuring the reliability of designed mechanical systems is one of the basic problems in mechanical engineering, instrument making, aircraft, space-rocket engineering and many other industries. It is explained by the continuous growth of reliability and life requirements to new equipment that should normally function in severe operation conditions. By convention all problems of estimating the reliability of mechanical systems can be divided into three classes.

1. The estimation of the reliability of a structure or structure components under a single or low-cycle loading. The problems of estimating reliability under single loading arise both at static and dynamic loading, for example, at impact or pulse loading [22].

No damage accumulation occurs in these problems or, if it does at a small number of load cycles, this damage accumulation can be ignored.

2. Problems involving the accumulation of residual macroscopic strain within a limited time interval $(0, T)$ under stationary or quasi-stationary random loads [9], when it is possible to consider single overloads resulting in the failure of a system as highly improbable. During analysis pertaining to these problems the task is to determine probability distributions for residual strains at $t = T$. The quasi-stationary loads (processes) are considered to mean processes whose probability characteristics change in time slowly compared with the variation of random functions.

3. Problems related to accumulation of fatigue damage under the action of stationary or quasi-stationary random loads, when the probability that stresses arising in a structure will exceed the elastic limit is very low, and the structure fails as a result of gradual development of fatigue cracks [9, 19].

9.2 Elementary Problems of Reliability Theory

Before proceeding to methods of finding the numerical values of the probability that inequalities (9.1)–(9.6) will be fulfilled, we shall tackle elementary problems of determining a systems reliability when that of its elements is known.

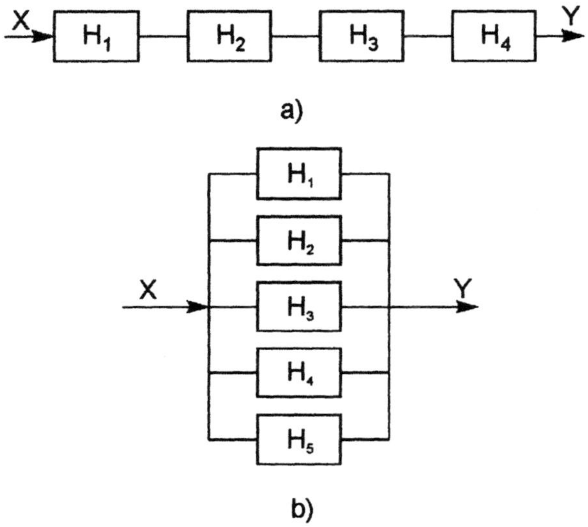

Fig. 9.8.

Let us consider an example of determining the reliability of a system that comprises separate elements whose reliability is equal to H_k (Fig. 9.8). Two mechanical systems with the known probabilities P_k ($P_k = H_k$) of no-failure operation of each of the members of the system are shown in **Fig.** 9.8 *a, b* as block diagrams. The system presented in Fig. 9.8 *a* consists of members connected in series; while that drawn in Fig. 9.8 *b* includes members connected in parallel. Let us consider the case of members being connected in series (Fig. 9.8 *a*) and interacting in such a way that their failures ($R_k = 1 - P_k$) are independent. It is required to determine the reliability of the system as a whole. This system maintains serviceability only when all of its members connected in series operate failure-free. It is to be recalled that the probability of occurrence of a joint event consisting of n independent events is equal to the product of the probabilities of occurrence of each of the n events. Therefore the probability P of no-failure operation of the system as whole is equal to

$$P = H = \prod_{j=1}^{\rho} H_j. \tag{9.16}$$

For example, if $\rho = 4$, $H_1 = 0.9$, $H_2 = 0.8$, $H_3 = 0.7$, $H_4 = 0.6$, the reliability of the system is equal to $H = 0.3024$, i.e. the reliability of any system made of members connected in series is lower than the reliability of its components.

Let us consider a system whose components are connected in parallel (Fig. 9.8 *b*) and duplicate each another. The failure of the system will occur only if all of its components fail.

The probability that each of its members fail is equal to

$$R_i = 1 - H_i.$$

The probability that the whole system fails (theorem of product of probabilities for independent events) is equal to

$$R = \prod_{i=1}^{5} (1 - H_i). \tag{9.17}$$

The probability of no-failure operation of the system (the reliability of the system) is

$$H = 1 - \prod_{i=1}^{5} (1 - H_i). \tag{9.18}$$

For example, if $H_i = 0.5$, the total reliability of a given block is equal to $H = 0.97$. The reliability of a system with its components being connected in parallel is higher than that of its components, i.e. if such system includes low-reliability components, we can substantially increase its reliability by substituting a block of several low reliability components connected in parallel for one low reliability component. For example, there is one low-reliability component (H_2) in a system of three members connected in series (Fig. 9.9). If the reliabilities of these components are equal to: $H_1 = 0.9$, $H_2^{(1)} = 0.3$, $H_3 = 0.8$, the total reliability of the system is $H = H_1 H_2^{(1)} H_3 = 0.216$. If we replace the component $H_2^{(1)}$ with a block of three exactly the same

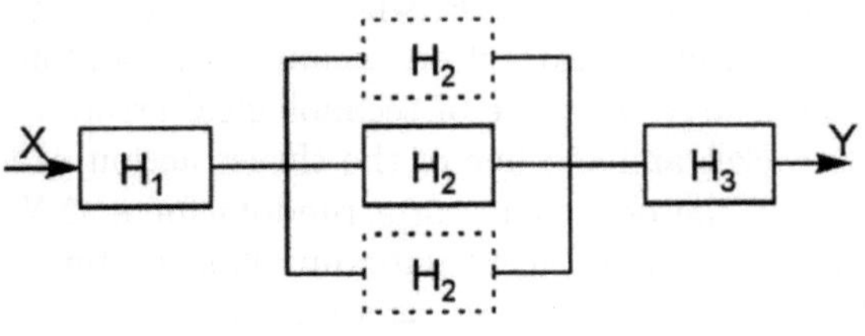

Fig. 9.9.

Fig. 9.10.

members connected in parallel (in Fig. 9.9 they are shown by dashed lines), the reliability of this unit will be $H_2^{(1)} = 1 - \left(1 - H_2^{(1)}\right)^3 = 0.657$ and the total probability of no-failure operation will equal $H = 0.9 \cdot 0.657 \cdot 0.8 = 0.473$, i.e. the reliability of the system will increase more than 100 percent!

An four-engine airliner is shown in Fig. 9.10. We may consider its system of engines as that of members connected in parallel (Fig. 9.10 *b*). Let us assume that a reliability H_j of all engines is the same and equal to $H_j = 0.9$. The airliner can fly if three of its engines fail. In this case the failure of the system will occur, when all four engines fail. At the independent work of the engines the probability of failure is equal to

$$R = (1 - H_j)^4 = (0.1)^4 = 0.0001,$$

therefore, $H = 1 - R = 0.9999$.

If the airliner can fly only with two of its engines working, the reliability is equal to $H = 1 - (0.1)^3 = 0.999$.

9.3 Possible Causes of Failures

Among the principal causes of failures of mechanical systems are: incomplete agreement between the load used in the analysis and the real load; intensive loads resulting in an excess of the load carrying capacity of a structure; plastic deformations of elastic elements; wear of parts; excess of allowable deformations; loss of stability, etc. A failure can result from random defects present in structural elements and developed in them while in service (development of cracks), as well as from damage accumulation and changes in the mechanical properties of material (for example, due to irradiation). For example, a rocket thrust R (Fig. 0.2) will have a scatter ($R = R_0 + \Delta R$, where R_0 is the design value of the thrust and ΔR is the random scatter) on account of a random change in the temperature of the charge. Because of technological errors at assembling the axial line of the rocket and the line of the thrust action can be out of alignment that results in the occurrence of a random force ΔN_c and a random moment ΔM_c and eventually in an impermissible scatter of the hit point of the rocket, which may be considered as a failure.

During the flight of flying vehicles in a rough (turbulent) atmosphere they are subjected to the action of random aerodynamic forces that can substantially change their flight trajectories with the latter going beyond the allowable "tube" (Fig. 9.7) and this is also a failure. Similar failure can result from a random cross wind. Figure 0.1 shows a vehicle moving along a road with random irregularities that account for random components in the forces of interaction of the vehicles wheels with the road. Depending on their "magnitudes" these components can cause such failures as break-down of the suspension or suspension fatigue failure.

Failures can also occur owing to a discrepancy between the design scheme (mathematical model) and the real structure, for example, because of ignoring non-linearities, gaps, friction and the scatter of the mechanical properties of the material in the mathematical model.

9.4 Determination of Numerical Values of No-Failure Operation Probability (Reliability)

It will be recalled that the reliability is defined as the probability of nofailure operation (9.14)

$$H = P\left[(S - F) > 0\right], \tag{9.19}$$

where S is the function defining the limiting capabilities of a structure (the function of load carrying capacity of a structure), F is the function defining the current state of a system. This section is devoted to problems of determining reliability at the action of single loading or a small number of sequential loadings. When tackling these problems we can ignore the accumulation of damage in the structure. In order to determine the probability P we must know the joint distribution law $f(z)$ of the random quantity $Z = S - F$ provided that the distribution laws of S and F are known. If we know $f(z)$, the probability P is equal to

$$P(Z > 0) = \int\limits_{0}^{\infty} f(z)\,\mathrm{d}z = \iint\limits_{\Omega(S-F>0)} f(S, F)\,\mathrm{d}S\,\mathrm{d}F. \tag{9.20}$$

In order to obtain the numerical values of H we must know (determine) the distribution law of the random quantity Z that is functionally dependent on two continuous random quantities X and Y

$$Z = \varphi(X, Y), \tag{9.21}$$

at their known joint probability density distribution law $f(x, y)$ (See Chap. 1).

Generally the distribution law $F(z)$ of the random quantity Z is

$$F(z) = P(Z < z) = P\left[(X, Y) \subset D\right] = \iint\limits_{D(Z<z)} f(x, y)\,\mathrm{d}x\,\mathrm{d}y. \tag{9.22}$$

The mathematical problem of determining $F(z)$ has reduced to taking the following double integral:

$$F(z) = \iint\limits_{D(Z<z)} f(x,y)\, dx\, dy.$$

For the case of $Z = X + Y$ the integral is taken over the area D shown in Fig. 9.11, where

$$x + y < z,$$

therefore, fixing z and assuming that $y = z - x$, we obtain the specific limits of integration

$$F(z) = \iint\limits_{D} f(x,y)\, dx\, dy = \int\limits_{-\infty}^{\infty} \left[\int\limits_{-\infty}^{z-x} f(x,y)\, dy \right] dx. \tag{9.23}$$

We obtain the density of distribution $f(z)$ by differentiating (9.23) with respect to z that enters into the upper limit of the integral as a parameter

$$f(z) = \frac{\partial F(z)}{\partial z} = \int\limits_{-\infty}^{\infty} \left\{ \frac{\partial}{\partial z} \int\limits_{-\infty}^{z-x} f(x,y)\, dy \right\} dx = \int\limits_{-\infty}^{\infty} f(x, z-x)\, dx. \tag{9.24}$$

As the random quantities X and Y are equivalent, we could exclude not y, but x and obtain another expression for the probability density distribution law $f(z)$

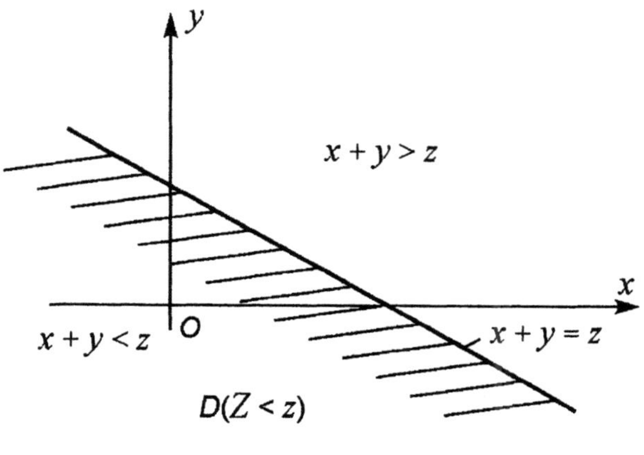

Fig. 9.11.

$$f(z) = \int_{-\infty}^{\infty} f(z-y, y)\, dy. \tag{9.25}$$

If X and Y are independent, the density of Z distribution is equal to

$$f(z) = \int_{-\infty}^{\infty} f_1(x)\, f_2(z-x)\, dx \tag{9.26}$$

or

$$f(z) = \int_{-\infty}^{\infty} f_1(z-y)\, f_2(y)\, dy.$$

Integrating the right-hand side of (9.26) over x, we obtain a function dependent on z.

When determining the no-failure operation probability, we must determine the distribution function $F(z)$ of the random quantity z that is equal to the difference of two random quantities s and f_0

$$z = s - f_0$$

at the known joint distribution law $f(s, f_0)$, i.e.

$$F(z) = \iint_{D(Z<z)} f(s, f_0)\, ds\, df_0.$$

In the shaded area shown in Fig. 9.12 $s - f_0 < z$. Therefore, the integration over the area D as in the previous case we can replace by integration over f_0 at a fixed z $(f_0 = s - z)$, and then over s

$$\iint_D f(s, f)\, ds\, df = \int_{-\infty}^{\infty} \left[\int_{s-z}^{\infty} f(s, f_0)\, df_0 \right] ds. \tag{9.27}$$

Differentiating (9.27) with respect to z, we obtain

$$f(z) = \int_{-\infty}^{\infty} \left[\frac{\partial}{\partial z} \int_{s-z}^{\infty} f(s, f_0)\, df_0 \right] ds = \int_{-\infty}^{\infty} f(s, s-z)\, ds.$$

If the random quantities s and f_0 are independent, we have

$$f(z) = \int_{-\infty}^{\infty} f_1(s)\, f_2(s-z)\, ds \tag{9.28}$$

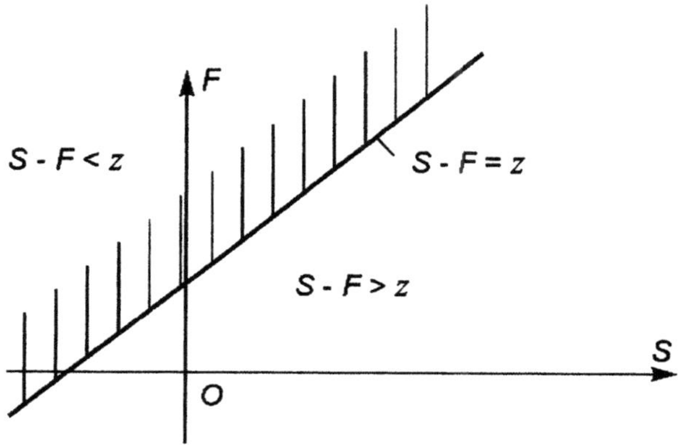

Fig. 9.12.

or, changing the order of integration,

$$f(z) = \int_{-\infty}^{\infty} \frac{\partial}{\partial z} \left[\int_{-\infty}^{z+f} f(s, f_0)\, \mathrm{d}s \right] \mathrm{d}f_0 = \int_{-\infty}^{\infty} \frac{\partial}{\partial z} f(z + f_0, f_0)\, \mathrm{d}f$$

$$+ \int_{-\infty}^{\infty} f_1(z + f_0)\, f_2(f_0)\, \mathrm{d}f_0. \tag{9.29}$$

Let us consider the case of s and f_0 having normal distributions

$$f(s) = \frac{1}{\sqrt{2\pi}\sigma_s} \exp\left\{ -\frac{(s - m_s)^2}{2\sigma_s^2} \right\}, \tag{9.30}$$

$$f(f) = \frac{1}{\sqrt{2\pi}\sigma_{f_0}} \exp\left\{ -\frac{(f - m_{f_0})^2}{2\sigma_{f_0}^2} \right\}. \tag{9.31}$$

Therefore in this case the distribution law of the random quantity z (9.28) is equal to

$$f(z) = \frac{1}{2\pi\sigma_s\sigma_{f_0}} \int_{-\infty}^{\infty} \exp\left\{ -\frac{(s - m_s)^2}{2\sigma_x^2} \right\} \exp\left\{ -\frac{(s - z - m_{f_0})^2}{2\sigma^2{}_{f_0}} \right\} \mathrm{d}s \tag{9.32}$$

or

$$f(z) = \frac{1}{2\pi\sigma_s\sigma_{f_0}} \int_{-\infty}^{\infty} \exp\left\{ -As^2 \pm 2B(z)\, s - C(z) \right\} \mathrm{d}s,$$

where

$$A = \frac{\sigma_s^2 + \sigma_{fo}^2}{2\sigma_s^2 \sigma_{fo}^2}; \quad B = \frac{m_s}{2\sigma_s^2} + \frac{z + m_{fo}}{2\sigma_{fo}^2}; \quad C = \frac{m_s^2}{2\sigma_s^2} + \frac{\left(z + m_{fo}\right)^2}{2\sigma_{fo}^2}.$$

Using the tabulated values of certain integrals, we obtain

$$f(z) = \frac{1}{2\pi\sigma_s\sigma_{fo}} \sqrt{\frac{\pi}{A}} \exp\left\{-\frac{AC - B^2}{A}\right\}$$

or by manipulations we get

$$f(z) = \frac{1}{\sqrt{2\pi}\sigma_z} \exp\left\{-\frac{\left(z - m_z\right)^2}{2\sigma_z^2}\right\}, \tag{9.33}$$

where $m_z = m_s - m_{fo}$, $\sigma_z = \sqrt{\sigma_s^2 + \sigma_{fo}^2}$.

Having determined $f(z)$, we find the probability of no-failure operation (reliability)

$$H = P(Z > 0) = \int_0^\infty f(z)\, dz. \tag{9.34}$$

Introducing a new notation

$$\beta = \frac{z - (m_s - m_{fo})}{\sigma_z},$$

we obtain

$$H = \frac{1}{2\pi} \int_{-\beta_0}^{\infty} \exp\left\{-\frac{\beta^2}{2}\right\} d\beta, \quad (\beta_0 = \beta\,|_{z=0}). \tag{9.35}$$

The plot of an integrand is shown in Fig. 9.13. The function $f(\beta)$ is symmetric about the vertical axis, therefore

$$\frac{1}{\sqrt{2\pi}} \int_{-\beta_0}^{\infty} \{\ldots\} = \frac{1}{\sqrt{2\pi}} \int_{-\infty}^{\beta_0} \{\ldots\}$$

$$= \frac{1}{\sqrt{2\pi}} \int_{-\infty}^{0} \{\ldots\} + \frac{1}{\sqrt{2\pi}} \int_{0}^{\beta_0} \{\ldots\} = 0,5 + \frac{1}{\sqrt{2\pi}} \int_{0}^{\beta_0} \{\ldots\}.$$

We obtain the final numerical value of reliability

$$H = 0,5 + \frac{1}{\sqrt{2\pi}} \int_0^{\frac{m_s - m_F}{\sigma_z}} \exp\left\{-\frac{\beta^2}{2}\right\} d\beta. \tag{9.36}$$

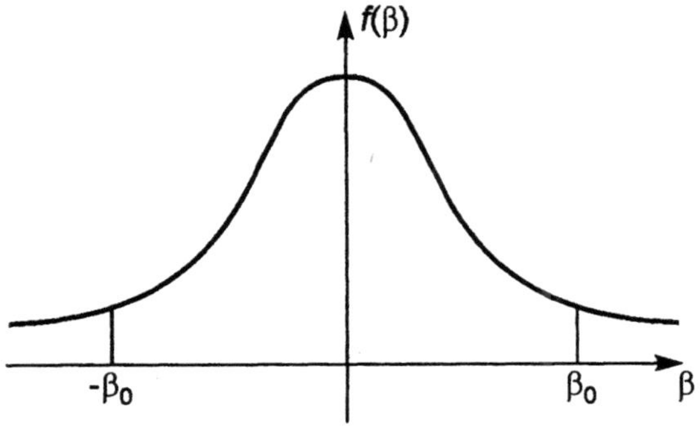

Fig. 9.13.

The integral entering the right-hand side of (9.41) is a tabulated integral (a "probability integral").

Example 9.1. Let us return to the problem of estimating the "strength" of the rod tensioned by the force P (Fig. 9.1), using determinate (9.10) and probability (9.14) criteria. At deterministic σ_y and $\sigma = \frac{P}{F}$ the "strength" is estimated by a factor of safety n

$$n = \frac{\sigma_y}{\sigma}. \tag{9.37}$$

Given the random scatters of σ_y and σ a factor $\tilde{n}$, equal to the ratio of mathematical expectations of σ_y and σ, is an analogue of the factor n

$$\tilde{n} = \frac{m_{\sigma_y}}{m_\sigma}. \tag{9.38}$$

But the factor $\tilde{n}$ does not take into account the root-mean-square scatters of σ_y and σ. As mentioned above, when estimating "strength", we must use the "probabilistic" criterion of "strength", i.e. the probability of no-failure operation, in order to take into consideration all probability characteristics of σ_y and σ.

We assume that the random quantities σ_y and σ have normal distributions (Fig. 9.14), i.e. the numerical values of the mathematical expectations (m_{σ_y} and m_σ) and standard deviations σ_{σ_y} and σ_σ are known.

If random quantities s and f are normally distributed, the reliability (the probability of no-failure operation) is equal to

$$H = P\left[(S - F) > 0\right] = 0.5 + \frac{1}{\sqrt{2\pi}} \int\limits_0^{\frac{m_{\sigma_T} - m_\sigma}{\sigma_z}} \exp\left\{-\frac{\beta^2}{2}\right\} \mathrm{d}\beta,$$

where

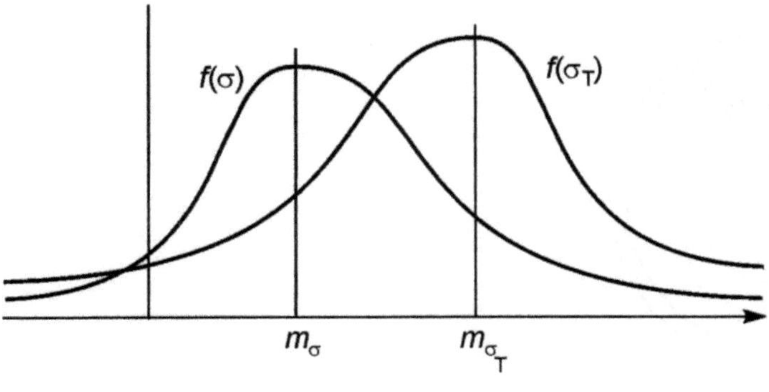

Fig. 9.14.

$$\sigma_z = \sqrt{\sigma_{\sigma_y}^2 + \sigma_\sigma^2}.$$

The results of analysis at various values of the probability characteristics σ_y and σ are presented in Table 9.1.

Table 9.1.

Mathematical expectation of yield limit, m_{σ_r}	Mathematical expectation of stress, m_σ	Mean square deviation, σ_{σ_r}	Mean square deviation of stress, σ_σ	Safety factor, $\tilde{n} = \frac{m_{\sigma_t}}{m_\sigma}$	Probability of no-failure operation, $P = H$
$5 \cdot 10^4$	$2 \cdot 10^4$	$5 \cdot 10^3$	$2 \cdot 10^3$	2.5	0.9999
$5 \cdot 10^4$	$2.5 \cdot 10$	$5 \cdot 10^3$	$2.5 \cdot 10^3$	2	0.9999
$5 \cdot 10^4$	$3.33 \cdot 10$	$5 \cdot 10^3$	$3.33 \cdot 10^3$	1.5	0.9974
$5 \cdot 10^4$	$3.8 \cdot 10^4$	$5 \cdot 10^3$	$3.8 \cdot 10^3$	1.3	0.9719
$5 \cdot 10^4$	$4.17 \cdot 10^4$	$5 \cdot 10^3$	$4.17 \cdot 10^3$	1.2	0.8997
$5 \cdot 10^4$	$4.35 \cdot 10^4$	$5 \cdot 10^3$	$4.37 \cdot 10^3$	1.15	0.8365
$5 \cdot 10^4$	$4.55 \cdot 10^4$	$5 \cdot 10^3$	$4.55 \cdot 10^3$	1.1	0.7486

When determining the reliability H, it was assumed that the root-mean-square values of the scatters σ_y and σ_σ were equal to 10 % of their mathematical expectations. It follows from Table 9.1 that the probability of failure R ($R = 1 - H$) is not zero for all values of the safety factor. As one would expect the probability of no-failure operation is higher at large factors of safety $\tilde{n}$ and lower at small $\tilde{n}$. How will the probabilities of no-failure operation corresponding to the factors of safety $\tilde{n} = 1.15$ and $\tilde{n} = 1.1$ change, if we take a higherquality material and reduce the possible scatter of load?

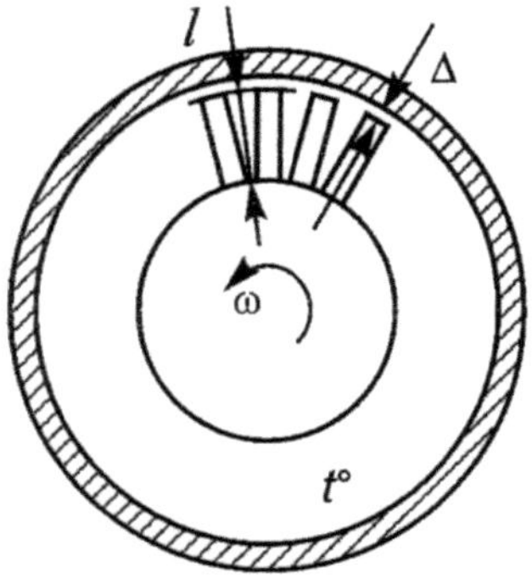

Fig. 9.15.

For example, the root-mean-square scatters of σ_y and σ (dependent on external load) are equal to 5 % of their average values. In this case the probabilities of nofailure operation are equal to: 1) at $n = 1.15$, $H = 0.9726$; and 2) at $n = 1.1$, $H = 0.9082$, which is respectively 16 % and 20 % higher than their values at the 10 % scatter (Table 9.1) of the root-mean-square values of σ_{σ_y} and σ_σ at the same factors of safety. Numerical values of the probability of no-failure operation enable us to investigate the sensitivity of a structure to the possible scatters of σ_y and the random force P.

As stated above, the absolute values of probabilities of no-failure operation are of little use, but the possibility of finding out how changes in the probability characteristics of allowable stresses and external load influence serviceability of a structure allows us to make an objective estimation of the structures quality. By comparing the probabilities of no-failure operation, for instance, at $\tilde{n} = 1.1$, we see, that a 5 % reduction in the root-mean-square values of σ_y and σ has resulted in the 20 % higher reliability. Let us determine, by way of example, the probability of no-failure operation of the bladed disk of a stage of the compressor of a turbojet engine (Fig. 9.15). During the operation of the engine the blades are elongated due to the axial distributed centrifugal forces resulting from the rotation of the disk and because of a temperature t°. The angular velocity of the disks rotation ω and the temperature t° have a scatter; as does a gap Δ, therefore, in order to estimate the reliability of the disk it is necessary to determine the probability that the inequality $(\Delta - \Delta l) > 0$ will be fulfilled

$$H = P\left[(\Delta - \Delta l) > 0\right].$$

Let us assume that distribution law of the gap Δ is equiprobabilistic (Fig. 9.16)

$$f(\Delta) = \begin{cases} a & \Delta_1 < \Delta < \Delta_2 \\ 0 & \Delta < \Delta_1 \text{ or } \Delta > \Delta_2 \end{cases} \qquad \left(a = \frac{1}{\Delta_2 - \Delta_1}\right), \tag{9.39}$$

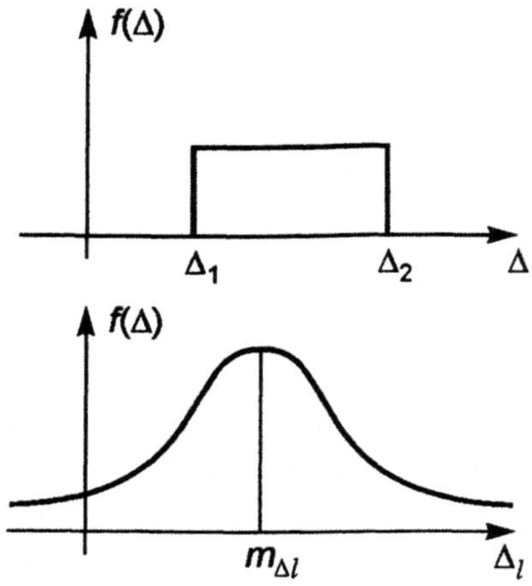

Fig. 9.16.

and Δl has the normal distribution (Fig. 9.16)

$$f(\Delta l) = \frac{1}{\sqrt{2\pi}\sigma_{\Delta l}} \exp\left\{-\frac{(\Delta l - m_{\Delta l})^2}{2\sigma_{\Delta l}^2}\right\}. \tag{9.40}$$

Having introduced a random quantity $z = \Delta - \Delta l$, we find a distribution function $F(z)$

$$F(z) = \iint\limits_{\Omega(Z<z)} f(\Delta, \Delta l)\, \mathrm{d}\Delta\, \mathrm{d}\Delta l$$

$$= \int\limits_{\Delta_1}^{\Delta_2} \mathrm{d}\Delta \left\{ \int\limits_{\Delta - z}^{\infty} \left(\frac{1}{\Delta_2 - \Delta_1}\right) \frac{1}{\sqrt{2\pi}\sigma_{\Delta l}} \exp\left\{-\frac{(\Delta l - m_{\Delta l})^2}{2\sigma_{\Delta l}^2}\right\}\right\} \mathrm{d}\Delta l.$$

Differentiating with respect to z, we obtain

$$\frac{\mathrm{d}F(z)}{\mathrm{d}z} = f(z) = \frac{1}{\Delta_2 - \Delta_1} \cdot \frac{1}{\sqrt{2\pi}\sigma_{\Delta l}} \int\limits_{\Delta_1}^{\Delta_2} \exp\left\{-\frac{[(\Delta - z) - m_{\Delta l}]^2}{2\sigma_{\Delta l}^2}\right\} \mathrm{d}\Delta. \tag{9.41}$$

Having introduced a new variable t

$$t = \frac{\Delta - (z + m_{\Delta l})}{\sigma_{\Delta l}},$$

we obtain

$$f(z) = \frac{1}{(\Delta_2 - \Delta_1)\sqrt{2\pi}} \int_{t_1}^{t_2} \exp\left\{-\frac{t^2}{2}\right\} \mathrm{d}t,$$

where

$$t_1 = \frac{\Delta_1 - (z + m_{\Delta l})}{\sigma_{\Delta l}};$$

$$t_2 = \frac{\Delta_2 - (z + m_{\Delta l})}{\sigma_{\Delta l}}.$$

For each given z_j the integral is expressed in terms of functions $\Phi(t_1)$ and $\Phi(t_2)$, that are presented in handbooks on higher mathematics

$$f(z_j) = \frac{1}{\Delta_2 - \Delta_1}\left[\Phi(t_2, z_j) - \Phi(t_1, z_j)\right]. \tag{9.42}$$

By discretely changing z, we obtain the numerical values of the distribution law $f(z)$ of the random quantity z at discrete points (Fig. 9.17).

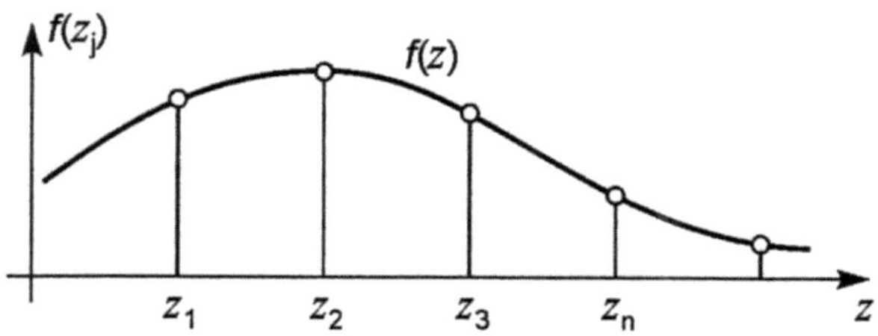

Fig. 9.17.

Having used spline functions, we can obtain, confining ourselves to some finite value z_n, which of the requirements on an accuracy a continuous function $f(z)$, and then determine also the no-failure operation probability on the interval $(0, z_n)$

$$P(z > 0) = \int_0^{\infty} f(z)\,\mathrm{d}z \approx \int_0^{z_n} f(z)\,\mathrm{d}z.$$

Having taken, for example, the following numerical values:

$$\Delta_1 = 0.4\,\mathrm{cm}, \quad \Delta_2 = 0.8\,\mathrm{cm}, \quad m_{\Delta l} = 0.6\,\mathrm{cm}, \quad \sigma_{\Delta l} = 0.06\,\mathrm{cm}$$

$$\left(t_1 = \frac{(0.1 - z)\,10^2}{12}, \quad t_2 = \frac{(0.5 - z)\,10^2}{12}\right),$$

we obtain as a result of calculations

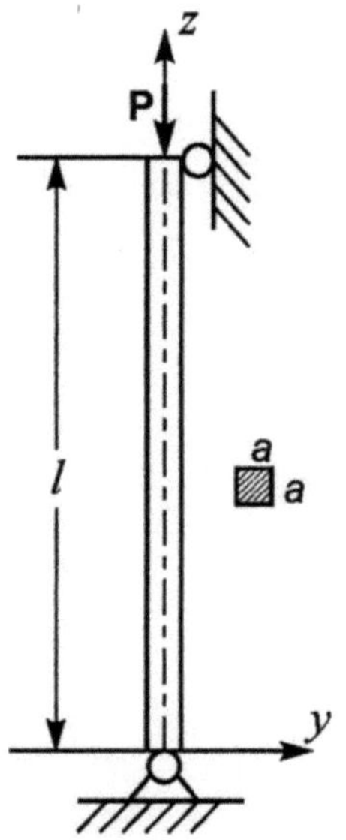

Fig. 9.18.

$$P\,(z > 0) = 0.972.$$

Let us estimate the reliability of the performance of a straight rod loaded with a compressive force P (Fig. 9.18). The limiting state of the rod in this case is caused by the loss of stability. Therefore the condition

$$P_{cr} - P > 0, \tag{9.43}$$

should be satisfied for the normal operation of the rod, where P_{cr} is the critical compressive force

$$P_{cr} = \frac{\pi^2 EJ}{l^2}.$$

As P_{cr} and P have random scatters, we must determine the probability of no-failure operation

$$P\left[(P_{cr} - P) > 0\right].$$

Let us take the following distribution laws for P_{cr} and P

$$f_1\,(P_{cr}) = \delta\,(P_{cr} - P_{cr0})\,;$$

$$f_2\,(P) = \frac{1}{\sqrt{2\pi}\sigma_p} \exp\left\{-\frac{(P - m_p)^2}{2\sigma_p^2}\right\},$$

where δ is the delta function.

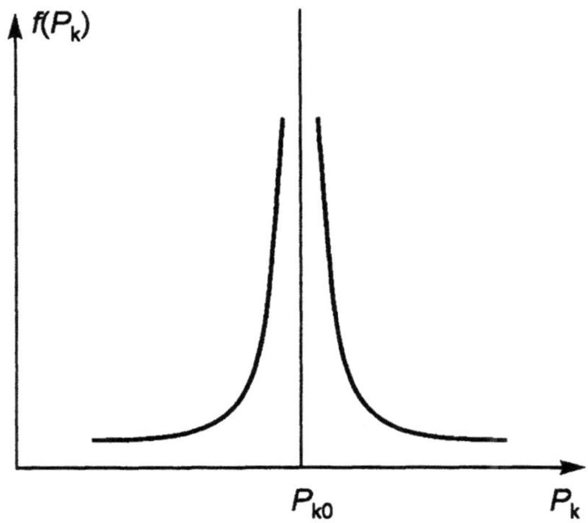

Fig. 9.19.

The distribution law $f(P_{cr})$ is shown in Fig. 9.19, i.e. P_{cr} is a deterministic quantity equal to P_{cr0}.

Let us introduce a random quantity z

$$z = P_{cr} - P > 0. \tag{9.44}$$

In order to determine the distribution law $f(z)$ let us find the distribution function $F(z)$, dependent on the joint distribution law of P_{cr} and P. Therefore at independent P_{cr} and P we can present this joint distribution law as

$$f(P_{cr}, P) = \delta(P_{cr} - P_{cr0}) f(P). \tag{9.45}$$

The distribution law $F(z)$ is

$$F(z) = \iint\limits_{D(Z<z)} \delta(P_{cr} - P_{cr0}) f(P) \, \mathrm{d}P_{cr} \, \mathrm{d}P. \tag{9.46}$$

Fixing z, we obtain

$$P = P_{cr} - z,$$

therefore,

$$F(z) = \int\limits_{-\infty}^{\infty} \left\{ \int\limits_{P_{cr}-z}^{\infty} f(P) \, dP \right\} \delta(P_{cr} - P_{cr0}) \, \mathrm{d}P_{cr}. \tag{9.47}$$

Differentiating $F(z)$ with respect to z, we obtain

$$f(z) = \frac{\mathrm{d}F(z)}{\mathrm{d}z} = \int_{-\infty}^{\infty} f(P_{cr} - z)\,\delta(P_{cr} - P_{cr0})\,\mathrm{d}P_{cr}$$

or

$$f(z) = f(P_{cr0} - z).$$

Therefore

$$f(z) = \frac{1}{\sqrt{2\pi}\sigma_p} \exp\left\{-\frac{(P_{cr0} - z - m_p)^2}{2\sigma_p^2}\right\}$$

or

$$f(z) = \frac{1}{\sqrt{2\pi}\sigma_p} \exp\left\{-\frac{[z - (P_{cr0} - m_p)]^2}{2\sigma_p^2}\right\}. \tag{9.48}$$

Assuming that

$$t = \frac{z - (P_{cr} - m_p)}{\sigma_p},$$

we obtain the probability of no-failure operation

$$P(z > 0) = \frac{1}{\sqrt{2\pi}} \int_{-\frac{P_{cr0} - m_p}{\sigma_p}}^{\infty} \exp\left\{-\frac{t^2}{2}\right\} \mathrm{d}t. \tag{9.49}$$

9.5 Determination of Reliability at the Linear Dependence of a Stress State on Random Loads

Let us consider a more general case, when a random quantity F, characterizing the loaded state of a structure linearly depends on concentrated and distributed forces (Fig. 9.20)

$$F = k_1 P + k_2 g, \tag{9.50}$$

where P is the concentrated force and g is the distributed force. For instance, the maximum stress at the clamped end for the rod shown in Fig. 9.20 linearly depends on P and g

$$\sigma_{\max} = \frac{M(0)}{W} = \frac{\beta_1 P + \beta_2 g}{W} = k_1 P + k_2 g. \tag{9.51}$$

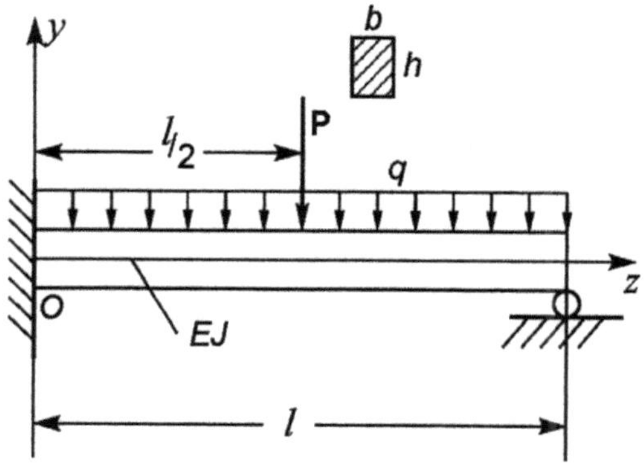

Fig. 9.20.

The probability of no-failure operation in this case is equal to

$$P\left(\sigma_y - \sigma_{\max} > 0\right) = P\left[\left(\sigma_y - \left(k_1 P + k_2 g\right)\right) > 0\right] . \tag{9.52}$$

Depending on specific conditions both dependent and independent P and g are possible. Let us first consider the case of P and g being independent and having normal distribution laws. As a preliminary let us consider the general case where random quantity Y is

$$Y = \sum_{i=1}^{n} a_i X_i \tag{9.53}$$

and it is required to find its distribution law $f(y)$ when the (normal) laws of X_i distribution are known. The operation of determining the distribution law of a sum of the independent random quantities X_i, having normal distribution is referred to as a composition of normal laws [13]. If we have a composition of normal laws, we obtain a normal law. Therefore the distribution law $f(y)$ takes the form

$$f(y) = \frac{1}{\sqrt{2\pi}\,\sigma_y} \exp\left\{-\frac{\left(y - m_y\right)^2}{2\sigma_y^2}\right\}, \tag{9.54}$$

where

$$m_y = \sum_{i=1}^{n} a_i m_{x_i}, \qquad \sigma_y = \sqrt{\sum_{i=1}^{n} a_i^2 \sigma_{x_i}^2}.$$

For the considered problem $y = k_1 P + k_2 g$, therefore, we have

$$f(y) = \frac{1}{\sqrt{2\pi}\,\sigma_y} \exp\left\{-\frac{\left(y - m_y\right)^2}{2\sigma_y^2}\right\},$$

where $m_y = k_1 m_P + k_2 m_g, \quad \sigma_y = \sqrt{k_1^2 \sigma_P^2 + k_2^2 \sigma_g^2}.$

Considering that the random quantities S and Y are independent and S has normal distribution, we obtain a distribution law $f(z)$ for $(Z = S - Y)$

$$f(z) = \frac{1}{\sqrt{2\pi}\sigma_z} \exp\left\{-\frac{(z-m_z)^2}{2\sigma_z^2}\right\},$$

where $m_z = k_1 + m_P + k_2 m_g - m_s$, $\sigma_z = \sqrt{k_1^2\sigma_P^2 + k_2^2\sigma_g^2 + \sigma_s^2}$.

The probability of the systems no-failure operation is

$$H = P\left[(S - (k_1 P + k_2 g)) > 0\right] = \int\limits_0^\infty f(z)\,\mathrm{d}z.$$

Going on to a new variable $t = \dfrac{Z - m_Z}{\sigma_Z}$, we obtain

$$P(z > 0) = H = \frac{1}{\sqrt{2\pi}} \int\limits_{-\frac{m_Z}{\sigma_Z}}^{\infty} \exp\left\{-\frac{t^2}{2}\right\} \mathrm{d}t.$$

The algorithm of determining the probability of no-failure operation at a single or low-cycle loading presented in this section allows the solution of problems pertaining to the optimization of structure, as well as to the analysis of structures with a given reliability.

9.6 Determination of the Probability of No-Failure Operation at the Nonlinear Dependence of the Random Quantity *F* on External Loads

The problems of determining the probability of no-failure operation considered in the previous sections referred to the case where the function F linearly depended on loads and the state of stress of structural elements was uniaxial. Let us consider more complex cases where the function F describing the real state of a system non-linearly depends on external loads. It occures , for example, when the state of stress of structural elements is two-dimensional. A straight, rectangular cross-section rod loaded with forces P_1 and P_2 and a twisting moment M is shown in Fig. 9.21 *a*. The laws of distribution of P_1, P_2 and M are considered to be known. There is a biaxial state of stress at the points A and B of the rod's section (see Fig. 9.21 *b, c*). The dangerous section is the section at $Z = 0$.

At point A (Fig. 9.21 *a*) we have

$$\sigma_{P_1} = \frac{P_1}{bh}, \quad \tau_A = \frac{M}{\alpha h b^2}, \quad \sigma_{P_2} = 0,$$

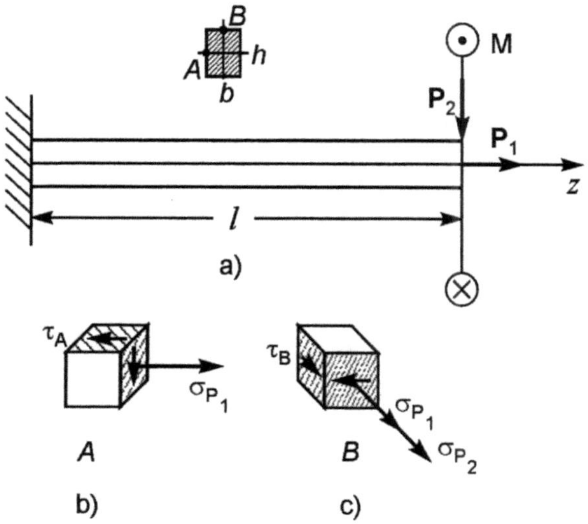

Fig. 9.21.

where α is the Saint-Venant factor (Table 9.2). At point B (Fig. 9.21 *b*) we have

$$\sigma_{P_1} = \frac{P_1}{bh}; \quad \sigma_{P_2} = \frac{P_2 l}{\frac{bh^2}{6}}; \quad \tau_B = \eta \tau_A .$$

Let us determine the probability of the equivalent stress at points A and B being less than the yield strength, i.e.

$$P\left[(\sigma_y - \sigma_e) > 0\right] . \tag{9.55}$$

Table 9.2.

$\frac{h}{b}$	1	1.5	1.75	2	2.5	3
α	0.208	0.231	0.239	0.246	0.258	0.264
η	1	0.859	0.820	0.795	0.766	0.753

$\frac{h}{b}$	4	6	8	10	∞
α	0.282	0.299	0.207	0.313	0.333
η	0.745	0.743	0.742	0.743	0.742

The probability P is the probability of the elements no-failure operation. Equivalent stress according to the distortion energy hypothesis is equal to

$$\sigma_e = \sqrt{\sigma^2 + 3\tau^2}.$$

At points A and B the equivalent stress is equal respectively to

$$\sigma_e^{(A)} = \sqrt{a_1 P_1^2 + a_2 M^2}, \tag{9.56}$$

$$\sigma_e^{(B)} = \sqrt{\left(a_1 P_1 + a_3 P_2\right)^2 + a_4 M^2}, \tag{9.57}$$

where $a_1 = (bh)^{-2}$, $a_2 = 3/\left(\alpha h b^2\right)^2$, $a_3 = \left(\dfrac{6l}{bh^2}\right)$, $a_4 = 3\eta/\left(\alpha h b^2\right)^2$. If $h > b$, the shear stress (τ_A) at point A is greater than that at point B, therefore, depending on the rods parameters $(l,\ b,\ h)$ and the numerical values of P_1, P_2 and M, the maximum equivalent stress in the section can be either at point A or at point B. Therefore the probability of no-failure operation of the rod will be equal to the least of the probabilities

$$P\left[\left(\sigma_y - \sigma_e^{(A)}\right) > 0\right],$$
$$P\left[\left(\sigma_y - \sigma_e^{(B)}\right) > 0\right].$$

The main difficulty in finding the probabilities of no-failure operation consists in the determination of the laws of distribution of $\sigma_e^{(A)}$ and when equivalent stresses non-linearly depend on the external load.

Let us consider the algorithm of approximate solution of this problem. Let us assume that

$$P_1 = P_{10} + \Delta P_1, \quad P_2 = P_{20} + \Delta P_2, \quad M = M_0 + \Delta M,$$

where P_{10}, P_{20}, M_0 are mathematical expectations of random quantities and ΔP_1, ΔP_2 and ΔM are random scatters that follow certain distribution laws, for example, normal distribution laws (Fig. 9.22) or Rayleigh distribution laws (Fig. 9.23). We consider that ΔP_1, ΔP_2 and ΔM are "small" random quantities (compared with mathematical expectations). For instance, if the random scatters follow normal distribution laws, then, using the three sigma rule, we can find their maximum values at zero mathematical expectations $\max \Delta P_1 = \left|3\sigma_{\Delta P_1}\right|$, $\max \Delta P_2 = \left|3\sigma_{\Delta P_2}\right|$, $\max \Delta M = \left|3\sigma_{\Delta M}\right|$. Therefore, if $\left|\max \Delta P_j\right| \ll P_{0j}\ (j = 1, 2)$ and $\left|\max \Delta M\right| \ll M_0$, the random quantities $\Delta P_j\ (j = 1, 2)$ and ΔM can be considered small. If the random quantities follow Rayleigh distribution laws, we can obtain their maximum values from conditions

$$P = \int\limits_0^{\max \Delta P_j} f\left(\Delta P_j\right) \mathrm{d}\Delta P_j = 0.98. \tag{9.58}$$

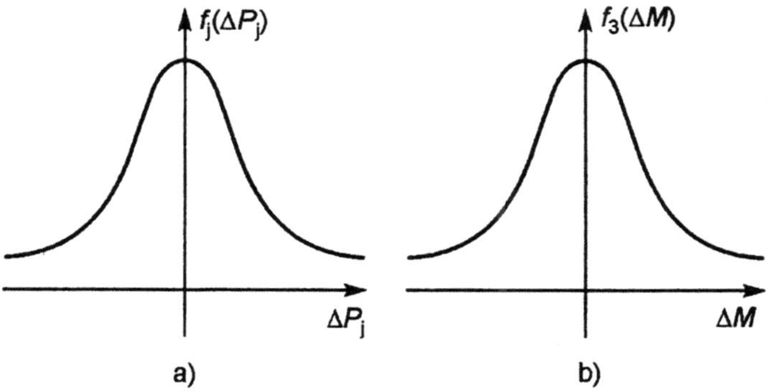

Fig. 9.22.

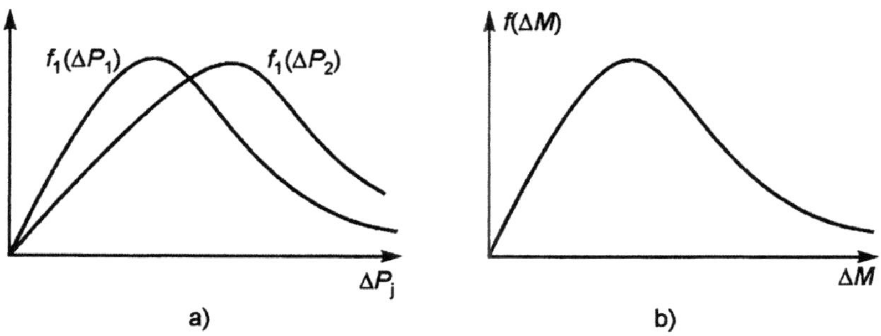

Fig. 9.23.

Similarly, we can determine the maximum values of random scatters for other distribution laws, which is necessary to justify the assumption that ΔP_1, ΔP_2, ΔM are small and to linearize nonlinear functions (9.56) and (9.57).

Let us consider the general case of the function F depending on a finite number of random quantities x_j :

$$H = P\left[(S - F(x_j)) > 0\right], \tag{9.59}$$

where $x_j = m_{0j} + \Delta x_j$; ($\Delta x_j$ are small quantities).

To find an approximate solution we must expand the function F in a series. A Taylor series for the function F in the vicinity of the point $F(m_{x_1}, m_{x_2}, \ldots, m_{x_n})$ is

$$F = F_0\left(m_{x_j}\right) + \sum_{j=1}^{n} \left.\frac{dF}{dx_j}\right|_{x_j = m_{x_j}} \Delta x_j + \frac{1}{2}\sum_{j=1}^{n}\sum_{i=1}^{n}\left(\left.\frac{\partial^2 F}{\partial x_j \partial x_i}\right|_{\substack{x_j = m_{x_j} \\ x_i = m_{x_i}}}\right)\Delta x_j \Delta x_i + \ldots .$$

If we restrict ourselves by the linear part of the expansion, we obtain

$$F = F_0 + \sum_{j=1}^{n} c_j \Delta x_j, \tag{9.60}$$

where

$$c_j = \left.\frac{\mathrm{d}F}{\mathrm{d}x_i}\right|_{x_i = m_{x_i}}.$$

The probability characteristics of the stochastic function include:
1). The mathematical expectation of F is equal to

$$m_F = F_0. \tag{9.61}$$

2). The variance of the random quantity F (9.60) is

$$D_F = M\left[\overset{\circ}{F}{}^2\right] = M\left[\left(\sum_{i=1}^{n} c_i \Delta x_i\right)^2\right] = \sum_{i=1}^{n} c_i^2 D_{x_i} + \sum_{j \neq i} c_i c_j k_{x_i x_j}. \tag{9.62}$$

For the independent random quantities Δx_j we have respectively

$$\begin{aligned} m_F &= F_0 \\ D_F &= \sum_{i=1}^{n} c_i^2 D_{x_i} \end{aligned} \tag{9.63}$$

As a result we obtain

$$H = P\left[(S - F) > 0\right] = P\left[\left(S - \left(F_0 + \sum_{i=1}^{n} c_i \Delta x_i\right)\right) > 0\right]. \tag{9.64}$$

For the further solution we must obtain the distribution law of the following random quantity

$$Y = F_0 + \sum_{i=1}^{n} c_i \Delta X_i, \tag{9.65}$$

where ΔX_i are independent random quantities whose distribution laws are considered known. Let us confine ourselves to the case of the distribution laws of ΔX_i being normal. Then, using characteristic functions, we obtain the distribution law $f(y)$. The characteristic function g_y of the random quantity y is equal to (t is a parameter)

$$g_y(t) = M\left[e^{itY}\right] = M\left[\exp\left\{it\left(F_0 + \sum_{i=1}^{n} c_i \Delta x_i\right)\right\}\right]$$
$$= e^{itF_0} M\left[\prod_{i=1}^{n} e^{itc_i \Delta x_i}\right] = e^{itF_0} \prod_{i=1}^{n} g_{\Delta x_i}(c_i t)\,. \tag{9.66}$$

The characteristic function of the random quantity $x_j = c_j \Delta x_j$ having normal distribution is [13, 29]

$$g_{x_i}^{(t)} = g_{\Delta x_i}(t) = \exp\left\{itm_{x_j} - \frac{t^2\sigma_{x_j}^2}{2}\right\} = \exp\left\{it\left(c_j m_{\Delta x_j}\right) - \frac{t^2 c_j^2 \sigma_{\Delta x_j}^2}{2}\right\}.$$

Therefore, for the characteristic function $g_y(t)$ we obtain

$$g_y(t) = e^{itF_0} \prod_{j=1}^{n} \exp\left\{it\left(c_j m \Delta x_j\right) - \frac{t^2 c_j \sigma \Delta x_j}{2}\right\}$$
$$= \exp\left\{i\left(tF_0 + \sum_{j=1}^{n} c_j m \Delta x_j\right) - \frac{t^2\left(\sum_{j=1}^{n} c_j^2 \sigma_{\Delta x_j}^2\right)}{2}\right\}$$

or

$$g_y(t) = \exp\left\{itm_y - \frac{t^2\sigma_y^2}{2}\right\}, \tag{9.67}$$

We can show that the characteristic function $g_y(t)$ corresponding to the normal distribution law [29]

$$f(y) = \frac{1}{\sqrt{2\pi}\sigma_y} \exp\left\{-\frac{(y-m_y)^2}{2\sigma_y^2}\right\}$$

is equal to

$$g_y(t) = \exp\left\{itm_y - \frac{t^2\sigma_y^2}{2}\right\}.$$

The functions $f(y)$ and $g_y(t)$ are connected by a Fourier transform, therefore the following normal distribution law

$$f(y) = \frac{1}{\sqrt{2\pi}\sigma_y} \exp\left\{-\frac{(y-m_y)^2}{2\sigma_y^2}\right\}, \tag{9.68}$$

where $m_y = F_0 + \sum_{i=1}^{n} c_i m_{\Delta x_i}$; $\sigma_y^2 = \sum_{i=1}^{n} c_i^2 \sigma_{\Delta x_i}^2$ corresponds to characteristic function (9.67).

As regards the considered problem, for example, for the point B after the linearization of the equivalent stress we obtain

$$Y = \sigma_e^{(B)} = \sigma_{e0}^{(B)} + c_1 \Delta P_1 + c_2 \Delta P_2 + c_3 \Delta M, \tag{9.69}$$

where

$$\sigma_e^{(B)} = \sqrt{(a_1 P_{10} + a_3 P_{20})^2 + a_4 M_0^2},$$

$$c_1 = \frac{\partial \sigma_e^{(B)}}{\partial P_1} = \frac{a_1 (a_1 P_{10} + a_3 P_{20})}{\sqrt{(a_1 P_{10} + a_3 P_{20})^2 + a_4 M_0^2}},$$

$$c_2 = \frac{\partial \sigma_e^{(B)}}{\partial P_2} = \frac{a_3 (a_1 P_{10} + a_3 P_{20})}{\sqrt{(a_1 P_{10} + a_3 P_{20})^2 + a_4 M_0^2}},$$

$$c_2 = \frac{\partial \sigma_e^{(B)}}{\partial M} = \frac{a_4 M}{\sqrt{(a_1 P_{10} + a_3 P_{20})^2 + a_4 M_0^2}}$$

At the normally distributed ΔP_1, ΔP_2 and ΔM (Fig. 9.22) the distribution law $f\left(\sigma_e^{(B)}\right)$ is normal

$$f\left(\sigma_e^{(B)}\right) = \frac{1}{\sqrt{2\pi}\sigma_{\sigma_e^{(B)}}} \exp\left\{-\frac{\left(\sigma_e^{(B)} - m_{\sigma_e^{(B)}}\right)^2}{2\sigma_{\sigma_e^{(B)}}}\right\},$$

where

$$m_{\sigma_e^{(B)}} = \sigma_{e0}^{(B)};$$

$$\sigma_{\sigma_e^{(B)}}^2 = c_1^2 \sigma_{\Delta P_1}^2 + c_2^2 \sigma_{\Delta P_2}^2 + c_3^2 \sigma_{\Delta M}^2.$$

Therefore the distribution law $f(z)$ of the random quantity z equal to $z = S - F = \sigma_y - \sigma_e^{(B)}$ at a normal distribution law is

$$f(z) = \frac{1}{\sqrt{2\pi}\sigma_z} \exp\left\{-\frac{(z - m_z)^2}{2\sigma_z^2}\right\},$$

where

$$m_z = m_{\sigma_e} - \sigma_{e0}^{(B)};$$

$$\sigma_z^2 = c_1^2 \sigma_{\Delta P_1}^2 + c_2^2 \sigma_{\Delta P_2}^2 + c_3^2 \sigma_{\Delta m}^2 + \sigma_{\sigma_y}^2.$$

The probability of no-failure operation is equal to

$$P\left[(\sigma_y - \sigma_e) > 0\right] = \frac{1}{\sqrt{2\pi}} \int\limits_{-\frac{m_z}{\sigma_z}}^{\infty} \exp\left\{-\frac{t^2}{2}\right\} \mathrm{d}t. \tag{9.70}$$

Let us consider the case of the random quantities ΔP_1, ΔP_2 having normal distributions (Fig. 9.22 *a*), and ΔM having the Rayleigh distribution (Fig. 9.23 *b*).

It is required to determine the reliability (9.64) for the case, where random quantities entering in the function $F(x_j)$

$$F(x_j) = \sum_{j=1}^{n} c_j x_j + b \tag{9.71}$$

have different distribution laws. For example, x_j $(j = 1, 2, \ldots, n-1)$ have normal distribution laws, and x_n has the Rayleigh distribution. Let us present (9.71) as

$$F(x_j) = y + c_n x_n = y + x_n^{(1)}, \tag{9.72}$$

where

$$y = \sum_{i=1}^{n-1} c_i x_i + b.$$

Having used characteristic functions, we obtain the distribution law of y that is a normal distribution law (9.68).

Then, we find the composition of the distribution laws $f(y)$ and $f_n\left(x_n^{(1)}\right)$.

Let us introduce a random quantity y_1 equal to

$$y_1 = y + x_n^{(1)}. \tag{9.73}$$

According to the general algorithm that determines the distribution law of the sum of independent random quantities (9.23)–(9.26), we have

$$f(y_1) = \int\limits_{0}^{\infty} f_n\left(x_n^{(1)}\right) \left[\frac{\mathrm{d}}{\mathrm{d}y_1} \int\limits_{-\infty}^{y_1 - x_n^{(1)}} f(y)\,\mathrm{d}y\right] \mathrm{d}x_n^{(1)} \tag{9.74}$$

or

$$f(y_1) = \int\limits_{0}^{\infty} f_n\left(x_n^{(1)}\right) f\left(y_1 - x_n^{(1)}\right) \mathrm{d}x_n^{(1)}. \tag{9.75}$$

The limits of the y_1 variation are from $-\infty$ to ∞. As a result we obtain

$$H = P\left[(S - Y_1) > 0\right].$$

Assuming that $z = s - Y_1$, we find the distribution law $f(z)$

$$f(z) = \int_{-\infty}^{\infty} f_s(s)\, f(s - z)\, \mathrm{d}s. \tag{9.76}$$

Having determined $f(z)$, we find the reliability of no-failure operation of a system at the action of independent random excitations that have different distribution laws

$$H = \int_{0}^{\infty} f(z)\, \mathrm{d}z. \tag{9.77}$$

If the integral is not tabulated, it can be approximately determined numerically. We must obtain in advance the values of $f(z)$ in a set of discrete points and then using spline functions we must obtain continuous function $f(z)$ on a bounded interval of z variation and to obtain H numerically.

10. Random Processes at the Action of Random Functions Bounded in Absolute Value

10.1 Introduction

During the presentation of the theory of random vibrations (see Chap.6–8) it was considered that all necessary information on random excitations (distribution laws or probability characteristics of random functions) is known, which allowed us to obtain the probability characteristics of the output based on the known probability characteristics of the input.

Besides, a necessary condition for applying methods of the theory of random processes is the repeated occurrence of a random event in practically homogeneous conditions. The application of the probability methods of analysis is worthwhile only at mass events. Very often, however, in the process of analyzing specific problems of the dynamics of mechanical systems, the necessary information on random excitations is either absent or its obtainment represents a problem that is incommensurably more complicated and labour-consuming than the subsequent solution of equations of motion.

The absence of necessary information on random forces complicates the solution of applied problems, while introduced assumptions and suggestions lead to highly approximate and poorly authentic numerical results.

When solving problems related to the random vibrations of a motor vehicle at the movement along a road with random irregularities (see Chap. 6), we considered that the necessary probability characteristics of the irregularities of the road are known. In order to obtain these probability characteristics a very large experimental work is required, therefore these characteristics are available only for a limited number of roads. Designing a structure (for example, a motor vehicle) demands that we should take into account the conditions of its operation. Only with due account of these conditions can we pose the problem of optimizing a structure and increasing its reliability. When external conditions are known (for example, the probability characteristics of the external forces which will act on a structure), we can use methods of the theory of random processes to determine internal forces (stresses) arising in the structure that allow us to judge its possible reliability. However, if the probability characteristics of external excitations are not known, these methods are of little use [32].

The service reliability of a structure substantially depends on how accurately the external excitations or a structures operation conditions that have

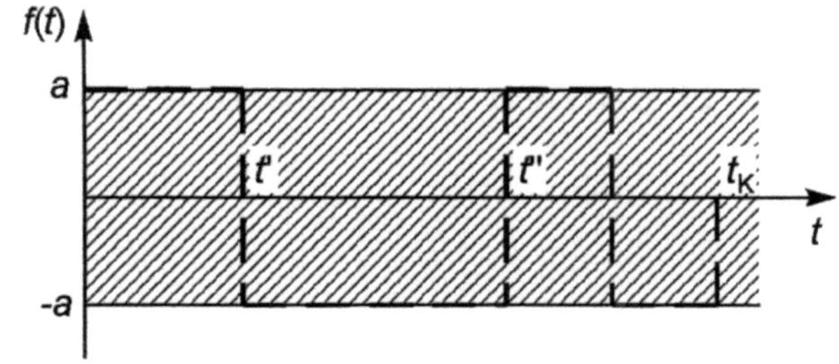

Fig. 10.1.

been taken in an analysis represent the actual excitations or operation conditions. In order to obtain the probability characteristics random functions we must have a large number of process realizations, which can entail great difficulties in experimental investigations or large material expenses. This gives rise to the problem of devising methods that would allow us to estimate the impact of random excitations on a mechanical system at an easily obtainable minimum of possible information about them.

The easiest thing in experimental investigations of the random processes involved is determining the tolerance of a random quantity or the area of possible values of a random function. The area of possible values of a random function $f(t)$, when the extreme values of the function $f(t)$ are equal in absolute value and time constant is shown in Fig. 10.1. In what follows a random function given by the area of possible values will be referred to as a random function bounded in absolute value. Generally, the boundary of the area of possible values of the function $f(t)$ can vary in time (Fig. 10.2). As information on the behaviour of the function $f(t)$ inside the area of possible

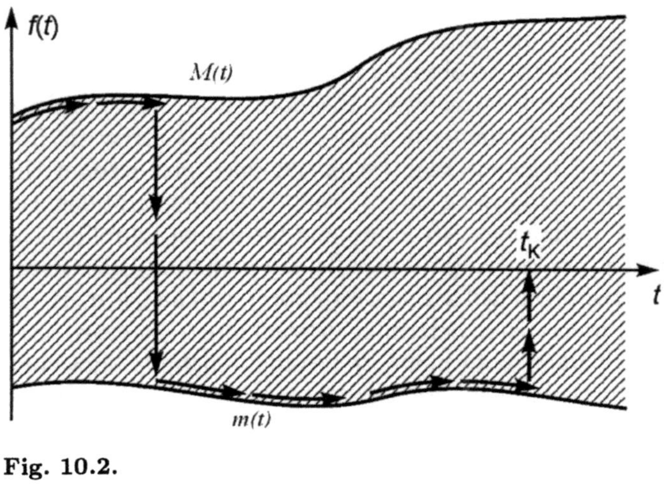

Fig. 10.2.

values is absent, we may assume that this behaviour can be arbitrary, i.e. the class of functions characterizing the possible behaviour of the random function $f(t)$ includes, among other things, the class of sectionally-continuous functions with discontinuities of the first kind. The possible discontinuous behaviour of the function $f(t)$ in time is shown in Figures 10.1 and 10.2 by the dash line. In physical terms a discontinuous behaviour of an excitation in time means an inertialess behaviour. Real random excitations have an inertia. For example, a random wind load can not change its direction instantly, which is tantamount to an instant change in the velocity of the wind. Therefore, the assumptions to the effect that any discontinuous behaviour of a random excitation is possible, are an idealization of the real behaviour of $f(t)$. In what follows it will be shown that this idealization leads to the worst possible actions on a system.

In order to elucidate the basic features of dynamic problems at the action of discontinuous excitations bounded in absolute value, let us consider the following example.

Example 10.1. A body of a mass m moves horizontally under the action of a force R (Fig. 10.3). The force R has a scatter $\Delta R(t)$, with $|\Delta R_{\max}| = a = $ $=$ const (i.e. $\Delta R(t)$ is a random function bounded in absolute value (Fig. 10.4). We can present the force R as

$$R(t) = R_0(t) \pm \Delta R(t), \tag{10.1}$$

where $R_0(t)$ is the design value of the force R.

If we ignore all forces other than R, the equation of motion of the mass m takes the form

$$\ddot{x}(t) = \frac{1}{m} R. \tag{10.2}$$

Fig. 10.3.

It is required to determine the greatest possible scatters of x and $\dot{x}$, which the body can have at the given time instant t_k at the action of the random force ΔR given by the area of its possible values. The displacement x can be presented as

$$x(t) = x_0(t) + \Delta x(t), \tag{10.3}$$

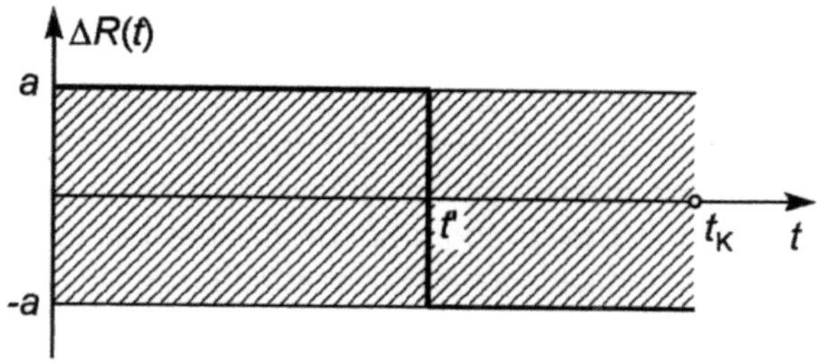

Fig. 10.4.

where x_0 is the displacement corresponding to the design value R_0; $\Delta x\,(t)$ is the possible scatter of the displacement caused by the random force $\Delta R\,(t)$.

The equation of the body's disturbed motion is

$$\Delta \ddot{x}\,(t) = \frac{1}{m}\Delta R. \tag{10.4}$$

Assuming that

$$\Delta \dot{x} = x_1; \quad \Delta x = x_2, \tag{10.5}$$

we obtain a system of equations

$$\begin{aligned} \dot{x}_1 &= f\,(t)\,; \\ \dot{x}_2 &= x_1, \end{aligned} \tag{10.6}$$

where $f\,(t) = \dfrac{\Delta R}{m}$.

Certain values Δx and $\dot{x}_k$ at the time instant t_k correspond to each possible behaviour of $\Delta R\,(t)$. As the time of the process and the force ΔR are bounded, Δx_k and $\Delta \dot{x}_k$ will be bounded as well. Therefore, when using the phase plane, its points $(\Delta x_k, \Delta x_k)$ will correspond to each possible behaviour of $\Delta R\,(t)$ with the points being localized in the bounded area of the phase plane. In Fig. 10.5 the area of possible values of Δx_k and $\Delta \dot{x}_k$ is shaded. The displacement x_k and the velocity $\dot{x}_k$ correspond to the design condition of the bodys motion at the time instant t_k. It follows from the boundedness of the area of possible values that there is a limit curve covering this area (methods of determining the area of possible values of the solutions of the equations of motion of a system will be presented below). The limit curve divides the phase plane into two areas: the area of Δx_k and $\Delta \dot{x}_k$ values whose realization is possible (the shaded area) and that of Δx_k and $\Delta \dot{x}_k$ values which will

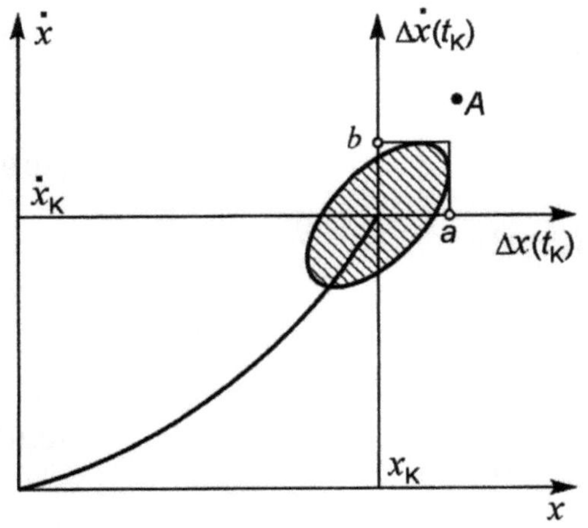

Fig. 10.5.

never be realized (at the given area of possible values of ΔR), for example the values of Δx_k and $\Delta \dot{x}_k$ corresponding to the point A in Fig. 10.5.

The knowledge of the area of possible values of Δx_k and $\Delta \dot{x}_k$ allows us to answer a number of practically important questions, in particular, what maximum values of Δx_k (Point a) and $\Delta \dot{x}_k$ (Point b) are possible at the most unfavourable laws of $\Delta R\,(t)$ variation, in what manner the sizes of the area depend on the parameters of a system and how, in this context, we can choose the parameters of a system, at which the area reaches its minimal sizes.

Let us determine the area of possible values of the solution of system (10.6) which (at zero initial data) may be presented as

$$x_1 = \int_0^{t_k} f\,(\tau)\,\mathrm{d}\tau; \quad x_2 = \int_0^{t_k}\int_0^{t_k} f\mathrm{d}\tau = \int_0^{t_k} (t_k - \tau)\,f\,(\tau)\,\mathrm{d}\tau. \tag{10.7}$$

Relationships (10.7) do not take into account the discontinuous behaviour of $f\,(\tau)$, therefore let us obtain the expression for x_1 and x_2 with due account of the possible discontinuous behaviour of $f\,(t)$ on the time interval $(0, t_k)$. Suppose, for example, that at the time instant t' $(t' < t_k)$ the function $f\,(t)$ reverses sign (dash line in Fig. 10.4). For the sake of definiteness we may consider that it was equal to $+a$ up to t', and to $-a$ after it, i.e.

$$f\,(t) = \begin{cases} a & 0 \le t \le t'; \\ -a & t < t \le t_k. \end{cases} \tag{10.8}$$

At such variation of the function $f(t)$ we obtain

$$\begin{aligned} x_1 &= \int_0^{t'} a\mathrm{d}\tau - \int_0^{t_k} a\mathrm{d}\tau, \\ x_2 &= a\int_0^{t'} (t_k - \tau)\,\mathrm{d}\tau - a\int_{t'}^{t_k} (t_k - \tau)\,\mathrm{d}\tau \end{aligned} \tag{10.9}$$

or after integration

$$\begin{aligned} x_1 &= (2t' - t_k)\,a; \\ x_2 &= -\frac{a}{2}\left[(t_k - t')^2 - t_k^2\right] + \frac{a}{2}\left[-(t_k - t')^2\right]. \end{aligned} \tag{10.10}$$

The function f may also behave in somewhat different manner, being $f = -a$, from zero to t' and $f = +a$ from t' to t_k. At such law of f variation the signs in front of the integrals in the right-hand side of expressions (10.9) reverse, i.e. the values x_1 and x_2 can be of opposite signs. The area of possible values of x_1 and x_2 at $t = t_k$ is limited by curves (10.10) given in parametric form. If we put $t_k = 1$, the expressions for x_1 and x_2 will be

$$x_1 = (2t' - 1)\,a; \quad x_2 = a\left(2t' - \frac{1}{2} - t'^2\right) \quad \text{at} \quad 0 \le t' \le 1. \tag{10.11}$$

Expressions (10.11) give only one branch of the limit curve, the second one being symmetric (with respect to the origin of coordinates), as the values of x_1 and x_2 equal in magnitude but opposite in sign are possible. In Fig. 10.6 the area of possible values is shaded, occupying only part of the area plotted on

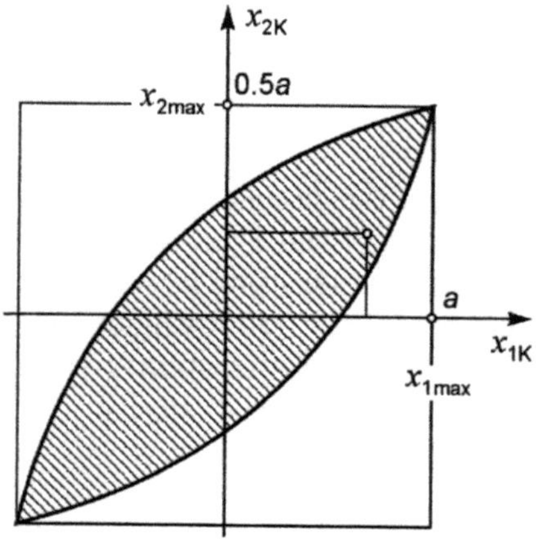

Fig. 10.6.

the maximum values of x_1 and x_2 if the function f had only one discontinuity on the interval $(0,\ t_k)$. What values will the functions x_1 and x_2 take if the function f has two or more discontinuities? It may be that we shall obtain a point located on the plane $(x_1,\ x_2)$ and lying outside the shaded area. Let us consider the case of the function f having two discontinuities at the instants t' and t'' $(t'' > t')$.

$$f = \begin{cases} a & 0 \le \tau \le t'; \\ -a & t' < \tau < t''; \\ a & t'' < \tau < t_k. \end{cases} \tag{10.12}$$

On integrating the expressions for x_1 and x_2 will be as follows (at $t_k = 1$):

$$\begin{aligned} x_1 &= (2t' - 2t'' + 1)\, a; \\ x_2 &= \frac{a}{2}\left[1 + 2\,(1 - t'')^2 - 2\,(1 - t')^2\right]. \end{aligned} \tag{10.13}$$

Let us take two arbitrary time instants $t' = 0.4$; $t'' = 0.6$ and substituting in the values of x_1 and x_2, we obtain $x_1 = 0.6a$; $x_2 = 0.3a$.

Judging from Fig. 10.6 where this point is placed in the shaded area, no worse case has occurred. The law of functions variation (with one discontinuity) that was taken when plotting the area turned out to be the most objectionable. In what follows it will be shown that for this system of equations the worst law of function variation is, in fact, that with one discontinuity. In this particular example such law of the function f $(f = \text{const})$ variation on the whole time interval is possible when the simultaneous maximum of the functions x_1 and x_2 is reached at $t = t_k$ (with simultaneous maxima of only one sign being possible). What will happen with the area if the function varies in wider bounds, for example, inside a band limited by straight lines $\pm 2a$ (instead of $\pm\, a$, as in the example)?

It is easy to establish from expression (10.13) that the functions x_1 and x_2 would be two times greater, i.e. the area would become larger, but its form and position inside the rectangular area would not change. Such increase or decrease in the tolerance for the functions perturbation leads to a similar change in the area of possible values of the solutions of a system of differential equations. In the considered example the area of possible values of x_1 and x_2 proved symmetric with respect to the origin of coordinates, but this is not always the case. The position of the area of possible values of the solutions of a system (for the considered example of the functions x_1 and x_2) depends on the boundaries of possible excitations (of the function f in the given example). Let us plot the area of possible values of x_1 and x_2 for the function f, specified inside a band shown in Fig. 10.7. Let us first consider a case of the function f variation shown in Fig. 10.7 by a solid line. In this case the function f is given as

$$f = \begin{cases} a & \text{at } \ 0 \le \tau \le t'; \\ 0 & \text{at } \ t' \le \tau \le t_k. \end{cases} \tag{10.14}$$

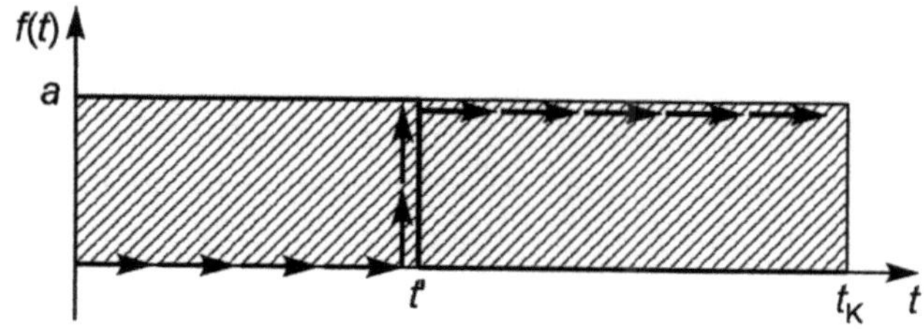

Fig. 10.7.

Having made the necessary calculations, we obtain at $t_k = 1$

$$x_1 = t'a; \quad x_2 = \frac{a}{2}\left[2t' - t'^2\right]. \tag{10.15}$$

At a change in the function f

$$f = \begin{cases} 0 & \text{at } 0 \le \tau \le t'; \\ a & \text{at } t' < \tau \le t_k. \end{cases}$$

x_1 and x_2 are of the form

$$x_1 = 1 - t'; \quad x_2 = \frac{1}{2}\left(1 - t'\right)^2. \tag{10.16}$$

Systems (10.15) and (10.16) define the equation of the limit curve (two branches) with the corresponding area being shown in Fig. 10.8. In the considered examples the boundaries of the area within which the function f varied were constant in time. Similarly, we can determine the area when its borders depend on time t, as, for example, is shown in Fig. 10.2, i.e.

$$m(t) \le f(t) \le M(t). \tag{10.17}$$

Generally, several random excitations $f_k(t)$ bounded in absolute value and satisfying the conditions

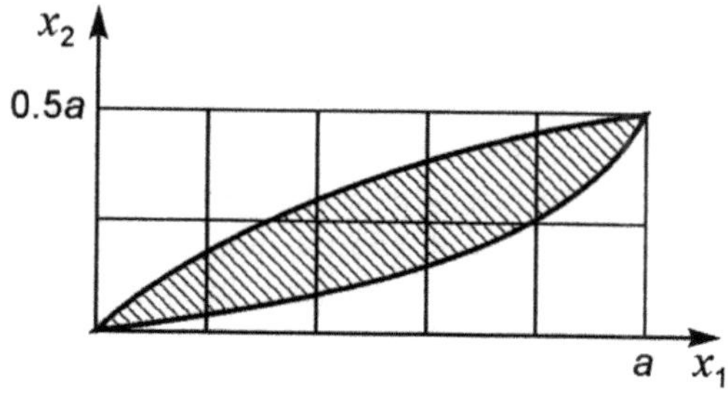

Fig. 10.8.

$$m_i\,(t) \le f_i\,(t) \le M_i\,(t) \quad (i = 1, 2, ..., k)\,, \tag{10.18}$$

can act on a system, where $m_i\,(t)$, $M_i\,(t)$ are known functions of time that represent the bottom and top boundaries of the possible values area within which the function f_i can take any values. The limiting values of the function f_i (m_i and M_i) can be independent on time t. The set of functions satisfying conditions (10.18) includes, among other things, deterministic functions varying in time according to known laws but remaining within the area of possible values. The functions satisfying condition (10.18) can embrace those retaining constant random values during the whole process, for example, random initial data. Let us introduce a notation f_i^0 for random functions retaining a constant numerical value during the process. The functions f_i^0 can take any value within the interval

$$m_i^0 \le f_i^0 \le M_i^0. \tag{10.19}$$

Restrictions on excitations of the (10.18) or (10.19) type have a definitive physical meaning and characterize the utmost possible deviations of excitations that can occur in a given system. As regards technical problems involving excitations f_i that evaluate the scatters of the forces or the parameters of a system, the tolerances for the deviations of these quantities from their nominal values can be determined with a sufficient accuracy.

Random functions f_i and random quantities f_i^0 that satisfy conditions (10.18) and (10.19) respectively, are independent. For example, if two random functions $f_1\,(t)$ and $f_2\,(t)$ meeting conditions (10.18) act on a system, their possible values are inside rectangular area (Fig. 10.9) at any time instant. In this case the components of the vector of excitations are independent.

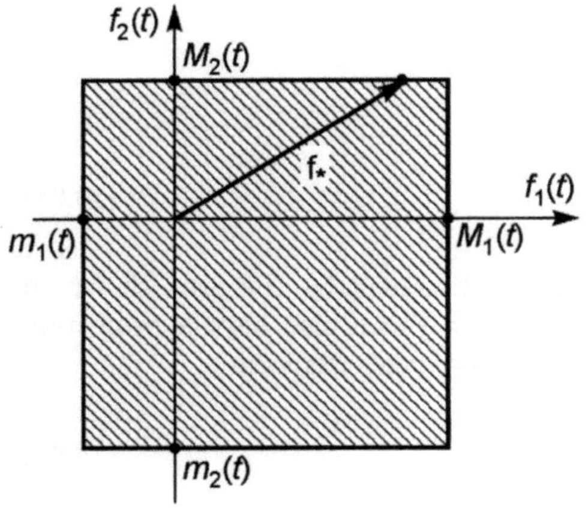

Fig. 10.9.

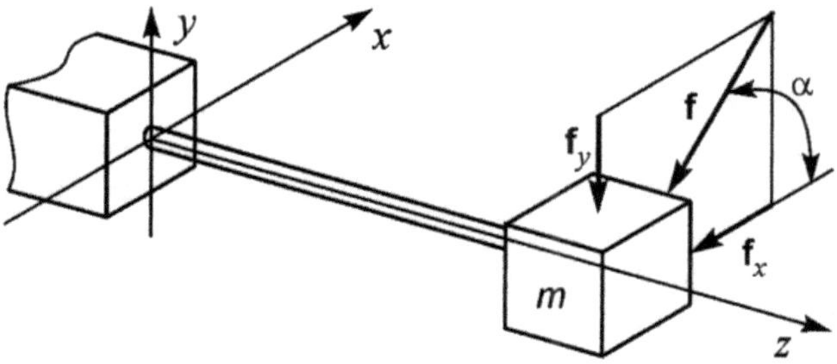

Fig. 10.10.

Along with independent random excitations, dependent random excitations limited in absolute value can occur. Let us consider as an example a mass m (Fig. 10.10) subjected to the action of a force $\mathbf{f}$ that is randomly directed (angle α is random) and limited in absolute value $|\mathbf{f}| \leq a$. The mass is attached to an inertialess elastic rod. Let us confine ourselves to deriving a condition that should be satisfied by the projections of the force $\mathbf{f}$ onto axes X and Y. Force $\mathbf{f}$ projections, f_x, and f_y, equal to

$$f_x = f \cos \alpha; \quad f_y = f \sin \alpha. \tag{10.20}$$

will enter in the equations of motion of the mass m written in the projections on the axes X and Y.

From (10.20) we can obtain a relationship

$$f_x^2 + f_y^2 = f^2,$$

but since f is limited in absolute value

$$|f(t)| \leq a = f_m$$

we have

$$\left(\frac{f_x}{f_m}\right)^2 + \left(\frac{f_y}{f_m}\right)^2 \leq 1, \tag{10.21}$$

i.e. the possible values of excitations are bounded by the area shown in Fig. 10.11. In a more general case, the force f can also have a projection onto the axis Z, therefore relationship (10.21) takes the form

$$\left(\frac{f_x}{f_m}\right)^2 + \left(\frac{f_y}{f_m}\right)^2 + \left(\frac{f_z}{f_m}\right)^2 \leq 1. \tag{10.22}$$

Condition (10.22) can be written in a more general form

$$(C\mathbf{f} \cdot \mathbf{f}) \leq 1, \tag{10.23}$$

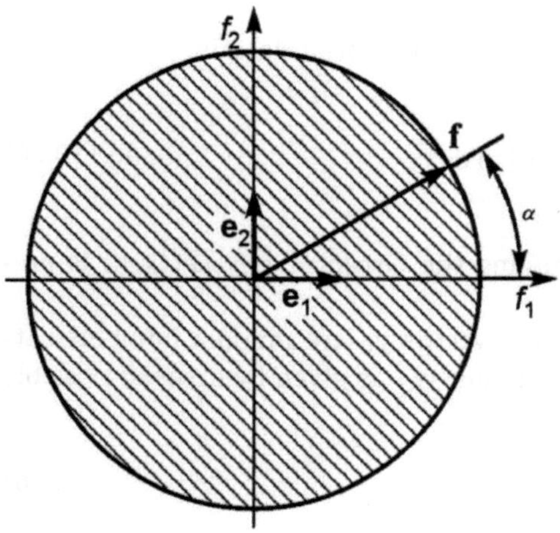

Fig. 10.11.

where C is a completely positive, self-adjoint square matrix. The left-hand side of condition (10.23) represents a completely positive quadratic form.

Two types of random excitations $f_i\left(t\right)$ satisfying (10.23) are possible: a) $f_i\left(t\right)$ do not depend on time, i.e. they are the excitations of the f_i^0 type; b) the random excitations $f_i\left(t\right)$ can be presented as $f_i = h_i\left(t\right) f_i^0$, where $h_i\left(t\right)$ are known functions dependent on time. In the b) case random excitations f_i meeting (10.23) vary in time. The principal problem connected with the analysis of the non-stationary vibrations of a mechanical system at the action of excitations that meet limitations (10.18) or (10.19) and (10.23) can be formulated as follows. It is required to determine the area of possible values of a vector $\mathbf{y}$ that describes the disturbed state of a system at a fixed time instant t_k, if the components of the vector of excitations are random functions satisfying conditions (10.18) or (10.19) and (10.23).

The special cases of the formulated problem are: the problem of determining the projection of the $n-$dimensional area of the possible values of the system state vector at non-stationary vibrations onto two-dimensional planes (areas similar to that presented in Fig. 10.6) and the problem of determining the greatest possible values of each of the components y_i of the state vector y at a fixed time instant.

10.2 Determining the Maximum Values of the Components of the Systems State Vector

Let us consider the vector equation of motion of a system that has the form

$$\dot{\mathbf{y}} + A\left(t\right)\mathbf{y} = B\left(t\right)\mathbf{f}\left(t\right). \tag{10.24}$$

The solution of equation (10.24) at non-zero initial data is

$$\mathbf{y} = K(t_k)\,\mathbf{y}_0 + \int_0^{t_k} G(t_k,\tau)\,B(\tau)\,\mathbf{f}(\tau)\,\mathrm{d}\tau. \tag{10.25}$$

The numerical method of determining the Green matrix $G(t_k,\tau)$ is presented in Chap. 5.

The components of the initial values vector $\mathbf{y}_0$ are random numbers satisfying the following conditions (considering $\mathbf{y}_{0j}$ as given by areas of possible values)

$$(y_{0j})_{\min} \le y_{0j} \le (y_{0j})_{\max}, \tag{10.26}$$

where $(y_{0j})_{\min}$ and $(y_{0j})_{\max}$ are known values.

At first, let us consider the case of the components of the vector $\mathbf{f}$ being independent, i.e. satisfying conditions (10.18)

$$m_i(t) \le f_i(t) \le M_i(t).$$

In scalar form the m–th component of the vector is

$$y_m(t_k) = \sum_{j=1}^{n} k_{mj} y_{0j} + \sum_{\nu=1}^{\rho} \int_0^{t_k} d_{m\nu} f_\nu \mathrm{d}\tau, \tag{10.27}$$

where ρ is the number of the components of the vector $\mathbf{f}$, which is not necessarily equal to n; $d_{m\nu}$ are the elements of the matrix $D = G(\tau)\,B(\tau)$. The maximum value of y_m is

$$\max(y_m) = \sum_{j=1}^{n} \max(k_{mj} y_{0j}) + \sum_{\nu=1}^{\rho} \max \int_0^{k} d_{m\nu} f_\nu \mathrm{d}\tau. \tag{10.28}$$

The maximum values of the terms dependent on the components of the initial values vector are reached at the following values

$$y_{0j} = \begin{cases} y_{0j\,\max} & \text{at } k_{mj} > 0; \\ y_{0j\,\min} & \text{at } k_{mj} < 0. \end{cases} \tag{10.29}$$

The maximum values of the integrals dependent on f_ν correspond to the following variation laws of excitations $f_\nu(\tau)$: $f_\nu(\tau) = M_\nu(\tau)$ on time τ intervals, where $d_{m\nu}(\tau) > 0$ and $f_\nu(\tau) = m_\nu(\tau)$ on time τ intervals, where $d_{m\nu}(\tau) < 0$, i.e.

$$f_\nu = \begin{cases} M_\nu & \text{at } d_{m\nu} > 0; \\ m_\nu & \text{at } d_{m\nu} < 0. \end{cases} \tag{10.30}$$

For example, if τ_j are the values of time where the functions $d_{m\nu}$ reverse sign, the maximum values of the integrals entering the right-hand side of (10.27) (for the sake of definiteness we assume that on the first interval from 0 to $\tau^{(\nu)}$ $d_{m\nu} > 0$) are

$$\max \int_0^{t_k} d_{m\nu} f_\nu \mathrm{d}\tau = \int_0^{\tau_1^{(\nu)}} d_{m\nu} M_\nu \mathrm{d}\tau + \int_{\tau_1^{(\nu)}}^{\tau_2^{(\nu)}} d_{m\nu} m_\nu \mathrm{d}\tau + \ldots \tag{10.31}$$

In the particular case of

$$|y_{0j\,\min}| = |y_{0j\,\max}| = a_j; \quad |m_j| = |M_j| = b_j,$$

the maximum value of the component y_m is

$$\max(y_m) = \sum_{j=1}^{n} |k_{mj}|\, a_j + \sum_{j=1}^{k} b_j \int_0^{t_k} |d_{mj}|\, \mathrm{d}\tau. \tag{10.32}$$

As a result, we determine the greatest possible values of each of the components y_m of the system state vector (10.28) at a fixed time instant, for example, at the time instant corresponding to the loss of contact between the rocket and the guide (Fig. 0.2).

The presented method makes it possible to determine not only the greatest possible values of the components of the system state vector but also the corresponding worst laws of the variation in time of random excitations bounded in absolute value. The worst laws of the variation in time of the excitations f_i which impart maximum values to each of the component y_m of the state vector are different, i.e. these laws cannot occur at a single realization of the process. The knowledge of the greatest possible values of the components of the system state vector is very useful because these values are guaranteed.

10.3 Areas of Possible Values of the System State Vector at the Action of Independent Excitations

The modulus and direction of a state vector $\mathbf{y}$ will change depending on the variation in time of the components of the vector $\mathbf{f}$. Owing to the boundedness of the components of the vectors $\mathbf{f}$ and $\mathbf{y}_0$ and conditions (10.18) and (10.19), the vector $\mathbf{y}$ will also be bounded at a finite time interval t_k (Fig. 10.12). Therefore, all points of the $n-$dimensional phase space can be divided into two sets: the set of points, which can be reached by the vector $\mathbf{y}$ at all possible variations of the vector $\mathbf{f}$, and the set of points which vector $\mathbf{y}$ can not reach.

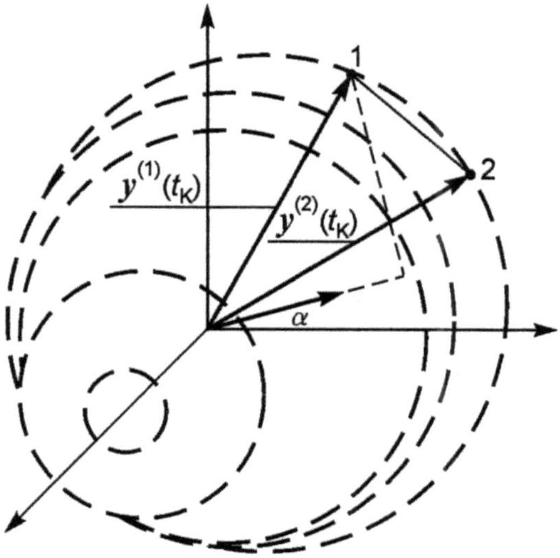

Fig. 10.12.

Let us assume that the maximum values of the projection of the vector $\mathbf{y}$ onto the directions determined by a unit vector $\boldsymbol{\alpha}$ are known. Drawing hyperplanes through the ends of these projections at different $\boldsymbol{\alpha}$ and perpendicularly to them, we obtain some closed area contained within the planes. As the unit vector $\boldsymbol{\alpha}$ depends on $n-1$ parameters (projections onto coordinate axes), we obtain the $n-1$ parametric family of hyperplanes. The earlier described geometrical method of obtaining the area is essentially related to the assumption that the area is convex, i.e., if any two points belong to the boundary of the area, all points of the straight-line segment connecting the two points belong to the area. For example, if a vector $\mathbf{f}^{(1)}$ defines point 1 (it corresponds to a state vector $\mathbf{y}^{(1)}$ resulting from the action of the vector $\mathbf{f}^{(1)}$, while a vector $\mathbf{f}^{(2)}$ defines point 2 (vector $\mathbf{y}^{(2)}$)), than vector $\mathbf{f}$

$$\mathbf{f} = \lambda\mathbf{f}^{(1)} + (1-\lambda)\,\mathbf{f}^{(2)} \tag{10.33}$$

defines all points of the straight-line segment connecting points *1* and *2*, when λ changes from 0 to 1. The system state vector at a given $\mathbf{f}$ can be presented as

$$\mathbf{y} = \lambda\mathbf{y}^{(1)} + (1-\lambda)\,\mathbf{y}^{(2)}, \tag{10.34}$$

and at the continuous variation of λ from 0 to 1 the vector $\mathbf{y}$ runs through all points of the straight line connecting the ends of the vectors $\mathbf{y}^{(1)}$ and $\mathbf{y}^{(2)}$. This testifies to the fact that all vectors $\mathbf{y}$ (at $0 \le \lambda \le 1$) lie inside the area of possible values, i.e. the area is convex.

Having determined the maximum values of the projections of the vector $\mathbf{y}$ for each direction of the vector $\boldsymbol{\alpha}$, we can proceed to the determination of a limit surface. The projection of the vector $\mathbf{y}$ onto the direction determined by the unit vector $\boldsymbol{\alpha}$ with components $(\alpha_1, \alpha_2, \ldots \alpha_n)$ can be written in the following form (Fig. 10.12):

$$y_{pr\alpha} = (\mathbf{y} \cdot \boldsymbol{\alpha}) \tag{10.35}$$

or

$$y_{pr\alpha} = \sum_{i=1}^{n} y_i \alpha_i. \tag{10.36}$$

By substituting the expressions obtained for y_m (10.27) in condition (10.36) we get

$$y_{pr\alpha} = \sum_{j=1}^{n} r_j y_{0j} + \sum_{j=1}^{n} \int_0^{t_k} q_j(t_k, \tau) f_j \mathrm{d}\tau, \tag{10.37}$$

where

$$r_j = \sum_{i=1}^{n} k_{ij} \alpha_i; \tag{10.38}$$

$$q_j(t_k, \tau) = \sum_{i=1}^{n} d_{ij} \alpha_i. \tag{10.39}$$

The components of the vector $\mathbf{f}$ can, among other things, include excitations (f_i^0) that do not vary in time, therefore let us present expression (10.37) as one explicitly dependent on f_j and f_ν^0

$$y_{pr\alpha} = \sum_{j=1}^{n} r_j y_{0j} + \sum_{j=1}^{n} \int_0^{k} q_j(t_k, \tau) f_j \mathrm{d}\tau + \sum_{\nu=1}^{k} Q_\nu^0 f_\nu^0, \tag{10.40}$$

where

$$Q_\nu^0 = \int_0^{t_k} q_\nu(t_k, \tau)\, \mathrm{d}\tau, \quad (\rho + k = n).$$

The problem of finding the area of possible values of a systems disturbed state vector $\mathbf{y}$ has reduced to the determination of the greatest possible value of expression (10.40) at restrictions met by the functions y_{j0}, f_j and f_ν^0 that maximize the afore-mentioned expression (10.40) with the functions f_j belonging to a closed set. This condition means that these functions may also have their limit values, i.e. on some interval of time (t_1, t_2) f_j can be equal to its limit values, for example

$$f_j = m_j(t) \quad (t_1 \le t \le t_2),$$
$$f_j = M_j(t) \quad (t_1 \le t \le t_2). \tag{10.41}$$

Let us show that the extreme values of vectors $\mathbf{y}(t_k)$ (Fig. 10.12) forming the limit surface are reached on the class of discontinuous functions f_j that satisfy conditions (10.18). Then we separately determine the maximum value of each of the sums entering the right-hand side of expression (10.40). Next we find the maximum value of the first sum

$$\sum_{j=1}^{n} r_j(t_k)\, y_{j0} \tag{10.42}$$

with the conditions $y_{im} \le y_{j0} \le y_{jM}$ being fulfilled, where y_{jm}, y_{jM} are the minimum and maximum possible values of y_{j0} respectively. Sum (10.42) takes the maximum value, if y_{j0} are equal to y_{jM} at $r_j(t_k) > 0$ or y_{j0} are equal to y_{jm} at $r_j < 0$.

The maximum value of the second sum in (10.40) can be obtained if we take $f_j(\tau)$ according to the band upper bound in those intervals of the variation of τ, where $q_j(t_k, \tau) > 0$ and according to the lower bound where $q_j(t_k, \tau) < 0$, i.e. the extreme value of the sum

$$\sum_{j=1}^{n} \int_0^{t_k} q_j(t_k, \tau)\, f_j \mathrm{d}\tau$$

is reached within the functions of the following type (discontinuous functions)

$$f_{j\alpha}^* = \begin{cases} M_j(\tau) & \text{if } q_j > 0; \\ m_j(\tau) & \text{if } q_j < 0, \end{cases} \tag{10.43}$$

as was to be shown since the laws of the variation of f_j that give an extremum to a scalar product $(\mathbf{y} \cdot \boldsymbol{\alpha})$, also define the extreme values of the vector $\mathbf{y}(t_k)$.

In order to determine the discontinuity points of the function f_j let us determine τ_{jk} where $q_j(t_k, \tau)$ reverses sign. In order to determine τ_{jk} we have an equation

$$q_j(t_k, \tau) = 0.$$

Knowing τ_{jk}, we determine the variation laws of $f_j(\tau)$ giving a maximum to the appropriate term in expression (10.40). The maximum value of the third sum can be obtained, if we take f_ν^0 equal to its maximum value at $Q_\nu^0 > 0$ and to its minimum value at $Q_\nu^0 < 0$. Thus, for each direction $\boldsymbol{\alpha}$ such values of y_{j0}, f_ν^0 and the function f_j are determined at which the projection of the vector $\mathbf{y}(t_k)$ on this direction takes its maximum value. Knowing y_{j0}, f_ν^0 and $f_{j\alpha}^*$, we determine the components of the vector $\mathbf{y}(t_k)$ that are coordinates of points of the vector $\mathbf{y}(t_k)$ region of possible values

$$y_i = \sum_{j=1}^{n} k_{ij} y_{j0} + \sum_{j=1}^{n} \int_0^{t_k} d_{ij} f_{j\alpha}^* \mathrm{d}\tau + \sum_{\nu=1}^{k} V_{i\nu} f_\nu^0, \tag{10.44}$$

where

$$V_{i\nu} = \int_0^{t_k} d_{i\nu} \mathrm{d}\tau.$$

The integral $\int_0^{t_k} d_{ij} f_{j\alpha}^* \mathrm{d}\tau$ with due account of the variation law of the function $f_{j\alpha}^*$ can be calculated in terms of antiderivatives

$$\begin{aligned} F_{ij}^M(\tau) &= \int_{\tau_{j,k}}^{\tau_{j,k+1}} d_{ij} M_j \mathrm{d}\tau; \\ F_{ij}^m(\tau) &= \int_{\tau_{j,k}}^{\tau_{j,k+1}} d_{ij} m_j \mathrm{d}\tau. \end{aligned} \tag{10.45}$$

Then

$$\int_0^{t_k} d_{ij} f_j^* \mathrm{d}\tau = F_{ij}^M(\tau_{j1}) + \left[F_{ij}^m(\tau_{j2}) - F_{ij}^m(\tau_{j1})\right] + \ldots, \tag{10.46}$$

if $q_j(t_k, \tau) > 0$ at $0 \le \tau < \tau_{j1}$ and

$$\int_0^{t_k} d_{ij} f_j^* \mathrm{d}\tau = F_{ij}^m(\tau_{j1}) + \left[F_{ij}^m(\tau_{j2}) - F_{ij}^M(\tau_{j1})\right] + \ldots, \tag{10.47}$$

if $q_j(t_k, \tau) < 0$ at $0 \le \tau \le \tau_{j1}$.

Let us consider the special case of the boundaries of the area of possible values of the random excitations f_j being constant in time

$$|f_j| \le a_j. \tag{10.48}$$

For this class of excitations the maximum value of the second term in expression (10.40) will be equal to

$$\sum_{j=1}^{\rho} a_j \int_0^{t_k} |q_j(t_k, \tau)| \mathrm{d}\tau. \tag{10.49}$$

In spite of the fact that the initial equations are linear, the problem involving the determination of the area of possible deviations of the solutions of a system at a fixed instant of time, is nonlinear, since the worst actions of excitations (the time instants of the transition of excitations from one boundary curve of the area of possible values to another) occur at time instants dependent on the properties of a system rather than externally given time instants. The area of possible deviations retains its configuration (in the similarity context) given a proportional variation of all excitations. At a similar variation of excitations we have

$$\mathbf{f}_1 = \beta \mathbf{f}; \quad \mathbf{f}_1^0 = \beta \mathbf{f}^0, \tag{10.50}$$

where β is a scalar multiplier.

Conditions (10.18) and (10.19) change accordingly and expressed as

$$\beta m_i \leq f_{i1} \leq \beta M_i; \quad \beta m_i^0 \leq f_{i0}^0 \leq \beta M_i^0.$$

In consequence of the linear dependence of the solution of equation (10.25) on the vector of excitations, the vectors $\mathbf{y}\,(t_k)$ that determine the limit surface also change by the factor β, i.e. a similar variation of the vectors $\mathbf{f}$ and $\mathbf{f}^0$ results in a similar variation of the area of possible deviations.

Example 10.2. The dynamically stable motion of a rocket can be accompanied by its small vibrations on a trajectory that are caused by a scatter in the thrust of the engine, the linear and angular misalignments of the thrust and a number of other factors. The equation of the small angular vibrations of the rocket (Fig. 10.13) in pitch is of the form

$$J_0 \ddot{\varphi} + a\varphi = M_b, \tag{10.51}$$

where J_0 is the rocket's moment of inertia with respect to the axis going through the center of gravity (perpendicular to the plane of the drawing);

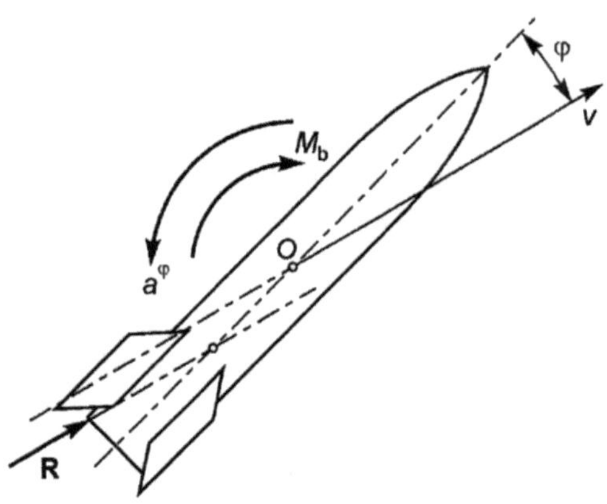

Fig. 10.13.

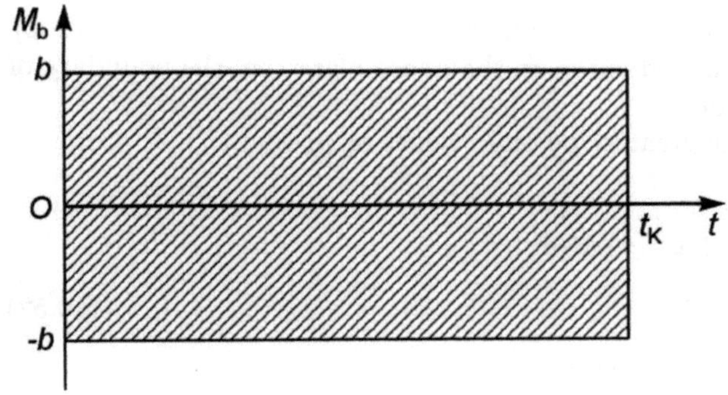

Fig. 10.14.

M_b is the random exciting moment (caused, in particular, by the gasdynamic misalignments of the thrust) whose value is usually known in the form of some field of scatter (Fig. 10.14) that has the boundaries of $\pm b$; $a\varphi$ is the restoring moment. It is required to determine the area of possible values of the angle φ and the angular velocity $\dot{\varphi}$(the area on the phase plane $(\dot{\varphi}, \varphi)$) at the time instant $t_k = 5\,\mathrm{s}$. When solving the problem, we consider that at $t = 0$ $\varphi(0) = \dot{\varphi}(0) = 0$ $J_0 = 10^4\ \mathrm{kgm}^2$, $a = 1{,}6{\cdot}10^4\,\mathrm{N{\cdot}m}$, $|b| = 200\,\mathrm{N{\cdot}m}$.

We may present the solution of equation (10.51) at the time instant t_k as a vector $\boldsymbol{\varphi}_k$ on the phase plane $\dot{\varphi}0\varphi$ (Fig. 10.15). The vector of solution $\boldsymbol{\varphi}_k$ corresponds to each variation law of the exciting moment M_b.

There is such law of variation of the moment M_b at which the modulus of the vector $\boldsymbol{\varphi}_k$ directed at an angle ω to the axis φ reaches its greatest value (Fig. 10.15). If we change the value of the angle α within the limit $0 \div 2\pi$ and

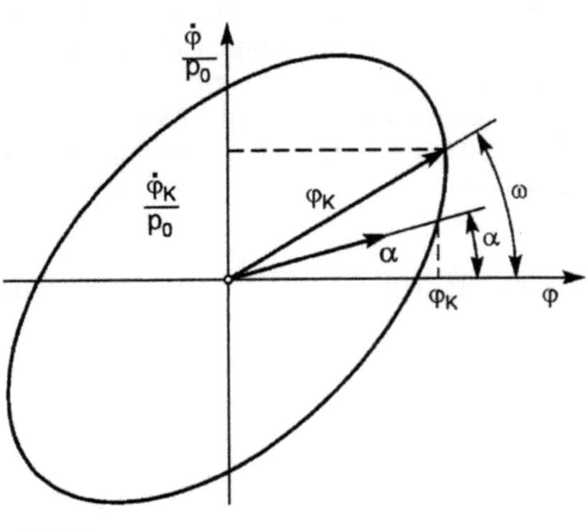

Fig. 10.15.

determine $|\varphi|_{\max}$, for each intermediate value of α, the end of the vector $\boldsymbol{\varphi}_k$ will circumscribe some closed curve on the phase plane, i.e. the boundary of the area of possible values.

Let us determine the greatest possible values of φ_k and $\dot{\varphi}_k$:

$$\begin{aligned} \varphi_k &= \frac{1}{J_0 p_0} \int_0^{t_k} \sin p_0 \,(t_k - \tau)\, M_b \mathrm{d}\tau; \\ \dot{\varphi}_k &= \frac{1}{J_0} \int_0^{t_k} \cos p_0 \,(t_k - \tau)\, M_b \mathrm{d}\tau. \end{aligned} \tag{10.52}$$

Since $|M_b| \leq b$,

$$\begin{aligned} \varphi_{k\,\max} &= \frac{b}{J_0 P_0} \int_0^{t_k} |\sin p_0 \,(t_k - \tau)|\, \mathrm{d}\tau, \quad \left(p_0 = \sqrt{\frac{a}{J_0}}\right); \\ \dot{\varphi}_{k\,\max} &= \frac{b}{J_0} \int_0^{t_k} |\cos p_0 \,(t_k - \tau)|\, \mathrm{d}\tau. \end{aligned} \tag{10.53}$$

In order to determine the boundaries of the area of possible solutions let us project the vector $\boldsymbol{\varphi}_k$ onto the straight line determined by the unit vector $\boldsymbol{\alpha}$

$$\varphi_\alpha = (\boldsymbol{\varphi}_k \cdot \boldsymbol{\alpha}) = \left(\varphi_k \cos\alpha + \frac{\dot{\varphi}_k}{p_0} \sin\alpha\right). \tag{10.54}$$

Having determined the maximum value of the projection of the vector of solution φ_α for a fixed value of the angle α and the variation law of M_b that corresponds to $\varphi_{\alpha\,\max}$, we find the values of the components of the vector $\boldsymbol{\varphi}_k$, i.e. the values of φ_k and $\dot{\varphi}_k$, which are the coordinates of one of a point of the boundary of the area of possible values of solutions. As the solution of equation (10.51) takes the form of (10.52), we obtain the following expression for projection of the vector $\boldsymbol{\varphi}_k$ (10.54)

$$\varphi_\alpha = \frac{b}{J_0 p_0} \int_0^{t_k} \sin p_0 \,(t_k - \tau + \alpha)\, M_b \, \mathrm{d}\tau. \tag{10.55}$$

For example, at $\alpha = \dfrac{\pi}{6}$ expression (10.55) is

$$\varphi_\alpha = \frac{b}{J_0 p_0} \int_0^{t_k} \sin \left[p_0 \,(t_k - \tau) + \frac{\pi}{6}\right] M_b \, \mathrm{d}\tau.$$

Assuming that $\sin\left[p_0(t_k-\tau)+\frac{\pi}{6}\right]=0$, let us determine τ_i, the instants of the sign reversal of M_b.

Taking into account that $p_0\approx 0.4\pi$, at $t_k=5\,\mathrm{s}$, we determine $\tau_1 = = 0.42\,\mathrm{s}$; $\tau_2=0.29\,\mathrm{s}$.

Hence, the variation law of the moment M_b, at which φ_α reaches the maximum value at the moment t_k, can be expressed as

$$M_b=\begin{cases}+b & \text{at}\quad 0\le\tau\le 0.42;\\ -b & \text{at}\quad 0.42\le\tau\le 2.9;\\ +b & \text{at}\quad 2.9\le\tau\le 5.0.\end{cases}$$

Let us calculate the value of φ_k for the obtained variation law of the moment M_b :

$$\varphi_k=\frac{b}{J_0p_0}\left[\int\limits_0^{0.42}\sin p_0(t_k-\tau)\,\mathrm{d}\tau-\int\limits_{0.42}^{2.9}\sin p_0(t_k-\tau)\,\mathrm{d}\tau\right.$$
$$\left.+\int\limits_{2.9}^{5.0}\sin p_0(t_k-\tau)\,\mathrm{d}\tau\right]=3.494\frac{b}{J_0p_0^2}.$$

Similarly, we find $\dot{\varphi}_k$ for $\alpha=\frac{\pi}{6}$

$$\dot{\varphi}_k=1.94\frac{b}{J_0p_0}.$$

Having made similar calculations for various values of the angle , we obtain the following data given in the table:

Table 10.1.

α	0	30°	60°	90°
φ_k	$5\cdot10^{-2}$	$4.3\cdot10^{-2}$	$2.4\cdot10^{-2}$	0
$\dot{\varphi}_k/p_0$	0	$2.4\cdot10^{-2}$	$4.3\cdot10^{-2}$	$5\cdot10^{-2}$

The area of possible values of solutions in non-dimensional coordinates plotted in accordance with the given data is shown in Fig. 10.16.

Let us consider problem 10.2, using the theory of random processes. For this purpose, let us supplement the available information on the random moment (field of possible values) with the probability characteristics of M_b, by relating them to the restriction taken for M_b $(|M_b|<b)$. Let us assume that M_b is a stationary random function with a time-constant normal distribution law (Fig. 10.17) and a correlation function of the following type

$$K_{M_b}=D_M\mathrm{e}^{-\alpha|\tau|}.$$

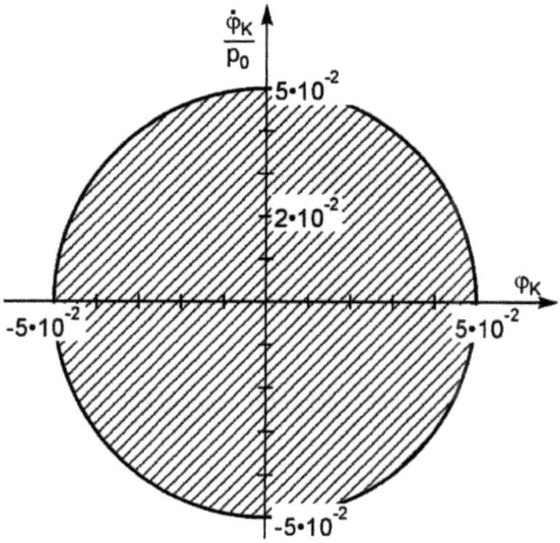

Fig. 10.16.

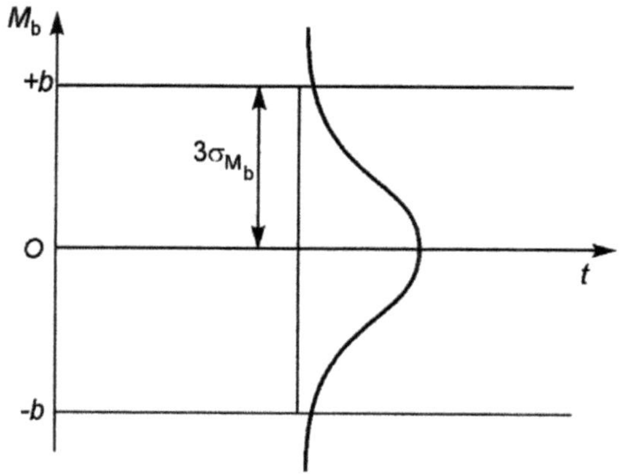

Fig. 10.17.

It would appear reasonable to assume that the greatest possible values of M_b are connected with the variance σ_M by the relationship (three sigma rule) $b = 3\sigma_M$, hence, $D_M = b^2/9$.

The correlation function of the solution of equation (10.51) (confining ourselves to the angle φ) is (see Chap. 5)

$$K_\varphi = \frac{D_M}{J_0^2 p_0^2} \int_0^{t_1} \int_0^{t_2} \sin p_0 (t_1 - \tau_1) \sin p_0 (t_2 - \tau_2) \, \mathrm{e}^{-\alpha|\tau_1 - \tau_2|} \mathrm{d}\tau_1 \mathrm{d}\tau_2 . \quad (10.56)$$

For the purpose of integration let us present expression (10.56) without the modulus sign $K_\varphi = K_\varphi^{(')} + K_\varphi^{('')}$:

$$K_\varphi^{(')} = \frac{D_M}{J_0^2 p_0^2} \int_0^{t_1} \sin p_0 (t_1 - \tau_1) \left[\int_0^{\tau_1} \sin p_0 (t_2 - \tau_2) \, \mathrm{e}^{-\alpha(\tau_1 - \tau_2)} \mathrm{d}\tau_2 \right.$$
$$\left. + \int_{\tau_1}^{t_2} \sin p_0 (t_2 - \tau_2) \, \mathrm{e}^{-\alpha(\tau_2 - \tau_1)} \mathrm{d}\tau_2 \right] \mathrm{d}\tau_1, \quad \text{at} \quad t_2 > t_1; \qquad (10.57)$$

$$K_\varphi'' = \frac{D_M}{J_0^2 p_0^2} \int_0^{t_2} \sin p_0 (t_2 - \tau_2) \left[\int_0^{\tau_2} \sin p_0 (t_1 - \tau_1) \, \mathrm{e}^{-\alpha(\tau_2 - \tau_1)} \mathrm{d}\tau_1 \right.$$
$$\left. + \int_{\tau_2}^{t_1} \sin p_0 (t_1 - \tau_1) \, \mathrm{e}^{-\alpha(\tau_1 - \tau_2)} \mathrm{d}\tau_2 \right] \mathrm{d}\tau_2 \quad \text{at} \quad t_1 > t_2. \qquad (10.58)$$

By way of integration and manipulation we obtain the expression for the variance of the solution

$$D_\varphi = \frac{D_M}{\alpha^2 + p_0^2} \left\{ \alpha t_k - \frac{\alpha}{2p_0 \sin 2p_0 t_k} \right.$$
$$+ \frac{1}{\alpha^2 + p_0^2} [p_0^2 - 2p_0 \mathrm{e}^{-\alpha t_k} (\alpha \sin p_0 t_k + p_0 \cos p_0 t_k)$$
$$\left. + (p_0^2 \cos^2 p_0 t_k - \alpha^2 \sin^2 p_0 t_k)] \right\} \frac{1}{J_0^2 p_0^2}. \qquad (10.59)$$

Having made the necessary calculations, we obtain the root-mean-square value of the angle φ

$$\sigma_\varphi = 0.59 \frac{b}{J_0 p_0^2}. \qquad (10.60)$$

Considering that for the angle φ a normal distribution is true, we obtain

$$\varphi_{\max} = 3\sigma_\varphi = 1.77 \frac{b}{J_0 p_0^2} = 2.2 \cdot 10^{-5}, \qquad (10.61)$$

that is approximately 2 times less than the value for φ in the table at $\alpha = 0$. The presented method with the use of discontinuous random functions bounded in absolute value gives higher values of the maximum deviations of systems from the undisturbed mode, as the worst variation law of the excitation M_b, is unlikely if we estimate it in the context of probability. The obtained value of $\varphi_{\max}$ (10.61), according to the correlation theory, is meaningful as a characteristic of the process only in the case of mass realization, i.e. when, for example, the disturbed motion of a large number of rockets is estimated on the average. Naturally, the average characteristics of the process (for example, the maximum values of the angle $\varphi = 3\sigma_{\varphi}$) are smaller than values φ (10.53) that can occur for individual rockets. If we have only one or two realizations of the process, average characteristics are of no avail, because it is impossible to reveal the statistical properties of the given process at such a small number of realizations. The value φ, found with the use of discontinuous random functions bounded in absolute value is guaranteed.

10.4 Projections of the Area of Possible Values of the System State Vector Onto Two-Dimensional Planes

The area of possible values is n-dimensional for a set of equations of the n-th order. In practical analysis it is often sufficient to know the area of possible values of two components of the system state vector, i.e. it is enough to know the projection of an n-dimensional area onto a two-dimensional plane.

Let us consider the scalar product of the vector of solutions $\mathbf{y}$ and the vector $\boldsymbol{\alpha}$ lying in a plane defined by two unit vectors $\boldsymbol{\omega}_1$ and $\boldsymbol{\omega}_2$ (Fig. 10.18).

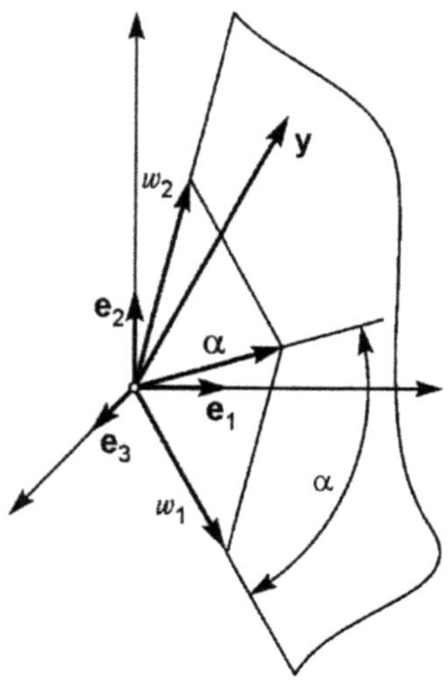

Fig. 10.18.

We consider the most general case where the plane defined by the vectors $\boldsymbol{\omega}_1$, $\boldsymbol{\omega}_2$ does not coincide with the coordinate planes. It is possible to present the vector $\boldsymbol{\alpha}$ as

$$\boldsymbol{\alpha} = \cos\alpha\,\boldsymbol{\omega}_1 + \sin\alpha\,\boldsymbol{\omega}_2. \tag{10.62}$$

The projection of the vector $\mathbf{y}$ onto the direction of the vector $\boldsymbol{\alpha}$ is

$$\begin{aligned} y_{pr\alpha} &= (\mathbf{y},\boldsymbol{\alpha}) = (\mathbf{y}\,\boldsymbol{\omega}_1)\cos\alpha + (\mathbf{y}\,\boldsymbol{\omega}_2)\sin\alpha = y_{\omega_1}\cos\alpha + y_{\omega_2}\sin\alpha \\ &= \cos\alpha\left[\sum_{i=1}^{n}(y_i\omega_{1i})\right] + \sin\alpha\left[\sum_{i=1}^{n}(y_i\omega_{21i})\right], \end{aligned} \tag{10.63}$$

or

$$\begin{aligned} y_{pr\alpha} &= \sum_{j=1}^{n}(r_{1j}\cos\alpha + r_{2j}\sin\alpha)\,y_{j0} \\ &+ \sum_{j=1}^{\rho}\int_{0}^{t_k}[q_{1j}\cos\alpha + q_{2j}\sin\alpha]\,f_1\,\mathrm{d}\tau \\ &+ \sum_{\nu=1}^{k}[V_{1\nu}\cos\alpha + V_{2\nu}\sin\alpha]\,f_\nu^0, \end{aligned} \tag{10.64}$$

where

$$\begin{aligned} r_{1j} &= \sum_{i=1}^{n}k_{ij}(t_k)\,\omega_{1i};\quad q_{1j} = \sum_{i=1}^{n}Q_{ij}\omega_{1i};\quad V_{1j} = \sum_{i=1}^{n}V_{ij}\omega_{1i}; \\ r_{2j} &= \sum_{i=1}^{n}k_{ij}(t_k)\,\omega_{2i};\quad q_{2j} = \sum_{i=1}^{n}Q_{ij}\omega_{2i};\quad V_{2j} = \sum_{i=1}^{n}V_{ij}\omega_{2i}. \end{aligned} \tag{10.65}$$

In the specific case of the vectors $\boldsymbol{\omega}_1$ and $\boldsymbol{\omega}_2$ coinciding with the coordinate axes, i.e. $\boldsymbol{\omega}_1 = \mathbf{e}_k$; $\boldsymbol{\omega}_2 = \mathbf{e}_\nu$, we obtain

$$\begin{aligned} r_{1j} &= K_{kj}(t_k);\quad q_{1j} = Q_{kj};\quad q_{2j} = Q_{\nu j}; \\ r_{2j} &= K_{\nu j}(t_k);\quad V_{1j} = V_{nj};\quad V_{2j} = V_{\nu j}. \end{aligned}$$

The values of y_{j0} are determined depending on the sign of the expression

$$r_{1j}\cos\alpha + r_{2j}\sin\alpha.$$

The maximum value of the sum of integrals is reached at the following variation laws of random functions:

$$\begin{aligned} f_j &= M_j, \quad \text{at } [q_{1j}\cos\alpha + q_{2j}\sin\alpha] > 0; \\ f_j &= m_j, \quad \text{at } [q_{1j}\cos\alpha + q_{2j}\sin\alpha] < 0. \end{aligned}$$

The maximum value of the sum dependent on time-constant random excitations is reached at values f_ν^0 equal to

$$f_\nu^0 = M_\nu^0, \quad \text{at } [V_{1\nu}\cos\alpha + V_{2\nu}\sin\alpha] > 0;$$
$$f_\nu^0 = m_\nu^0, \quad \text{at } [V_{1\nu}\cos\alpha + V_{2\nu}\sin\alpha] < 0.$$

After determining y_j and f_j^* for each α ($f_{j\alpha}^*$ we derive $y_i(t_k)$) (10.44) and than y_{w_1} and y_{w_2} that are coordinates of two-dimension region border points.

10.5 Determination of the Maximum Values of Dynamic Reactions

When analyzing the motion of a system of bodies it is often required to determine not only the system "state" $\mathbf{y}$ (coordinates and velocities), but also the force interaction (reactions) between the separate bodies of a system. We may present the following problem as the simplest physical model of similar problems. A weight of a mass m moves at a velocity v on an absolutely rigid beam with an elastic fixation (Fig. 10.19 *a*). The right support of the beam represents a spring with a stiffness c and a viscous friction damper (coefficient of friction is α). A random force $f(t)$, bounded in absolute value acts on the mass m. A reaction force N that depends on the time behaviour of a function f within the area of possible values takes place between the beam and the mass m. It is required to determine in analysis the greatest possible value of the dynamic reaction N arising between the mass m and the beam.

To carry out a strength design we must know the greatest possible values of the reaction force for each instant of time. The value of the reaction at each time instant depends on the behaviour of the function $f(t)$ from the beginning of the motion up to the given moment. A certain value of the reaction will correspond to each possible behaviour of the function $f(t)$ in an interval $(0, t_1)$, where t_1 is the arbitrary instant of time corresponding to the position of the mass m, shown in Fig. 10.19 *a*. This being so, amongst infinite set of possible behaviours of the function f, including discontinuous variations in time, there is a variation law of f, that gives a maximum value to the reaction at the moment t_1. If we use the kinetostatics method during the derivation of the equations of motion of a system, we shall obtain equations that algebraically (i.e. without derivatives) embrace unknown reactions. Let us consider the general case of the motion of the system, assuming that the equations of the systems motion contain k unknown reactions entering in these equations algebraically. Here the complete system of n equations can be divided into two subsystems (including equations of constraints, with the constraints being considered bilateral) that can be presented in a vector-matrix form as

$$\begin{aligned} &A_{10}\ddot{\mathbf{y}}_1 + A_{11}\dot{\mathbf{y}}_1 + A_{12}\mathbf{y}_1 + A_{13}\mathbf{z} = B_1\mathbf{f}_1; \\ &\ddot{\mathbf{y}}_1 + A_{21}\dot{\mathbf{y}}_1 + A_{22}\mathbf{y}_1 + A_{23}\mathbf{z} = B_2\mathbf{f}_2. \end{aligned} \tag{10.66}$$

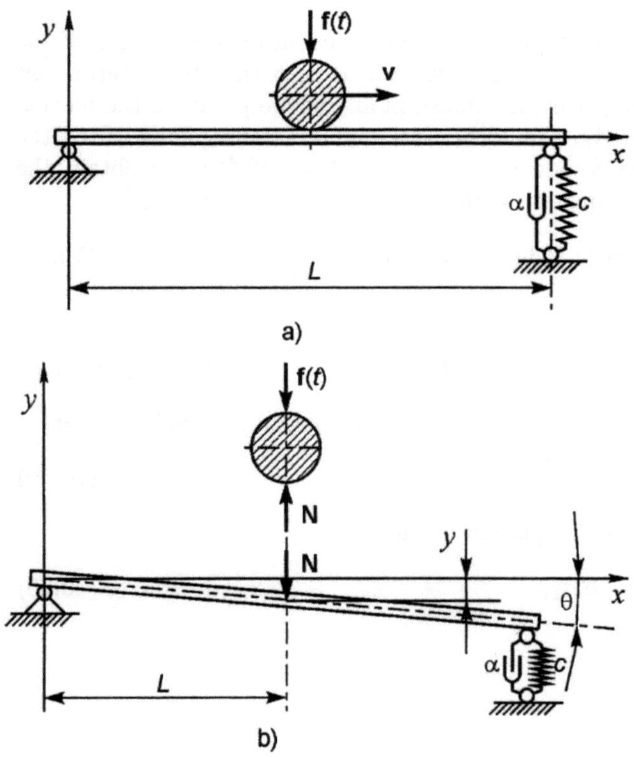

Fig. 10.19.

where $\mathbf{y}_1$ is a vector with $n - k$ components that are the generalized coordinates of the system; $\mathbf{z}$ is a vector with k components formed from unknown reactions; $\mathbf{f}_1$, $\mathbf{f}_2$ are vectors whose components are random excitations bounded in absolute value and satisfying condition (10.18); A_{ij}, B_1, B_2 are matrixes whose elements in the general case can be functions of time.

In what follows we consider that the dimensionality of the vectors $\mathbf{f}_1$ and $\mathbf{f}_2$ is equal to $n - k$. In system of equations (10.66) the matrixes A_{10}, A_{11}, A_{12}, A_{23}, B_1 and B_2 are rectangular, except for the matrix A_{13}, that should be square. For the matrix A_{13} to be square it is sufficient to take from the general system of equations the number of equations equal to the dimensionality of the vector $\mathbf{z}$ (first equation of system (10.66)). The second vector equation of system (10.66) has the dimensionality equal to $n - k$. Multiplying the first equation of system (10.66) by the matrix A_{13}^{-1}, we obtain

$$\mathbf{z} = A_{13}^{-1} B_1 \mathbf{f}_1 - A_{13}^{-1} A_{10} \ddot{\mathbf{y}}_1 - A_{13}^{-1} A_{11} \dot{\mathbf{y}}_1 - A_{13}^{-1} A_{12} \mathbf{y}_1. \tag{10.67}$$

The break-down of the general set of equations into two equations of the system is, in general, arbitrary, but it should be such that the determinant of the formed matrix A_{13} does not vanish in any of the points of the motion interval, i.e. the matrix A_{13} should be non-singular. Having substituted the expression for $\mathbf{z}$ in the second equation of system (10.66), we obtain the following vector-matrix equation with eliminated algebraic unknowns

$$A_4\ddot{\mathbf{y}}_1 + A_5\dot{\mathbf{y}}_1 + A_6\mathbf{y}_1 = B_2\mathbf{f}_2 - A_{23}A_{13}^{-1}B_1\mathbf{f}_1, \tag{10.68}$$

where

$$A_4 = E - A_{23}A_{13}^{-1}A_{10}; \quad A_5 = A_{21} - A_{23}A_{13}^{-1}A_{11}; \quad A_6 = A_{22} - A_{23}A_{13}^{-1}A_{12}.$$

Let us lower the order of equation (10.68) having introduced a vector

$$\dot{\mathbf{y}}_1 = \dot{\mathbf{y}}_2, \tag{10.69}$$

and transform (10.68) to an equation of the form

$$\dot{\mathbf{y}} + A(t)\,\mathbf{y} = B_3\mathbf{f}_2 - B_4\mathbf{f}_1, \tag{10.70}$$

where

$$\mathbf{y} = \begin{bmatrix} \mathbf{y}_2 \\ \mathbf{y}_1 \end{bmatrix}.$$

The dimensionality of the vectors $\mathbf{f}_1$ and $\mathbf{f}_2$ entering equation (10.70) is $2(n-k)$ (the first $(n-k)$ components of the vectors $\mathbf{f}_1$ and $\mathbf{f}_2$ are non-zero and the remaining $(n-k)$ components are equal to zero). The solution of equation (10.70) can be presented as

$$\mathbf{y} = K(t)\,\mathbf{y}_0 + \int_0^t G(t,\tau)\,(B_3\mathbf{f}_2 - B_4\mathbf{f}_1)\,d\tau. \tag{10.71}$$

The matrix $K(t)$ satisfies the equation

$$\dot{K}(t) + A(t)\,K(t) = 0. \tag{10.72}$$

In determining the vector $\mathbf{z}$, we are concerned only with the first derivative of the vector $\mathbf{y}$ with respect to t:

$$\dot{\mathbf{y}} = \dot{K}(t)\,\mathbf{y}_0 + \int_0^t \frac{\partial}{\partial t}G(t,\,\tau)\,(B_3\mathbf{f}_2 - B_4\mathbf{f}_1)\,d\tau + B_3(t)\,\mathbf{f}_2(t) - B_4(t)\,\mathbf{f}_1(t). \tag{10.73}$$

The partial derivative of the Green matrix with respect to t entering into the integral in expression (10.73) can be transformed to the form

$$\frac{\partial}{\partial t} G(t, \tau) = \dot{K}(t) K^{-1}(\tau),$$

or with due account of equation (10.72)

$$\frac{\partial}{\partial t} G(t, \tau) = -A(t) K(t, \tau). \tag{10.74}$$

Having substituted expression (10.75) in (10.73), we obtain

$$\dot{\mathbf{y}} = -A(t)K(t)\mathbf{y}_0 - \int_0^t A(t) G(t, \tau) B_3(t) \mathbf{f}_2 \mathrm{d}\tau$$
$$+ \int_0^t A(t) G(t, \tau) B_4(t) \mathbf{f}_1 \mathrm{d}\tau + B_3(t) \mathbf{f}_2(t) - B_4(t) \mathbf{f}_1(t). \tag{10.75}$$

It is possible to present the expression for the vector $\mathbf{z}$ (10.67) as (since $\ddot{\mathbf{y}}_1 = \dot{\mathbf{y}}_2$),

$$\mathbf{z} = A_8 \mathbf{f}_1 - A_9 \dot{\mathbf{y}} - A_{10} \mathbf{y}, \tag{10.76}$$

where A_8, A_9, A_{10} are rectangular matrixes. The matrixes A_9, A_{10} have $2(n-k)$ columns and k rows. By the substitution of expressions for $\mathbf{y}$, $\dot{\mathbf{y}}$ (10.71), (10.75) in (10.76) and manipulations we obtain

$$\mathbf{z} = C\mathbf{f}_1 - D\mathbf{f}_2 + G\mathbf{y}_0 + \int_0^t K'(t, \tau) \mathbf{f}_1 \mathrm{d}\tau + \int_0^t K''(t, \tau) \mathbf{f}_2 \mathrm{d}\tau, \tag{10.77}$$

where $C = A_9 B_4 + A_8$; $D = A_9 B_3$; $G = A_9 A K_0 - A_{10} K$; $K' = [A_9 A - A_{10}] G B_3$; $K'' = [A_{10} - A_9 A] G B_4$.

In scalar form (10.77) is

$$z_i = \sum_{j=1}^{n-k} \left[c_{ij} f_{1j} - d_{ij} f_{2j} + g_{ij}^0 y_{0j} \right] + \sum_{j=1}^{n-k} k'_{ij} f_{1j} \mathrm{d}\tau$$
$$+ \sum_{j=1}^{n-k} k''_{ij} f_{2j} \mathrm{d}\tau, \quad (i = 1, 2, \ldots, k). \tag{10.78}$$

When finding the maximum value of each of the reactions the variation laws of f_{1j} and f_{2j} are determined by the method presented in Sect. 10.2. Let us find, for example, the variation laws of f_{1j} and f_{2j}, when z_ρ reaches its maximum value. The maximum value of the sum $\sum_{j=1}^{n-k} c_{\rho j} f_{1j}$ is reached in the following way:

If $c_{\rho j} > 0$, then $f_{1j} = M_j(t)$;
If $c_{\rho j} < 0$, then $f_{1j} = m_j(t)$.

Similarly we determine the maximum value of the sum $\sum_{j=1}^{n-k} d_{\rho j} f_{2j}$:

If $d_{\rho j} > 0$, then $f_2 = M_j(t)$;
If $d_{\rho j} < 0$, then $f_2 = m_j(t)$.

The maximum values of the integrals entering in the right-hand side of relationship (10.78) are determined at a fixed time instant in the following manner. If $k'_{\rho j} > 0$ (or $k''_{\rho j} > 0$) on the interval $(\tau_\nu, \tau_{\nu+1})$, then $f_{1j} = M_j(\tau)$ (or $f_{2j} = M_j(\tau)$), and if $k'_{\rho j} < 0$ on the interval$(\tau_\nu, \tau_{\nu+1})$, then $f_{1j} = m_j(\tau)$.
If the initial data are also given as

$$y_{0j\,\min} \le y_{0j} \le y_{0j\,\max},$$

then the "worst" values of y_{0j}, at which z_ρ, reaches maximum are determined from the conditions:

If $g^0_{\rho j} \gg 0$, then $y_{oj} = y_{0j\,\max}$;
If $g^0_{\rho j} \ll 0$, then $y_{oj} = y_{0j\,\min}$.

Example 10.3. Let us determine the maximum value of the reaction $N(t)$ (Fig. 10.19 *b*), if $|f(t)| \le a$. With the use of d'Alembert's principle we obtain the equations of motion of the beam and the mass m :

$$\begin{aligned} &J_0\ddot{\theta} + \alpha L^2\dot{\theta} + cL^2\theta = -Nl; \\ &m\ddot{y} + N - f = 0. \end{aligned} \tag{10.79}$$

Having eliminated the reaction N, from system (10.79), we obtain (as $y = l\theta$)

$$\left(J_0 + ml^2\right)\ddot{\theta} + \alpha L^2\dot{\theta} + cL^2\theta = fl. \tag{10.80}$$

At zero initial data the solution of equation (10.80) takes the form

$$\theta = \frac{1}{(J_0 + ml^2)} \int_0^t \mathrm{e}^{-n(t-\tau)} \sin p_1 (t-\tau)\, f \mathrm{d}\tau, \tag{10.81}$$

where

$$p_1 = \sqrt{p_0^2 - n^2}; \quad n = \frac{\alpha L^2}{2\left(J_0 + ml^2\right)}; \quad p_0^2 = \frac{cL^2}{J_0 + ml^2}.$$

By manipulations we can obtain from the second equation of system (10.79) the following expression for the reaction N as

$$N = \left(1 - \frac{ml^2}{J_0 + ml^2}\right) f + \frac{ml^2 c}{(J_0 + ml^2)\, p_1} \times \int_0^t e^{-n(t-\tau)} \sin\left[p_1 (t - \tau) + \beta\right] f \, d\tau, \tag{10.82}$$

where $c = \sqrt{p_0^4 + 4n^2 p_1^2}$; $\mathrm{tg}\beta = \frac{2np_1}{p_0^2}$.

The maximum value of expression (10.82) is basically determined by the integral, which reaches its extreme value, if the function $f(t)$ changes its values from $+a$ to $-a$ at time instants τ_n found from a condition

$$\sin\left[p_1 (t - \tau_n) + \beta\right] = 0.$$

The greatest possible value of the reaction N is reached at $t \to \infty$, therefore let us go into expression (10.82) to the independent variable $\varepsilon = t - \tau$, that varies from infinity to zero. In this case we determine the instants of discontinuity ε_k of a function $f(\varepsilon)$ from a condition

$$\sin\left[p_1 \varepsilon + \beta\right] = 0. \tag{10.83}$$

The maximum value of the reaction N is

$$N_{\max} = \frac{J_0 a}{J_0 + ml^2} + c_1 \left[\sin(\beta + \beta_1) + 2 \sin\beta_1 \sum_{k=1}^{\infty} \exp\left\{\frac{-n(k\pi - \beta)}{p_1}\right\}\right], \tag{10.84}$$

where

$$c_1 = \frac{cml^2 a}{p_0 p_1 (J_0 + ml^2)}; \quad \mathrm{tg}\beta_1 = \frac{p_1}{n}.$$

As the sum in the right-hand side of (10.84) is

$$\sum_{k=1}^{\infty} \exp\left\{\frac{-n(k\pi - \beta)}{p_1}\right\} = \frac{\exp\left\{\frac{-n(\pi-\beta)}{p_1}\right\}}{1 - \exp\left\{\frac{-n\pi}{p_1}\right\}}, \tag{10.85}$$

we finally obtain

$$N_{\max} = \frac{J_0 a}{J_0 + ml^2} + c_1 \left[\sin(\beta + \beta_1) + 2 \sin\beta_1 \frac{\exp\left\{\frac{-n(\pi-\beta)}{p_1}\right\}}{1 - \exp\left\{\frac{-n\pi}{p_1}\right\}}\right], \tag{10.86}$$

where

$$\sin\beta = \frac{2np_1}{\sqrt{p_0^2 + 4n^2 p_1^2}}; \quad \sin\beta_1 = \frac{p_0}{\sqrt{p_0^2 + n^2}}.$$

10.6 Areas of Possible Values of the System State Vector in the Case of Several Sections of Motion

In the previous sections we examined problems involving cases where the number of a systems degrees of freedom on the given interval of motion time remained unchanged as did the number of random excitations. Dealing with real systems, however, we run into violations of these conditions when during the motion of a system its number of degrees of freedom or the number and even the type of random excitations can change. A system with one degree of freedom is shown in Fig. 10.20. A mass m moves on an absolutely rigid beam with a velocity v. A random force f bounded in absolute value acts on the mass m on an interval $(0, l_1)$ and a random twisting moment bounded in absolute value ΔM (the mass m begins to rotate) acts on this mass on an interval (l_1, L). The area of possible values of θ and $\dot{\theta}$ at the moment of the mass m losing contact with the beam will depend on two sections of motion, on each of which different random excitations occur.

A system with varying number of degrees of freedom is shown in Fig. 10.21. Ring *1* can freely move on rigid guides *2* connected with disk *3*. The instants t_i of the ring's contact with the disk *4* are known. When there is a gap between ring *1* and disk *4*, the given system has two degrees of freedom. In the case of

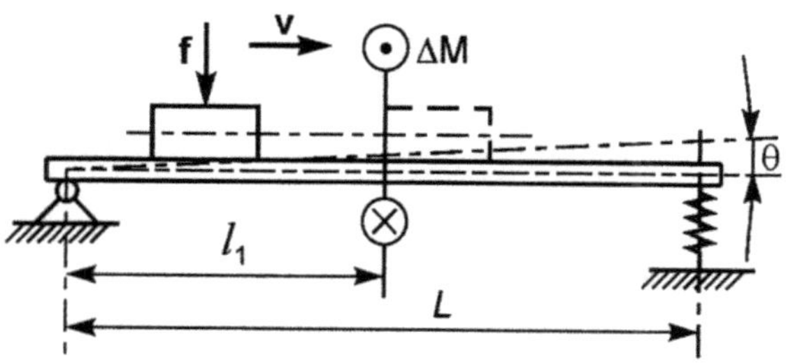

Fig. 10.20.

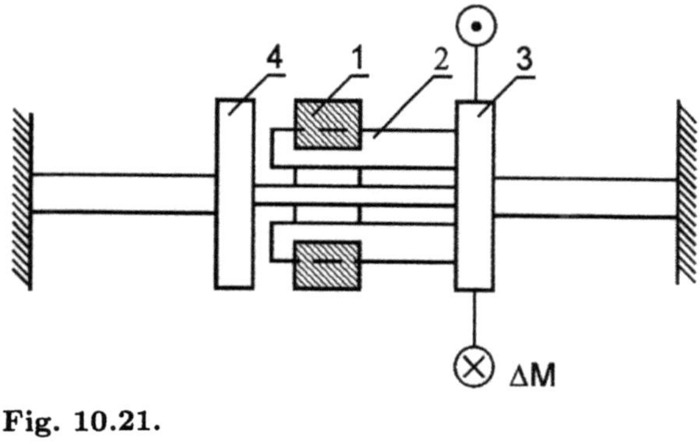

Fig. 10.21.

the ring being tightly forced against disk *4* (it is supposed that relative slip of ring *1* and disk *4* is excluded), the system has one degree of freedom.

If it is required to determine the possible angular displacements of disk *3* at a fixed time instant during the action of a random moment ΔM bounded in absolute value, we must consider the motion of the system with due account of the changes in the number of degrees of freedom.

Let us, first, consider a case where there are two sections of motion and the dimensionality of the state vector does not vary. On each of its sections of motion the system is described by its own vector-matrix equation expressed as

$$\dot{\mathbf{y}}_1 + A_1\mathbf{y}_1 = B_1\mathbf{f}_{11}; \quad \mathbf{y}_1(0) = \mathbf{y}_{10}, \quad (0 \le t \le t'_k); \tag{10.87}$$

$$\dot{\mathbf{y}}_2 + A_2\mathbf{y}_2 = B_2\mathbf{f}_{12}; \quad \mathbf{y}_2(t'_k) = \mathbf{y}_{20}, \quad (t'_k \le t \le t_k), \tag{10.88}$$

where $\mathbf{y}_1$ and $\mathbf{y}_2$ are the n-dimensional vectors of the disturbed state of the system on the first and second sections of its motion respectively; $\mathbf{f}_{11}$ is the vector of random excitations that act on the system on the first section of its motion; $\mathbf{f}_{12}$ is the vector of random excitations that act on the system on the second section of its motion. The components of the vectors $\mathbf{f}_{11}$ and $\mathbf{f}_{12}$ follow conditions of the form

$$m_{ij} \le f_{ij} \le M_{ij},$$

where m_{ij} and M_{ij} are some known functions.

The solutions of equations (10.87) and (10.88) are

$$\mathbf{y}_1(t'_k) = K_1(t'_k)\,\mathbf{y}_{10} + \int\limits_0^{t_k} G_1(t'_k, \tau)\, B_1\mathbf{f}_{11}\mathrm{d}\tau; \tag{10.89}$$

$$\mathbf{y}_2(t_k - t'_k) = K_2(t_k - t'_k)\,\mathbf{y}_{20} + \int\limits_0^{t_k - t'_k} G_2\left[(t_k - t'_k), \tau\right] B_2\mathbf{f}_{12}\mathrm{d}\tau. \tag{10.90}$$

Since $\mathbf{y}_1(t'_k) = \mathbf{y}_{20}$, then, substituting the expression for $\mathbf{y}_1(t'_k)$ in (10.90) we obtain

$$\mathbf{y}_2(t_k - t'_k) = K_2(t_k - t'_k)\, K_1(t'_k)\,\mathbf{y}_{10} + \int\limits_0^{t_k} K_2(t_k - t'_k)$$
$$\times\, G_1(t'_k, \tau)\, B_1\mathbf{f}_{11}\mathrm{d}\tau + \int\limits_0^{t_k - t'_k} G_2\left[(t_k - t'_k), \tau\right] B_2\mathbf{f}_{12}\mathrm{d}\tau. \tag{10.91}$$

In the case of k sections the expression for $\mathbf{y}_k$ will include k integral terms with nonintersecting limits of integration.

The maximum value of the projection of the vector $\mathbf{y}_2$ onto an arbitrary direction is determined in the same manner as for one section of motion.

Example 10.4. There are the following equations of a systems motion on a time interval $0 \le t \le t_1$

$$\dot{x}_{11} = f_1; \quad \dot{x}_{21} = x_{11}. \tag{10.92}$$

The equations of motion on a time interval $t_1 \le t \le t_2$ are

$$\dot{x}_{12} = f_2; \quad \dot{x}_{22} = -x_{12}. \tag{10.93}$$

Random excitations f_1 and f_2 satisfy the condition

$$1 \le f_i \le 1.$$

The area of possible values of x_{11}^* and x_{21}^* at the end of the first section of motion is the area of initial data for system (10.93). The solution of system (10.93) is

$$\begin{aligned} x_{12} &= \int_0^{t_2-t_1} f_2 \mathrm{d}\tau + x_{11}^*; \\ x_{22} &= -\int_0^{t_2-t_1} (t_2 - t_1 - \tau) f_2 \mathrm{d}\tau + x_{11}^* (t_2 - t_1) + x_{21}^*, \end{aligned} \tag{10.94}$$

where x_{11}^*, x_{21}^* are the values of x_{11} and x_{21} at $t = t_1$.

In order to plot the area it is more convenient to present solutions (10.94) in the form

$$\begin{aligned} x_{12} &= \int_0^{t_1} f_1 \mathrm{d}\tau + \int_0^{t_2-t_1} f_2 \mathrm{d}\tau; \\ x_{22} &= \int_0^{t_1} (t_1 - \tau) f_1 \mathrm{d}\tau - \int_0^{t_2-t_1} [(t_2 - t_1) - \tau] f_2 \mathrm{d}\tau. \end{aligned} \tag{10.95}$$

Let us determine the maximum value of the projection of the vector $\mathbf{x}$ (x_{12}, x_{22}) onto the direction determined by the unit vector $\mathbf{e}$:

$$\begin{aligned} (\mathbf{x} \cdot \mathbf{e}) = &\int_0^{t_1} [\cos\alpha + (2t_1 - t_1 - \tau)\sin\alpha] f_1 \mathrm{d}\tau \\ &+ \int_0^{t_2-t_1} [\cos\alpha - (t_2 - t_1 - \tau)\sin\alpha] f_2 \mathrm{d}\tau. \end{aligned}$$

The maximum value of the projection $(\mathbf{x} \cdot \mathbf{e})$ is equal to the sum of the maximum values of the terms which can be determined separately (for each x we determine the instant τ_i of discontinuity of the functions f_1 and f_2). As the integrands in (10.95) linearly depend on τ on the intervals $(0, t_1)$ and (t_1, t_2), they can reverse sign only once, i.e. the extreme values of each of the integrals are reached at the variation laws of f_1 and f_2 with one discontinuity. The area from f_2 at $t = t_2$ shown by the dashed line corresponds to each point of the boundary of the area at $t = t_1$ (Fig. 10.22). Envelope 1 bounds the full area at the moment t_2 on the plane $x_{22}x_{12}$. When the dimensionality of the system state vector $\mathbf{y}_1$ varies on different sections of motion, as, for example, in the system shown in Fig. 10.21, we have systems of equations of different dimensionality on different sections of motion. For the purpose of greater definiteness let the dimensionality of the vector $\mathbf{y}_1$ be equal to n, and the dimensionality of the vector $\mathbf{y}_2$ be equal to ν $(\nu < n)$, i.e.

$$\mathbf{y}_1 = \begin{bmatrix} y_{11} \\ y_{12} \\ \vdots \\ y_{1\nu} \end{bmatrix}; \quad \mathbf{y}_2 = \begin{bmatrix} y_{21} \\ y_{22} \\ \vdots \\ y_{2\nu} \end{bmatrix}.$$

The components $y_{21}, \ldots, y_{2\nu}$ are the continuation of the components $y_{11}, \ldots, y_{1\nu}$ on the second section of motion, i.e. the components of the vectors $\mathbf{y}_1$ and $\mathbf{y}_2$ $(y_{11}, y_{12}) \ldots (y_{1\nu}, y_{2\nu})$ are physically identical. The vector $\mathbf{y}_1(t'_k)$

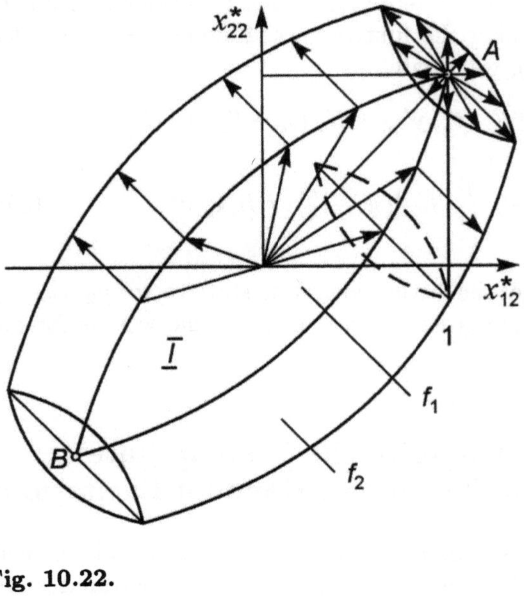

Fig. 10.22.

can be presented as

$$\mathbf{y}_1(t'_k) = \begin{bmatrix} \mathbf{y}_{20} \\ \mathbf{y'}_{20} \end{bmatrix} = K_1(t'_k)\begin{bmatrix} \mathbf{y'}_{10} \\ \mathbf{y''}_{10} \end{bmatrix} + \int_0^{t'_k} K_1 B_1 \mathbf{f}_{11} \mathrm{d}\tau, \tag{10.96}$$

where $\mathbf{y'}_{20}$ contains the components of the vector y_1 that have no continuation on the second section of motion. It is possible to derive the following expression for $\mathbf{y}_{20}$ from equation (10.96)

$$\mathbf{y}_{20} = K'_1(t'_k)\,\mathbf{y'}_{10} + K''_1(t'_k)\,\mathbf{y''}_{10} + \int_0^{t_k} P\mathbf{f}_{11}\mathrm{d}\tau, \tag{10.97}$$

where

$$K'_1(t'_k) = \begin{bmatrix} k_{11} & k_{12} & \dots & k_{1\nu} \\ k_{21} & k_{22} & \dots & k_{2\nu} \\ \vdots & \vdots & \ddots & \vdots \\ k_{\nu 1} & k_{\nu 2} & \dots & k_{\nu\nu} \end{bmatrix}; \quad K''_1(t'_k) = \begin{bmatrix} k_{1,\nu+1} & \dots\dots & k_{1n} \\ k_{2,\nu+1} & \dots\dots & k_{2n} \\ \vdots & \vdots\ \ddots & \vdots \\ k_{\nu,\nu+1} & \dots\dots & k_{\nu n} \end{bmatrix};$$

$$P = \begin{bmatrix} p_{11} & p_{12} & \cdots & p_{1n} \\ p_{21} & p_{22} & \cdots & p_{2n} \\ \dots & \dots & \dots & \dots \\ p_{\nu 1} & p_{\nu 2} & \cdots & p_{\nu n} \end{bmatrix}.$$

The matrix K'_1 is a square matrix (of the ν-th order); the matrixes K''_1 and P are rectangular. The matrix P represents a matrix obtained from the matrix $K_1 B_1$ by way of retaining the first ν rows. Having substituted the expression for $\mathbf{y}_{20}$ into (10.90), we obtain

$$\mathbf{y}_2(t_k - t'_k) = K_2(t_k - t'_k)\,K'_1(t'_k)\,\mathbf{y'}_{10} + K_2(t_k - t'_k)\,K''_1(t'_k)\,\mathbf{y''}_{10} + \int_0^{t'_k} K_2(t_k - t'_k)\,P\mathbf{f}_{11}\mathrm{d}\tau + \int_0^{t_k - t'_k} K_2\left[(t_k - t'_k), \tau\right] B_2 \mathbf{f}_{12}\mathrm{d}\tau. \tag{10.98}$$

The area of possible values of the disturbed system state vector $\mathbf{y}_2(t_k - t'_k)$ at the end of the second section is determined in the same way as for one section of motion.

10.7 Areas of Possible Values of the System State Vector at the Action of Dependent Random Excitations

Not infrequently we come across random excitations that cannot be considered independent. They include, among others, a scatter of coordinates and their first two derivatives at some time instant and projections of a random force onto coordinate axes. An example of a mechanical system subjected to the action of dependent random excitations bounded in absolute value is presented in Sect. 10.1 (see Fig. 10.10). The dependence between the excitations has the same effect as the additional restrictions imposed on these functions. One of such restrictions is condition (10.23)

$$(C\mathbf{f} \cdot \mathbf{f}) \leq 1, \tag{10.99}$$

where C is a square self-adjoint positively definite matrix.

In the general case, the elements c_{ij} of the matrix C can depend on time. As is known, for a matrix to be positively definite it is necessary and sufficient that its elements satisfy the Sylvester criterion (for any time instant t)

$$\Delta_1 = c_{11} > 0, \quad \Delta_2 = \begin{vmatrix} c_{11} & c_{12} \\ c_{21} & c_{22} \end{vmatrix} > 0, \ldots, \Delta_n > 0.$$

The vector of random initial data $\mathbf{y}_0$ can satisfy the similar condition

$$(\mathrm{C}_0\mathbf{y}_0 \cdot \mathbf{y}_0) \leq 1, \tag{10.100}$$

where C_0 is the square self-adjoint positively definite matrix with constant elements.

The solution of equation (10.24) is

$$\mathbf{y} = K(t)\,\mathbf{y}_0 + \int_0^{t_k} G(t_k, \tau)\, B(\tau)\, \mathbf{f} \mathrm{d}\tau. \tag{10.101}$$

Let us consider a case where the vector of random excitations can be presented as

$$\mathbf{f} = H(t)\, \mathbf{f}^0, \tag{10.102}$$

where $H(t)$ is the diagonal matrix with continuously time-dependent elements and $\mathbf{f}^0$ is the vector whose components are random numbers limited in absolute value.

Having substituted expression (10.102) in equation (10.101), we obtain

$$\mathbf{y} = K\mathbf{y}_0 + \int_0^{t_k} P(t_k, \tau)\, \mathbf{f}^0 \mathrm{d}\tau, \tag{10.103}$$

where $P(t_k, \tau) = G(t_k, \tau) \cdot B(\tau)\, H(\tau)$.

Condition (10.99) is modified into the following form

$$(\mathrm{C}_1\mathbf{f}^0 \cdot \mathbf{f}^0) \leq 1, \tag{10.104}$$

where $\mathrm{C}_1 = HCH$.

In order to obtain the area of deviations of the vector $\mathbf{y}$ let us determine the maximum projection of the vector $\mathbf{y}$ onto the unit vector α:

$$y_{pre} = (\mathbf{y} \cdot \boldsymbol{\alpha}) = (K(t)\,\mathbf{y}_0 \cdot \boldsymbol{\alpha}) + \int_0^{t_k} P(t_k, \tau)\,\mathbf{f}^0 \mathrm{d}\tau \cdot \boldsymbol{\alpha}. \tag{10.105}$$

Let us determine the maximum of expression (10.105) at additional conditions (10.99) and (10.100) which must be satisfied by the vectors $\mathbf{y}_0$ and $\mathbf{f}^0$. Having used Lagrangian multipliers, we obtain the following functional:

$$J = (K(t)\,\mathbf{y}_0 \cdot \boldsymbol{\alpha}) + \int_0^{t_k} P(t_k, \tau)\,\mathbf{f}^0 \mathrm{d}\tau \cdot \boldsymbol{\alpha}$$
$$- \frac{\mu}{2}\left[(\mathrm{C}_0\mathbf{y}_0 \cdot \mathbf{y}_0) - 1\right] - \frac{\lambda}{2}\left[(\mathrm{C}_0\mathbf{f}^0 \cdot \mathbf{f}^0) - 1\right],$$

or

$$J = (\mathbf{y}_0 \cdot K^*(t)\,\boldsymbol{\alpha}) + \mathbf{f}^0 \int_0^{t_k} P^*(t_k, \tau)\,\mathrm{d}\tau \cdot \boldsymbol{\alpha}$$
$$- \frac{\mu}{2}\left[(\mathrm{C}_0\mathbf{y}_0 \cdot \mathbf{y}_0) - 1\right] - \frac{\lambda}{2}\left[(\mathrm{C}_0\mathbf{f}^0 \cdot \mathbf{f}^0) - 1\right]. \tag{10.106}$$

The functional J depends on y_{0i} and f_i^0, therefore y_{0i} and f_i^0 at which J reaches its extreme values are determined from the conditions

$$\frac{\partial J}{\partial y_{0i}} = 0; \quad \frac{\partial J}{\partial f_i^0} = 0, \quad (i = 1, 2, \ldots, n). \tag{10.107}$$

Conditions (10.107) can be obtained in a more compact form, by differentiating (10.106) with respect to the vectors $\mathbf{y}_0$ and $\mathbf{f}^0$:

$$\begin{aligned} \frac{\partial J}{\partial \mathbf{y}_0} &= K^*\boldsymbol{\alpha} - \mu\mathrm{C}_0\mathbf{y}_0 = 0; \\ \frac{\partial J}{\partial \mathbf{f}^0} &= \int_0^{t_k} P^* \mathrm{d}\tau \cdot \boldsymbol{\alpha} - \lambda\mathrm{C}_1\mathbf{f}^0 = 0. \end{aligned} \tag{10.108}$$

The vectors $\mathbf{y}_0$ and $\mathbf{f}^0$ that give extreme values to expression (10.106) are determined from the following equations

$$K^*(t_k)\,\boldsymbol{\alpha} = \mu\mathrm{C}_0\mathbf{y}_0; \tag{10.109}$$

$$\int_0^{t_k} P^*(t_k, \tau)\,\mathrm{d}\tau\,\boldsymbol{\alpha} = \lambda\mathrm{C}_1\mathbf{f}^0. \tag{10.110}$$

Multiplying relationships (10.109) and (10.110) by matrixes C_0^{-1} and C_1^{-1} respectively, we obtain

$$C_0^{-1}K^*(t_k)\boldsymbol{\alpha} = \mu\mathbf{y}_0; \tag{10.111}$$

$$C_1^{-1}\int_0^{t_k} P^*(t_k,\tau)\,\mathrm{d}\tau\,\boldsymbol{\alpha} = \lambda\mathbf{f}^0. \tag{10.112}$$

Let us scalarly multiply expression (10.109) by expression (10.111)

$$\left[K^*(t_k)\boldsymbol{\alpha}\cdot C_0^{-1}K^*(t_k)\,\boldsymbol{\alpha}\right] = \mu^2,$$

or

$$\left(KC_0^{-1}K^*\boldsymbol{\alpha}\cdot\boldsymbol{\alpha}\right) = \mu^2. \tag{10.113}$$

We shall deal with expressions (10.110) and (10.112) in a similar way:

$$\left(\int_0^{t_k} P^*(t_k,\tau)\,\mathrm{d}\tau\boldsymbol{\alpha}\cdot C_1^{-1}\int_0^{t_k} P^*(t_k,\tau)\,\mathrm{d}\tau\,\boldsymbol{\alpha}\right) = \lambda^2,$$

or

$$\left(\int_0^{t_k} P\mathrm{d}\tau C_1^{-1}\int_0^{t_k} C^*\mathrm{d}\tau\,\boldsymbol{\alpha}\cdot\boldsymbol{\alpha}\right) = \lambda^2. \tag{10.114}$$

Expressions (10.113) and (10.114) define the introduced Lagrangian multipliers μ and λ for each direction in space.

Let us determine the maximum value of each of the terms defining the vector $\mathbf{y}$. Then, we present the vector $\mathbf{y}$ as a sum of two vectors:

$$\mathbf{y} = \mathbf{y}_1 + \mathbf{y}_2, \tag{10.115}$$

where

$$\mathbf{y}_1 = K(t_k)\,\mathbf{y}_0,$$

$$\mathbf{y}_2 = \int_0^{t_k} P(t_k,\tau)\,\mathbf{f}^0\mathrm{d}\tau = \int_0^{t_k} P\,\mathrm{d}\tau\,\mathbf{f}^0.$$

Let us prove the following theorem: the maximum value of the projections of the vectors $\mathbf{y}_1$ and $\mathbf{y}_2$ onto the arbitrary direction determined by the unit vector $\boldsymbol{\alpha}$ is equal to the value of the Lagrangian multiplier for this direction.

Let us multiply both parts of equality (10.111) from the left by the matrix $K(t_k)$:

$$K(t_k)\,C_0^{-1}K^*(t_k)\,\boldsymbol{\alpha} = \mu K(t_k)\,\mathbf{y}_0 = \mu \mathbf{y}_1. \tag{10.116}$$

Let us determine the maximum value of the projection of the vector $\mathbf{y}_1$ onto the direction α from the following relationship

$$\mathbf{y}_{1pr\boldsymbol{\alpha}} = (\mathbf{y}_1 \cdot \boldsymbol{\alpha}) = (K(t_k)\,\mathbf{y}_0 \cdot \boldsymbol{\alpha}) = (\mathbf{y}_0 K^*(t_k) \cdot \boldsymbol{\alpha}),$$

or, having used relationship (10.109),

$$y_{1pr\boldsymbol{\alpha}} = (\mathbf{y}_0 \cdot \mu C_0 \mathbf{y}_0) = \mu\,(\mathbf{y}_0 \cdot C_0 \mathbf{y}_0) = \mu. \tag{10.117}$$

Similarly we may show that

$$y_{1pr\boldsymbol{\alpha}} = \lambda. \tag{10.118}$$

The formulated theorem is proved. The maximum value of the functional J or of the projection of the vector onto the direction $\boldsymbol{\alpha}$ is

$$J_{\max} = y_{pr\alpha\,\max} = \mu + \lambda, \tag{10.119}$$

where μ and λ are determined from relationships (10.113) and (10.114).

Let us determine the vectors $\mathbf{y}_0$ and $\mathbf{f}^0$ that give maximum values to $\mathbf{y}_1$ and $\mathbf{y}_2$ for the given direction $\boldsymbol{\alpha}$. From relationships (10.109) and (10.110) we obtain

$$\mathbf{y}_0 = \frac{1}{\mu} C_0^{-1} K^*(t_k)\,\boldsymbol{\alpha}; \tag{10.120}$$

$$\mathbf{f}^0 = \frac{1}{\lambda} C_1^{-1} \int_0^{t_k} P^*(t_k,\tau)\,\mathrm{d}\tau\,\boldsymbol{\alpha}. \tag{10.121}$$

The vectors $\mathbf{y}_1$ and $\mathbf{y}_2$ for the given direction $\boldsymbol{\alpha}$ are determined from the following relationships

$$\mathbf{y}_1 = \frac{1}{\mu} K(t_k)\,C_0^{-1} K^*(t_k)\,\boldsymbol{\alpha}; \tag{10.122}$$

$$\mathbf{y}_2 = \frac{1}{\lambda} \int_0^{t_k} P(t_k,\tau)\,\mathrm{d}\tau\, C_0^{-1} \int_0^{t_k} P^*(t_k,\tau)\,\mathrm{d}\tau\,\boldsymbol{\alpha}, \tag{10.123}$$

or

$$\mathbf{y}_1 = \frac{K C_0^{-1} K^* \boldsymbol{\alpha}}{\sqrt{(K C_0^{-1} K^* \boldsymbol{\alpha} \cdot \boldsymbol{\alpha})}}, \tag{10.124}$$

$$\mathbf{y}_2 = \frac{\int\limits_0^{t_k} P\mathrm{d}\tau \mathrm{C}_1^{-1} \int\limits_0^{t_k} P^* \mathrm{d}\tau \boldsymbol{\alpha}}{\sqrt{\int\limits_0^{t_k} P\mathrm{d}\tau \mathrm{C}_1^{-1} \int\limits_0^{t_k} P^* \mathrm{d}\tau \boldsymbol{\alpha} \cdot \boldsymbol{\alpha}}} = \frac{\mathbf{A}}{a}. \tag{10.125}$$

Formulas (10.124) and (10.125) enable us to determine for each unit vector $\boldsymbol{\alpha}$ a vector $\mathbf{y}$, that connects the point of origin with some point of the boundary of the area of possible values of the vector $\mathbf{y}$. Formulas (10.120)–(10.123) are true provided that $\mu \neq 0$ and $\lambda \neq 0$. If the Lagrangian multipliers for some directions of the vector $\boldsymbol{\alpha}$ vanish, formulas (10.120)-(10.123) are not applicable.

It follows from relationships (10.116) and (10.117) that in this case the projections of the vectors $\mathbf{y}_1$ and $\mathbf{y}_2$ onto some direction are equal to zero (it is not always the case that both projections of the vectors $\mathbf{y}_1$ and $\mathbf{y}_2$ vanish simultaneously). The vanishing of one of the projections of the vectors (for example, that of $\mathbf{y}_1$) means that the area of deviations of the vector $\mathbf{y}_1$ degenerates, i.e. has a smaller dimensionality. The Lagrangian multipliers can vanish only when

$$K^* (t_k) \boldsymbol{\alpha} = 0; \tag{10.126}$$

$$\int\limits_0^{t_k} P^* (t_k, \tau) \, \mathrm{d}\tau = 0. \tag{10.127}$$

The dimensionality of the subspace of the vector $\boldsymbol{\alpha}$ for which the Lagrangian multipliers vanish is equal to the following rank of matrixes

$$K^* (t_k); \quad \int\limits_0^{t_k} P^* (t_k, \tau) \, \mathrm{d}\tau.$$

If the rank of the matrixes is equal to n, the Lagrangian multipliers do not vanish. The state-of-the-system vector at a fixed time instant (the greatest possible value of the vector $\mathbf{y}$) is

$$\mathbf{y}_* = \mathbf{y}_{1*} + \mathbf{y}_{2*} = \left(\frac{K_1}{\sqrt{(K_1 \boldsymbol{\alpha} \cdot \boldsymbol{\alpha})}} + \frac{P_1}{\sqrt{(P_1 \boldsymbol{\alpha} \cdot \boldsymbol{\alpha})}} \right) \boldsymbol{\alpha}, \tag{10.128}$$

where

$$K_1 = K \mathrm{C}_0^{-1} K^*; \quad P_1 = \int\limits_0^{t_k} P\mathrm{d}\tau \mathrm{C}_1^{-1} \int\limits_0^{t_k} P^* \mathrm{d}\tau.$$

Relationship (10.128) allows us to establish the dependence of the complete area on the scatter of initial data (vector $\mathbf{y}_{1*}$) and on random excitations that act during the system motion (vector $\mathbf{y}_{2*}$).

Let us find the equations of the areas of possible values of each of the vectors $\mathbf{y}_1$ and $\mathbf{y}_2$ of the general solution $\mathbf{y}$. Let us consider the expression for $\mathbf{y}_1$ (10.122):

$$\mathbf{y}_1 = K_1 (t_k) \boldsymbol{\alpha}. \tag{10.129}$$

Let us show, that the matrix K_1 is self-adjoint, i.e. satisfies the condition $(K_1)^* = (K_1)$:

$$\begin{aligned}(K_1)^* &= \frac{1}{\mu} \left(K \mathrm{C}_0^{-1} K^*\right)^* = \frac{1}{\mu} \left(K^*\right)^* \left(\mathrm{C}_0^{-1}\right)^* \left(K^*\right) \\ &= \frac{1}{\mu} K \left(\mathrm{C}_0^{-1}\right)^* K^* = \frac{1}{\mu} K \left(\mathrm{C}_0^*\right)^{-1} K^* = \frac{1}{\mu} K \mathrm{C}_0^{-1} K^* = K_1.\end{aligned}$$

In the process of derivation we used the condition of the self-adjointness of the matrix C_0.

From relationship (10.129) we have

$$K_1^{-1} \mathbf{y}_1 = \boldsymbol{\alpha}. \tag{10.130}$$

Scalarly multiplying the left- and right-hand sides of equation (10.130) by $K_1^{-1}\mathbf{y}_1$ and $\boldsymbol{\alpha}$ respectively, we obtain

$$\left(K_1^{-1} \mathbf{y}_1 \cdot K_1^{-1} \mathbf{y}_1\right) = (\boldsymbol{\alpha} \cdot \boldsymbol{\alpha}) = 1,$$

or

$$\left(\left(K_1^{-1}\right)^2 \mathbf{y}_1 \cdot \mathbf{y}_1\right) = 1. \tag{10.131}$$

Equations (10.131) represent an equation of a second order surface in an $n-$dimensional space. It follows from the boundedness of the area of deviations that this surface is an ellipsoid. Similarly we can show that the vector $\mathbf{y}_2$ satisfies the following equation

$$\left(\left(P_1^{-1}\right)^2 \mathbf{y}_2 \cdot \mathbf{y}_2\right) = 1, \tag{10.132}$$

where P_1 is the self-adjoint matrix.

Relationships (10.124), (10.125) define the greatest possible value of the vector $\mathbf{y}$ (10.115) as a function of the direction of the arbitrary unit vector $\boldsymbol{\alpha}$. When the direction of the vector $\boldsymbol{\alpha}$ continuously changes in space, the end of the vector $\mathbf{y}_*$ describes the limit surface (the boundary of the area of possible values of the state vector). It is possible to present the vector $\boldsymbol{\alpha}$ in terms of projections in the initial coordinate system:

$$\boldsymbol{\alpha} = \sum_{i=1}^{n} h_i \mathbf{e}_i, \tag{10.133}$$

where h_i are the projections of the vector $\boldsymbol{\alpha}$ onto the coordinate axes and $\mathbf{e}_i$ are unit vectors of the coordinate axes.

The component of the vector $\mathbf{y}_*$ is

$$\mathbf{y}_{*i} = \sum_{j=1}^{n} a_{ij} h_j, \quad (i = 1, 2, \ldots, n). \tag{10.134}$$

Let us consider a special case where it is required to determine the projection of the area onto a two-dimensional plane. Here

$$\boldsymbol{\alpha} = h_1 \mathbf{e}_1 + h_2 \mathbf{e}_2 = \cos\alpha\, \mathbf{e}_1 + \sin\alpha\, \mathbf{e}_2.$$

From relationships (10.134) we obtain

$$\begin{aligned} y_{*1} &= a_{11} h_1 + a_{12} h_2, \\ y_{*2} &= a_{21} h_1 + a_{22} h_2. \end{aligned} \tag{10.135}$$

Changing α from 0 to 360°, we obtain from relationships (10.135) the coordinates of the boundary points of the projection of the area onto a two-dimensional plane.

Example 10.5. Let us consider the free vibrations of a mass m (Fig. 10.23). Without considering damping forces the equation of the small vibrations of the mass is

$$\ddot{y} + p_0^2 y = 0, \tag{10.136}$$

where $p_0^2 = \dfrac{c}{m}$; $c = \dfrac{3EJ_x}{l^3}$.

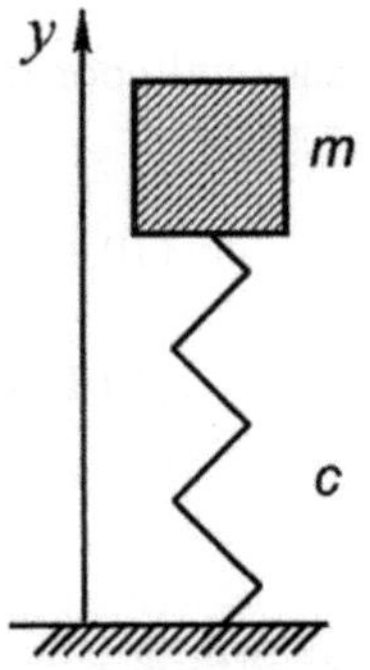

Fig. 10.23.

The initial data are random dependent quantities that satisfy the condition

$$\left(\frac{y_0}{y_{0m}}\right)^2 + \left(\frac{\dot{y}_0}{\dot{y}_{0m}}\right)^2 \le 1, \tag{10.137}$$

where y_{0m}, $\dot{y}_{0m}$ are the greatest possible values of the scatter of the initial data.

In the considered case condition (10.137) has a simple physical meaning. At the initial time instant the system can have a random deviation y_0 and a random velocity $\dot{y}_0$, i.e. the system obtains a random energy equal to

$$\frac{cy_0^2}{2} + \frac{m\dot{y}_0^2}{2} = W. \tag{10.138}$$

The greatest possible energy that can be obtained by the system is limited and equal to W_m, it can be expressed in terms of the greatest possible deviation and the greatest possible velocity:

$$W_m = \frac{cy_{0m}^2}{2} = \frac{m\dot{y}_{0m}^2}{2}. \tag{10.139}$$

At arbitrary deviations

$$\frac{cy_0^2}{2} + \frac{m\dot{y}_0^2}{2} \le W_m \tag{10.140}$$

or, having divided the inequality by W_m, we obtain

$$\left(\frac{y_0}{y_{0m}}\right)^2 + \left(\frac{\dot{y}_0}{\dot{y}_{0m}}\right)^2 \le 1. \tag{10.141}$$

Let us pass to non-dimensional quantities, putting

$$tp_0 = \tau; \quad y = zy_{0m}.$$

By transformations we obtain (having passed to non-dimensional coordinate and non- dimensional time)

$$\ddot{z} + z = 0; \tag{10.142}$$

$$z_0^2 + \left(\frac{\dot{z}_0}{\dot{z}_{0m}}\right)^2 \le 1, \tag{10.143}$$

where $\dot{z}_{0m} = \dfrac{\dot{y}_{0m}}{(y_{0m}p_0)}$.

Condition (10.143) can be presented as the following scalar product

$$(\mathbf{C}_0\mathbf{z}_0 \cdot \mathbf{z}_0) \le 1, \tag{10.144}$$

where

$$\mathrm{C}_0 = \begin{bmatrix} \frac{1}{z_{0m}^2} & 0 \\ 0 & 1 \end{bmatrix}; \qquad \mathbf{z}_0 = \begin{bmatrix} \dot{z}_0 \\ z_0 \end{bmatrix}.$$

The solution of equation (10.142) can be presented as

$$\mathbf{z} = \begin{bmatrix} \dot{z} \\ z \end{bmatrix} = \begin{bmatrix} \cos\tau & -\sin\tau \\ \sin\tau & \cos\tau \end{bmatrix} \cdot \begin{bmatrix} \dot{z}_0 \\ z_0 \end{bmatrix} = K(\tau)\,\mathbf{z}_0. \tag{10.145}$$

At the time instant $t = t_k$ or $\tau_k = t_k p_0$ (for the purpose of simplification let us assume that $\tau_k = 1$)

$$\mathbf{z} = K(1)\,\mathbf{z}_0. \tag{10.146}$$

Lagrangian multiplier (10.113) for the arbitrary direction of the unit vector $\boldsymbol{\alpha}$ (see Fig. 10.18) is

$$\mu = \sqrt{\left(K\mathrm{C}_0^{-1}K^{*}\boldsymbol{\alpha}\cdot\boldsymbol{\alpha}\right)}, \tag{10.147}$$

where

$$K = \begin{bmatrix} \cos 1 & -\sin 1 \\ \sin 1 & \cos 1 \end{bmatrix}; \quad \mathrm{C} = \begin{bmatrix} \dot{z}_{0m}^2 & 0 \\ 0 & 1 \end{bmatrix}; \quad K^{*} = \begin{bmatrix} \cos 1 & \sin 1 \\ -\sin 1 & \cos 1 \end{bmatrix}.$$

By transformations we obtain

$$\mu = \sqrt{k_{11}\cos^2\alpha + k_{12}\sin 2\alpha + k_{22}\sin^2\alpha}$$

or

$$\mu = \sqrt{\dot{z}_{0m}^2\cos^2(\alpha-1) + \sin^2(\alpha-1)}, \tag{10.148}$$

where α is the angle expressed in radians;

$$k_{11} = \dot{z}_{0m}^2\cos^2 1 + \sin^2 1;$$
$$k_{12} = \frac{1}{2}\sin 2\left(-1 + \dot{z}_{0m}^2\right);$$
$$k_{22} = \dot{z}_{0m}^2\sin^2 1 + \cos^2 1.$$

Let us find a vector $\mathbf{z}_0$, at which the projection of the vector $\mathbf{z}$ on the direction determined by the vector $\boldsymbol{\alpha}$, reaches its maximum value, from the expression

$$\mathbf{z}_0 = \frac{1}{\mu}\mathrm{C}_0^{-1}K^{*}(1)\,\boldsymbol{\alpha},$$

or

$$\mathbf{z}_0 = \frac{1}{\mu}\left[\dot{z}_{0m}^2 \cos(\alpha - 1)\,\mathbf{e}_1 + \sin(\alpha - 1)\,\mathbf{e}_2\right]. \tag{10.149}$$

The components of the vector $\mathbf{z}_0$ are

$$\begin{aligned} z_{01} &= \dot{z}_0 = \frac{z_{0m}^2 \cos(\alpha - 1)}{\mu}; \\ z_{02} &= z_0 = \frac{1}{\mu}\sin(\alpha - 1). \end{aligned} \tag{10.150}$$

The obtained expressions for $\dot{z}_0$, z_0 and μ should satisfy condition (10.144) (the test of the solution). It is easy to verify that this is true. The system state vector $\mathbf{z}$ at a time instant $\tau = 1$ (10.124) is

$$\mathbf{z} = \frac{1}{\mu} K \mathrm{C}_0^{-1} K^* \boldsymbol{\alpha},$$

or

$$\begin{aligned} \mathbf{z} = {} & \frac{1}{\mu}\left[\left(\dot{z}_{0m}^2 \cos^2 1 + \sin^2 1\right)\cos\alpha + \frac{1}{2}\sin 2\left(\dot{z}_{0m}^2 - 1\right)\sin\alpha\right]\mathbf{e}_1 \\ & + \frac{1}{\mu}\left[\frac{1}{2}\left(\dot{z}_{0m}^2 - 1\right)\sin 2\cos\alpha + \left(\dot{z}_{0m}^2 \sin^2 1 + \cos^2 1\right)\sin\alpha\right]\mathbf{e}_2. \end{aligned}$$

The components of the system state vector are:

$$\begin{aligned} \dot{z}(1) &= \frac{1}{\mu}\left[\left(\dot{z}_{0m}^2 \cos^2 1 + \sin 1\right)\cos\alpha + \frac{1}{2}\sin 2\left(\dot{z}_{0m}^2 - 1\right)\right]; \\ z(1) &= \frac{1}{\mu}\left[\frac{1}{2}\cos\alpha\sin 2\left(\dot{z}_{0m}^2 - 1\right) + \left(\dot{z}_{0m}^2 \sin^2 1 + \cos^2 1\right)\sin\alpha\right]. \end{aligned} \tag{10.151}$$

Changing the angle α from 0 to 2π, we obtain from (10.151) the coordinates of the boundary of the area of possible values of the system state vector at a fixed time instant.

Let us consider the special case of $\dot{z}_{om}^2 = \mu = 1$

$$\dot{z}(1) = \cos\alpha; \quad z(1) = \sin\alpha. \tag{10.152}$$

Eliminating α from (10.151), we obtain the equation of the boundary of the area of possible values of solutions $\dot{z}^2 + z^2 = 1$ which is true for any time τ.

In the given example the free vibrations of the conservative system, for which the total energy remains unaltered are considered, therefore the area of possible values of solutions also remains unaltered in time.

Let us consider the case of a force random in its direction and bounded in its absolute value $\mathbf{f}$ ($|\mathbf{f}| \leq \mathbf{f}_m$) acting on a mass m (Fig. 10.10).

The equations of motion of the mass m are

$$\ddot{y} + p_1^2 y = \frac{1}{m} f_y; \quad \ddot{x} + p_2^2 y = \frac{1}{m} f_x, \tag{10.153}$$

where

$$p_1^2 = \frac{c_1}{m} = \frac{EJ_x}{3l^3 m}; \quad p_2^2 = \frac{c_2}{m} = \frac{EJ_y}{3l^3 m}.$$

Going on to non-dimensional parameters $tp_1 = \tau$, $y = y_0 l$, $x = x_0 l$, by way of transformations we obtain

$$\begin{aligned} \ddot{y}_0(\tau) + y_0(\tau) &= \frac{f_y}{mp_1^2 l} = f_{y0}; \\ \ddot{x}_0(\tau) + p_0^2 x_0(\tau) &= \frac{f_x'}{mp_1^2 l} = f_{x0}. \end{aligned} \tag{10.154}$$

The non-dimensional components of the vector $\mathbf{f}$ satisfy the condition

$$\left(\frac{f_{x0}'}{f_{m0}'}\right)^2 + \left(\frac{f_{y0}'}{f_{m0}'}\right)^2 \leq 1; \quad \left(f_{m0} = \frac{f_m'}{mp_1^2 l}\right)$$

or

$$(\mathrm{C}_1 \mathbf{f} \cdot \mathbf{f}) \leq 1,$$

where matrix C_1 is

$$\mathrm{C}_1 = \begin{bmatrix} \frac{1}{f_m^2} & 0 & 0 & 0 \\ 0 & 1 & 0 & 0 \\ 0 & 0 & \frac{1}{f_m^2} & 0 \\ 0 & 0 & 0 & 1 \end{bmatrix}.$$

It is required to determine the area of possible values of the non-dimensional coordinates x_0, y_0 of the mass m at a fixed time instant t_k ($\tau_k = p_1 t_k$) at zero initial data.

From system (10.154) we obtain the following set of four equations of the first order (omitting index 0 in non-dimensional quantities)

$$\dot{y}_1 + y_2 = f_1; \quad \dot{y}_1 - y_2 = 0; \tag{10.155}$$

$$\dot{y}_3 + p_0^2 y_4 = f_3; \quad \dot{y}_4 - y_3 = 0, \tag{10.156}$$

where

$$\begin{aligned} y_1 &= \dot{y}; \quad y_2 = y; \quad f_1 = f_y; \\ y_3 &= \dot{x}; \quad y_4 = x; \quad f_3 = f_x. \end{aligned}$$

In vector form we have

$$\dot{y} + A\mathbf{y} = \mathbf{f}, \tag{10.157}$$

where

$$\mathbf{y} = \begin{bmatrix} y_1 \\ y_2 \\ y_3 \\ y_4 \end{bmatrix}; \quad A = \begin{bmatrix} 0 & 1 & 0 & 0 \\ -1 & 0 & 0 & 0 \\ 0 & 0 & 0 & p_0^2 \\ 0 & 0 & -1 & 0 \end{bmatrix}; \quad \mathbf{f} = \begin{bmatrix} f_1 \\ 0 \\ f_3 \\ 0 \end{bmatrix}.$$

As follows from formula (10.125), we must know the matrix $P(\tau_k, \tau)$ in order to determine the state-of-the-system vector $\mathbf{y}$. In the considered case the matrix P is the Green matrix of equation (10.157):

$$P(\tau_k, \tau) = G(\tau_k, \tau).$$

As the matrix $G(\tau_k, \tau)$ is determined from the solution of homogeneous equations (10.155) and (10.156) that are independent, we can obtain it, having determined Green matrixes for each of the systems of equations (10.155) and (10.156). The Green matrixes for these equations are

$$G_1(\tau_k, \tau) = \begin{bmatrix} \cos(\tau_k - \tau) & \sin(\tau_k - \tau) \\ -\sin(\tau_k - \tau) & \cos(\tau_k - \tau) \end{bmatrix};$$

$$G_2(\tau_k, \tau) = \begin{bmatrix} \cos p_0(\tau_k - \tau) & \sin p_0(\tau_k - \tau) \\ -\sin p_0(\tau_k - \tau) & \cos p_0(\tau_k - \tau) \end{bmatrix},$$

therefore, we have

$$G(\tau_k, \tau) = \begin{bmatrix} G_1 & 0 \\ 0 & G_2 \end{bmatrix}.$$

Integrating the matrix $G(\tau_k, \tau)$, we obtain

$$\int_0^{t_k} G(\tau_k, \tau)\, d\tau = \begin{bmatrix} G_1' & 0 \\ 0 & G_2' \end{bmatrix},$$

where

$$G_1' = \begin{bmatrix} \sin\tau_k & 1 - \cos\tau_k \\ -(1 - \cos\tau_k) & \sin\tau_k \end{bmatrix},$$

$$G_2' = \begin{bmatrix} \frac{1}{p_0}\sin p_0\tau_k & \frac{1}{p_0}(1 - \cos p_0\tau_k) \\ -\frac{1}{p_0}(1 - \cos p_0\tau_k) & \frac{1}{p_0}\sin p_0\tau_k \end{bmatrix}.$$

By transformations we obtain the matrix that enters in the numerator of expression (10.125)

$$B = \int_0^{t_k} \mathrm{P}\mathrm{d}\tau \mathrm{C}_1^{-1} \int_0^{t_k} P^* \mathrm{d}\tau = \begin{bmatrix} B_{11} & 0 \\ 0 & B_{22} \end{bmatrix},$$

where

$$B_{11} = \begin{bmatrix} f_m^2 \sin^2 \tau_k + (\cos \tau_k - 1)^2 & (f_m^2 - 1) \sin \tau_k (\cos \tau_k - 1) \\ (f_m^2 - 1) \sin \tau_k (\cos \tau_k - 1) & f_m^2 (1 - \cos \tau_k)^2 + \sin^2 \tau_k \end{bmatrix};$$

$$B_{22} = \begin{bmatrix} f_m^2 \sin^2 p_0 \tau_k + (\cos p_0 \tau_k - 1)^2 & (f_m^2 - 1) \sin p_0 \tau_k (\cos p_0 \tau_k - 1) \\ (f_m^2 - 1) \sin p_0 \tau_k (\cos p_0 \tau_k - 1) & f_m^2 (1 - \cos p_0 \tau_k)^2 + \sin^2 p_0 \tau_k \end{bmatrix}.$$

Under the conditions of the problem it is required to find the area of possible values in the plane $x0y$, i.e. the projection of a four-dimensional area onto the plane $x0y$, therefore vector $\boldsymbol{\alpha}$ entering in formula (10.125) lies in this plane, i.e.

$$\boldsymbol{\alpha} = \begin{bmatrix} 0 \\ \sin \alpha \\ 0 \\ \cos \alpha \end{bmatrix} = 0\, \mathbf{e}_1 + \sin \alpha\, \mathbf{e}_2 + 0\, \mathbf{e}_3 + \cos \alpha\, \mathbf{e}_4;$$

$$\begin{aligned} \mathbf{A} = & (f_m^2 - 1) \sin \tau_k (\cos \tau_k - 1) \sin \alpha\, \mathbf{e}_1 \\ & + \left[f_m^2 (1 - \cos \tau_k)^2 + \sin^2 \tau_k \right] \sin \alpha\, \mathbf{e}_2 \\ & + \frac{1}{p_0^2} (f_m^2 - 1) \sin p_0 \tau_k (\cos p_0 \tau_k - 1) \cos \alpha\, \mathbf{e}_3 \\ & + \frac{1}{p_0^2} \left[f_m^2 (1 - \cos p_0 \tau_k)^2 + \sin^2 p_0 \tau_k \right] \cos \alpha\, \mathbf{e}_4, \end{aligned}$$

where $\mathbf{e}_i$ are the unit vectors of the coordinate axes.

The denominator of formula (10.125) for the considered example is

$$a = \sqrt{\left[f_m^2 (1 - \cos \tau_k)^2 + \sin^2 \tau_k \right] \sin^2 \alpha + \Rightarrow}$$

$$\Rightarrow \overline{+ \frac{1}{p_0^2} \left[f_m^2 (1 - \cos p_0 \tau_k)^2 + \sin^2 p_0 \tau_k \right] \cos^2 \alpha}.$$

For plotting the projection of the area onto the plane $x0y$ we need the following projections

$$y_2 = \frac{1}{a}\left[f_m^2(1-\cos\tau_k)^2 + \sin^2\tau_k\right]\sin\alpha;$$
$$y_4 = \frac{1}{p_0^2 a}\left[f_m^2(1-\cos p_0\tau_k)^2 + \sin^2 p_0\tau_k\right]\cos\alpha.$$

Changing α from 0 to 360°, we obtain the area of possible values of the coordinates of the point at the time instant τ_k.

Let us consider the special case of $p_0 = 1$. By manipulations we obtain

$$y_2 = \sqrt{f_m^2(1-\cos\tau_k)^2 + \sin^2\tau_k}\cdot\sin\alpha,$$
$$y_4 = \sqrt{f_m^2(1-\cos\tau_k)^2 + \sin^2\tau_k}\cdot\cos\alpha.$$

Eliminating the angle α, we obtain the following equation of a circle

$$\left(\frac{y_2}{a}\right)^2 + \left(\frac{y_4}{a}\right)^2 = 1.$$

Example 10.6. Let us consider the mechanical system shown in Fig. 10.24. A body of a mass m_1 moves under the action of a force $\mathbf{R}$ in an absolutely rigid tube elastically fixed on the right end. The direction of the force $\mathbf{R}$ is characterized by two random quantities $\mathbf{e}$ and ε, where $\mathbf{e}$ is the linear displacement and ε is the angular displacement. The random displacements are bounded in the absolute value

$$|\mathbf{e}| \le e_{\max}; \quad |\varepsilon| \le \varepsilon_{\max}.$$

We may consider that both random quantities are small. The projections of the force $\mathbf{R}$ onto the planes $y0z$ and $x0z$ and its position relative to the points 0 and 0_1 that depends on the projections of the random displacements e_x, e_y, ε_y and ε_x are shown in Fig. 10.25. For the purpose of simplification we take that at the initial time instant the coordinate of the point 0_1 is equal to $l_1/2$. Due to the random misalignments of the force $\mathbf{R}$ two random moments acting in the planes $y0z$ and $x0z$ take place:

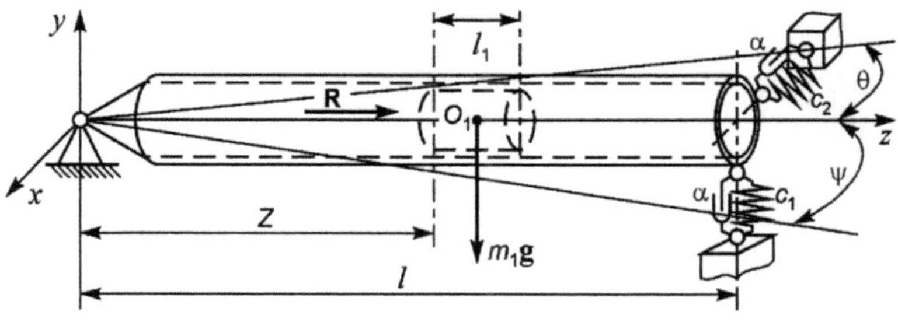

Fig. 10.24.

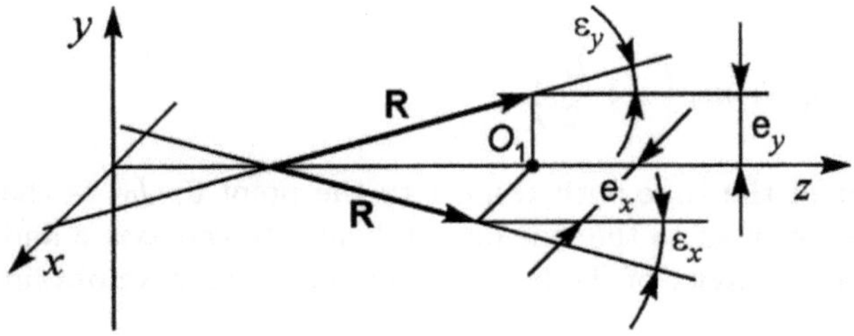

Fig. 10.25.

$$M_x = (z\varepsilon_y - e_y)\,R; \tag{10.158}$$

$$M_y = (z\varepsilon_x - e_x)\,R. \tag{10.159}$$

The moments M_x and M_y will cause a disturbed motion of the tube in two planes. This motion is characterized by small angular deviations θ and ψ (Fig. 10.24). Let us obtain the equations of the tubes disturbed motion about the equilibrium state in which the system is up to the moment of application of the force **R**.

At the moment of the body leaving the tube the axis of the body will obtain angular deviations θ_k, ψ_k and also angular velocities $\dot{\theta}_k$ and $\dot{\psi}_k$ that form a four-dimensional area in the phase space. As the projections are $e_x = |\mathbf{e}|\cos\alpha$ and $e_y = |\mathbf{e}|\sin\alpha$, we obtain a condition satisfied by the random quantities e_x and e_y :

$$\left(\frac{e_x}{e_{\max}}\right)^2 + \left(\frac{e_y}{e_{\max}}\right)^2 \leq 1. \tag{10.160}$$

The random quantities ε_x and ε_y meet a similar condition

$$\left(\frac{\varepsilon_x}{\varepsilon_{\max}}\right)^2 + \left(\frac{\varepsilon_y}{\varepsilon_{\max}}\right)^2 \leq 1. \tag{10.161}$$

When analyzing the disturbed motion of the system under the action of the random quantities e_x, e_y, ε_x, ε_y that meet conditions (10.160), (10.161), it is required to determine the areas of possible values of the system state vector at the instant of the body's exit. Assuming that the angles are small, we obtain the following equations of motion:

$$\frac{\mathrm{d}}{\mathrm{d}t}\left[J_x\dot{\theta}\right] + \alpha l^2\dot{\theta} + c_1 l^2\dot{\theta} = M_x - m_1 g\, z; \tag{10.162}$$

$$\frac{\mathrm{d}}{\mathrm{d}t}\left[J_y\dot{\psi}\right] + \alpha l^2\dot{\psi} + c_1 l^2\dot{\psi} = M_y, \quad (J_x = J_y)\,. \tag{10.163}$$

Since

$$J = J_x = J_y = J_0 + J_{01} + m_1 \left(z + \frac{l_1}{2} \right)^2 ,$$

where J_0 is the moment of the tube with respect to the point 0; J_{01} is the moment of the body with respect to the axis perpendicular to the axis z and going through the center of gravity of the body. By manipulations we obtain

$$\ddot{\theta} + a_{11}\dot{\theta} + a_{12}\theta = b_0 + b_{11}\varepsilon_y + b_{12}e_y; \tag{10.164}$$

$$\ddot{\psi} + a_{21}\dot{\psi} + a_{22}\psi = b_{21}\varepsilon_x + b_{22}e_x, \tag{10.165}$$

where

$$a_{11} = \frac{2m_1 \left(z + \frac{l_1}{2}\right) \dot{z} + \alpha l^2}{J}; \quad a_{12} = \frac{c_1 l^2}{J};$$

$$a_{21} = a_{11}; \quad a_{22} = \frac{c_2 l^2}{J}; \quad b_0 = \frac{m_1 g z}{J};$$

$$b_{11} = b_{21} = \frac{Rz}{J}; \quad b_{12} = b_{22} = -\frac{R}{J}.$$

Considering that the small vibrations of the system and the small deviations of the line of action of the force from the axis z practically do not influence the motion of the body along the axis z, we obtain at zero initial data

$$z = \frac{R}{2m_1} t^2. \tag{10.166}$$

Equations (10.164) and (10.165) are connected through the right-hand sides that embrace the random quantities satisfying conditions (10.160) and (10.161). If we introduce the vector

$$\mathbf{y} = \begin{bmatrix} \dot{\theta} \\ \dot{\psi} \\ \theta \\ \psi \end{bmatrix} = \begin{bmatrix} y_1 \\ y_2 \\ y_3 \\ y_4 \end{bmatrix}, \tag{10.167}$$

equations (10.164) and (10.165) can be written as

$$\dot{\mathbf{y}} + A(t)\,\mathbf{y} = B_1(t)\,\mathbf{f}_1 + B_1(t)\,\mathbf{f}_2 + \mathbf{b}_0(t), \tag{10.168}$$

where

$$A(t) = \begin{bmatrix} a_{11} & 0 & a_{12} & 0 \\ 0 & a_{21} & 0 & a_{22} \\ -1 & 0 & 0 & 0 \\ 0 & -1 & 0 & 0 \end{bmatrix}, \quad B_1(t) = \begin{bmatrix} b_{11} & 0 & b_{12} & 0 \\ 0 & b_{12} & 0 & b_{12} \\ 0 & 0 & 0 & 0 \\ 0 & 0 & 0 & 0 \end{bmatrix};$$

$$\mathbf{f}_1 = \begin{bmatrix} \varepsilon_y \\ \varepsilon_x \\ 0 \\ 0 \end{bmatrix}, \quad \mathbf{f}_2 = \begin{bmatrix} 0 \\ 0 \\ e_y \\ e_x \end{bmatrix}, \quad \mathbf{b}_0 = \begin{bmatrix} b_0 \\ 0 \\ 0 \\ 0 \end{bmatrix}.$$

We can obtain the solution of equation (10.168) only by using numerical methods because the elements of matrix A(t) are varying in time. The general form of the solution at zero initial data is

$$\mathbf{y} = \int\limits_0^{t_k} G\left(t_k, t\right) \left[B_1 \left(\mathbf{f}_1 + \mathbf{f}_2\right) + \mathbf{b}_0\right] \mathrm{d}t. \tag{10.169}$$

The matrix $G\left(t_k, t\right)$ can be determined from the solution of the homogeneous equation adjoint with the following homogeneous equation

$$\frac{\mathrm{d}\mathbf{z}}{\mathrm{d}\tau} + A^* \mathbf{z} = 0, \quad \left(\tau = t_k - t\right). \tag{10.170}$$

The fundamental matrix of solution (10.170) $K\left(\tau\right)$ is connected with the matrix $K\left(t_k, t\right)$ by the relationship

$$K^*\left(\tau\right) = G\left(t_k, t\right). \tag{10.171}$$

Let us confine ourselves to the determination of the projection of the four-dimensional area onto the phase planes $\left(\dot{\theta}_k, \dot{\psi}_k\right)$ and $\left(\theta_k, \psi_k\right)$.

Let us take the arbitrary vector $\boldsymbol{\alpha}_* = \boldsymbol{\alpha}_* \left(\cos\beta,\ \sin\beta,\ 0,\ 0\right)$, lying in the plane $\left(\dot{\theta},\ \dot{\psi}\right)$, and determine the projection of the state-of-the-system vector $\mathbf{y}$ onto the direction defined by the vector $\boldsymbol{\alpha}_*$:

$$y_{\boldsymbol{\alpha}_*} = \left(\mathbf{y} \cdot \boldsymbol{\alpha}_*\right) = \int\limits_0^{t_k} K\left(t_k, t\right) \left[B_1 \left(\mathbf{f}_1 + \mathbf{f}_2\right) + \mathbf{b}_0\right] \boldsymbol{\alpha}_* \, \mathrm{d}t. \tag{10.172}$$

To plot the area it is required to determine the maximum of the projection $y_{\boldsymbol{\alpha}_*}$ at additional conditions (10.160) and (10.161) imposed on the vectors $\mathbf{f}_1$ and $\mathbf{f}_2$, i.e. we must find the maximum of a functional of the form

$$J = y_{\boldsymbol{\alpha}_*} - \frac{\lambda_1}{2}\left[\left(C_1 \mathbf{f}_1 \cdot \mathbf{f}_1\right) - 1\right] - \frac{\lambda_2}{2}\left[\left(C_2 \mathbf{f}_2 \cdot \mathbf{f}_2\right) - 1\right], \tag{10.173}$$

where λ_1 and λ_2 are Lagrangian multipliers;

$$C_1 = \begin{bmatrix} \frac{1}{\varepsilon^2_{\max}} & 0 & 0 & 0 \\ 0 & \frac{1}{\varepsilon^2_{\max}} & 0 & 0 \\ 0 & 0 & 1 & 0 \\ 0 & 0 & 0 & 1 \end{bmatrix}, \quad C_2 = \begin{bmatrix} 1 & 0 & 0 & 0 \\ 0 & 1 & 0 & 0 \\ 0 & 0 & \frac{1}{\varepsilon^2_{\max}} & 0 \\ 0 & 0 & 0 & \frac{1}{\varepsilon^2_{\max}} \end{bmatrix}.$$

According to the theory presented in Sect. 4.3, the maximum value of the functional J is equal to the following sum of Lagrangian multipliers:

$$J_{\max} = \lambda_1 + \lambda_2, \tag{10.174}$$

that are determined from the relationships

$$\begin{aligned} \left(C_1^{-1} D\boldsymbol{\alpha}_* \cdot D\boldsymbol{\alpha}_*\right) &= \lambda_1^2; \\ \left(C_2^{-1} D\boldsymbol{\alpha}_* \cdot D\boldsymbol{\alpha}_*\right) &= \lambda_2^2, \end{aligned} \tag{10.175}$$

where

$$D = \int_0^{t_k} (KB_1)^* \, \mathrm{d}\tau.$$

It follows from expressions (10.175) that in order to obtain the values λ_i for any direction of the vector $\boldsymbol{\alpha}_*$ it is sufficient to determine the numerical values of the matrix D once. The vectors $\mathbf{f}_1$ and $\mathbf{f}_2$ that give a maximum value to the projections of the vector $\mathbf{y}$ onto the direction of the vector $\boldsymbol{\alpha}_*$ are:

$$\mathbf{f}_{1m} = \frac{1}{\lambda_1} C_1^{-1} D\boldsymbol{\alpha}_*; \quad \mathbf{f}_{2m} = \frac{1}{\lambda_2} C_2^{-1} D\boldsymbol{\alpha}_*. \tag{10.176}$$

Having substituted the values $\mathbf{f}_{1m}$ and $\mathbf{f}_{2m}$ into solution (10.169), we obtain

$$\mathbf{y_m} = D^* \mathbf{f}_{1m} + D^* \mathbf{f}_{2m} + \int_0^{t_k} K\mathbf{b}_0 \mathrm{d}\tau. \tag{10.177}$$

To plot the projection of the area onto the plane $\left(\dot{\theta}_k, \dot{\psi}_k\right)$ it is sufficient to determine two projections of the vector $\mathbf{y}_m$ (y_{1m} and y_{2m}) :

$$\begin{aligned} \dot{\theta}_k &= y_{1m} = d_{11}^{(*)} \varepsilon_{ym} + d_{12}^{(*)} \varepsilon_{xm} + d_{13}^{(*)} e_{ym} + d_{14}^{(*)} e_{xm} + y_{1k}; \\ \dot{\psi}_k &= y_{2m} = d_{21}^{(*)} \varepsilon_{ym} + d_{22}^{(*)} \varepsilon_{xm} + d_{23}^{(*)} e_{ym} + d_{24}^{(*)} e_{xm} + y_{2k}, \end{aligned} \tag{10.178}$$

where ε_{ym}, ε_{xm}, e_{ym}, e_{xm} are the components of the vectors $\mathbf{f}_{1m}$ and $\mathbf{f}_{2m}$ (10.176).

A point of the boundary of the projection of the area corresponds to each pair of the numbers y_{1m} and y_{2m} on the plane $\left(\dot{\theta}_k, \dot{\psi}_k\right)$. Expressions for projections y_{1m} and y_{2m} (10.178) embrace the non-random components y_{1k} and y_{2k} that can be ignored when plotting the area, because they do not influence its form. The numerical values of parameters that enter into equations of motion (10.164), (10.165) and conditions (10.160), (10.161) are

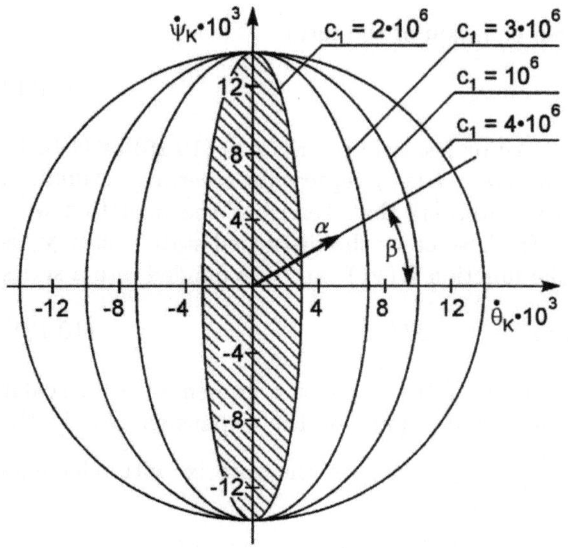

Fig. 10.26.

$l = 6$ m; $l_1 = 1$ m; $m_1 g = 600$ N; $J_0 = 4 \cdot 10^3$ kg·m^2; $J_{01} = 50$ kg·m^2; $c_2 = 4 \cdot 10^6$ N/m; $\alpha_1 = 0$; $e_{\max} = 5 \cdot 10^{-3}$ m; $\varepsilon_{\max} = 10^{-3}$ rad; $R = 10^6$ N; $t_k = 0.49$ s.

Changing β (Fig. 10.26) from 0 to 360° and determining y_{1m} and y_{2m} for each value of β (practically for a number of discrete values of β), we obtain the projection of the area of possible values (Fig. 10.26) of the angular velocities of the axis of the body at the instant of its exit from the tube.

An area for a number of rigidity values c_1 is given in Fig. 10.26. It substantially depends on the parameters of the system, in this case – on its rigidity (at a fixed value of c_2). There is such value of c_1, at which the area becomes minimal. This value may be considered optimal (as, when solving technical problems, it is desirable to have the minimal area of possible values of the vector of the systems disturbed motion).

10.8 Determination of the Maximum Values of Linear Functionals at Independent Excitations

In analyzing the motion of a dynamic system it is often required to determine the extreme values of some function J that characterizes the quality of a process and depends on the vector $\mathbf{y}$ of the disturbed state of a system at a fixed time instant t_k, for example

$$J = L(\mathbf{y}_k), \tag{10.179}$$

where the vector $\mathbf{y}$ satisfies an equation of the form

$$\dot{\mathbf{y}} + A\mathbf{y} = B\mathbf{f}, \tag{10.180}$$

and the components f_i of the vector $\mathbf{f}$ satisfy restrictions (10.18) or (10.23).

In applied problems the function $L(\mathbf{y}_k)$ represents a certain estimation characteristic (process quality criterion) that describes the deviation of a process from the given state. In these cases the disturbed-state vector $\mathbf{y}_k$ is considered a small one and the function $L(\mathbf{y}_k)$ can be expanded into a series

$$L(\mathbf{y}_k) = (\mathbf{L}_k \cdot \mathbf{y}_k) + (L_0\mathbf{y}_k \cdot \mathbf{y}_k) + \Delta L', \tag{10.181}$$

where $(\mathbf{L}_k, \mathbf{y}_k)$ is the linear part of the expansion presented as a scalar product; $(L_0\mathbf{y}_k, \mathbf{y}_k)$ is the square-law part of the expansion; $\mathbf{L}_k$ is the vector with components $\left(\dfrac{\partial L}{\partial y_{ki}}\right)_{y_{ki}=0}$; L_0 is the matrix with elements $\left.\dfrac{\partial^2 L}{\partial y_{ki}\partial y_{kj}}\right|_{\substack{y_{ki}=0\\ y_{kj}=0}}$ =0; and $\Delta L'$ is the remaining sum of a series.

For example, Sect. 10.3 dealt with the motion of a rocket subjected to the action of a random exciting moment M_b limited in absolute value (see Fig. 10.13). Such a moment can take place at the action of random wind gusts or at random misalignments of thrust. Although it is usually very difficult to obtain information on the random action of a wind, which is indispensable when applying methods of the theory of random processes, at the same time we cannot ignore the effect of the winds possible action. Therefore there is a necessity for estimating its greatest possible effect with a limited volume of information on this score, including data on the greatest possible wind velocities in a given geographical place, which allows us to determine the area of possible values of the exciting moment M_n (Fig. 10.14).

Example 10.2 involves the obtainment of the area of possible values of the angle φ and $\dot{\varphi}$ at the time instant t_k. These values can be considered as the angular deflection of the axial line of a rocket (and its first derivative) from its nominal value, i.e. φ_k and $\dot{\varphi}_k$ can be considered small quantities. If by the time instant t_k we shall basically mean the instant of termination of the action of the force R, the subsequent motion of the body will follow a trajectory differing from the estimated one. The estimated trajectory that corresponds to the distance x_0 on the axis x, terminates in point A (Fig. 10.27). The possible trajectories at all other conditions being equal depend on φ_k and $\dot{\varphi}_k$. Let Δx denote the deviations of the body's fall points from point A. Then Δx becomes a function of φ_k and $\dot{\varphi}_k$. Owing to the boundedness of the area of possible values of $(\varphi_k, \dot{\varphi}_k)$ and of the time of motion on the passive segment of the trajectory the distance dispersion $\Delta x(\varphi_k, \dot{\varphi}_k)$ is also a bounded function. Let us take that $\Delta x_{\max} \ll x$. Expanding the function $x(\varphi_k, \dot{\varphi}_k)$ into a Taylor series and confining ourselves to the linear part of the expansion, we obtain

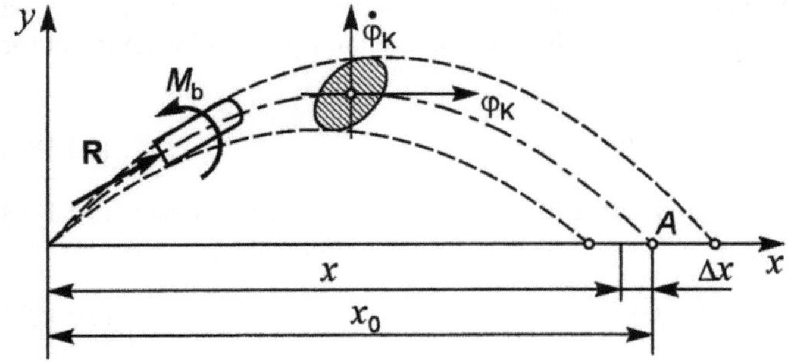

Fig. 10.27.

$$\Delta x\left(\varphi_k,\ \dot{\varphi}_k\right)=\frac{\partial x}{\partial \varphi_k}\varphi_k+\frac{\partial x}{\partial \dot{\varphi}_k}\dot{\varphi}_k \tag{10.182}$$

or

$$J=\left(\mathbf{L}_k\cdot\boldsymbol{\phi}_k\right), \tag{10.183}$$

where $\mathbf{L}_k=\begin{bmatrix}\dfrac{\partial x}{\partial \varphi_k}\\ \dfrac{\partial x}{\partial \dot{\varphi}_k}\end{bmatrix}$; $\boldsymbol{\varphi}_k=\begin{bmatrix}\varphi_k\\ \dot{\varphi}_k\end{bmatrix}$ are the components of the vector $\mathbf{L}_k$ that are considered known.

It follows from (10.183) that Δx is a linear functional dependent on the vector $\boldsymbol{\varphi}_k$. It is required to determine the extreme value of Δx at the known area of the possible values of the random exciting moment M_b and the possible effect (the worst one) of the random wind action.

In the general case, this problem is stated in the following way: at the fixed time instant t_k it is required to determine the extreme value of a linear functional of the form

$$J=\left(\mathbf{L}_k\cdot\mathbf{y}_k\right),\quad \mathbf{y}_k=\mathbf{y}\left(t_k\right) \tag{10.184}$$

provided that the vector $\mathbf{y}$ satisfies the following vector-matrix equation

$$\dot{\mathbf{y}}+A\mathbf{y}=B\mathbf{f},\quad \left(\mathbf{y}\left(0\right)=\mathbf{y}_0\right), \tag{10.185}$$

and the components of the vector $\mathbf{f}$ satisfy conditions (10.18) and (10.23). The vector $\mathbf{L}_k$ is specified with its components being linearly independent, i.e. the relationship $\sum_{i=1}^{n} L_{ki}\alpha_i$ can be satisfied only if $\alpha_1=\alpha_2=\cdots=\alpha_n=0$.

Let us consider the case of excitations f_j being independent, i.e. satisfying conditions (10.18). Then, we multiply scalar equation (10.185) by some vector $\mathbf{L}$:

$$(\mathbf{L} \cdot \dot{\mathbf{y}}) + (\mathbf{L} \cdot A\mathbf{y}) = (\mathbf{L} \cdot B\mathbf{f}) . \tag{10.186}$$

Let us transform the first term in the left-hand side of equation (10.186) to the form

$$(\mathbf{L} \cdot \dot{\mathbf{y}}) = (\mathbf{L} \cdot \mathbf{y})^{\dot{}} - \left(\dot{\mathbf{L}} \cdot \mathbf{y}\right) . \tag{10.187}$$

Equation (10.186) with the use of (10.187) takes the form

$$(\mathbf{L} \cdot \mathbf{y})^{\dot{}} - \left(\dot{\mathbf{L}} \cdot \mathbf{y}\right) + (\mathbf{L} \cdot A\mathbf{y}) = (\mathbf{L} \cdot B\mathbf{f}) . \tag{10.188}$$

Let us integrate both parts of equation (10.188) with respect to t:

$$(\mathbf{L} \cdot \mathbf{y})\Big|_{t=t_k} - (\mathbf{L} \cdot \mathbf{y})\Big|_{t=0} + \int\limits_0^{t_k} \left[(\mathbf{L} \cdot A\mathbf{y}) - \left(\dot{\mathbf{L}} \cdot \mathbf{y}\right)\right] \mathrm{d}\tau = \int\limits_0^{t_k} (\mathbf{L} \cdot B\mathbf{f})\, \mathrm{d}\tau . \tag{10.189}$$

The expression under the integral sign can be transformed to the form

$$(\mathbf{L} \cdot A\mathbf{y}) - \left(\dot{\mathbf{L}} \cdot \mathbf{y}\right) = (A^*\mathbf{L} \cdot \mathbf{y}) - \left(\dot{\mathbf{L}} \cdot \mathbf{y}\right) = \left(A^*\mathbf{L} - \dot{\mathbf{L}}\right) \cdot \mathbf{y}. \tag{10.190}$$

Let us demand that the vector $\mathbf{L}$ at the time instant t_k be equal to the vector $\mathbf{L}_k$. From relationship (10.189) we finally obtain

$$J = (\mathbf{L}_k \cdot \mathbf{y}_k) = \int\limits_0^{t_k} (\mathbf{L} - A^*\mathbf{L} \cdot \mathbf{y})\, \mathrm{d}\tau + \int\limits_0^{t_k} (\mathbf{L} \cdot B\mathbf{f})\, \mathrm{d}\tau + (\mathbf{L}_0 \cdot \mathbf{y}_0) . \tag{10.191}$$

Let us require that the introduced vector $\mathbf{L}$ satisfy the equation

$$\dot{\mathbf{L}} - A^*\mathbf{L} = 0. \tag{10.192}$$

Finally, we obtain the expression for the functional

$$J = \int\limits_0^{t_k} (\mathbf{L} \cdot B\mathbf{f})\, \mathrm{d}\tau + (\mathbf{L}_0 \cdot \mathbf{y}_0) . \tag{10.193}$$

For solving equation (10.192) it is better to pass to a variable $\tau_1 = t_k - \tau$ (at $\tau_1 = 0$ $K(0) = E$):

$$\frac{\mathrm{d}\mathbf{L}}{\mathrm{d}\tau_1} + A^*\mathbf{L} = 0. \tag{10.194}$$

The solution of equation (10.194) is

$$\mathbf{L} = K(\tau_1)\,\mathbf{L}_k. \tag{10.195}$$

(at $\tau_1 = 0 \quad K(0) = E$).

By manipulations we obtain

$$J_1 = \int_0^{t_k} \left(K(\tau_1)\,\mathbf{L}_k \cdot B\mathbf{f}\right) \mathrm{d}\tau_1 + \left(\mathbf{L}_0 \cdot \mathbf{y}_0\right),$$

or

$$J_1 = \int_0^{t_k} \left(B^* K(\tau_1)\,\mathbf{L}_k \cdot \mathbf{f}\right) \mathrm{d}\tau_1 + \left(\mathbf{L}_0 \cdot \mathbf{y}_0\right). \tag{10.196}$$

We obtain the vector $\mathbf{L}_0$ from (10.195) at $\tau_1 = t_k$, i.e.

$$\mathbf{L}_0 = K(t_0)\,\mathbf{L}_k. \tag{10.197}$$

If the vector of initial values $\mathbf{y}_0$ is zero, we have

$$J_1 = \int_0^{t_k} \left(B^* K(\tau_1)\,\mathbf{L}_k \cdot \mathbf{f}\right) \mathrm{d}\tau_1. \tag{10.198}$$

At random functions with time-constant boundaries ($f_{i1} = \pm a$) we have

$$J_1 = \int_0^{t_k} \left|\left(B^* K(\tau_1)\right)\right| \mathrm{d}\tau_1\,\mathbf{L}_k \cdot \mathbf{a}. \tag{10.199}$$

From relationship (10.198) it is easy to find the variation law of the components of the vector $\mathbf{f}$ that give a maximum value to the functional J. In scalar form expression (10.196) is

$$J_1 = \sum_{i=1}^{n} \int_0^{t_k} \left(\sum_{j=1}^{n} p_{ij} L_{kj} f_i\right) \mathrm{d}\tau_1 + \sum_{i=1}^{n} \left(L_{0i} y_{0i}\right), \tag{10.200}$$

where p_{ij} are the elements of a matrix $P = B^* K(\tau_1)$.

We find the maximum of expression $\sum_{i=1}^{n} (L_{0i} y_{0i})$, having determined y_{0i} from the following conditions:

$$y_{0i} = y_{iM}, \quad \text{if} \quad L_{0i} > 0,$$
$$y_{0i} = y_{im}, \quad \text{if} \quad L_{0i} < 0.$$

Let τ_{ik} be the values of time in which the integrand $\sum_{j=1}^{n} p_{ij} L_{kj}$ reverses sign.

The maximum value of the first term in right-hand side of (10.200) is determined from the formula

$$\sum_{i=1}^{n} \int_{0}^{t_k} \left(\sum_{j=1}^{n} p_{ij} L_{kj} f_{1i} \right) \mathrm{d}\tau_1 = \sum_{i=1}^{n} \left\{ \sum_{k=1}^{\rho} \int_{0}^{t_{i,k+1}} \left[\left(\sum_{j=1}^{n} p_{ij} L_{kj} \right) \begin{Bmatrix} M_i \\ m_i \end{Bmatrix} \right] \mathrm{d}\tau_1 \right\}. \tag{10.201}$$

Either the function M_i or m_i depending on the sign of the function

$$\left(\sum_{j=1}^{n} p_{ij} L_{kj} \right) = \delta_i \tag{10.202}$$

will be under the integral on each of the intervals from $t_{i,k}$ to $t_{i,k+1}$.

If on the interval $(t_{i,k}, t_{i,k+1})$ $\delta_i > 0$, there will be M_i under the integral in the right-hand side of expression (10.201), and if $\delta_i < 0$, m_i will be there.

The presented method enables us to determine not only the maximum value of J_1, but also the worst variation law of the function f_i. If the boundaries of the areas of possible values of f_i are independent of time, i.e. $f_i = \pm a_i$, the expression for the maximum value of J_1 takes the form

$$J_1 = \sum_{i=1}^{n} a_i \int_{0}^{t_k} |\delta_i| \, \mathrm{d}\tau_i + m \sum_{i=1}^{n} L_{0i} y_{0i}. \tag{10.203}$$

When determining the maximum value of J_1, we established the variation laws of the functions f_i that give a maximum value to J_1. If to include these laws in expression (10.25), some vector $\mathbf{y}$ will be obtained. Will it be inside the area of possible values or only touch its surface?

If the vector $\mathbf{y}$ for the found laws f_i is inside the area of possible values, it means that the maximum of the functional J_1 is reached on the internal vectors $\mathbf{y}$; if this vector only touches the surface bounding the area, the maximum is reached on the vectors which determine the limit surface of the deviations area.

Let us show that the maximum value of the linear functional $J = (\mathbf{L}_k \cdot \mathbf{y}_k)$ is reached on the vectors which define the limit surface of the area of possible values of the system state vector. To make it clearer let us consider the particular case of the vector $\mathbf{y}_k$ having two components. Here the area of possible values is bounded by a flat curve on the phase plane (Fig. 10.28).

Let us assume that the maximum $J =$ corresponds to the vector $\mathbf{y}_k$ (Fig. 10.28):

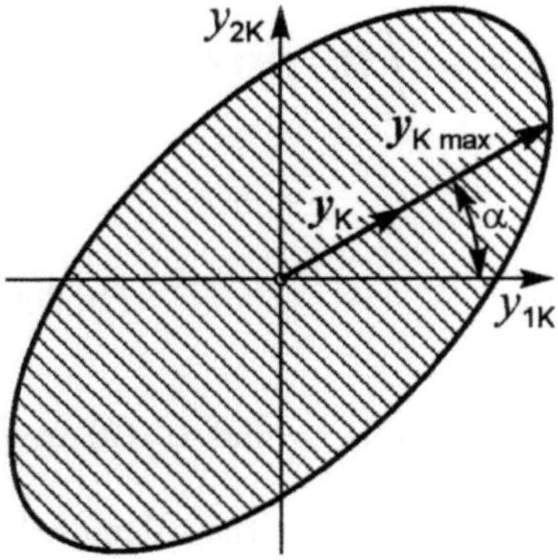

Fig. 10.28.

$$J = (\mathbf{L}_k \cdot \mathbf{y}_k) = \max . \tag{10.204}$$

Any vector $\mathbf{y}_k$ directionally coinciding in direction with the vector $\mathbf{y}_{k\,\max}$ can be written as

$$\mathbf{y}_k = k\mathbf{y}_{k\,\max}, \tag{10.205}$$

where k is the scalar multiplier that varies from zero to unity.

Substituting $\mathbf{y}_k$ into expression (10.204), we obtain

$$J = k\,(\mathbf{L}_k \cdot \mathbf{y}_{k\,\max}) . \tag{10.206}$$

It follows from expression (10.206) that for any vector $\mathbf{y}_k$ directionally coinciding with the vector $\mathbf{y}_{k\,\max}$ the value of the functional J for a given angle can only be smaller than that for the vector $\mathbf{y}_{k\,\max}$that determines the boundary of the area, as was to be shown.

Let us determine the maximum value of the linear functional in the case of several sections of motion being available. We shall consider the motion of a system on the interval $(0, t_k)$ when there are several of time intervals (t_i, t_{i+1}) on which the action of different random excitations $\mathbf{f}_i$ occurs while the dimensionality of the vectors $\mathbf{f}_i$ may differ (their dimensionality considered to be smaller than that of the equations of motion of the system). Let us, first, consider a case where the dimensionality of the equations of motion of the system is the same on all intervals of motion. We shall begin our consideration with determining the maximum value of the functional J at the end of motion (at $t = t_k$), with this value being depends on all stages of motion. On each of the time intervals embracing the action of the vector of random excitations $\mathbf{f}_i$, the equation of motion is

$$\dot{\mathbf{y}}_i + A_i\mathbf{y}_i = B_i\mathbf{f}_i \quad (i = 1, 2, \ldots, n) , \tag{10.207}$$

where n is the number of sections.

Let us connect linear functional (10.196) $J = \int\limits_0^{t_k} (\mathbf{L} \cdot B\mathbf{f})\, d\tau + (\mathbf{L}_0 \cdot \mathbf{y}_0)$ with the n–th (last) section of motion:

$$J_n = \int\limits_0^{t_{k_n}} (\mathbf{L}_n \cdot B_n \mathbf{f}_n)\, d\tau + (\mathbf{L}_{0n} \cdot \mathbf{y}_{0n}). \tag{10.208}$$

The value of $\mathbf{L}_n\,(t_{k_n})$ is known, therefore it is possible to find $\mathbf{L}_n\,(0)$ from equation (10.134).

In its turn, the scalar product $(\mathbf{L}_n\,(0) \cdot \mathbf{y}_n\,(0))$ represents the value of the functional on the previous section of motion, i.e.

$$(\mathbf{L}_{0n} \cdot \mathbf{y}_{0n}\,(0)) = \int\limits_0^{t_{k_{n-1}}} (\mathbf{L}_{n-1} \cdot B_{n-1} \mathbf{f}_{n-1})\, d\tau + (\mathbf{L}_{0n-1} \cdot \mathbf{y}_{0n-1}\,(0)), \tag{10.209}$$

and so on, therefore it is possible to obtain the expression for the functional dependent on the motion of the system on the whole interval $(0, t_k)$:

$$J = \sum_{i=1}^{n} \int\limits_0^{t_{k_i}} (\mathbf{L}_i \cdot B_i \mathbf{f}_i)\, d\tau + (\mathbf{L}_{01} \cdot \mathbf{y}_1\,(0)), \tag{10.210}$$

where $\mathbf{L}_{01}$, $\mathbf{y}_1\,(0)$ are the values of the vectors $\mathbf{L}_1$ and $\mathbf{y}_1$ in the beginning of the first interval of motion.

The maximum value of functional (10.210) is

$$J_{\max} = \sum_{i=1}^{n} \left(\int\limits_0^{t_{k_i}} (\mathbf{L}_i \cdot B_i \mathbf{f}_i)\, d\tau \right)_{\max} + (\mathbf{L}_{01} \cdot \mathbf{y}_1\,(0))_{\max}, \tag{10.211}$$

where $\sum\limits_{i=1}^{n} t_{k_i} = t_k$.

The method of determining the maximum value of each of the terms entering into the right-hand side of expression (10.210) is presented above.

A mechanical system with the varying dimensionality of a state vector $\mathbf{y}$ is shown in Fig. 10.29. A body moves along a rod and comes into contact with it in two points: A and B. In the beginning of the motion of the system the latter has one degree of freedom (Fig. 10.29 a), and after the loss of the contact of the forward support (point) the system has two degrees of freedom (Fig. 10.29 b). There are also mechanical systems with the dimension of the vector $\mathbf{y}$ on the first section of motion being greater than that on the second one (Fig. 10.21).

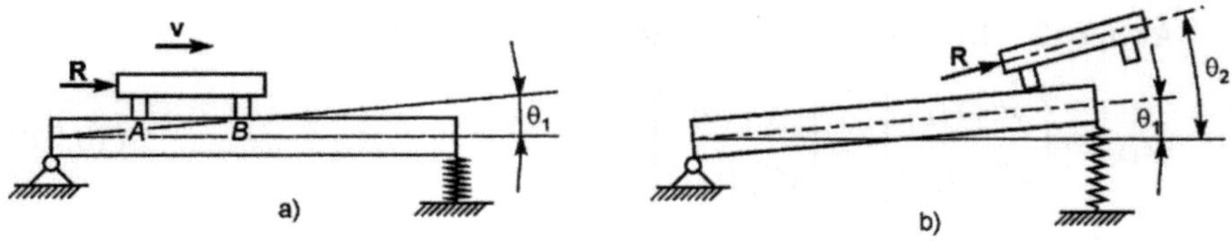

Fig. 10.29.

Let us consider a system involving two sections of motion, and for the purpose of definiteness we assume that the dimensionality n of the state vector $\mathbf{y}_1$ on the first section of motion is greater than the dimensionality ν of the vector $\mathbf{y}_2$. Then we shall consider the second section of the system's motion of and some vector $\mathbf{L}_2$ connected with this section and having the dimensionality of the vector $\mathbf{y}_2$. The vector $\mathbf{L}_2$ satisfies adjoint equation (10.194). On the second section of motion there is a functional

$$J_2 = \int_0^{t_k - t_{k-1}} (\mathbf{L}_2 \cdot B_2 \mathbf{f}_2)\,\mathrm{d}\tau + (\mathbf{L}_{02} \cdot \mathbf{y}_2(0)), \tag{10.212}$$

where t_{k_1} is the time of the systems motion on the first section.

The vector $\mathbf{y}_2(0)$ is equal to the vector $\mathbf{y}_{11}$ at the end of the first section, where $\mathbf{y}_{11}$ is a vector with components that retain their value on the second section, i.e. the vector $\mathbf{y}_2$ represents a continuation of the vector $\mathbf{y}_{11}$. Since, under the condition of the problem, the vector $\mathbf{y}_1$ has the dimensionality n (that is greater than that of the vector $\mathbf{y}_2$), then at $t = t_{k_1}$

$$\mathbf{y}_1(t_{k_1}) = \begin{bmatrix} \mathbf{y}_{11}(t_{k_1}) \\ \mathbf{y}_{12}(t_{k_1}) \end{bmatrix} = \begin{bmatrix} \mathbf{y}_2(0) \\ \mathbf{y}_{12}(t_{k_1}) \end{bmatrix}. \tag{10.213}$$

The vector $\mathbf{y}_{12}$ has the dimensionality $n - \nu$.

The vector $\mathbf{L}_2(0)$ is determined from the solution of the following equation

$$\frac{\mathrm{d}\mathbf{L}_2}{\mathrm{d}\tau} + A_2^* \mathbf{L}_2 = 0, \quad (\tau = (t_k - t_{k_1}) - t). \tag{10.214}$$

At $\tau = 0$ the vector $\mathbf{L}_2$ is given.

Let us introduce the vector $\mathbf{L}_{1k}$ with n components

$$\mathbf{L}_{1k} = \begin{bmatrix} \mathbf{L}_2(0) \\ 0 \end{bmatrix}. \tag{10.215}$$

The vector $\mathbf{L}_{1k}$ is initial values vector for the equation

$$\frac{\mathrm{d}\mathbf{L}_1}{\mathrm{d}\tau} + A_1^* \mathbf{L}_1 = 0, \quad (\tau = t_{k_1} - t). \tag{10.216}$$

At $\tau = 0$ $\mathbf{L}_1 = \mathbf{L}_{1k}$. Let us present the vector $\mathbf{y}_1\,(t_{k1})$ as

$$\mathbf{y}_{1k} = \begin{bmatrix} \mathbf{y}''_1\,(t_k) \\ \mathbf{y}''_1\,(t_{k_1}) \end{bmatrix} = \begin{bmatrix} \mathbf{y}_2\,(0) \\ \mathbf{y}''_1\,(t_{k_1}) \end{bmatrix} \tag{10.217}$$

and consider the scalar product $(\mathbf{L}_{1k} \cdot \mathbf{y}_{1k})$ which, according to (10.215) and (10.217), is

$$(\mathbf{L}_{1k} \cdot \mathbf{y}_1\,(t_{k_1})) = (\mathbf{L}_{02} \cdot \mathbf{y}_2\,(0))\,. \tag{10.218}$$

In its turn, scalar product (10.218) is a functional connected with the first section of motion, i.e.

$$J_1 = (\mathbf{L}_{1k} \cdot \mathbf{y}_{1k}) = \int\limits_0^{t_{k_1}} (\mathbf{L}_1 \cdot B_1\mathbf{f}_1)\,\mathrm{d}\tau + (\mathbf{L}_{01} \cdot \mathbf{y}_1\,(0)), \tag{10.219}$$

therefore, we finally obtain the expression for the functional J for two sections of motion

$$J = \int\limits_0^{t_{k_1}} (\mathbf{L}_1 \cdot B_1\mathbf{f}_1)\,\mathrm{d}\tau + \int\limits_0^{t_k - t_{k_1}} (\mathbf{L}_2 \cdot B_2\mathbf{f}_2)\,\mathrm{d}\tau + (\mathbf{L}_{01} \cdot \mathbf{y}_1\,(0)). \tag{10.220}$$

10.9 Maximum Value of a Linear Functional at Dependent Excitations

Let us determine the maximum value of a linear functional at dependent random excitations satisfying condition (10.23).

It is required to find the maximum of the following functional

$$J = \int\limits_0^{t_k} (\mathbf{L} \cdot B\mathbf{f})\,\mathrm{d}\tau + (\mathbf{L}_0 \cdot \mathbf{y}_0). \tag{10.221}$$

A vector of excitations $\mathbf{f}$ can be written as

$$\mathbf{f} = H\,(t)\,\mathbf{f}_0, \tag{10.222}$$

where $H\,(t)$ is the diagonal matrix with continuously time-dependent elements; $\mathbf{f}_0$ is the vector, whose components satisfy the condition

$$(C_1\mathbf{f}_0 \cdot \mathbf{f}_0) \le 1.$$

The components of the vector $\mathbf{y}_0$ satisfy a similar condition

$(C_0 \mathbf{y}_0 \mathbf{y}_0) \leq 1.$

Having used Lagrangian multipliers, we obtain a functional of the form

$$J_1 = \int_0^{t_k} (\mathbf{L} \cdot BH\mathbf{f}_0)\, \mathrm{d}\tau + (\mathbf{L}_0 \cdot \mathbf{y}_0) - \frac{\lambda}{2}\left[(C_1\mathbf{f}_0 \cdot \mathbf{f}_0) - 1\right] - \frac{\mu}{2}\left[(C_0\mathbf{y}_0 \cdot \mathbf{y}_0) - 1\right] = 0. \tag{10.223}$$

Since $\mathbf{L} = K\mathbf{L}_0$, we have

$$J_1 = \int_0^{t_k} \left[(BH)^* K\mathbf{L}_0 \cdot \mathbf{f}_0\right] \mathrm{d}\tau + (\mathbf{L}_0 \cdot \mathbf{y}_0) - \frac{\lambda}{2}\left[(C_1\mathbf{f}_0 \cdot \mathbf{f}_0) - 1\right] - \\ - \frac{\mu}{2}\left[(C_0\mathbf{y}_0 \cdot \mathbf{y}_0) - 1\right] = 0. \tag{10.224}$$

Functional (10.224) is similar to expression (10.106) for the projection of the state-of-the-system vector onto the arbitrary direction determined by a vector $\mathbf{e}_1$. Therefore its maximum value is

$$J_{1\,\max} = \lambda + \mu, \tag{10.225}$$

where

$$\lambda^2 = \left(C_1^{-1} D\mathbf{L}_0 \cdot D\mathbf{L}_0\right); \quad D = \int_0^{t_k} (BH)^* K\, \mathrm{d}\tau; \quad \mu^2 = \left(C_1^{-1}\mathbf{L}_0 \cdot \mathbf{L}_0\right).$$

Let us consider the case of the components f_i of the vector of excitations being random discontinuous functions bounded in absolute value, as, for example, independent random excitations (see Fig. 10.1). Here it is possible to present the vector $\mathbf{f}$ as

$$\mathbf{f} = H(t)\, S(t)\, \mathbf{f}_0, \tag{10.226}$$

where $H(t)$ is the diagonal matrix similar to the matrix entering in (10.222); S is the diagonal matrix whose elements are unit discontinuous functions $s_{ii}(t)$.

The latter can take values ± 1 with the instants of their discontinuity being random. Physically it means that the components f_i can instantaneously reverse the sign at random time instants, while the components of vector (10.222) vary in time continuously. Condition (10.23) for vectors (10.226) is modified to the form

$$(C_1(t)\, \mathbf{f}_0,\ \mathbf{f}_0) \leq 1, \tag{10.227}$$

where

$$C_1(t) = (HS)^* C(t)(HS).$$

If the matrix $C(t)$ is diagonal, the matrix $C_1(t)$ is diagonal too. For example, for the case of the vector $\mathbf{f}$ having three components the matrix $C_1(t)$ is

$$C_1(t) = \begin{bmatrix} \frac{h_{11}^2}{|\mathbf{f}|_{\max}^2} & 0 & 0 \\ 0 & \frac{h_{22}^2}{|\mathbf{f}|_{\max}^2} & 0 \\ 0 & 0 & \frac{h_{33}^2}{|\mathbf{f}|_{\max}^2} \end{bmatrix},$$

where h_{ii} are the elements of the matrix H.

In this special case (when the matrix $C(t)$ is diagonal) condition (10.227) does not depend on s_{ii}, as s_{ii}^2, that are equal to unity enter into the elements of the matrix $C_1(t)$.

Linear functional (10.223) with due account of (10.226) takes the form

$$J_1 = \int_0^{t_k} (\mathbf{L} \cdot BHS\mathbf{f}_0)\mathrm{d}\tau + (\mathbf{L}_0 \cdot \mathbf{y}_0) - \frac{\lambda}{2}\left[(C_1\mathbf{f}_0 \cdot \mathbf{f}_0) - 1\right]$$
$$- \frac{\mu}{2}\left[(C_0\mathbf{y}_0 \cdot \mathbf{y}_0) - 1\right]. \tag{10.228}$$

The maximum value of J_1 (10.228) is

$$J_{1\max} = \lambda_m + \mu, \tag{10.229}$$

but as opposed to similar expression (10.225) the Lagrangian multiplier λ_m is determined with due account of the possible discontinuous variation of the components of the vector $\mathbf{f}$ in time.

Depending on the discontinuous variation of $\mathbf{f}_j$ in time, the Lagrangian multiplier λ can take different values. Among the possible discontinuous variations of $\mathbf{f}_j$, however, there are such laws of variation, at which λ reaches its maximum value. For this value of λ the notation λ_m is introduced. Let us determine λ_m considering expressions for the square of the Lagrangian multiplier (explanation to (10.225))

$$\lambda_m^2 = \left(C_1^{-1}\mathbf{b} \cdot \mathbf{b}\right), \quad \left(\mathbf{b} = D^{(0)}\mathbf{L}_0\right), \tag{10.230}$$

where

$$D^{(0)} = \int_0^{t_k} (BHS)^* K \,\mathrm{d}\tau.$$

The elements $b_{ij}^{(1)}$ and d_{ij} of the matrixes $B^{(1)} = (BHS)^*$ and $D = BK$ are respectively equal to:

$$b_{ij}^{(1)} = h_{ii}s_{ii}b_{ji}; \quad d_{ij} = h_{ii}s_{ii}\left(\sum_{\rho=1}^{n} b_{\rho i}k_{\rho j}\right).$$

Therefore, the components of the vector $\mathbf{b}$ are

$$b_k = \int_0^{t_k} \left(\sum_{\nu=1}^{n} d_{k\nu}l_{0\nu}\right) h_{kk}s_{kk}\mathrm{d}\tau; \tag{10.231}$$

as the matrix C_1^{-1} is diagonal and its elements are positive, the maximum value of quadratic form (10.230), or, what is the same, of the square of the Lagrangian multiplier λ_m, is reached when the components of the vector $\mathbf{b}$ take their maximum values, which occurs at the following variation laws of functions:

$$s_{kk}^{(m)} = \begin{cases} 1, & \text{at} \quad g_k > 0, \\ -1, & \text{at} \quad g_k < 0, \end{cases} \tag{10.232}$$

where

$$g_k = \left(\sum_{\nu=1}^{n} d_{k\nu}l_{0\nu}\right) h_{kk}.$$

The time instants τ_ρ, when the functions s_{kk} reverse the sign (the points of discontinuity), are determined from conditions

$$g_k = 0. \tag{10.233}$$

Having determined $s_{kk}^{(m)}$, we find a vector $\mathbf{f}_0^{(m)}$ that corresponds to these laws. When determining the maximum value of J_1 condition (10.107) is used

$$\frac{\partial J_1}{\partial \mathbf{f}_0} = 0,$$

which in the considered problem results in the following relationship

$$\int_0^{t_k} (BHS)^* K\mathbf{L}\,\mathrm{d}\tau_0 - \lambda_m C_1 \mathbf{f}_0^{(m)} = 0. \tag{10.234}$$

Having determined the maximum value of λ_m and the variation laws of the unit functions $s_{kk}^{(m)}$ (10.232), we find the vector $\mathbf{f}_0^{(m)}$

$$\mathbf{f}_0^{(m)} = \frac{1}{\lambda_m} C_1^{-1} \int_0^{t_k} (BHS)^* K\mathbf{L}_0\,\mathrm{d}\tau. \tag{10.235}$$

The vector $\mathbf{f}^{(m)}$ at which functional J (10.221) with due account of the discontinuous variation of components $f_j^{(m)}$ in time reaches its maximum value is

$$\mathbf{f}^{(m)} = HS^{(m)}\mathbf{f}_0^{(m)}. \tag{10.236}$$

Let us consider a case where during the motion of the system on different time intervals act different random excitations bounded in absolute value and satisfying the following conditions

$$\left(C^{(j)}\mathbf{f}_0^{(j)} \cdot \mathbf{f}_0^{(j)}\right) \leq 1; \quad \left(C^{(j)} = (HS)^* C_j HS\right), \tag{10.237}$$

where the index j corresponds to the $j-$ th interval of the motion of the system. The time instants of the beginning (t_{0j}) and end (t_{kj}) of the intervals are known. Let us confine ourselves to the case of the dimensionality of the system state vector remaining unchanged on all intervals of motion. Let us determine the maximum value of the linear functional J for the time instant t_k $\left(t_k = j = 1\sum^{m} t_{kj}\right)$ with due account of the excitations $f^{(j)}$ $(j = 1, 2\ \ldots\ m)$ that act on the system and satisfy conditions (10.237).

For each of the intervals of motion $(t_{0j},\ t_{kj})$ we can obtain a functional similar to (10.221)

$$J^{(j)} = \int_0^{t_k} \left(S^{(j)}D^{(j)*}\mathbf{L}^{(j)} \cdot \mathbf{f}_0^{(j)}\right) d\tau_1 + \left(\mathbf{L}^{(j)} \cdot \mathbf{y}_0^{(j)}\right), \tag{10.238}$$

where

$$D^{(j)*} = \left(B^{(j)}H^{(j)}\right)^*.$$

It is possible to present the expression for the functional $J^{(j)}$ as

$$J^{(j)} = \int_0^{t_{kj}} \left(S^{(j)}D^{(j)*}\mathbf{L}^{(j)} \cdot \mathbf{f}_0^{(j)}\right) d\tau_1 + J^{(j-1)}, \tag{10.239}$$

where

$$J^{(j-1)} = \int_0^{t_{kj-1}} \left(S^{(j-1)}D^{(j-1)*}\mathbf{L}^{(j-1)} \cdot \mathbf{f}_0^{(j-1)}\right) d\tau_1 + \left(\mathbf{L}^{(j-1)} \cdot \mathbf{y}_0^{(j-1)}\right). \tag{10.240}$$

Let us confine ourselves to the case of one random vector $\mathbf{f}^{(j)}$ acting on each of the sections.

The vectors $\mathbf{L}_0^{(j)}$, entering into (10.238), are determined from the following equations (at $\tau_1 = t_{kj}$)

$$\frac{\mathrm{d}\mathbf{L}^{(j)}}{\mathrm{d}\tau_1} + A_j^* \mathbf{L}^{(j)} = 0; \quad \left(\mathbf{L}^{(j)} = K_j(\tau_1)\,\mathbf{L}_k^{(j)}\right). \tag{10.241}$$

The value of the vector $\mathbf{L}_k^{(j)}$ is known (at the end of the jth section of motion), which allows us to determine the value of the vector $\mathbf{L}_0^{(j)}$

$$\mathbf{L}_0^{(j)} = K_j(t_{kj})\,\mathbf{L}_k^{(j)}, \tag{10.242}$$

representing the value of the vector $\left(\mathbf{L}^{(j-1)}\right)$ at the end of the previous section of motion

$$\mathbf{L}_k^{(j-1)} = \mathbf{L}_0^{(j)}. \tag{10.243}$$

For the last section of motion we have

$$J^{(m)} = \int_0^{t_{km}} \left(S^{(m)} D^{(m)*} \mathbf{L}^{(m)} \cdot \mathbf{f}_0^{(m)}\right) \mathrm{d}\tau_1 + J^{(m-1)}. \tag{10.244}$$

Sequentially eliminating $J^{(m-1)}$ from (10.244) (with the use of recurrent relationship (10.239)), we obtain

$$J = \sum_{j=1}^{m} \int_0^{t_{kj}} \left(S^{(j)} D^{(j)*} \mathbf{L}^{(j)} \cdot \mathbf{f}_0^{(j)}\right) \mathrm{d}\tau_1 + \left(\mathbf{L}_0^{(1)} \cdot \mathbf{y}_0^{(1)}\right). \tag{10.245}$$

The integrals entering into the right-hand side of (10.245) contain the vectors $\mathbf{L}^{(j)}$ which satisfy equations (10.241). The solution of these equations can be obtained, if we know the value of the vector $\mathbf{L}^{(j)}$ at $\tau_1 = 0$, i.e. if the values of $\mathbf{L}^{(j)}$ are known. Only the vector $\mathbf{L}_k^{(m)}$ is known among the vectors $\mathbf{L}^{(j)}$. This is quite sufficient, however, for the obtainment, by sequential calculations, of the values of the integrals for each of the sections, beginning from the last one. Solving equation (10.241) at $j = m$, we obtain $\mathbf{L}_0^{(m)}$ that is equal to the initial values vector $\mathbf{L}_k^{(m-1)}$intended for the solution of this equation on the previous interval of the systems motion, etc.

As on each of the sections of motion of the system the vectors $\mathbf{f}_0^{(j)}$ satisfy conditions (10.237)

$$\left(C^{(j)} \mathbf{f}_0^{(j)} \cdot \mathbf{f}_0^{(j)}\right) 1 \quad (j = 1, 2, \cdots, m),$$

then, without performing intermediate transformations during the determination of the maximum value of J_1 (similar to those considered earlier in the case of one section of the systems motion), we obtain

$$J_{1\,\max} = \mu + \sum_{j=1}^{m} \lambda_j, \tag{10.246}$$

where

$$\begin{aligned} \lambda_j^2 &= \left[\left(C^{(j)}\right)^{-1} \mathbf{b}^{(j)} \cdot \mathbf{b}^{(j)}\right], \\ \mathbf{b}^{(j)} &= \int_0^{t_{kj}} S^{(j)} D^{(j)*} \mathbf{L}^{(j)} \mathrm{d}\tau, \quad D^{(j)} = \left(B^{(j)} H^{(j)}\right)^* K_j. \end{aligned} \tag{10.247}$$

The greatest possible value of the functional J_1 will be reached at maximum λ_j, which occurs at the following variation laws of the unit discontinuous functions $s_{kk}^{(j)}$

$$s_{kk}^{(j)} = \begin{cases} 1 & \text{at} \quad \left(\sum\limits_{\nu=1}^{m} d_{k\nu} l_{0\nu} h_{kk}\right) > 0; \\ -1 & \text{at} \quad \left(\sum\limits_{\nu=1}^{m} d_{k\nu} l_{0\nu} h_{kk}\right) < 0. \end{cases} \tag{10.248}$$

At the variation laws of the matrixes elements $S^{(j)}$, (10.248), the components of the vectors $\mathbf{b}^{(j)}$ are equal to their greatest possible values and with the diagonal matrix $\left(C^{(j)}\right)^{-1}$ having positive elements we obtain the maximum values of $\lambda_{(j)}^2$. Therefore we finally obtain

$$J_{1\,\max} = \mu + \sum_{j=1}^{m} \lambda_{j\,\max}. \tag{10.249}$$

10.10 Vibration Protection of Mechanical Systems

In the introduction to this textbook we considered an automobile moving along a road with random irregularities and an airplane running on an aerodrome pavement characterized by random asperities (Fig 0.1 *a, b*) to illustrate the impact of vibrations that arise in the context of randomness. The acceleration and braking of the automobile and the airplane cause random vibrations that can be quite strong. These vibrations are fraught with the failure of control systems and to prevent it different measures of vibration protection are used in them for instruments.

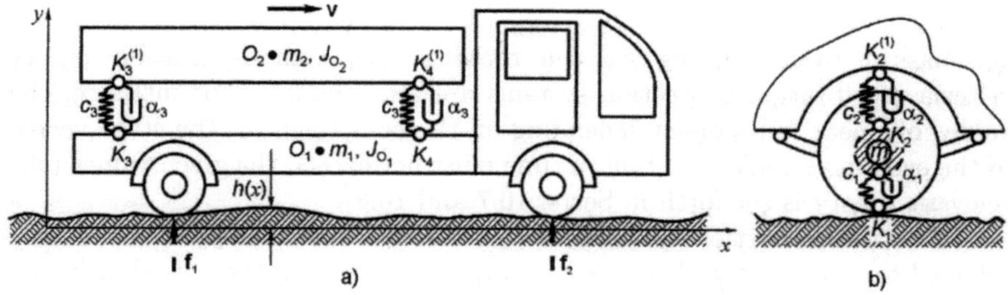

Fig. 10.30.

Figure 10.30 *a* shows a special-freight automobile very vulnerable to great overloads or shock loads resulting from a breakdown in the suspension during acceleration or braking.

That is why an additional vibration protection system is used between the vehicle and the freight to diminish overloads and do not let the masses m_1 and m_2 collide with each other at the suspension's break-down.

The need for the protection of instruments also arises during the active leg of the flight trajectory of a rocket (Fig. 10.31). Random perturbations that act on the rocket include those depending on the operation of the engine, e.g., the random angular misalignment of the thrust (ε), the linear misalignment of the thrust (e) and the scatter of the thrust (ΔR), which cause random inertia forces components acting on the mass m (instrument). As a result random vibrations of the mass m turn up. For the instrument to operate normally it is necessary that the displacements of the mass m do not lead to its collisions with the body of the rocket For example the displacement of the mass must be within the area shown in Fig. 10.31 by the dotted line. Methods making it possible to determine the two-dimensional areas of the possible values of state vector components at a given instant of time are presented in Sects. 10.4, 10.5 and 10.6. Therefore the vibration protection system of the instrument should be designed in such a way as to prevent

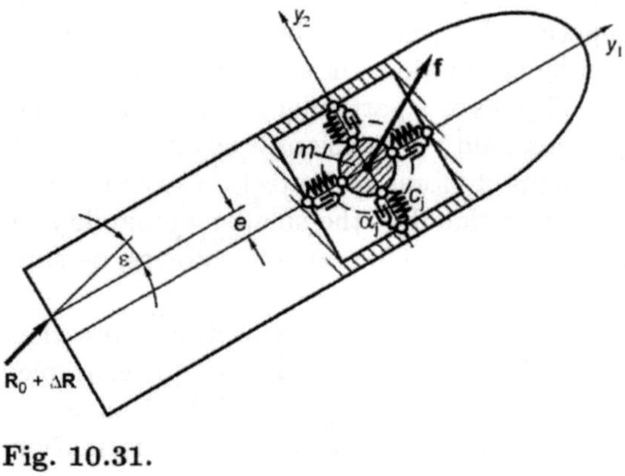

Fig. 10.31.

the mass m from colliding with the rocket's body. Other operation quality criteria of vibration protection systems are also possible. For instance, the values of linear functionals depending on the component of the state vector of the system at a given instant of time must not exceed the given values (the relevant theory is set forth in Sects. 10.7 and 10.8).

Vibration protection systems that do not use power sources are usually referred to as passive and embrace absorption systems without feedback. Below we shall discuss mechanical systems with a passive vibration protection of instruments and transported freights. Passive vibration protection systems have become extremely popular because they constitute very simple structures. As a rule, they consist of elastic and dissipative elements. Fig. 10.30 shows passive vibration protection devices (shock absorbers) made of elastic elements and of viscous-friction ones (stiffness of elastic elements is denoted by c and viscous-friction coefficient is indicated by α).

However, passive vibration protection systems have a disadvantage: they function effectively (at c and α chosen by calculations) only at certain perturbations.

If the perturbations change, e.g. the frequency range for determined perturbations or spectral densities for random perturbations change, we have to determine new appropriate values of c and α to make the performance of shock absorbers effective.

Vibration protection systems using feedback (active vibration protection systems at determined perturbations) are free of this drawback as the control system of the vibration protection device realizes a continuous variation in c and α at a continuous variation in perturbations, which allows the vibration protection system to operate effectively all the time.

It is very difficult to analyze vibration protection systems (passive and particularly active ones) at random perturbations during non-steady (non-stationary) processes even in cases when we have all necessary information about random perturbations. In addition to this, probabilistic estimations of vibration protection systems are not always acceptable. For example, when transporting dangerously explosive freights, where collisions between the freight and the vehicle are impermissible, the probability of motion of the system without collisions obtained by calculations, however small it may be, does not guarantee the absence of collisions at a particular process realization. The probabilistic estimation of the possibility of collisions is advantageous for making comparisons of different designs of vibration protection systems. The design of a vibration protection system that has the smaller probability of collisions can be considered more appropriate. One cannot, however, be sure of the absence of collisions.

Therefore, the probabilistic methods of analysis of vibration protection systems that must guarantee the fulfilment of quality criteria, for example, the absence of collisions between the instrument (freight) and the foundation are not always acceptable. If we have no necessary statistical information about random perturbations (this is most abundant case) it is impossible to analyze vibration protection system using methods of statistical dynamics. Therefore we must use a theory and methods that allow us to determine by calculation numerical values of parameters of vibration protection design (for example, c and α) in conditions of uncertainty of acting perturbations. The methods presented in the above mentioned paragraphs are some of such methods of analysis of dynamic processes that are used when there is no statistical information about random perturbations.

Let us formulate a problem of passive vibration protection, bearing in mind mechanical systems presented on Fig. 10.30 and 10.31. Random contact forces $\mathbf{f}_j$ induced by the random irregularities are acting on a moving vehicle with a freight (Fig. 10.30). To make it easy to grasp let us confine ourselves by a case where vibrations occur in the plane of the drawing. The mass m (Fig. 10.31) is acted by a random inertia force linearly dependent on the small misalignments ε and e and on the small scatter of the thrust ΔR. In both cases we have mechanical systems with a finite number of degrees of freedom. In the problem of transportation (Fig. 10.30) the vibration protection system of the freight must guarantee the absence of collisions between the freight and the vehicle during the time intervals corresponding to acceleration and braking. This means that the mutual approach of points $\left(K_3,\ K_3^{(1)}\right)$ and $\left(K_4,\ K_4^{(1)}\right)$ must not exceed the allowable value. An indispensable condition for a rigidly fixed freight is that the mutual approach of points K_2 and $K_2^{(1)}$ of the mass m and the body of the vehicle (Fig. 10.30 *b*) also does not exceed allowable values.

We need to obtain such numerical values of parameters of the vibration protection system $(c_j,\ \alpha_j)$ at which there are no collisions during the time intervals corresponding to acceleration or braking, when we have certain limitations on the random irregularities of the road and on the velocity of transportation v. As to the vibration protection of the mass m (Fig. 10.31), our task is to determine such values c_j, α_j at which the displacements of the point O would be inside the given area during the interval of the motion on the active leg under the action of random inertia forces depending on ε, e and ΔR.

Let us consider a mechanical system with a finite number of degrees of freedom whose motion is described by a non-homogeneous linear equation of the following type

$$A_1\ddot{\mathbf{Y}} + A_2\dot{\mathbf{Y}} + A_3\mathbf{Y} = D^{(1)}\mathbf{f}^{(1)}, \tag{10.250}$$

where $\mathbf{Y} = (y_1,\ y_2,\ \ldots,\ y_n)^T$ is a vector whose components are generalized linear and angular coordinates; $\mathbf{f}^{(1)}$ is a vector whose components are random forces and moments; A_j and $D^{(1)}$ are matrices whose elements in the general case can be both constant numbers and known functions depending on time (e.g., for mechanical systems with a variable mass). The term $A_2\dot{\mathbf{Y}}$

takes forces of viscous friction into account. The elements of the matrix A_3 depend on the characteristics of elastic elements and on the parameters of the vibration protection system (c_j, α_j).

Assuming $\dot{\mathbf{Y}} = \mathbf{Z}_1$ and $\mathbf{Y} = \mathbf{Z}_2$ we obtain an equation similar to that of (10.24),

$$\dot{\mathbf{Z}} + A(t, \boldsymbol{\lambda})\mathbf{Z} = D(t)\,\mathbf{f}, \tag{10.251}$$

where

$$\mathbf{Z} = (\mathbf{Z}_1, \mathbf{Z}_2)^{\mathrm{T}} = (\dot{y}_1,\ \dot{y}_2, \ldots, \dot{y}_n;\ y_1,\ y_2, \ldots, y_n)^{\mathrm{T}};$$

$$A = \begin{bmatrix} A_1^{-1} & A_1^{-1}A_3 \\ 0 & -E \end{bmatrix}, \quad D = \begin{bmatrix} A_1^{-1}D^{(1)} & 0 \\ 0 & 0 \end{bmatrix}, \quad \mathbf{f} = \begin{bmatrix} \mathbf{f}^{(1)} \\ 0 \end{bmatrix},$$

$\boldsymbol{\lambda}$ is a vector whose components are the parameters of the vibration protection system.

The information about random functions (i.e., about components $f_j^{(1)}$ of vector $\mathbf{f}^{(1)}$) necessary for the implementation of statistical dynamics is not available. We know only the areas of the possible values of random functions, i.e. $f_j^{(1)}$ are random functions bounded in absolute value. Random perturbations $f_j^{(1)}$ can be independent or dependent.

For independent perturbations the components of the vector $f_j^{(1)}$ satisfy conditions (10.18) (Fig. 10.32)

$$\min f_j^{(1)}(t) \le f_j^{(1)}(t) \le \max f_j^{(1)}(t). \tag{10.252}$$

In a specific case $\max f_j$ and $\min f_j$ can be constant, including an equality in absolute value

$$|\max f_j| = |\min f_j| = \alpha_j. \tag{10.253}$$

During the motion of an automobile on a road with random irregularities the forces f_j depend on the height of the irregularities $h(x)$ $\quad (x = vt)$. If we

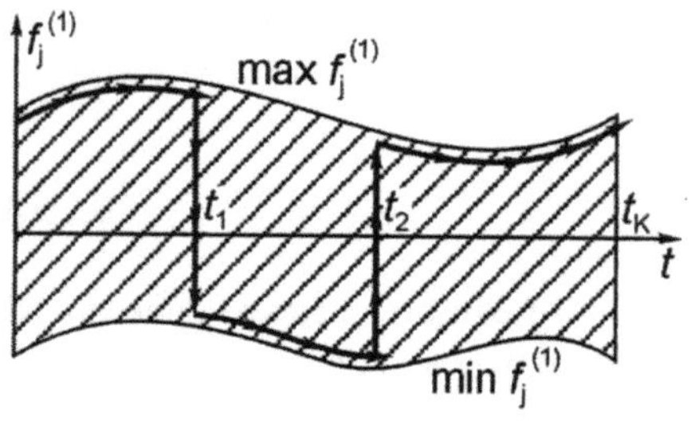

Fig. 10.32.

consider the irregularities of a road relative to its average state, then the area of possible values $h(vt)$ is similar to that presented in Fig. 10.32.

If we tackle a more complex problem where spatial vibrations arise at the motion of a vehicle, the equations of the small vibrations of a vehicle with a freight will contain projections of contact forces that are dependable and therefore meet conditions (10.22)

$$\left(\frac{f_{jx}^{(1)}}{\max|\mathbf{f}_j|}\right)^2+\left(\frac{f_{jy}^{(1)}}{\max|\mathbf{f}_j|}\right)^2+\left(\frac{f_{jz}^{(1)}}{\max|\mathbf{f}_j|}\right)^2\leq 1. \tag{10.254}$$

The equations of small vibrations of the mass m (Fig. 10.31) relative to the body of the rocket will include projections of the random inertia force $\mathbf{f}$ that are equal to the sum of three terms dependent on each of the random perturbations ε, e, ΔR $(\mathbf{f}_z,\ \mathbf{f}_e, \mathbf{f}_{\Delta R})$ meeting conditions similar to those of (10.254). The equation of small vibrations of the mass m relative to the body of the rocket is similar to equation (10.251). For the point mass m vector $\mathbf{Y}$ has six components.

In order to determine the numerical values of parameters c_i and α_i of a vibration protection system at which collisions will be absent, let us determine the projections of area D of the possible displacements of the mass m at an instant t_K on three coordinate planes $(y_1 0 y_2)$, $(y_1 0 y_3)$, $(y_2 0 y_3)$. The algorithm of determination of two-dimensional areas for dependent and independent perturbations is presented in paragraphs 10.4 and 10.6 respectively.

If the two-dimensional projections of area D are inside of the projection of area D_0 of the allowable values of the state vector $\mathbf{Y}(t_k)$ on coordinate planes, then c_i and α_i guarantee the absence of collisions during the interval $(0, t_K)$ of the system's motion.

We can minimize area D by varying parameters c_j, α_j of the vibration protection system. This area is denoted by D^* and is presented in Fig. 10.33 by dotted line. The appropriate values of c_j^* and α_j^* can be considered optimal.

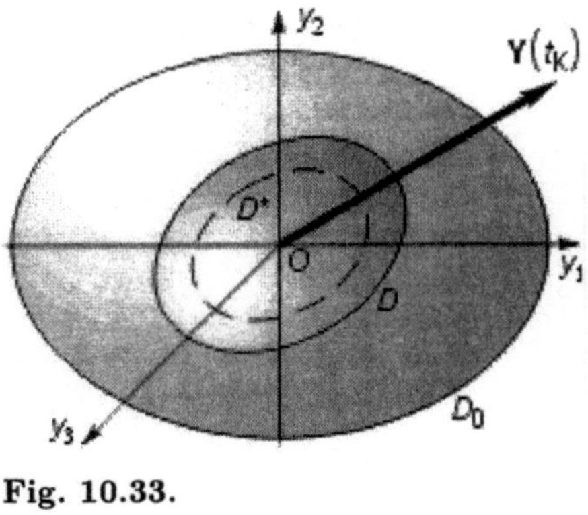

Fig. 10.33.

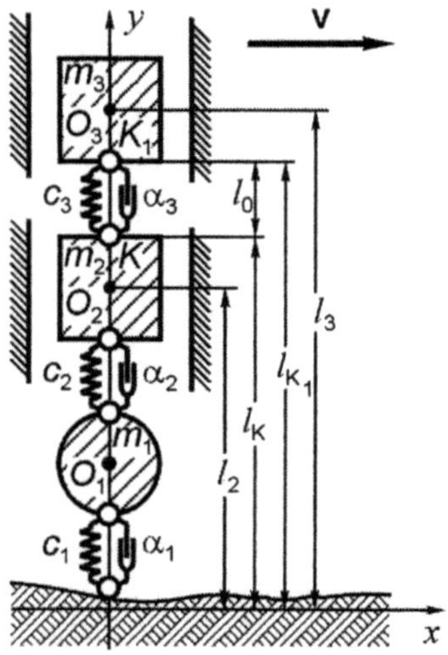

Fig. 10.34.

Let us consider a mechanical system with three degrees of freedom shown in Fig. 10.34 as an example of the numerical determination of values c_3, α_3 of the parameters of a passive vibration protection system of the weight (mass m_3). This is not very important in the practical plane but it allows us to show the algorithm of the numerical analysis applied to the passive vibration protection of objects under non- stationary vibrations of systems with a finite number of degrees of freedom at random perturbations bounded in absolute value.

For the example considered we obtain an equation of small vibrations

$$\ddot{\mathbf{Y}} + A^{(1)}\dot{\mathbf{Y}} + A^{(2)}\mathbf{Y} = D^{(1)}\mathbf{f}^{(1)} + D^{(2)}\mathbf{f}^{(2)}, \tag{10.255}$$

where

$$\mathbf{Y} = (y_1,\ y_2,\ y_3)^{\mathrm{T}}, \quad \mathbf{f}^{(1)} = (h,\ 0,\ 0)^{\mathrm{T}}, \quad \mathbf{f}^{(2)} = \left(0,\ \dot{h},\ 0\right)^{\mathrm{T}},$$

$$A^{(1)} = \begin{bmatrix} \dfrac{\alpha_1+\alpha_2}{m_1} & -\dfrac{\alpha_2}{m_2} & 0 \\ -\dfrac{\alpha_2}{m_2} & \dfrac{(\alpha_2+\alpha_3)}{m_2} & -\dfrac{\alpha_3}{m_2} \\ 0 & -\dfrac{\alpha_3}{m_3} & \dfrac{\alpha_3}{m_3} \end{bmatrix},$$

$$A^{(2)} = \begin{bmatrix} \frac{c_1 + c_2}{m_1} & -\frac{c_2}{m_2} & 0 \\ -\frac{c_2}{m_2} & \frac{c_1 + c_2}{m_2} & -\frac{c_3}{m_2} \\ 0 & -\frac{c_3}{m_3} & \frac{c_3}{m_3} \end{bmatrix},$$

$$D^{(1)} = \begin{bmatrix} \frac{c_1}{m_1} & 0 & 0 \\ 0 & 0 & 0 \\ 0 & 0 & 0 \end{bmatrix}, \quad D^{(2)} = \begin{bmatrix} 0 & \frac{\alpha_1}{m_1} & 0 \\ 0 & 0 & 0 \\ 0 & 0 & 0 \end{bmatrix}.$$

Letting $\dot{\mathbf{Y}} = \mathbf{Z}_1$, $\mathbf{Y} = \mathbf{Z}_2$ we obtain

$$\dot{\mathbf{Z}} + A\mathbf{Z} = D\mathbf{f} + D_2\mathbf{f}_2, \tag{10.256}$$

where

$$\mathbf{Z} = \left(\dot{\mathbf{Z}}_1, \mathbf{Z}\right)^{\mathrm{T}} = (\dot{y}_1, \dot{y}_2, \dot{y}_3, y_1, y_2, y_3)^{\mathrm{T}},$$

$$\mathbf{f}_1 = \left(f^{(1)}, 0\right)^{\mathrm{T}}, \quad \mathbf{f}_2 = \left(f^{(2)}, 0\right)^{\mathrm{T}}.$$

$$A = \begin{bmatrix} A^{(1)} & A^{(2)} \\ -E & 0 \end{bmatrix}, \quad D_1 = \begin{bmatrix} D^{(1)} & 0 \\ 0 & 0 \end{bmatrix}, \quad D_2 = \begin{bmatrix} D^{(2)} & 0 \\ 0 & 0 \end{bmatrix}.$$

At the initial instant of time the system was at rest. Then it began to move with an increasing velocity v (acceleration) up to the value $v = v_K$ at an instant $t = t_K$. It is necessary to determine c_3 and α_3 at which masses m_3 and m_2 would not collide with each other during the interval of time $(0, t_K)$ corresponding to the acceleration.

The solution of equation (10.256) is of the form

$$\mathbf{Z} = \int_0^t G(t,\tau)D_1\mathbf{f}_1 \mathrm{d}\tau + \int_0^t G(t,\tau)D_2\mathbf{f}_2 \mathrm{d}\tau. \tag{10.257}$$

From (10.257) we obtain $\mathbf{Z}_1$ and $\mathbf{Z}_2$ at an instant t_K

$$\mathbf{Z}_1 = \int_0^{t_K} G_{11}D^{(1)}\mathbf{f}^{(1)}\mathrm{d}\tau + \int_0^{t_K} G_{11}D^{(2)}\mathbf{f}^{(2)}\mathrm{d}\tau, \tag{10.258}$$

$$\mathbf{Z}_2 = \int_0^{t_K} G_{21}D^{(1)}\mathbf{f}^{(1)}\mathrm{d}\tau + \int_0^{t_K} G_{21}D^{(2)}\mathbf{f}^{(2)}\mathrm{d}\tau, \tag{10.259}$$

where G_{11} and G_{21} are block matrices entering in matrix G.

After transformations we obtain from (10.259)

$$z_5 = y_2 = \int_0^{t_k} g_{51}\frac{c_1}{m_1}h\,\mathrm{d}\tau + \int_0^{t_k} g_{51}\frac{\alpha_1}{m_1}\dot{h}\,\mathrm{d}\tau,$$
$$z_6 = y_3 = \int_0^{t_k} g_{61}\frac{c_1}{m_1}h\,\mathrm{d}\tau + \int_0^{t_k} g_{61}\frac{\alpha_1}{m_1}\dot{h}\,\mathrm{d}\tau. \tag{10.260}$$

The derivative of road irregularities with respect to time ($\dot{h}$) enters in expression (10.260) and we have no information about this derivative. Therefore, using integration by parts, let us present terms containing $\dot{h}$ in the form

$$\int_0^{t_k} g_{51}\frac{\alpha_1}{m_1}\dot{h}\,\mathrm{d}\tau = g_{51}\frac{\alpha_1}{m_1}h\Big|_0^{t_k} - \int_0^{t_k} \dot{g}_{51}\frac{\alpha_1}{m_1}h\,\mathrm{d}\tau,$$
$$\int_0^{t_k} g_{61}\frac{\alpha_1}{m_1}\dot{h}\,\mathrm{d}\tau = g_{61}\frac{\alpha_1}{m_1}h\Big|_0^{t_k} - \int_0^{t_k} \dot{g}_{61}\frac{\alpha_1}{m_1}h\,\mathrm{d}\tau. \tag{10.261}$$

As a result we obtain

$$y_2 = \int_0^{t_k} (g_{51}\frac{c_1}{m_1} - \dot{g}_{51}\frac{\alpha_1}{m_1})h\,\mathrm{d}\tau + g_{51}\frac{\alpha_1}{m_1}h\Big|_0^{t_k},$$
$$y_2 = \int_0^{t_k} (g_{61}\frac{c_1}{m_1} - \dot{g}_{61}\frac{\alpha_1}{m_1})h\,\mathrm{d}\tau + g_{61}\frac{\alpha_1}{m_1}h\Big|_0^{t_k}. \tag{10.262}$$

The Green matrix at $t = t_k$ is equal to identity matrix, therefore

$$g_{51}\frac{\alpha_1}{m_1}h\Big|_0^{t_k} = -g_{51}(0)\frac{\alpha_1}{m_1}h(0),$$
$$g_{61}\frac{\alpha_1}{m_1}h\Big|_0^{t_k} = -g_{61}(0)\frac{\alpha_1}{m_1}h(0).$$

Expressions (10.262) contain derivatives of Green matrix elements ($\dot{g}_{51}$ and $\dot{g}_{61}$) that can be obtained in the following way.

Let us consider an identity

$$K(t)\,K^{-1}(t) = E. \tag{10.263}$$

Taking derivatives with respect to t we obtain

$$\dot{K}K^{-1} + K\dot{K}^{-1} = 0.$$

By eliminating $\dot{K}$ (the matrix K meets equation $\dot{K} + AK = 0$) we obtain

$$-A + K\dot{K}^{-1} = 0. \tag{10.264}$$

Let us multiply equation (10.264) from left by matrix K^{-1} and by matrix $K(t_K)$

$$K(t_K)\,\dot{K}^{-1} - K(t_K)\,K^{-1}A = 0,$$

or

$$\dot{G}(t_k, t) = G(t_k, t)\,A, \tag{10.265}$$

where $G(t_k, t)$ is the Green matrix .

Therefore $\dot{g}_{51}$ and $\dot{g}_{61}$ are equal to

$$\dot{g}_{51} = \sum_{j=1}^{6} g_{5j}a_{j1}, \quad \dot{g}_{61} = \sum_{j=1}^{6} g_{6j}a_{j1}. \tag{10.266}$$

In order to determine the area of possible values of y_2 and y_3 at an instant t_K let us determine $h(\tau)$ at which functional J (10.63) attains maximum value for a given angle α

$$\max_h J = y_2 \cos\alpha + y_3 \sin\alpha = \int_0^{t_k} (d_1 \cos\alpha + d_2 \sin\alpha) h(\tau)\, d\tau$$
$$- (g_{51}(0)\cos\alpha + g_{61}(0)\sin\alpha)\frac{\alpha_1}{m_1} h(0),$$

where

$$d_1 = g_{51}\frac{c_1}{m_1} - \left(\sum_{j=1}^{6} g_{5j}a_{j1}\right)\frac{\alpha_1}{m_1},$$

$$d_2 = g_{61}\frac{c_1}{m_1} - \left(\sum_{j=1}^{6} g_{6j}a_{j1}\right)\frac{\alpha_1}{m_1}.$$

The maximum value of functional J is determined by an integral term. Therefore a law of $h(\tau)$ variation which for a given α produces values $y_2(t_k)$ and $y_3(t_k)$ corresponding to the boundary of the area is

$$h_*(\tau) = \begin{cases} \max h, & \text{at } (d_1 \cos\alpha + d_2 \sin\alpha) > 0, \\ \min h, & \text{at } (d_1 \cos\alpha + d_2 \sin\alpha) < 0. \end{cases} \tag{10.267}$$

In order to avoid collisions between the masses m_2 and m_3 (Fig. 10.35) the following condition

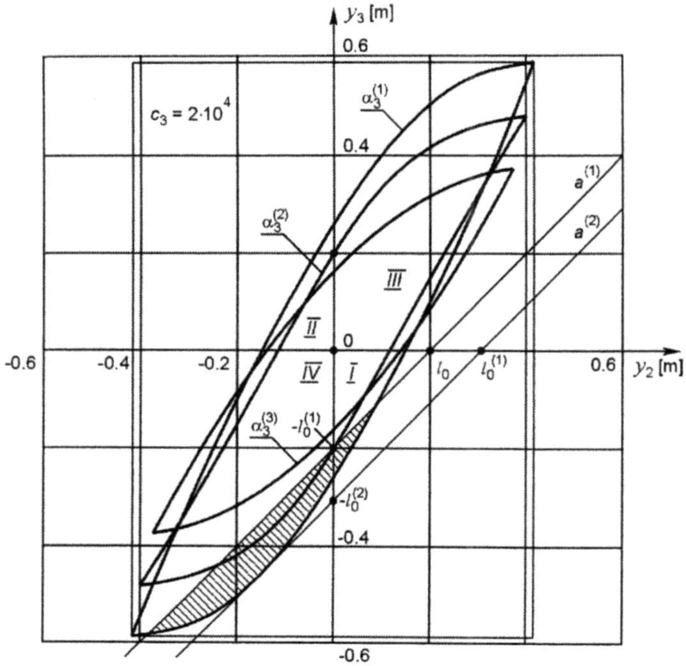

Fig. 10.35.

$$l_0 + y_3(t) - y_2(t) > 0. \tag{10.268}$$

must be fulfilled.

During the numerical solution the following quantities entering in equations were considered

$$m_1 = 20 \text{ kg}, \quad m_2 = 100 \text{ kg}, \quad m_3 = 80 \text{ kg}, \quad c_1 = 5 \cdot 10^4 \text{ N/m},$$
$$c_2 = 4 \cdot 10^4 \text{ N/m}, \ \ c_3 = 2 \cdot 10^4 \text{ N/m}, \quad \alpha_1 = 1.2 \cdot 10^3 \text{ Ns/m},$$
$$\alpha_2 = 2.4 \cdot 10^3 \text{ Ns/m}, \quad t_k = 4 \text{ s}, \quad |h_{\max}| = |h_{\min}| = 0.15 \text{ m}$$

The areas of the possible values of y_2 and y_3 at an instant t_k for a series of values of the coefficient of viscous friction force $\alpha_3^{(j)}$ ($\alpha_3^{(1)} = 760$, $\alpha_3^{(2)} = 1560$, $\alpha_3^{(3)} = 3040$) are presented in Fig. 10.35.

The boundary between the values y_2 and y_3 at which collisions are absent and values y_2 and y_3 at which collisions are possible is determined by an equation of a straight line

$$l_0 + y_3 - y_2 = 0. \tag{10.269}$$

For the case where $l_0 = 0.2$ m this straight line is shown in Fig. 10.35 (it is denoted by $a^{(1)}$). For a half-plane of the plane (y_2, y_3) that is situated to the left of the straight line $a^{(1)}$ an inequality

$$l_0 + y_3 - y_2 > 0,$$

fulfills, i.e. there are no collisions.

For a half-plane of the plane (y_2, y_3) that is situated to the right of the straight line $a^{(1)}$ an inequality

$$l_0 + y_3 - y_2 < 0$$

fulfills, i.e. collisions are possible.

In order to have no collisions the straight line $a^{(1)}$ must not intersect the area (y_2, y_3).

In Fig. 10.35 the straight line $a^{(1)}$ intersects the areas obtained for the coefficients of viscous friction $\alpha_3^{(1)} = 760$ and $\alpha_3^{(2)} = 1560$ (the area of values of y_2 and y_3 for which collisions are possible is shown by hatching). For the area at $\alpha_3^{(3)} = 3040$ possible values y_2 and y_3 are situated to the left of the straight line $a^{(1)}$ i.e. for values $c_3 = 2 \cdot 10^4$ and $\alpha_3 = 3040$ collisions are excluded.

The result obtained is true for any values $|\max h| = |\min h| < 0.15$ m. For the value $l_0^{(1)} = 0.28$ m (straight line $a^{(2)}$) there would not be collisions at all values of $\alpha_3^{(j)}$ because this straight line does not cross the area. And this result is guaranteed for the given irregularities of the road (h).

Appendices

A.1 Elementary Generalized Functions

Dirac Delta Function δ. Let us consider a function (Fig. A1)

$$\delta(t) = \frac{1}{\pi}\left(\frac{1}{1+t^2}\right), \tag{A.1}$$

having the maximum value at $t = 0$ and decreasing at modulus $|t|$ growth. Let us transform the function $\Phi(t)$ by increasing its value at $t = 0$ by the factor m

$$\Phi_1(mt) = \frac{1}{\pi}\frac{m}{\left[1+(mt)^2\right]}. \tag{A.2}$$

The functions $\Phi_1(mt)$ for several m are presented in Fig. A2. It can be shown that the integrals of the functions (1) and (2) are equal to unity, i.e.

$$\int_{-\infty}^{+\infty} \Phi(t)\,\mathrm{d}t = 1, \qquad \int_{-\infty}^{+\infty} \Phi_1(mt)\,\mathrm{d}t = 1.$$

As m increases indefinitely ($m \to \infty$) we obtain the function that is named as the Dirac delta function (Fig. A3):

$$\lim_{m\to\infty} \Phi_1(mt) = \delta(t).$$

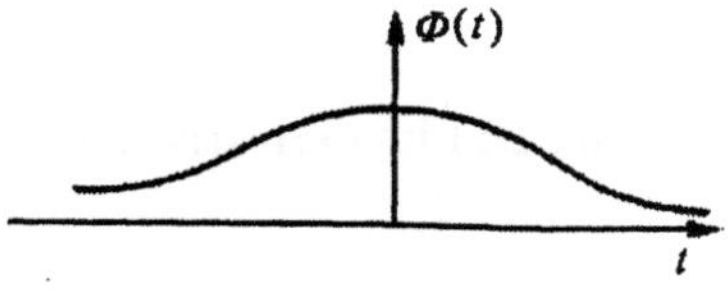

Fig. A.1.

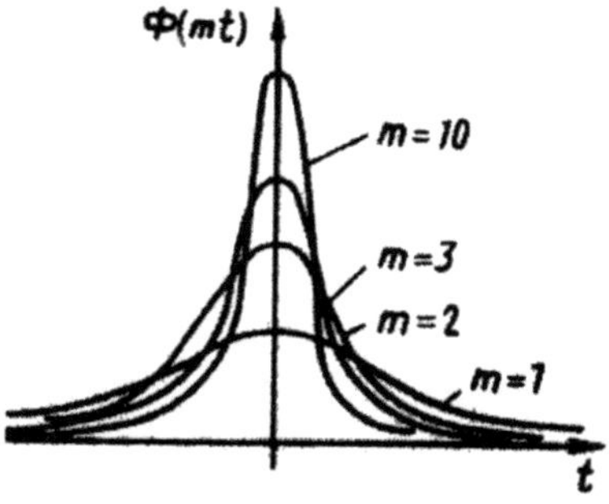

Fig. A.2.

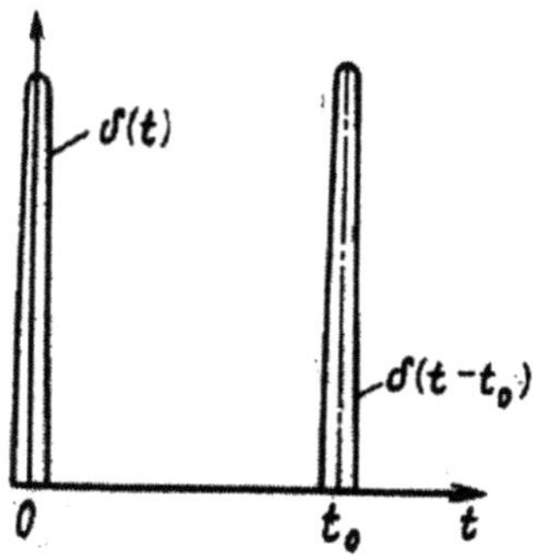

Fig. A.3.

The principal properties of the Dirac delta function are:

$$1)\ \delta(t) = \begin{cases} 0 & t < 0, \\ \infty & t = 0, \\ 0 & t > 0, \end{cases} \quad \delta(t - t_0) = \begin{cases} 0 & t < t_0, \\ \infty & t = t_0, \\ 0 & t > 0; \end{cases} \tag{A.3}$$

$$2)\ \delta(t) = \delta(-t);$$

$$3)\ \int\limits_{-\infty}^{\infty} \delta(t - t_0)\,\mathrm{d}t = 1;$$

$$4)\ \delta(t) = \delta(a\varepsilon) = \frac{1}{|a|}\delta(\varepsilon),$$

where ε is a dimensionless quantity.

The integrals containing the Dirac delta function and its derivatives are

$$\int\limits_{-\infty}^{\infty} \varphi\,(t)\delta\,(t - t_0)\ \mathrm{d}t = \varphi\,(t_0)\,;$$

$$\int_{-\infty}^{\infty} \varphi(t)\dot{\delta}(t-t_0)\,\mathrm{d}t = -\int_{-\infty}^{\infty} \dot{\varphi}(t)\delta(t-t_0)\,\mathrm{d}t = -\dot{\varphi}(t_0)\,,$$

where $\dot{\delta}$ is the derivative of the Dirac delta function.

For the derivative of n-th order we obtain

$$\int_{-\infty}^{\infty} \delta^n(t)\,\varphi(t)\,\mathrm{d}t = (-1)^n\,\varphi^n(t_0).$$

The derivatives of the Dirac delta function can be obtained (as easy-to-grasp presentation) as a limit of the function derivatives, i.e.

$$\lim_{m\to\infty} \Phi_1^m(mt) = \delta^n(t)\,.$$

The Heaviside Function. The integral of the Dirac delta function with variable upper limit is

$$\int_{-\infty}^{t} \delta(t-t_0)\,\mathrm{d}t = H(t-t_0), \quad H(t-t_0) = \begin{cases} 0 & t < t_0; \\ \frac{1}{2} & t = t_0; \\ 1 & t > t_0, \end{cases} \tag{A.4}$$

where $H(t)$ is the Heaviside function (Fig. A4).

Differentiating (4) with respect to t we obtain

$$\frac{\mathrm{d}H(t-t_0)}{\mathrm{d}t} = \delta(t-t_0)\,. \tag{A.5}$$

The Function sign(t) (Fig. A5) is

$$\mathrm{sign}(t) = \begin{cases} -1 & t < 0; \\ 0 & t = 0; \\ 1 & t > 0; \end{cases} \qquad \mathrm{sign}(t-t_0) = \begin{cases} -1 & t < t_0; \\ 0 & t = t_0; \\ 1 & t > t_0; \end{cases} \tag{A.6}$$

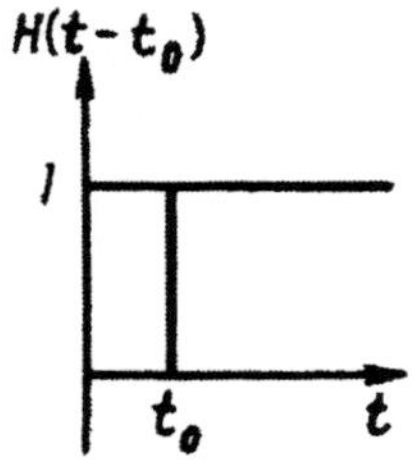

Fig. A.4.

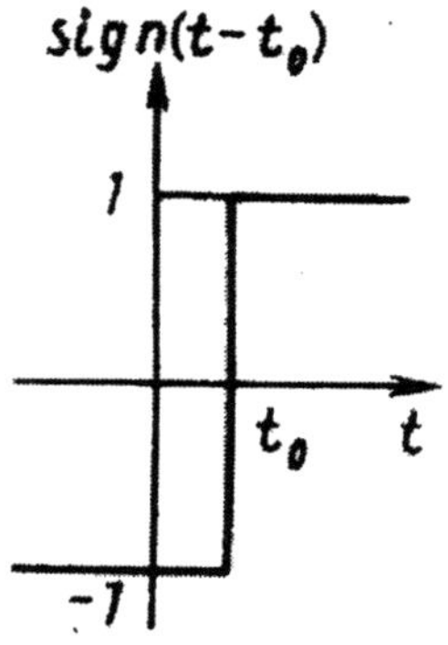

Fig. A.5.

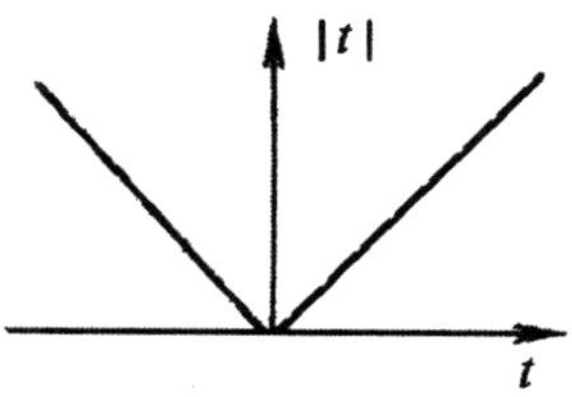

Fig. A.6.

$$\frac{\mathrm{dsign}t}{\mathrm{d}t} = 2\delta\left(t\right), \quad H(t) = \frac{1}{2}(1 + \mathrm{sign}t).$$

The function $f(t) = |t|$ (Fig. A6) is

$$|t| = t\mathrm{sign}t; \quad \frac{\mathrm{d}\,|t|}{\mathrm{d}t} = \mathrm{sign}t; \quad \frac{\mathrm{d}^2\,|t|}{\mathrm{d}t^2} = 2\delta\left(t\right). \tag{A.7}$$

A.2 Values of Integrals J_n

$$J_n = \frac{1}{2\pi} \int\limits_{-\infty}^{\infty} \frac{G(i\omega)}{\left|A\left(i\omega\right)\right|^2} \mathrm{d}\omega.$$

where

$$A\left(i\omega\right) = a_0\left(i\omega\right)^n + a_1\left(i\omega\right)^{n-1} + \cdots + a_n;$$
$$G\left(i\omega\right) = b_0\left(i\omega\right)^{2n-2} + b_1\left(i\omega\right)^{2n-4} + \cdots + b_{n-1};$$

$$J_1 = \frac{b_0}{2a_0a_1}; \quad J_2 = \frac{-b_0 + \frac{a_0b_1}{a_2}}{2a_0a_1}; \quad J_3 = \frac{-a_2b_0 + a_0b_1 - \dfrac{a_0a_1b_2}{a_3}}{2a_0(a_0a_3 - a_1a_2)};$$

$$J_4 = \frac{b_0(-a_1a_4 + a_2a_3) - a_0a_3a_1 + a_0a_1b_2 + \dfrac{a_0b_3}{a_4}(a_0a_3 - a_1a_2)}{2a_0(a_0a_3^2 + a_1^2a_4 - a_1a_2a_3)};$$

$$J_5 = \frac{M_5}{2a_0\Delta_5},$$

where

$$\begin{aligned} M_5 = b_0 - (a_0a_4a_5 + a_1a_4^2 + a_2^2a_5 - a_2a_3a_4) + a_0b_1(-a_2a_5 + a_3a_4) + \\ + a_0b_2(a_0a_5 - a_1a_4) + a_0b_3(-a_0a_3 + a_1a_2) + \\ + \frac{a_0b_4}{b_4a_5}\left(-a_0a_1a_5 + a_0a_3^2 + a_1^2a_4 - a_1a_2a_3\right); \end{aligned}$$

$$\Delta_5 = a_0^2a_5^2 - 2a_0a_1a_4a_5 - a_0a_3a_2a_5 + a_0a_3^2a_4 + a_1^2a_4 - a_1a_2a_3a_4.$$

A.3 Correlation Functions and Spectral Densities Corresponding to Them

Correlation functions	Spectral densities
$D_0 e^{-\alpha\|\tau\|}$ 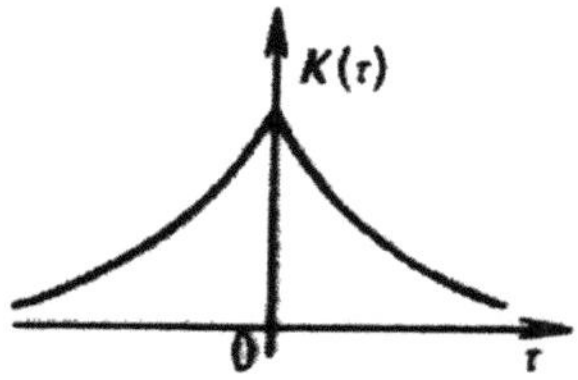	$\frac{2\alpha D_0}{2\pi(\alpha^2+\omega^2)}$ 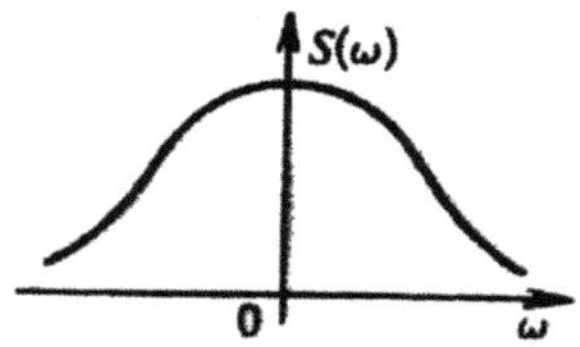
$D_0 e^{-\alpha\|\tau\|}\cos\beta\tau$ 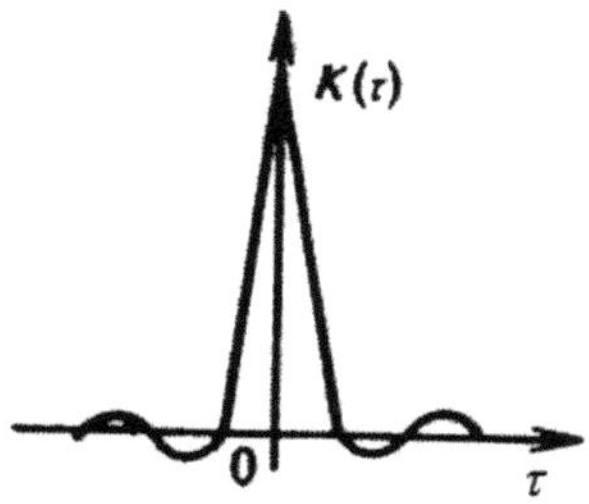	$\frac{2\alpha D_0}{\pi}\frac{\omega^2+\beta^2+\alpha^2}{(\omega^2-\beta^2-\alpha^2)^2-4\alpha^2\omega^2}$ 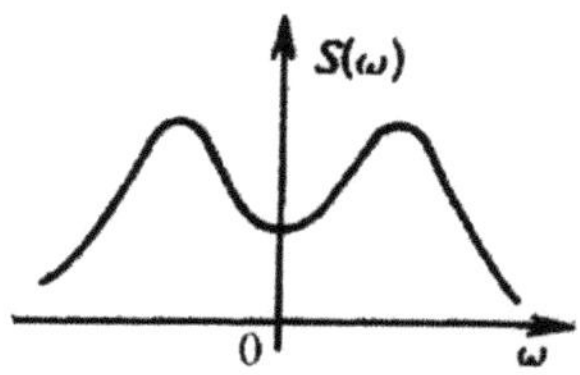
$D_0 e^{-\alpha\|\tau\|}\left(\cos\beta\tau+\frac{\alpha}{\beta}\sin\beta\|\tau\|\right)$ 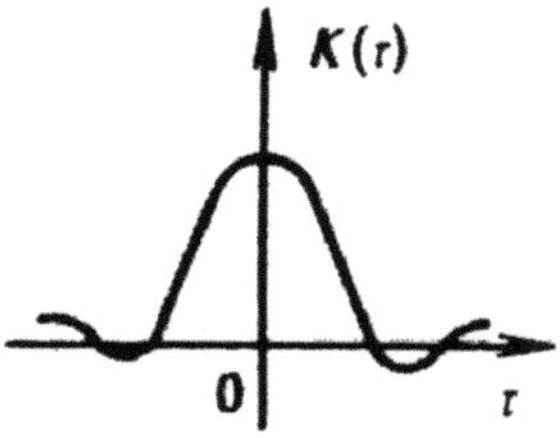	$\frac{2\alpha D_0}{\pi}\frac{\alpha^2+\beta^2}{(\omega^2-\beta^2-\alpha^2)^2-4\alpha^2\omega^2}$ 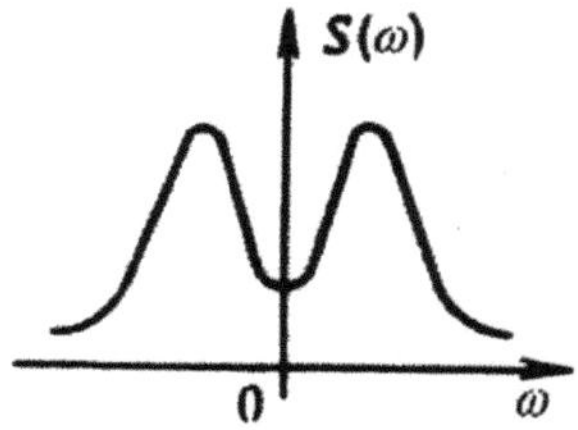

<table>
<tr><th>Correlation functions</th><th>Spectral densities</th></tr>
<tr><td>

$D_0 \frac{\sin \omega_\beta \tau}{\omega_\beta \tau}$

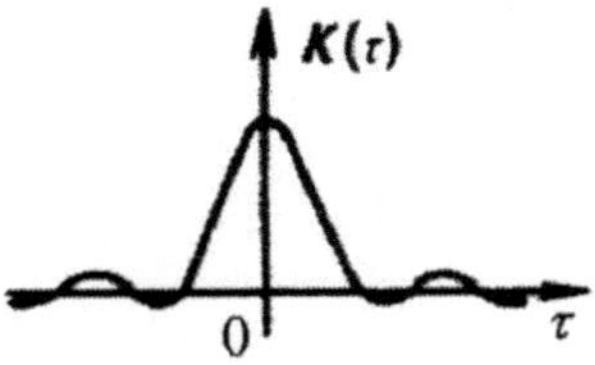

</td><td>

$\frac{D_0}{\omega_\beta}$ at $|\omega| < \omega_\beta$
0 at $|\omega| > \omega_\beta$

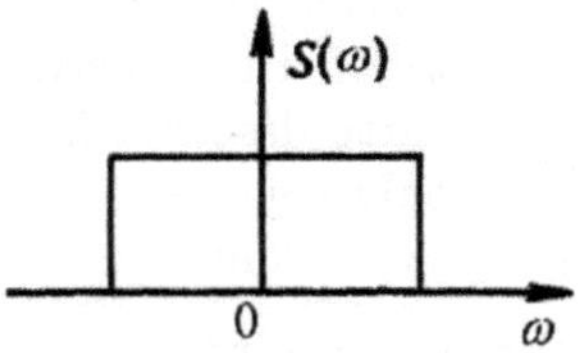

</td></tr>
<tr><td>

$K(\tau) = 2\pi s_0 \delta(\tau)$

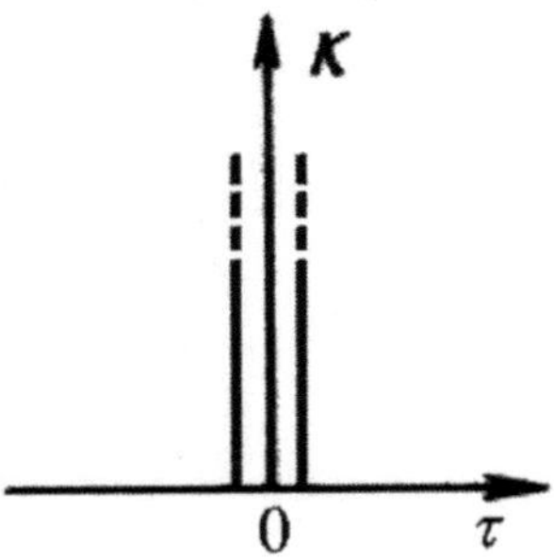

</td><td>

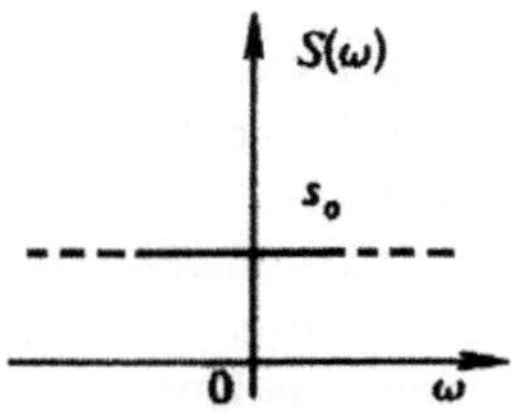

</td></tr>
<tr><td>

$D_0 \exp\left\{-\frac{\alpha^2 \tau^2}{4}\right\}$

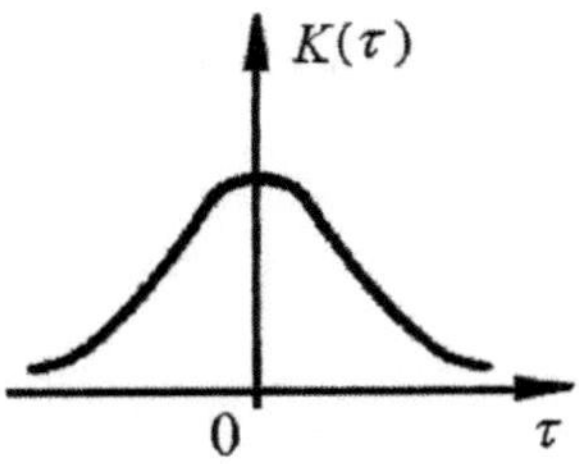

</td><td>

$\frac{2D_0}{\alpha\sqrt{\pi}} \exp\left\{-\frac{\omega^2}{\alpha^2}\right\}$

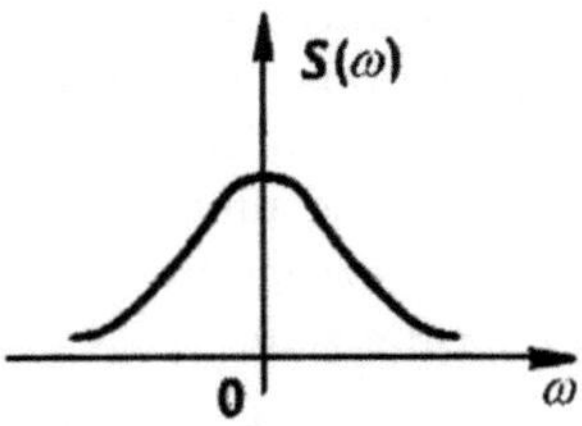

</td></tr>
</table>

A.4 Hiawatha Designs an Experiment

by Maurice G. Kendall

1. Hiawatha, mighty hunter
He could shoot ten arrows upwards
Shoot them witli such strength and swiftness
That the last had left the bowstring
Ere the first to earth descended.
This was commonly regarded
As a feat of skill and cunning.

2. One or two sarcastic spirits
Pointed out to him, however,
That it might be much more useful
If he sometimes hit the target.
Why not shoot a little straighter
And employ a smaller sample?

3. Hiawatha, who at college
Majored in applied statistics
Consequently felt entitled
To instruct his fellow men on
Any subject whatsoever.
Waxed exceedingly indignant
Talked about the law of error,
Talked about truncated normals.
Talked of loss of information,
Talked about his lack of bias
Pointed out that in the long run
Independent observations
Even though they missed the target
Had an average point of impact
Very near the spot he aimed at
With the possible exception
Of a set of measure zero.

4. This, they said, was rather doubtful.
Anyway, it didn't matter
What resulted in the long run:
Either he must hit the target
Much more often than at present
Or himself would have to pay for
All the arrows that he wasted.

5. Hiawatha, in a temper
Quoted parts of R.A. Fisher
Quoted Yates and quoted Finney
Quoted yards of Oscar Kempthorne
Quoted reams of Cox and Cochran
Quoted Anderson and Bancroft
Practically in extenso
Trying to impress upon them
That what actually mattered
Was to estimate the error.

6. One or two of them admitted
Such a thing might have its uses
Still, they said, he might do better
If he shot a little straighter.

7. Hiawatha, to convince them
Organized a shooting contest
Laid out in the proper manner
Of designs experimental
Recommended in the textbooks
(Mainly used for tasting tea, but
Sometimes used in other cases)
Randomized his shooting order
In factorial arrangements
Used in the theory of Galois
Fields of ideal polynomials
Got a nicely balanced layout
And successfully confounded
Second-order interactions.

8. All the other tribal marksmen
Ignorant, benighted creatures,
Of experimental set-ups
Spent their time of preparation
Putting in a lot of practice
Merely shooting at a target.

9. Thus it happened in the contest
That their scores were most impressive
With one solitary exception
This (I hate to have to say it)
Was the score of Hiawatha,
Who, as usual, shot his arrows
Shot them with great strength and swiftness
Managing to be unbiased
Not, however, with his salvo
Managing to hit the target.

10. There, they said to Hiawatha,
That is what we all expected.

11. Hiawatha, nothing daunted,
Called for pen and called for paper
Did analyses of variance
Finally produced the figures
Showing beyond peradventure
Everybody else was biased
And the variance components
Did not differ from each other
Or from Hiawatha's
(This last point, one should acknowledge
Might have been much more convincing
If he hadn't been compelled to
Estimate has own component
From experimental plots in
Which the values all were missing.
Still, they didn't understand it
So they couldn't raise objections
This is what so often happens
With analyses of variance).

12. All the same, his fellow tribesmen
Ignorant, benighted heathens.
Took away his bow and arrows.
Said that though my Hiawatha
Was a brilliant statistician
He was useless as a bowman,
As for variance components
Several of the more outspoken
Made primeval observations
Hurtfull to the finer feelings
Even of a statistician.

13. In a corner of the forest
Dwells alone my Hiawatha
Permanently cogitating
On the normal law of error
Wondering in idle moments
Whether an increased precision
Might perhaps be rather better
Even at the risk of bias
If thereby ore, now and then, could
Register upon the target.

The sense of the poem lies in the fact that in the case when it is required to obtain the guaranteed final result it is impossible to replace it by its probability estimation.
The theory and the numerical methods of the determination of guaranteed final results are presented in the last chapter

References

1. Arsalanov A.M. (1987) Design of Structure Elements with Preassigned Reliability at Random Actions. Mashinostroenie, Moscow (Russian)
2. Augusti G., Baratta A., Casciati F. (1985) Probabilistic Methods in Structural Engineering. Chapman and Hall, London, New York
3. Balabukh L.I., Alfutov N.A., Usyukin V.I. (1984) Structural Mechanics of Rockets. Vysshaya Shkola, Moscow (Russian)
4. Bellman R. (1961) Adaptative Control Processes: a Guided Tour. Princeton University Press, Princeton
5. Birger I.A. (1978) Technical Diagnostics. Mashinostroenie, Moscow (Russian)
6. Biderman V.L. (1980) Theory of Mechanical Vibrations. Vysshaya Shkola, Moscow (Russian)
7. Bendat J., Peerson A. (1989) Random Data Analysis and Measurement Procedures. John Wiley and Sons, New-York
8. Bolotin V.V. (1971) Application of Methods of the Theory of Probability and Theory of Reliability in Structural Design. Stroyizdat, Moscow (Russian)
9. Bolotin V.V. (1984) Prediction of Machines and Structures Life. Mashinostroenie, Moscow (Russian)
10. Bolotin V.V. (1979) Random Vibrations of Elastic Systems. Nauka, Moscow (Russian)
11. Born M. (1963) Physics in Life of my Generation. Inostrannaya literatura, Moscow (Russian)
12. Buchgolz N.N. (1939) Fundamentals of Theoretical Mechanics. Izdatelstvo technicoteoreticheskoy literatury, Moscow (Russian)
13. Ventzel E.S. (2001) Theory of Probability. Vysshaya Shkola, Moscow (Russian)
14. Ventzel E.S., Ovcharov L.A (2000) Theory of Random Processes and its Technical Applications. Vysshaya Shkola, Moscow (Russian)
15. Gladky V.F. (1969) Dynamics of Flying Vehicle Structure. Nauka, Moscow (Russian)
16. Gladky V.F. (1975) Strength, Vibrations and Reliability of Flying Vehicle Structure. Nauka, Moscow (Russian)
17. Gnedenko B.V. (1965) Theory of Probability. Nauka, Moscow (Russian)
18. Gusev A.S. (1989) Fatigue Strength and Life of Structures at Random Loads. Mashinostroenie, Moscow (Russian)
19. Gusev A.S., Svetlitsky V.A. (1984) Design of Structures at Random Loading. Mashinostroenie, Moscow (Russian)
20. Ermakov S.M., Mikhailov G.A. (1976) Fundamentals of Statistical Simulation. Nauka, Moscow (Russian)
21. Kasakov V.A. (1973) Introduction to the Theory of Markovian Processes and Some Radiotechnical Problems. Sovetskoye radio, Moscow (Russian)
22. Kapur K.C., Lamberson L.R. (1980) Reliability in Engineering Design. John Wiley and Sons, New-York

Index

23. Cooper G., McGillem K. (1989) Probabilistic Methods of Signals and Systems Analysis. Holt, Rinehart and Winston, New-York
24. Manley R. (1971) Waveform Analysis. John Wiley and Sons Inc., Chapman and Hall, London
25. Makarov B.P. (1983) Non-Linear Problems of Statistical Dynamics of Machines and Devices. Mashinostroenie, Moscow (Russian)
26. Nicolaenko N.A., Ulyanov S.V. (1977) Statistical Dynamics of Engineering Structures. Mashinostroenie, Moscow (Russian)
27. Panovko Ya. G. (1976) Fundamentals of Applied Theory of Vibrations and Impact. Mashinostroenie, Moscow (Russian)
28. Proscuryakov V.B. (1972) Dynamics and Strength of Frames and Bodies of Vehicles. Mashinostroenie, Moscow (Russian)
29. Pugachev V.S. (1962) Theory of Random Functions. Fizmatgiz, Moscow (Russian)
30. Pugachev V.S. (1968) Introduction to the Theory of Probability. Nauka. Moscow (Russian)
31. Rosin M.D. (1970) Statistical Dynamics and Theory of Control Systems Efficiency. Mashinostroenie, Moscow (Russian)
32. Svetlitsky V.A. (1986) Dynamics of Flying Vehicle Launch. Nauka, Moscow (Russian)
33. Svetlitsky V.A. (2000) Statics of Rods. Springer Verlag
34. Svetlitsky V.A. (2001) Dynamics of Rods. Springer Verlag (in print)
35. Sveshnikov A.A. (1968) Applied Methods of Random Functions Theory. Nauka, Moscow (Russian)
36. Silaev A.A. (1972) Spectral Theory of Vehicles Cushioning. Mashinostroenie, Moscow (Russian)
37. Skuchik E. (1968) Simple and Complex Vibratory Systems. The Pensylvania State University Press, University Park and London
38. Handbook of Probabilistic Analysis (1970). Voenizdat. Moscow (Russian)
39. Tikhonov V.I. (1970) Random Processes Overshoots. Nauka, Moscow (Russian)
40. Tikhonov V.I., Mironov M.A. (1970) Markovian Processes. Sovetskoye radio, Moscow (Russian)
41. Feodosiev V.I. (1995) Strength of Materials. BSTU Publishers, Moscow (Russian)
42. Kharkevich A.A. (1962) Spectrums and Analysis. Fizmatgiz, Moscow (Russian)
43. Chetaev N.G. (1987) Theoretical Mechanics. Nauka, Moscow (Russian)

Foundations of Engineering Mechanics

Series Editors: Vladimir I. Babitsky, Loughborough University
Jens Wittenburg, Karlsruhe University

Palmov	Vibrations of Elasto-Plastic Bodies (1998, ISBN 3-540-63724-9)
Babitsky	Theory of Vibro-Impact Systems and Applications (1998, ISBN 3-540-63723-0)
Skrzypek/ Ganczarski	Modeling of Material Damage and Failure of Structures Theory and Applications (1999, ISBN 3-540-63725-7)
Kovaleva	Optimal Control of Mechanical Oscillations (1999, ISBN 3-540-65442-9)
Kolovsky	Nonlinear Dynamics of Active and Passive Systems of Vibration Protection (1999, ISBN 3-540-65661-8)
Guz	Fundamentals of the Three-Dimensional Theory of Stability of Deformable Bodies (1999, ISBN 3-540-63721-4)
Alfutov	Stability of Elastic Structures (2000, ISBN 3-540-65700-2)
Morozov/ Petrov	Dynamics of Fracture (2000, ISBN 3-540-64274-9)
Astashev/ Babitsky/ Kolovsky	Dynamics and Control of Machines (2000, ISBN 3-540-63722-2)

Foundations of Engineering Mechanics

Series Editors: Vladimir I. Babitsky, Loughborough University
Jens Wittenburg, Karlsruhe University

Svetlitsky	Statics of Rods (2000, ISBN 3-540-67452-7)
Kolovsky/ Evgrafov/ Slousch/ Semenov	Advanced Theory of Mechanisms and Machines (2000, ISBN 3-540-67168-4)
Landa	Regular and Chaotic Oscillations (2001, ISBN 3-540-41001-5)
Muravskii	Mechanics of Non-Homogeneous and Anisotropic Foundations (2001, ISBN 3-540-41631-5)
Gorshkov/ Tarlakovsky	Transient Aerohydroelasticity of Spherical Bodies (2001, ISBN 3-540-42151-3)
Babitsky/ Krupenin	Vibration of Strongly Nonlinear Discontinuous Systems (2001, ISBN 3-540-41447-9
Manevitch/ Andrianov/ Oshmyan	Mechanics of Periodically Heterogeneous Structures (2002, ISBN 3-540-41630-7)
Lurie	Analytical Mechanics (2002, ISBN 3-540-42982-4)

Foundations of Engineering Mechanics

Series Editors: Vladimir I. Babitsky, Loughborough University
Jens Wittenburg, Karlsruhe University

Slepyan	Models and Phenomena in Fracture Mechanics (2002, ISBN 3-540-43767-3)
Nagaev	Dynamics of Synchronising Systems (2003, ISBN 3-540-44195-6)
Svetlitsky	Statistical Dynamics and Reliability Theory for Mechanical Structures (2003, ISBN 3-540-44297-9)

MIX
Papier aus verantwortungsvollen Quellen
Paper from responsible sources
FSC® C105338

If you have any concerns about our products,
you can contact us on
ProductSafety@springernature.com

In case Publisher is established outside the EU,
the EU authorized representative is:
Springer Nature Customer Service Center GmbH
Europaplatz 3, 69115 Heidelberg, Germany

Printed by Libri Plureos GmbH
in Hamburg, Germany